Lecture Notes in Mathematics Vol. 1111
ISBN 978-3-540-15195-1

F. Hirzebruch

Arbeitstagungen

Addendum

Lecture Notes Vol. 1111 (Springer) includes the programmes of all 25 Arbeitstagungen held between 1957 and 1984. The archives for AT 57 and AT 58 were very incomplete. However N. Kuiper had made a note for himself of the topics of the talks and communicated these to me in a letter of June 22nd 1991.

Arbeitstagung 1957

13.–20. Juli 1957

Atiyah:	1) Non-projective algebraic surfaces
	2) Chern classes and connections
Grauert:	Real cross-sections which are analytic
Grothendieck:	Kohärente Garben und verallgemeinerte Riemann-Roch-Hirzebruch-Formel auf algebraischen Mannigfaltigkeiten. Vier lange Vorträge *(4 long lectures)*
Kuiper:	A generalisation of convexity
Tits:	Geometrie von Ausnahmegruppen
Hirzebruch:	$c_n(\hat{\eta}) = (n-1)!$ generator $H^{2n}(S^{2n}, \mathbb{Z})$

Arbeitstagung 1958

9.–16. Juli 1958

Bott:	Homotopy groups of classical groups
Grauert:	Levi's problem
Grothendieck:	Rational sections in fibre spaces
Kervaire:	Parallelisability of spheres
Milnor:	Computations on cobordism
D. Puppe:	Non-additive functors and cross-effects
Remmert:	Plurisubharmonic functions
Serre:	Intersection theory
Stein:	Komplexe Basen zu holomorphen Abbildungen
Thom:	Combinatorics and C^{∞}-manifolds
Hirzebruch:	Applications of Milnor's results

Lecture Notes in Mathematics

Edited by A. Dold and B. Eckmann

Subseries: Mathematisches Institut der Universität und
Max-Planck-Institut für Mathematik, Bonn - vol. 5
Adviser: F. Hirzebruch

1111

Arbeitstagung Bonn 1984

Proceedings of the meeting
held by the Max-Planck-Institut für Mathematik, Bonn
June 15–22, 1984

Edited by F. Hirzebruch, J. Schwermer and S. Suter

Springer-Verlag
Berlin Heidelberg New York Tokyo

Herausgeber
Friedrich Hirzebruch
Joachim Schwermer
Silke Suter
Max-Planck-Institut für Mathematik
Gottfried-Claren-Str. 26
5300 Bonn 3, Federal Republic of Germany

AMS-Subject Classification (1980): 10D15, 10D21, 10F99, 12D30, 14H10, 14H40, 14K22, 17B65, 20G35, 22E47, 22E65, 32G15, 53C20, 57N13, 58F19

ISBN 3-540-15195-8 Springer-Verlag Berlin Heidelberg New York Tokyo
ISBN 0-387-15195-8 Springer-Verlag New York Heidelberg Berlin Tokyo

CIP-Kurztitelaufnahme der Deutschen Bibliothek. Mathematische Arbeitstagung <25. 1984. Bonn>: Arbeitstagung Bonn: 1984; proceedings of the meeting, held in Bonn, June 15–22, 1984 / [25. Math. Arbeitstagung]. Ed. by F. Hirzebruch ... – Berlin; Heidelberg; New York; Tokyo: Springer, 1985. (Lecture notes in mathematics; Vol. 1111: Subseries: Mathematisches Institut der Universität und Max-Planck-Institut für Mathematik Bonn; Vol. 5)
ISBN 3-540-15195-8 (Berlin ...)
ISBN 0-387-15195-8 (New York ...)
NE: Hirzebruch, Friedrich [Hrsg.]; Lecture notes in mathematics / Subseries: Mathematisches Institut der Universität und Max-Planck-Institut für Mathematik Bonn; HST

Printing and binding: Beltz Offsetdruck, Hemsbach/Bergstr.
2146/3140-543210

VORWORT

Die mathematische Arbeitstagung in Bonn hatte seit ihrem Anfang im Jahre 1957 bis hin zum 25. Treffen 1984 stets ihren eigenen unverwechselbaren Charakter. Das Programm wird nicht im voraus festgelegt, sondern Vortragende und Themen ergeben sich spontan aus einer Programmdiskussion aller Teilnehmer unter Leitung von F. Hirzebruch. Dies ermöglicht, auf neueste Entwicklungen in einzelnen mathematischen Gebieten unmittelbar einzugehen,und sie auf der Tagung vorzustellen. Darüber hinaus ist die Arbeitstagung nicht ausschließlich einem speziellen mathematischen Thema gewidmet, sondern es finden sich von der Zahlentheorie über die Topologie und Geometrie bis hin zur Analysis über die Jahre hinweg Beiträge aus fast allen Gebieten der Mathematik. Gerade diesem informellen und arbeitsmäßigem Charakter der Tagung hätte die Veröffentlichung von Ergebnisberichten widersprochen. Da jedoch immer wieder stark danach gefragt wurde, hat man, erst seit dem Jahre 1974, einen Kompromiß gefunden. Die Sprecher fassen ihre Vorträge - meist nur handschriftlich - kurz zusammen, und diese werden gesammelt zum Abschluß eines jeden Treffens an die Teilnehmer verteilt.

Anläßlich der 25. mathematischen Arbeitstagung im Jahre 1984 wurde und wird hiervon in mehrfacher Hinsicht abgewichen: Einige Mathematiker wurden im voraus gebeten, in Überblicksvorträgen Entwicklungen und Probleme in ihren eigenen Arbeitsgebieten darzustellen. Diese Beiträge finden sich im ersten Teil des vorliegenden Bandes. Den zweiten Teil bilden Ausarbeitungen der meisten der ad-hoc Vorträge, deren Sprecher erst während der Arbeitstagung ausgewählt wurden. Vielleicht spiegelt sich trotz dieser Änderungen im Procedere ein wenig der Charakter der Arbeitstagung wider. Im Anhang sind die Programme der Arbeitstagungen 1957 - 1984 wiedergegeben.

Die Mühe der Schreibarbeit fast aller Beiträge haben mit viel Geduld und Sorgfalt Frau K. Deutler und Frau C. Pearce auf sich genommen, denen wir dafür sehr dankbar sind.

Bonn, November 1984 Die Herausgeber

25. Mathematische Arbeitstagung 1984 (15. - 22. Juni)

J. Tits:	Groups and group functors attached to Kac-Moody data
M. Atiyah:	The eigenvalues of the Dirac operator
A. Connes:	K-theory, cyclic cohomology and operator algebras
G. Segal:	Loop groups
G. Harder:	Special values of Hecke L-functions and abelian integrals
H. Wente:	A counterexample in 3-space to a conjecture of H. Hopf
G. Faltings:	Compactification of $A_g/\mathbb{Z}$
C.T.C. Wall:	Geometric structures and algebraic surfaces
J. Harris:	Recent work on Hodge structures
Y.T. Siu:	Some recent results in complex manifold theory related to vanishing theorems for the semipositive case
W. Schmid:	Recent progress in representation theory
W. Ballmann:	Manifolds of non-positive curvature
B. Mazur - Ch. Soulé:	Conjectures of Beilinson on L-functions and K-theory
H.-O. Peitgen:	Morphology of Julia sets
S.S. Chern:	Some applications of the method of moving frames
S. Lang:	Vojta's conjecture on heights and Green's function
S. Donaldson:	4-manifolds with indefinite intersection form
D. Zagier:	Modular points, modular curves, modular surfaces and modular forms
G. van der Geer:	Schottky's problem
R. Bryant:	G_2 and Spin(7)-holonomy
S. Wolpert:	Homology of Teichmüller spaces
J.-P. Serre:	l-adic representations
M.F. Atiyah:	On Manin's manuscript "New dimensions in geometry"

Inhaltsverzeichnis

25. Mathematische Arbeitstagung 1984

Überblicksvorträge

INTRODUCTION TO NON COMMUTATIVE DIFFERENTIAL GEOMETRY

A. Connes
Institut des Hautes
Études Scientifiques
35, Route des Chartres
F-91440 Bures-Sur-Yvette
France

This is the introduction to a series of papers in which we shall extend the calculus of differential forms and the de Rham homology of currents beyond their customary framework of manifolds, in order to deal with spaces of a more elaborate nature, such as,

a) the space of leaves of a foliation,

b) the dual space of a finitely generated non-abelian discrete group (or Lie group),

c) the orbit space of the action of a discrete group (or Lie group) on a manifold.

What such spaces have in common is to be, in general, badly behaved as point sets, so that the usual tools of measure theory, topology and differential geometry lose their pertinence. These spaces are much better understood by means of a canonically associated algebra which is the group convolution algebra in case b). When the space V is an ordinary manifold, the associated algebra is <u>commutative.</u> It is an algebra of complex-valued functions on V, endowed with the pointwise operations of sum and product.

A smooth manifold V can be considered from different points of view such as

α) <u>Measure theory</u> (i.e. V appears as a measure space with a fixed measure class), β) Topology (i.e. V appears as a locally compact space),

γ) Differential geometry (i.e. V appears as a smooth manifold).

Each of these structures on V is fully specified by the corresponding algebra of functions, namely:

α) The commutative von Neumann algebra $L^\infty(V)$ of classes of essentially bounded measurable functions on V,

β) The C^*-algebra $C_0(V)$ of continuous functions on V which vanish at infinity,

γ) The algebra $C_c^\infty(V)$ of smooth functions with compact support.

It has long been known to operator algebraists that measure theory and topology extend far beyond their usual framework to:

A) The theory of weights and von Neumann algebras.

B) C^*- algebras, K theory and index theory.
Let us briefly discuss these two fields,

A) The theory of weights and von Neumann algebras.

To an ordinary measure space (X,μ) correspond the von Neumann algebra $L^\infty(X,\mu)$ and the weight φ :

$$\varphi(f) = \int_X f d\mu \quad \forall f \in L^\infty(x,\mu)^+ .$$

Any pair (M,φ) of a commutative von Neumann algebra M and weight φ is obtained in this way from a measure space (X,μ). Thus the place of ordinary measure theory in the theory of weights on von Neumann algebras is similar to that of commutative algebras among arbitrary ones. This is why A) is often called non-commutative measure theory.

Non-commutative measure theory has many features which are trivial in the commutative case. For instance to each weight φ on a von Neumann algebra M corresponds canonically a one-parameter group $\sigma_t^\varphi \in \mathrm{Aut}\, M$

of automorphisms of M, its <u>modular automorphism</u> group. When M is commutative, one has $\sigma_t^\varphi(x) = x$, $\forall x \in M$, and for any weight φ on M. We refer to [13] for a survey of non-commutative measure theory.

B) <u>C*-algebras, K theory and index theory.</u>

Gel'fand's theorem implies that the category of commutative C*-algebras and *-homomorphisms is dual to the category of locally compact spaces and proper continuous maps.

Non-commutative C*-algebras have first been used as a tool to construct von Neumann algebras and weights, exactly as in ordinary measure theory, where the Riesz representation theorem [38], Theorem 2.14, enables to construct a measure from a positive linear form on continuous functions. In this use of C*-algebras the main tool is positivity. The fine topological features of the "space" under consideration do not show up. These fine features came into play thanks to Atiyah's topological K-theory [2]. First the proof of the periodicity theorem of R. Bott shows that its natural set up is non-commutative Banach algebras (cf. [46]). Two functors K_0, K_1 (with values in the category of abelian groups) are defined and any short exact sequence of Banach algebras gives rise to an hexagonal exact sequence of K groups. For $A = C_0(X)$, the commutative C*-algebra associated to a locally compact space X, $K_j(A)$ is (in a natural manner) isomorphic to $K^j(X)$, the K theory with compact supports of X. Since (cf. [41]) for a commutative Banach algebra B, $K_j(B)$ depends only upon the Gel'fand spectrum of B, it is really the C*-algebra case which is most relevant.

Secondly, Brown, Douglas and Fillmore have classified (cf. [8]) short exact sequences of C*-algebras of the form:

$$0 \to \mathcal{K} \to A \to C(X) \to 0$$

where $\mathcal{K}$ is the C*-algebra of compact operators in Hilbert space, and X is a compact space. They have shown how to construct a group from such extensions. When X is a finite dimensional compact metric space, this group is naturally isomorphic to $K_1(X)$, the Steenrod K homology of X, cf. [19],[24].

Since the original classification problem of extensions did arise as an internal question in operator and C*-algebra theory, the work of Brown, Douglas and Fillmore made it clear that K theory is an indispensable tool even for studying C*-algebras per se. This fact was further emphasized by the role of K theory in the classification of C*-algebras which are inductive limits of finite dimensional ones (cf. [7] [22] [21]),and in the work of Cuntz and Krieger on C*-algebras associated to topological Markov chains [18].

Finally the work of the Russian school, of Miscenko and Kasparov in particular, ([30] [26] [27] [28]), on the Novikov conjecture, has shown that the K theory of non-commutative C*-algebras plays a crucial role in the solution of classical problems in the theory of non-simply-connected manifolds. For such a space X , a basic homotopy invariant is the Γ-equivariant signature σ of its universal covering $\tilde{X}$, where $\Gamma = \pi_1(X)$ is the fundamental group of X. This invariant σ lies in the K group, $K_0(C^*(\Gamma))$, of the group C*- algebra $C^*(\Gamma)$.

The K theory of C*-algebras, the extension theory of Brown, Douglas and Fillmore and the Ell theory of Atiyah ([1]) are all special cases of Kasparov's bivariant functor KK(A,B). Given two $\mathbb{Z}/2$ graded C*-algebras A and B, KK(A,B) is an abelian group whose elements are homotopy classes of Kasparov A-B bimodules (cf. [26] [27]).

After this quick overview of measure theory and topology in the non-commutative framework, let us be more specific about the algebras associated to the "spaces" occuring in a) b) c) above.

a) Let V be a smooth manifold, F a smooth foliation of V. The measure theory of the leaf space "V/F" is described by the von Neumann algebra of the foliation (cf.[10][11][12]). The topology of the leaf space is described by the C*-algebra $C^*(V,F)$ of the foliation (cf. [11] [12] [43]).

b) Let Γ be a discrete group. The measure theory of the (reduced) dual space $\hat{\Gamma}$ is described by the von Neumann algebra $\lambda(\Gamma)$ of operators in the Hilbert space $\ell^2(\Gamma)$ which are invariant under right translations. This von Neumann algebra is the weak closure of the group ring $\mathbb{C}\Gamma$ acting in $\ell^2(\Gamma)$ by left translations.

The topology of the (reduced) dual space $\hat{\Gamma}$ is described by the C*-algebra $C_r^*(\Gamma)$, the norm closure of $\mathbb{C}\Gamma$ in the algebra of bounded operators in $\ell^2(\Gamma)$.

b') For a Lie group G the discussion is the same, with $C_c^\infty(G)$ instead of $\mathbb{C}\Gamma$.

c) Let Γ be a discrete group acting on a manifold W. The measure theory of the "orbit space" W/Γ is described by the von Neumann algebra crossed product $L^\infty(W) \rtimes \Gamma$ (cf. [33]).

The situation is summarized in the following table:

Space	V	V/F	$\hat{\Gamma}$	$\hat{G}$	W/Γ
Measure theory	$L^\infty(V)$	v.N.algebra of (V,F)	$\lambda(\Gamma)$	$\lambda(G)$	$L(W) \rtimes \Gamma$
Topology	$C_0(V)$	$C^*(V,F)$	$C_r^*(\Gamma)$	$C_r^*(G)$	$C_0(W) \rtimes \Gamma$

It is a general principle (cf. [3] [14] [4]) that for families of elliptic operators $(D_y)_{y\in Y}$ parametrized by a "space" Y such as those occuring above, the index of the family is an element of $K_0(A)$, the K group of the C*-algebra associated to Y. For instance the Γ-equivariant signature of the universal covering X of a compact oriented manifold is the Γ-equivariant index of the elliptic signature operator on X. We are in case b) and $\sigma \in K_0(C_r^*(\Gamma))$. The obvious problem then is to compute K(A) for the C*-algebras of the above spaces, and then the index of families of elliptic operators.

After the breakthrough of Pimsner and Voiculescu ([34]) in the computation of K groups of crossed products, and under the influence of the Kasparov bivariant theory, the general program of computation of the K groups of the above spaces (i.e. of the associated C*-algebras) has undergone rapid progress in the last years ([12] [43] [31] [32] [45] [44]).

So far, each new result confirms the validity of the general conjecture formulated in [4]. In order to state it briefly, we shall deal only with case c) above. We also assume that Γ is discrete and <u>torsion free</u> , cf. [4] for the general case. By a familar construction of algebraic topology a space such as W/Γ, the orbit space of a discrete group action, can be realized as a simplicial complex, <u>up to homotopy</u>. One lets Γ act freely and properly on a contractible space $E\Gamma$, and forms the <u>homotopy quotient</u> $W \times_\Gamma E\Gamma$ which is a meaningful space even when the quotient topological space W/Γ is pathological. In case b) (Γ acting on $W = \{pt\}$) this yields the classifying space $B\Gamma$. In case a), see [12] for the analoguous construction. In [4] (using [12] and [14]) a map μ is defined from the twisted K homology $K_{*,\tau}(W \times_\Gamma E\Gamma)$ to the K group of the C*-algebra $C_0(W) \rtimes \Gamma$. The conjecture is that this map μ is always an ismorphism.

$$\mu : K_{*,\tau}(W \times_\Gamma E\Gamma) \to K_*(C_0(W) \rtimes \Gamma)$$

At this point it would be tempting to advocate that the space $W \times_\Gamma E\Gamma$ gives a sufficiently good description of the topology of W/Γ and that we can dispense with C^* algebras. However, it is already clear in the simplest cxamples that the C*-algebra $A = C_0(W) \rtimes \Gamma$ is a finer description of the "topological space" of orbits. For instance, with $W = S^1$ and $\Gamma = \mathbb{Z}$, the actions given by two irrational rotations $R_{\theta_1}, R_{\theta_2}$ yield isomorphic C*-algebras if and only if $\theta_1 = \pm\theta$ ([34] [35]) and Morita equivalent C*-algebras iff θ_1 and θ_2 belong to the same orbit of the action of $PSL(2,\mathbb{Z})$ on $P_1(\mathbb{R})$ [36]. On the contrary, the homotopy quotient is independent of θ (and is homotopic to the two torus).

Moreover, as we already mentioned, an important role of a "space" such as $Y = W/\Gamma$ is to parametrize a family of elliptic operators, $(D_y)_{y\in Y}$. Such a family has both a topological index $\mathrm{Ind}_t(D)$, which belongs to the twisted K homology group $K_{*,\tau}(W \times_\Gamma E\Gamma)$, and an analytic index $\mathrm{Ind}_a(D) = \mu(\mathrm{Ind}_t(D))$, which belongs to $K_*(C_0(W) \rtimes \Gamma)$ (cf. [4] [16]). But it is a priori only through $\mathrm{Ind}_a(D)$ that the analytic properties of the family $(D_y)_{y\in Y}$ are reflected. For instance, if each D_y is the Dirac operator on a Spin Riemannian manifold M_y of strictly positive scalar curvature, one has $\mathrm{Ind}_a(D) = 0$ (cf. [37][16]), but the equality $\mathrm{Ind}_t(D) = 0$ follows only if one knows that the map μ is injective (cf. [4][37][16]). The problem of injectivity of μ is an important reason for developing the analogue of de Rham homology for the above "spaces". Any closed de Rham current C on a manifold V yields a map φ_C from $K^*(V)$ to $\mathbb{C}$

$$\varphi_C(e) = \langle C, \mathrm{ch}\, e\rangle \quad \forall e \in K^*(V)$$

where $\mathrm{ch}: K^*(V) \to H^*(V,\mathbb{R})$ is the usual Chern character.

Now , any "closed de Rham current" C on the orbit space W/Γ should yield a map φ_C from $K_*(C_0(W) \rtimes \Gamma)$ to $\mathbb{C}$. The rational injectivity of μ would then follow from the existence, for each $\omega \in H^*(W \times_\Gamma E\Gamma)$, of a "closed current" $C(\omega)$ making the following diagram commutative,

$$\begin{array}{ccc} K_{*,\tau}(W \times_\Gamma E\Gamma) & \xrightarrow{\mu} & K_*((C_0(W) \rtimes \Gamma) \\ \downarrow \mathrm{ch}_* & & \downarrow \varphi_C(\omega) \\ H_*(W \times_\Gamma E\Gamma, \mathbb{R}) & \xrightarrow{\omega} & \mathbb{C} \end{array}$$

Here we assume that W is Γ-equivariantly oriented so that the dual Chern character $\mathrm{ch}_*: K_{*,\tau} \to H_*$ is well defined (See [16]). Also, we view $\omega \in H^*(W \times_\Gamma E\Gamma, \mathbb{C})$ as a linear map from $H_*(W \times_\Gamma E\Gamma, \mathbb{R})$ to $\mathbb{C}$.

This leads us to the subject to our series of papers which is;

1. <u>The construction of de Rham homology for the above spaces;</u>

2. <u>Its applications to K theory and index theory.</u>

The construction of the theory of currents, closed currents, and of the maps φ_C for the above "spaces", requires two quite different steps. The first is <u>purely algebraic:</u>

One starts with an algebra A over $\mathbb{C}$, which plays the role of $C^\infty(V)$, and one develops the analogue of de Rham homology, the pairing with the algebraic K theory groups $K_0(A)$, $K_1(A)$, and algebraic tools to perform the computations. This step yields a contravariant functor H^*_λ from non commutative algebras to graded modules over the polynomial ring $\mathbb{C}(\sigma)$ with a generator σ of degree 2. In the definition of this functor the finite cyclic groups play a crucial role, and this is why H^* is called <u>cyclic cohomology.</u> Note that it is a contravariant functor for algebras and hence a covariant one for "spaces". It is the subject of part II under the title,

De Rham homology and non-commutative algebra

The second step involves analysis:

The non-commutative algebra $\mathcal{A}$ is now a dense subalgebra of a C*-algebra A and the problem is, given a closed current C on $\mathcal{A}$ as above satisfying a suitable continuity condition relative to A, to extend $\varphi_C : K_0(\mathcal{A}) \to \mathbb{C}$ to a map from $K_0(A)$ to $\mathbb{C}$. In the simplest situation, which will be the only one treated in parts I and II, the algebra $\mathcal{A} \subset A$ is stable under holomorphic functional calculus (cf. Appendix 3 of part I) and the above problem is trivial to handle since the inclusion $\mathcal{A} \subset A$ induces an isomorphism $K_0(\mathcal{A}) \approx K_0(A)$. However, even to treat the fundamental class of W/Γ, where Γ is a discrete group acting by orientation preserving diffeomorphisms on W, a more elaborate method is required and will be discussed in part V (cf. [16]). In the context of actions of discrete groups we shall construct $C(\omega)$ and $\varphi_{C(\omega)}$ for any cohomology class $\omega \in H^*(W \times_\Gamma E\Gamma, \mathbb{C})$ in the subring R generated by the following classes:

a) Chern classes of Γ-equivariant (non unitary) bundles on W.

b) Γ-invariant differential forms on W.

c) Gel'fand Fuchs classes.

As applications of our construction we get (in the above context):

α) If $x \in K_{*,\tau}(W \times_\Gamma E\Gamma)$ and $\langle ch_* x, \omega\rangle \neq 0$ for some ω in the above ring R then $\mu(x) \neq 0$.

In fact we shall further improve this result by varying W; it will then apply also to the case $W = \{pt\}$, i.e. to the usual Novikov conjecture. All this will be discussed in part V, but see [16] for a preview.

β) For any $\omega \in R$ and any family $(D_y)_{y \in Y}$ of elliptic operators parametrized by $Y = W/\Gamma$, one has the index theorem.

$$\varphi_C(\mathrm{Ind}_a(D)) = \langle ch_* \mathrm{Ind}_t(D), \omega\rangle$$

When Y is an ordinary manifold, this is the cohomological form of the Atiyah Singer index theorem for families ([3]).

It is important to note that, in all cases, the right hand side is computable by a standard recipe of algebraic topology from the symbol of D. The left hand side carries the analytic information such as vanishing, homotopy invariance,...

All these results will be extended to the case of foliations (i.e. when Y is the leaf space of a foliation) in part VI.

As a third application of our analogue of de Rham homology for the above spaces we shall obtain index formulae for transversally elliptic operators; that is elliptic operators on the above "spaces" Y. In part IV we shall work out the pseudo-differential calculus for crossed products of a C*-algebra by a Lie group, (cf. [15]), thus yielding many non-trivial examples of elliptic operators on spaces of the above type. Let A be the C*-algebra associated to Y, any such elliptic operator on Y yields a finitely summable Fredholm module over the dense subalgebra $\mathcal{A}$ of smooth elements of A.

In part I we show how to construct canonically from such a Fredholm module a closed current on the dense subalgebra $\mathcal{A}$. The title of part I, the Chern character in K homology is motivated by the specialization of the above construction to the case when Y is an ordinary manifold. Then the K homology $K_*(V)$ is entirely described by elliptic operators on V ([6] [14]) and the association of a closed current provides us with a map,

$$K_*(V) \to H_*(V,\mathbb{C})$$

which is exactly the dual Chern character ch_*.

The explicit computation of this map ch_* will be treated in part III as an introduction to the asymptotic methods of computations of cyclic cocycles which will be used again in part IV. As a corollary we shall, in part IV give completely explicit formulae for indices of finite difference, differential operators on the real line.

If D is an elliptic operator on a "space" Y and C is the closed current $C = ch_* D$ (constructed in part I), the map $\varphi_C : K_*(A) \to \mathbb{C}$ makes sense and one has,

$$\varphi_C(x) = \langle x, [D] \rangle = \text{Index } D_x \quad \forall x \in K_*(A)$$

where the right hand side means the index of D with coefficients in x, or equivalently the value of the pairing between K homology and K cohomology. The <u>integrality</u> of this value, Index $D_x \in \mathbb{Z}$, is a basic result which will be already used in a very efficient way in part I, to control $K_*(A)$.

The aim of part I is to show that the construction of the Chern character ch_* in K homology dictates the basic definitions and operations - such as the suspension map S - in cyclic cohomology. It is motivated by the previous work of Helton and Howe [23], Carey and Pincus [9] and Douglas and Voiculescu [20].

There is another, equally important, natural route to cyclic cohomology. It was taken by Loday and Quillen ([29]) and by Tsigan ([42]). Since the latter work is independent from ours, cyclic cohomology was discovered from two quite different points of view.

There is also a strong relation with the work of I. Segal [39] on quantized differential forms, which will be discussed in part IV and with the work of M. Karoubi on secondary characteristic classes [25], which is discussed in part II, Theorem 33.

Our results and in particular the spectral sequence of part II were announced in the conference on operator algebras held in Oberwolfach in September 1981 ([17]).

Besides parts I and II, which will soon appear in the IHES Publications, our set of papers will contain:

I. The Chern character in K homology.
II. De Rham homology and non commutative algebra.
III. Smooth manifolds, Alexander Spanier cohomology and index theory.
IV. Pseudodifferential calculus for C^* dynamical systems, index

theorem for crossed products and the pseudo torus.

V. Discrete groups and actions on smooth manifolds.

VI. Foliations and transversally elliptic operators.

VII. Lie groups.

References

[1] M.F. Atiyah, Global theory of elliptic operators, Proc. Internat. Conf. on functional analysis and related topics, Univ. of Tokyo Press, Tokyo (1970).

[2] M.F. Atiyah, K theory, Benjamin (1967).

[3] M.F. Atiyah and I. Singer, The index of elliptic operators IV, Ann. of Math. 93 (1971) p. 119-138.

[4] P. Baum and A. Connes, Geometric K theory for Lie groups and Foliations, Preprint IHES, 1982.

[5] P. Baum and A. Connes, Leafwise homotopy equivalence and rational Pontrjagin classes, Preprint IHES, 1983.

[6] P. Baum and R. Douglas, K homotopy and index theory, Operator algebras and applications, Proc. Symposia Pure Math. 38 (1982), part I, p. 117-173.

[7] O. Bratteli, Inductive limits of finite dimensional C*-algebras, Trans. AMS 171 (1972), p. 195-234.

[8] L.G. Brown, R. Douglas and P.A. Fillmore, Extensions of C*-algebras and K homology, Ann. of Math. (2) 105 (1977) p. 265-324.

[9] R. Carey and J.D. Pincus, Almost commuting algebras, K theory and operator algebras, Lecture Notes in Math. N°575, Springer Berlin-New York (1977).

[10] A.Connes, The von Neumann algebra of a foliation, Lecture Notes in Physics N°80 (1978) p. 145-151, Springer Berlin-New York.

[11] A. Connes, Sur la théorie non commutative de l'intégration, Algèbres d'opérateurs, Lecture Notes in Math. N°725, Springer Berlin-New York (1979).

[12] A. Connes, A Survey of foliations and operator algebras, Operator algebras and applications, Proc. Symposia Pure Math. 38 (1982) Part I, p. 521-628.

[13] A. Connes, Classification des facteurs, Operator algebras and applications, Proc. Symposia Pure Math. 38 (1982) Part II, p. 43-109.

[14] A. Connes and G. Skandalis, The longitudinal index theorem for foliations, to appear in Publ. R.I.M.S. Kyoto.

[15] A. Connes, C*-algèbres et géométrie différentielle, C.R.Acad. Sci. Paris, tome 290, Série I (1980).

[16] A. Connes, Cyclic cohomology and the transverse fundamental class of a foliation, Preprint IHES M/84/7.

[17] A. Connes, Spectral sequence and homology of currents for operator algebras. Math. Forschungsinstitut Oberwolfach Tagungsbericht 42/81, Funktionalanalysis und C*-Algebren , 27.9. - 3.10.1981.

[18] J. Cuntz and W. Krieger, A class of C*-algebras and topological Markov chains, Invent. Math. 56 (1980) p. 251 - 268.

[19] R. Douglas, C*-algebra extensions and K homology, Annals of Math. Studies, N°95, Princeton University Press 1980.

[20] R. Douglas and D. Voiculescu, On the smoothness of sphere extensions, J. Operator Theory 6(1) (1981) p. 103.

[21] E.G. Effros, D.E. Handelman and C.L. Shen, Dimension groups and their affine representations, Amer. J. Math. 102 (1980) p. 385-407.

[22] G. Elliott, On the classification of inductive limits of sequences of semi-simple finite dimensional algebras, J. Alg. 38 (1976) p. 29-44.

[23] J. Helton and R. Howe, Integral operators, commutators, traces, index and homology, Proc. of Conf. on operator theory, Lecture Notes in Math. N°345, Springer Berlin-New York (1973).

[24] D.S. Kahn, J. Kaminker and C. Schochet, Generalized homology theories on compact metric spaces, Michigan Math. J. 24 (1977) p. 203-224.

[25] M. Karoubi, Connexions, courbures et classes caractéristiques en K theorie algébrique, Canadian Math. Soc. Proc. Vol.2, part I (1982) p. 19-27.

[26] G. Kasparov, K functor and extensions of C*-algebras, Izv. Akad. Nauk SSSR, Ser. Mat. 44(1980) p. 571-636.

[27] G. Kasparov, K theory, group C*-algebras and higher signature, Conspectus, Chernogolovka (1983).

[28] G. Kasparov, Lorentz groups: K theory of unitary representations and crossed products, preprint, Chernogolovka, 1983.

[29] J.L. Loday and D. Quillen, Cyclic homology and the Lie algebra of matrices, C.R. Acad. Sci. Paris, Série I, 296 (1983) p. 295-297.

[30] A.S. Miscenko, Infinite dimensional representations of discrete groups and higher signature, Math. USSR Izv. 8 (1974) p. 85-112.

[31] M.Penington, K theory and C*-algebras of Lie groups and Foliations, D. Phil. thesis, Michaelmas, Term. 1983, Oxford.

[32] M.Penington and R. Plymen, The Dirac operator and the principal series for complex semi-simple Lie groups, J. Funct. Analysis 53 (1983) p. 269-286.

[33] G. Pedersen, C*-algebras and their automorphism groups. Academic Press, New York (1979).

[34] M. Pimsner and D. Voiculescu, Exact sequences for K groups and Ext groups of certain cross-product C*-algebras. J. of operator theory 4 (1980), 93-118.

[35] M. Pimsner and D. Voiculescu, Imbedding the irrational rotation C* algebra into an AF algebra, J. of operator theory 4 (1980) 201-211.

[36] M. Rieffel, C*-algebras associated with irrational rotations, Pac. J. of Math. 93, N°2 (1981).

[37] J. Rosenberg, C* algebras, positive scalar curvature and the Novikov conjecture, Publ. Math. IHES, Vol. 58 (1984) p. 409-424.

[38] W. Rudin, Real and complex analysis, Mc. Graw Hill, New York (1966).

[39] I. Segal, Quantized differential forms, Topology 7 (1968) p. 147-172.

[40] I.M. Singer, Some remarks on operator theory and index theory, Lecture notes in Math. 575 (1977) p. 128-138, Springer New York.

[41] J.L. Taylor, Topological invariants of the maximal ideal space of a Banach algebra, Advances in Math. 19 (1976) N°2, p. 149-206.

[42] B.L. Tsigan, Homology of matrix Lie algebras over rings and Hochschild homology, Uspekhi Math. Nauk. Vol. 38 (1983) p. 217-218.

[43] A.M. Torpe, K theory for the leaf space of foliations by Reeb components, Mat. Institut, Odense Univ., Preprint (1982).

[44] A. Valette, K. Theory for the reduced C*-algebra of semisimple Lie groups with real rank one, Quarterly J. of Math., Oxford Série 2, Vol. 35 (1984) p. 334-359.

[45] A. Wasserman, Une démonstration de la conjecture de Connes-Kasparov, To appear in C.R. Acad. Sci. Paris.

[46] R. Wood, Banach algebras and Bott periodicity, Topology 4 (1965-66) p. 371-389.

SPECIAL VALUES OF HECKE L-FUNCTIONS AND ABELIAN INTEGRALS

G. Harder and N. Schappacher
Max-Planck-Institut
für Mathematik
Gottfried-Claren-Str. 26
5300 Bonn 3

In this article we attempt to explain the formalism of Deligne's rationality conjecture for special values of motivic L-functions (see [D1]) in the particular case of L-functions attached to algebraic Hecke characters ("Größencharaktere of type A_0"). In this case the conjecture is now a theorem by virtue of two complementary results, due to D. Blasius and G. Harder, respectively: see §5 below.

For any "motive" over an algebraic number field, Deligne's conjecture relates certain special values of its L-function to certain periods of the motive. Most of the time when motives come up in a geometric situation, we tend to know very little about their L-functions. In the special case envisaged here, however, the situation is quite different: The L-functions of algebraic Hecke characters are among those for which Hecke proved analytic continuation to the whole complex plane and functional equation. But the "geometry" of the corresponding motives has emerged only fairly recently - see §3 below.

The relatively good command we now have of the motives attached to algebraic Hecke characters reveals that many non-trivial period relations are in fact but reflections of character-identities. This point of view is systematically perused in [Sch], and we shall illustrate it here by the so-called formula of Chowla and Selberg: see § 6 .

This formula, in fact, goes back to the year 1897, as does the instance of Deligne's conjecture with which we start in § 1. Tying up these two relations in the motivic formalism, we hope to make it apparent that both results really should be viewed "comme les deux volets d'un même diptyque", as A. Weil has pointed out in [W III], p. 463.

Contents

§ 1. A formula of Hurwitz

In 1897, Hurwitz [Hu] proved that

$$(1) \qquad \sum_{a,b\in\mathbb{Z}}{}' \frac{1}{(a+bi)^{4\nu}} = \Omega^{4\nu} \times \text{(rational number)},$$

for all $\nu = 1,2,3,\ldots$, where

$$(2) \qquad \Omega = 2\int_0^1 \frac{dx}{\sqrt{1-x^4}} = 2.62205755\ldots = \frac{\Gamma(\frac{1}{4})^2}{2.\sqrt{2\pi}}$$

Notice the analogy of these identities with the well-known formula for the Riemann zeta-function at positive even integers:

$$(3) \qquad \sum_{a\in\mathbb{Z}}{}' \frac{1}{a^{2\nu}} = (2\pi i)^{2\nu} \times \text{(rational number)}.$$

Both formulas are special cases of Deligne's conjecture. To understand this in Hurwitz' case, we look at the elliptic curve A given by the equation

$$A : y^2 = 4x^3 - 4x .$$

A is defined over $\mathbb{Q}$, but we often prefer to look at it as defined over the field $k = \mathbb{Q}(i) \subset \mathbb{C}$. Over this field of definition, we can see that A admits complex multiplication by the same field k :

$$k \xrightarrow{\sim} \mathrm{End}(A) \otimes \mathbb{Q}$$

$$i \longmapsto \left\{ \begin{matrix} x \mapsto -x \\ \\ y \mapsto iy \end{matrix} \right\}$$

Deligne's account of Hurwitz' formula would start from the observation that both sides of (1) express information about the homology

$$H_1(A)^{\otimes 4\nu} \subset H_{4\nu}(A^{4\nu}) .$$

The left hand side of (1) carries data collected at the finite places of k , as does the right hand side for the infinite places.

In fact, look at the different cohomology theories:

<u>- Etale cohomology:</u> Fix a rational prime number ℓ , and denote, for $n \geq 1$, by $A[\ell^n]$ the group of ℓ^n-torsion points in $A(\overline{\mathbb{Q}})$, $\overline{\mathbb{Q}}$ being the algebraic closure of $\mathbb{Q}$ in $\mathbb{C}$. Then

$$V_\ell(A) = \left(\varprojlim_n A[\ell^n]\right) \otimes_{\mathbb{Z}_\ell} \mathbb{Q}_\ell$$

is the dual of the first ℓ-adic cohomology of $A \times_k \overline{\mathbb{Q}}$ with coefficients in $\mathbb{Q}_\ell$.

By functoriality, the isomorphism $k \cong \mathbb{Q} \otimes_{\mathbb{Z}} \mathrm{End}\, A$ makes $V_\ell(A)$ into a $k \otimes \mathbb{Q}_\ell$-module, free of rank 1. The natural continuous action of $\mathrm{Gal}(\overline{\mathbb{Q}}/k)$ on $V_\ell(A)$ is $k \otimes \mathbb{Q}_\ell$-linear, and therefore given by a continuous character

$$\Psi_\ell : \mathrm{Gal}(\overline{\mathbb{Q}}/k)_{ab} \longrightarrow GL_{k \otimes \mathbb{Q}_\ell}(V_\ell(A)) = (k \otimes \mathbb{Q}_\ell)^* .$$

This character was essentially determined - if from a rather different point of view - on July 6, 1814 by Gauss, [Ga]. The explicit analysis of the Galois-action on torsion-points of A was carried out (in a stunningly "modern" fashion) in 1850 by Eisenstein, [Ei]. - In any case, if π is a prime element of $\mathbb{Z}[i]$ not dividing 2ℓ , normalized so that $\pi \equiv 1 \pmod{(1+i)^3}$, and if $F_\pi \in \mathrm{Gal}(\overline{\mathbb{Q}}/k)_{ab}$ is a geometric Frobenius element at (π) (<u>i.e.</u>, $F_\pi^{-1}(x) \equiv x^{\mathbb{N}\pi} \pmod{\mathfrak{P}}$ for any prime $\mathfrak{P}$ of k_{ab} dividing (π), any algebraic integer $x \in k_{ab}$), then one finds

$$\psi_\ell(F_\pi) = \pi^{-1} \in k^* \hookrightarrow (k \otimes \mathbb{Q}_\ell)^* .$$

The characters ψ_ℓ all fit together to give an "algebraic Hecke character" ψ defined on the group I_2 of ideals of k that are prime to 2:

$$\begin{array}{ccccc} I_2 & \xrightarrow{\quad\psi\quad} & & & k^* \\ \uparrow & & & & \searrow \\ I_{2\ell} & \xrightarrow{F} & \mathrm{Gal}(\overline{\mathbb{Q}}/k)_{ab} & \xrightarrow{\psi_\ell} & (k \otimes \mathbb{Q}_\ell)^* \end{array}$$

Then for all $\nu \geq 1$, the character $\psi^{4\nu}$ can be defined on all ideals of k by $(\alpha) \mapsto \alpha^{-4\nu}$. Remember that k is embedded into $\mathbb{C}$, so that it makes sense to consider the L-functions

$$L(\psi^{4\nu}, s) = \prod_{\mathfrak{p}} \frac{1}{1 - \frac{\psi(\mathfrak{p})^{4\nu}}{\mathbb{N}\mathfrak{p}^s}} \qquad (\mathrm{Re}(s) > 1-2\nu) ,$$

where $\mathfrak{p}$ ranges over all prime ideals of $\mathbb{Z}[i]$. Then the left hand side of Hurwitz' formula (1) is simply $4.L(\psi^{4\nu},0)$. We have shown how this is a special value of the L-function afforded by the ℓ-adic cohomologies $V_\ell(A)^{\otimes_{k\otimes\mathbb{Q}_\ell} 4\nu}$.

<u>- Betti and de Rham cohomology.</u> Here we shall use the fact that the curve A (if not its complex multiplication) is already defined over $\mathbb{Q}$. Denote by $H_1^B(A) = H_1(A(\mathbb{C}),\mathbb{Q})$ the first rational singular homology of the Riemann surface $A(\mathbb{C})$, with the Hodge decomposition

$$H_1^B(A) \otimes_{\mathbb{Q}} \mathbb{C} = H^{-1,0} \oplus H^{0,-1} .$$

Complex conjugation on $A \times_{\mathbb{Q}} \mathbb{C}$ induces an endomorphism

$$F_\infty : H_1^B(A) \circlearrowleft \qquad \text{(the "Frobenius at } \infty\text{")}.$$

Call H_B^+ the fixed part of $H_1^B(A)$ under F_∞, and let η be a basis of this onedimensional $\mathbb{Q}$-vector space.

Let $H_1^{DR}(A) = H_{DR}^1(A)^\vee$ be the dual of the first algebraic de Rham cohomology of A over $\mathbb{Q}$, given with the Hodge filtration

$$H_1^{DR}(A) \supset F^+ \supset \{0\} ,$$

where $F^+ \otimes_{\mathbb{Q}} \mathbb{C} \cong H^{0,-1}$ under the GAGA isomorphism over $\mathbb{C}$:

$$I : H_1^B(A) \otimes_{\mathbb{Q}} \mathbb{C} \xrightarrow{\sim} H_1^{DR}(A) \otimes_{\mathbb{Q}} \mathbb{C} .$$

I induces an isomorphism of onedimensional $\mathbb{C}$-vector spaces

$$I^+ : H_B^+(A) \otimes_{\mathbb{Q}} \mathbb{C} \longrightarrow (H_1^{DR}(A)/F^+) \otimes_{\mathbb{Q}} \mathbb{C} .$$

Then, $\frac{1}{\Omega} \cdot I^+(\eta) \in H_1^{DR}(A)/F^+$, for Ω defined by (2). In fact, $\Omega = \int_1^\infty \frac{dx}{\sqrt{x^3-x}}$ is a real fundamental period of our curve, and so, up to $\mathbb{Q}^*$, Ω is the determinant of the integration-pairing

$$(H_B^+(A) \otimes_{\mathbb{Q}} \mathbb{C}) \times (H^0(A,\Omega^1) \otimes_{\mathbb{Q}} \mathbb{C}) \xrightarrow{\int} \mathbb{C} ,$$

calculated in terms of $\mathbb{Q}$-rational bases of both spaces. This determinant equals that of the map I^+ since $H^0(A,\Omega^1) \subset H_{DR}^1(A)$ is the dual of $H_1^{DR}(A)/F^+$.

Passing to tensor powers of the onedimensional vector spaces above we find the periods $\Omega^{4\nu}$ occuring in (1).

In a sense, we have cheated a little in deriving the period Ω from the cohomological setup: In the étale case we have used the action of k via complex multiplication to obtain a onedimensional situation (i.e., the k-valued character Ψ). In the calculation of the period, too, we should have considered $H_1^{DR}(A/k) = H_1^{DR}(A) \otimes_{\mathbb{Q}} k$, endowed with the further action of k via complex multiplication, and two copies of $H_1^B(A)$, indexed by the two possible embeddings of the base field k into $\mathbb{C}$.... But in the presence of an elliptic curve over $\mathbb{Q}$, this would have seemed too artificial, and the general procedure will be treated in § 4.

As a final remark about formula (1), it should be noted that it is proved fairly easily. Any lattice $\Gamma = \lambda . (\mathbb{Z} + \mathbb{Z} i)$ gives a Weierstraß $\wp$-function such that

$$\wp'(z,\Gamma) = 4\wp(z,\Gamma)^3 - g_2(\Gamma)\,\wp(z,\Gamma),$$

and for $\lambda = \Omega$ we get $g_2(\Gamma) = 4$. The rational numbers left unspecified in (1) are then essentially the coefficients of the z-expansion of $\wp(z,\Gamma)$. It is these numbers that Hurwitz studied in his papers.

§ 2. Algebraic Hecke Characters

Let k and E be totally imaginary number fields (of finite degree over $\mathbb{Q}$), and write

$$\Sigma = \mathrm{Hom}\,(k,\overline{\mathbb{Q}}) \quad \text{and} \quad T = \mathrm{Hom}\,(E,\overline{\mathbb{Q}})$$

the sets of complex embeddings of k and E. The group $\mathrm{Gal}(\overline{\mathbb{Q}}/\mathbb{Q})$ acts on $\Sigma \times T$, transitively on each individual factor. An <u>algebraic homomorphism</u>

$$\beta : k^* \longrightarrow E^*$$

is a homomorphism induced by a rational character

$$\beta : R_{k/\mathbb{Q}}(\mathbb{G}_m) \longrightarrow R_{E/\mathbb{Q}}(\mathbb{G}_m).$$

This means that, for all $\tau \in T$, the composite

$$\tau \circ \beta : k^* \longrightarrow \overline{\mathbb{Q}}^*$$

is given by

$$(4) \qquad \tau \circ \beta\,(x) = \prod_{\sigma \in \Sigma} \sigma(x)^{\,n(\sigma,\tau)} ,$$

for certain integers $n(\sigma,\tau)$, such that $n(\rho\sigma,\rho\tau) = n(\sigma,\tau)$ for all $\rho \in \mathrm{Gal}\,(\overline{\mathbb{Q}}/\mathbb{Q})$.

Let $k^*_{\mathbb{A},f} \hookrightarrow k^*_{\mathbb{A}}$ be the topological group of finite idèles of k - i.e., those idèles whose components at the infinite places are 1. For $x \in k^*$, let x also denote the corresponding principal idèle in $k^*_{\mathbb{A}}$,and x_f the finite idèle obtained by changing the infinite components of x to 1.

An algebraic Hecke character Ψ of k with values in E , of (infinity-) type β , is a continuous homomorphism

$$\Psi : k^*_{\mathbb{A},f} \longrightarrow E^*$$

such that, for all $x \in k^*$,

$$\Psi(x_f) = \beta(x) .$$

If β is the infinity-type of an algebraic Hecke character Ψ , then, by continuity, β has to kill a subgroup of finite index of the units of k. It follows that the integer

$$(5) \qquad w = n(\sigma,\tau) + n(c\sigma,\tau) = n(\sigma,\tau) + n(\sigma,c\tau)$$

(where c = complex conjugation on $\overline{\mathbb{Q}}$) is independent of σ,τ. It is called the weight of Ψ .

For any $\tau \in T$, we get a complex valued Größencharakter $\tau\circ\Psi$ which extends to a quasicharacter of the idèle-class-group:

$$\begin{array}{ccc}
k^*_{\mathbb{A},f} & \xrightarrow{\tau\circ\Psi} & \overline{\mathbb{Q}}^* \\
\downarrow & & \\
k^*_{\mathbb{A}} & & \downarrow \\
\downarrow & & \\
k^*_{\mathbb{A}}/k^* & \xrightarrow{\tau\circ\Psi} & \mathbb{C}^* .
\end{array}$$

Consider the array of L-functions, indexed by T:

$$L(\Psi,s) = (L(\tau\circ\Psi,s))_{\tau\in T} ,$$

where, for $\mathrm{Re}(s) > \frac{w}{2}+1$,

$$L(\tau\circ\Psi,s) = \prod_{\mathfrak{p}} \left(1 - \frac{(\tau\circ\varphi)(\pi_{\mathfrak{p}})}{\mathbb{N}\mathfrak{p}^s}\right)^{-1} ,$$

the product being over all prime ideals $\mathfrak{p}$ of k for which the value $\Psi(\pi_{\mathfrak{p}})$ does not depend on the choice of uniformizing parameter $\pi_{\mathfrak{p}}$ of $k_{\mathfrak{p}}$.

The point $s = 0$ is called critical for Ψ , if for any τ , no Γ-factor on either side of the functional equation of $L(\tau\circ\Psi,s)$ has a pole at $s = 0$. This is really a property of the infinity-type β of Ψ , for it turns out that $s = 0$ is critical for Ψ if and only if there is a disjoint decomposition

$$\Sigma\times T = \{(\sigma,\tau) \mid n(\sigma,\tau) < 0\} \dot{\cup} \{(\sigma,\tau) \mid n(c\sigma,\tau) < 0\} .$$

In other words, for every $\tau\in T$, there is a "CM-type" $\Phi(\tau\circ\beta)\subset\Sigma$ such that

(6)
- $\Phi(\alpha\tau\circ\beta) = \Phi(\tau\circ\beta)^{\alpha}$, for $\alpha\in \mathrm{Gal}(\overline{\mathbb{Q}}/\mathbb{Q})$
- $\Phi(\tau\circ\beta) \dot{\cup} \Phi(c\tau\circ\beta) = \Sigma$
- $\sigma\in\Phi(\tau\circ\beta) \Leftrightarrow n(\sigma,\tau) < 0 \Leftrightarrow n(c\sigma,\tau) \geq 0$.

For Ψ such that $s=0$ is critical Deligne defined an array of periods $\Omega(\Psi) = (\Omega(\Psi,\tau))_{\tau\in T} \in (\mathbb{C}^*)^T = (E\otimes_{\mathbb{Q}}\mathbb{C})^*$, and conjectured that

$$\frac{L(\Psi,0)}{\Omega(\Psi)} \in E \hookrightarrow E\otimes\mathbb{C} . \tag{7}$$

In other words, he conjectured that there is $x\in E$ such that for all $\tau : E \hookrightarrow \mathbb{C}$,

$$L(\tau\circ\Psi,0) = \tau(x) . \Omega(\Psi,\tau) .$$

The definition of $\Omega(\Psi)$ is discussed in § 4. It requires attaching a motive to an algebraic Hecke character.

§ 3. Motives

3.1 In the example of § 1, we constructed a "motive" for our Hecke characters $\Psi^{4\nu}$ by taking tensor powers of $H_1(A)$, i.e., a certain direct factor of $H_{4\nu}(A^{4\nu})$, in the various cohomology theories. This illustrates fairly well the general idea of what a motive should be: Starting from an algebraic variety over a number field, we have the right to consistently choose certain parts of its cohomology. Just what "consistenly" means constitutes the difference between various notions of motive. Here we shall be concerned with a fairly weak and therefore half way manageable version: motives defined using "absolute Hodge cycles" - see [DMOS], I and II. In this theory motives can often be shown to be isomorphic when their L - functions and periods coincide. A little more precisely, giving a homomorphism between two such motives M and N amounts to giving a family of homomorphisms

$$\begin{cases} H_\sigma(M) \longrightarrow H_\sigma(N) & \text{(Betti cohomology depends on the choice of } \sigma : k \longrightarrow \mathbb{C} \text{ yielding } M \longmapsto M\times_\sigma\mathbb{C}) \\ H_{DR}(M) \longrightarrow H_{DR}(N) & \\ H_\ell(M) \longrightarrow H_\ell(N) & \text{(for all } \ell) \end{cases}$$

compatible with all the natural structures on these cohomology groups: Hodge decomposition, Hodge filtration, $Gal(\bar{k}/k)$-action, as well as with the comparison isomorphisms between H_B and H_{DR}, H_B and the H_ℓ's .

3.2 Let us state more precisely what a <u>motive attached to an algebraic Hecke character</u> Ψ should be! - In the example of § 1, the curve $A/\mathbb{Q}$ defines the motive $H_1(A)$ <u>over</u> $\mathbb{Q}$ whose L-function is $L(\Psi,s)$. (<u>This</u> is really what Gauss observed in 1814; nowadays this follows from a result of Deuring, which has been further generalized by Shimura [Sh 1]...) But this is <u>not</u> what we are looking for. The complex multiplication of A and therefore the Hecke character Ψ are not visible over $\mathbb{Q}$. That is why we considered A <u>over</u> k in our treatment of the étale cohomology, and used the field of values of Ψ (which again happened to be k) to obtain onedimensional Galois-representations, and thus Ψ .

Given a general algebraic Hecke character Ψ like in § 2, a motive M for Ψ has to be a motive defined over the base field k such that the field E acts on all the realizations of M in the various cohomology theories, and such that for all ℓ, $H_\ell(M)$ is an $E\otimes\mathbb{Q}_\ell$-module of rank 1 with $\mathrm{Gal}(\overline{\mathbb{Q}}/k)$ acting via Ψ . The action of E on the various realizations of M should of course be compatible with their extra structures and with the various comparison isomorphisms. In other words (see 3.1), E should embed into $\mathrm{End}\, M$. Thus the rank-condition on $H_\ell(M)$ can also be stated by saying that Betti cohomology $H_\sigma(M)$ should form a onedimensional E-vector space.

<u>3.3</u> The typical example is $H_1(A)$, for an abelian variety A/k with $E \cong \mathbb{Q}\otimes_{\mathbb{Z}} \mathrm{End}_{/k}A$ and $2 \dim A = [E : \mathbb{Q}]$. The fact that these motives always give rise to an algebraic Hecke character was one of the main results of the theory of complex multiplication by Shimura and Taniyama. The Hecke characters occuring with abelian varieties of CM-type are precisely those of weight -1 such that $n(\sigma,\tau) \in \{-1,0\}$, for all $(\sigma,\tau) \in \Sigma\times T$.

In fact, given such an algebraic Hecke character Ψ of k with values in E , we can assume without loss of generality that E is the field generated by the values of Ψ on the finite idèles of k . Then E is a CM-field (i.e., quadratic over a totally real subfield), and a theorem of Casselman, [Sh 1], can be applied to get an abelian variety A defined over k such that:

- $2 \dim A = [E : \mathbb{Q}]$
- there is an isomorphism $E \xrightarrow{\sim} \mathbb{Q}\otimes_{\mathbb{Z}} \mathrm{End}_{/k}A$
- $H_1(A)$ is a motive for Ψ .

3.4. When Ψ has arbitray weight ($\neq 0$) the homogeneity condition (5) above still forces the infinity-type β to be of the form $\beta = \prod_i \beta_i$, with $|\text{weight}(\beta_i)| = 1$, $n_i(\sigma,\tau) \in \{\pm 1, 0\}$. Since twisting with finite order characters is easy to control motivically one would naively expect to be able to assemble a motive for any given algebraic Hecke character essentially as tensor product of constituents of the form $H_1(A)$ or $H^1(A)$ like in 3.3.

There is however the nasty problem of controlling the fields of values E. For example, if k is imaginary quadratic with class number $h > 1$, then a Hecke character of k with $\{n(\sigma,\tau)\} = \{-1,0\}$ or $\{n(\sigma,\tau)\} = \{1,0\}$ can never take all values in $E = k$, but its h-th power may.

Constructing a motive for the h-th power as an E-linear tensor power of a motive for the character of weight ± 1, one still has to show that the field of coefficients E can be "descended" to k in weight $\pm h$.

3.5 This "descent" of the field of coefficients can be dealt with directly. But we gain much more insight if we use a very elegant formalism due to Langlands, [La] § 5, and Deligne, [DMOS] IV. Langlands defined a group scheme over $\mathbb{Q}$, the "Taniyama group" T, of which Deligne was subsequently able to show that the category of its $\mathbb{Q}$-rational representations is equivalent to the category of those motives as can be obtained (eventually after twisting by a character of finite order) from abelian varieties over $\mathbb{Q}$ which admit complex multiplication over $\overline{\mathbb{Q}}$. Since the Taniyama group - along with many other beautiful properties - has, for every k, a certain subquotient S_k (isomorphic to a group scheme constructed by Serre in [Sℓ]) whose irreducible representations are given precisely by the algebraic Hecke characters Ψ of k, we "find" the motive attached to a given Ψ by lifting the corresponding representation of S_k back to the subgroup of T whose representations give the motives defined over k.

3.6 So, for every algebraic Hecke character Ψ of k with values in E, a motive over k equipped with an E-action can be constructed from CM-abelian varieties over k, whose ℓ-adic Galois representations are onedimensional $E \otimes \mathbb{Q}_\ell$-modules given by Ψ. Furthermore, the Tate-conjecture would imply that the ℓ-adic realizations determine a motive up to isomorphism - even in the strictest sense of "motives" (algebraic cycles). As we are dealing with motives for absolute Hodge cycles, it is perhaps not too surprising that one can actually prove: in the category of motives that can be obtained from all abelian varieties over k (not necessarily CM), any two motives attached to the same Hecke character are actually isomorphic - see [Sch], I. Still, this does not seem to be known in any larger category of motives. In fact, it hinges on Deligne's theorem that "every Hodge cycle on an abelian variety over an algebraically closed field is absolutely Hodge" - see [DMOS], I. Anyway, whenever we find two motives constructed from the cohomology of abelian varieties that belong to the same Hecke character they will have the same periods...

§ 4. Periods

As in the example of § 1, periods are going to arise from a comparison of the Betti and de Rham cohomology groups of our motive. So, let us first look at these cohomologies more closely in the case of a motive for an algebraic Hecke character. We are going to use some facts which are well-known for the cohomology of algebraic varieties, and which carry over to motives.

4.1 As in § 2, let k and E be totally imaginary number fields, and Ψ an algebraic Hecke character of k with values in E. Let M be a motive over k attached to Ψ (in the sense of 3.2 above). Then for any embedding $\sigma \in \Sigma$, the singular rational cohomology $H_\sigma(M)$ is an E-vector space of dimension 1. The E-action respects the Hodge-decomposition

$$H_\sigma(M) \otimes_{\mathbb{Q}} \mathbb{C} = \bigoplus_{p,q} H^{p,q} .$$

$H_\sigma(M) \otimes_{\mathbb{Q}} \mathbb{C}$ is an $E \otimes_{\mathbb{Q}} \mathbb{C} = \mathbb{C}^T$-module of rank 1. (Σ and T were defined at the beginning of § 2.)

Starting from the special case where $M = H_1(A)$ with an abelian variety A/k of CM-type, and using the uniqueness of the motive attached to a Hecke character (see 3.6), one finds that, for any embedding $\tau \in T$, the direct factor of $H_\sigma(M) \otimes_{\mathbb{Q}} \mathbb{C}$ on which E acts via τ lies in

$$H^{n(\sigma,\tau),\, w-n(\sigma,\tau)} .$$

(The $n(\sigma,\tau)$ are given by the infinity-type of Ψ : see § 2, formula (5).)

<u>4.2</u> Let us note in passing that, if $M(\Psi)$ and $M(\Psi')$ are motives for Hecke characters Ψ and Ψ' of k with values in E, then the following are equivalent:

- $M(\Psi) \cong M(\Psi')$ <u>over</u> $\overline{\mathbb{Q}}$.
- For some $\sigma\in\Sigma$, $H_\sigma(M(\Psi)) \cong H_\sigma(M(\Psi'))$, as rational Hodge-structures.
- Ψ and Ψ' have the same infinity-type β .

<u>4.3</u> Coming back to our motive M for Ψ , suppose now that $s = 0$ is critical for Ψ (see § 2, formula (6)), and consider the comparison isomorphism

$$I : \bigoplus_{\sigma\in\Sigma} H_\sigma(M) \otimes_{\mathbb{Q}} \mathbb{C} \xrightarrow{\sim} H_{DR}(M) \otimes_{\mathbb{Q}} \mathbb{C} .$$

Note that $H_{DR}(M)$ is by definition a k-vector space, and that $k \otimes_{\mathbb{Q}} \mathbb{C} \cong \mathbb{C}^\Sigma$. So, I is an isomorphism of $k \otimes E \otimes \mathbb{C}$ - modules of rank 1. For $\sigma\in\Sigma$, let e_σ be an E-basis of $H_\sigma(M)$, and put $e = (e_\sigma \otimes 1)_{\sigma\in\Sigma}$. On the right hand side, choose a basis ω of $H_{DR}(M)$ over $k \otimes_{\mathbb{Q}} E$, and decompose

$$\omega = \sum_{(\sigma,\tau)\in\Sigma\times T} \omega_{\sigma,\tau} ,$$

with $\omega_{\sigma,\tau} \in \tau$-eigenspace of $H_{DR}(M) \otimes_{k,\sigma} \mathbb{C}$. Writing $I(e_\sigma) = \sum_{\tau\in T} I(e_\sigma)_\tau$ for the corresponding decomposition of $I(e)$, we find for all $(\sigma,\tau)\in\Sigma\times T$ that

$$\omega_{\sigma,\tau} = p(\sigma,\tau) \,.\, I(e_\sigma)_\tau \quad , \text{ for some } \; p(\sigma,\tau) \in \mathbb{C}^* \;.$$

The unit

$$(p(\sigma,\tau))_{(\sigma,\tau)\in \Sigma\times T} \in (\mathbb{C}^*)^{\Sigma\times T} = (k \otimes E \otimes \mathbb{C})^*$$

gives the "matrix" of I and, up to multiplication by $(k\otimes E)^*$, depends only on Ψ .

4.4 Modulo such a factor one has the relation

$$(8) \qquad p(\sigma,\tau) \cdot p(c\sigma,\tau) \sim (2\,\pi i)^w \;.$$

This amounts essentially to Legendre's period relation, and can be proved in our context (using uniqueness of motives for Hecke characters) from the identity $\Psi\overline{\Psi} = \mathbb{N}^w$. - The motive $\mathbb{Q}(-1)$ attached to the norm character is discussed in more detail, e.g., in [D1], § 3. For (8), it is enough to know that $\mathbb{Q}(-1)$ is a motive defined over $\mathbb{Q}$, with coefficients in $\mathbb{Q}$ such that

$$H_B(\mathbb{Q}(-1)) = \frac{1}{2\pi i}\,\mathbb{Q} \quad \text{and} \quad H_{DR}(\mathbb{Q}(-1)) = \mathbb{Q} \;,$$

with trivial comparison isomorphism. Incidentally, $\mathbb{Q}(-1)$ has no critical s , if considered over a totally imaginary field k .

With (8) and 3.4, calculating the $p(\sigma,\tau)$'s (or their inverses) usually reduces to integrating holomorphic differentials on which E acts via τ or $c\tau$.

4.5 In terms of these $p(\sigma,\tau)$, Deligne's period $\Omega(\Psi)\in(E\otimes\mathbb{C})^*/E^*$ (see (7) above) can be defined componentwise by

$$(9) \qquad \Omega(\Psi,\tau) = D(\Psi)_\tau \cdot \prod_{\sigma\in\Phi(\tau\circ\beta)} p(\sigma,\tau)^{-1} \;.$$

For the definition of the "CM-types" $\Phi(\tau\circ\beta)$, see § 2, formula (6).

Note that the product in (9) is, in fact, well-defined up to a factor in $(E\otimes 1)^*$. - One definition of the "discriminant factor" $D(\Psi) = (D(\Psi)_\tau)_{\tau\in T}$ can be found in [D1], 8.15. This factor arises when one computes the cohomology of $R_{k/\mathbb{Q}}M$ by the Künneth formula: among other things, one has to choose an ordering of Σ. A definition of $D(\Psi)$ which was born out by these cohomological computations - cf. [Ha], esp. 2.4.1 and Cor. 5.7.2B - is as follows. Start with one $\tau\in T$, and let $K_\tau \subset E^\tau$ be the fixed field of

$$\{\rho\in \mathrm{Gal}(\bar{\mathbb{Q}}/\mathbb{Q}) \mid \rho\,\Phi(\tau\circ\beta) = \Phi(\tau\circ\beta)\}\ .$$

$\mathrm{Gal}(\bar{\mathbb{Q}}/K_\tau)$ permutes the set $\Phi(\tau\circ\beta)$. Let $L_\tau \supset K_\tau$ be the fixed field of the kernel of the character

$$\mathrm{Gal}(\bar{\mathbb{Q}}/K_\tau) \longrightarrow \mathfrak{S}(\Phi(\tau\circ\beta)) \xrightarrow{\mathrm{sgn}} \{\pm 1\}\ .$$

Then $[L_\tau : K_\tau] \leq 2$, and $L_\tau = K_\tau(D(\Psi)_\tau)$ for some $D(\Psi)_\tau$ with $D(\Psi)_\tau^{\,2}\in K_\tau^*$.

Now, any $\rho\in\mathrm{Gal}(\bar{\mathbb{Q}}/\mathbb{Q})$ induces a permutation of the set of infinite places of k: both $\Phi(\tau\circ\beta)$ and $\Phi(\rho\tau\circ\beta)$ are in bijection with this set. Call $\varepsilon(\rho)$ the sign of this permutation. Then we set

$$D(\Psi)_{\rho\tau} = \varepsilon(\rho)\ (D(\Psi)_\tau)^\rho$$

The array $(D(\Psi)_\tau)_{\tau\in T}$ is independent, up to a factor in $(E\otimes 1)^*$, of the choices made in defining its components.

Let us list some properties of $D(\Psi)$ - cf. also [Sch].

4.6 a) $D(\Psi)$ depends only on k, E, and the collection of "CM-types" $\{\Phi(\tau\circ\beta) \mid \tau\in T\}$.

b) $D(\Psi)^2 \in (E\otimes 1)^* \subset (E\otimes\mathbb{C})^*$.

c) If k is a CM-field, with maximal totally real subfield k_0, then $D(\Psi) \sim \sqrt{\mathrm{discr}(k_0)}$, up to a factor in $(E\otimes 1)^*$.

d) Let F/k be a finite extension of degree n. Then

$$\frac{D(\Psi\circ N_{F/k})}{D(\Psi)^n} \sim \left(\prod_{\sigma\in\Phi(\tau\circ\beta)} \sigma(\delta_{|\sigma|}) \right)_{\tau\in T} ,$$

up to a factor in $(E\otimes 1)^*$. Here, the right hand side means the following:

Let $d(k^*)^2 \in k^*/(k^*)^2$ be the relative discriminant of F/k . For any infinite place v of k , choose a square root $\delta_v = \sqrt{d} \in k_v^*$. For $\sigma \in \Sigma$, let $|\sigma|$ be the infinite place of k determined by σ and $c\sigma$, and denote by $\sigma(\delta_{|\sigma|}) \in \mathbb{C}^*$ the well-defined image of $\delta_{|\sigma|} \in k^*_{|\sigma|}$ under the continuous isomorphism $k_{|\sigma|} \xrightarrow{\sim} \mathbb{C}$ given by σ . - Note that changing the representative of d or the signs of δ_v , at some places v , multiplies the right hand side of our formula only by a factor in $(E\otimes 1)^*$.

Assume the situation of 4.6,d). From the very definition of the $p(\sigma,\tau)$, and the properties 4.6,a) and d), one finds the following formula for the behaviour of the periods under extension of the base field:

(10)
$$\Delta(F/k,\beta) := \frac{\Omega(\Psi\circ N_{F/k})}{\Omega(\Psi^n)} = \frac{D(\Psi\circ N_{F/k})}{D(\Psi^n)}$$
$$= \frac{D(\Psi\circ N_{F/k})}{D(\Psi)^n} \; \frac{D(\Psi)^n}{D(\Psi^n)} = \left(\prod_{\sigma\in\Phi(\tau\circ\beta)} \sigma(\delta_{|\sigma|}) \right)_{\tau\in T} \cdot D(\Psi)^{n-1} .$$

The array $\Delta(F/k,\beta) \in (E\otimes\mathbb{C})^*$ will reappear in the second theorem of § 5 below. Note that, if k is a CM-field, the second factor of $\Delta(F/k,\beta)$ can be evaluated by 4.6,c). Both factors of Δ are already present in [Ha], in the case $n = 2$, although the formalism there is still somewhat clumsier than the one employed here.

4.7 Let us close this section with a few words on the behaviour of our periods under twisting. For the Tate twist, one finds

(11) $$\Omega(\Psi\cdot\mathbb{N}^m) \sim (2\pi i)^m \, \Omega(\Psi) .$$

If ω is a character of finite order on $K^*_{\mathbb{A},f}/k^*$ with values in E^*, one passes from $\Omega(\Psi)$ to $\Omega(\omega\Psi)$ by leaving $D(\Psi)$ unchanged, and multiplying the $p(\sigma,\tau)$ by certain algebraic numbers with eigen-properties under ω. The details can be found in [Sch]. All we need to know is the following invariance lemma:

If F is a finite extension of k, χ a character of finite order on $F^*_{\mathbb{A},f}/F^*$ with values in E^*, and ω the restriction of χ to $k^*_{\mathbb{A},f}$ (in other words, considering χ and ω on $\mathrm{Gal}(\bar{k}/F)$, $\mathrm{Gal}(\bar{k}/k)$, resp., via class field theory, $\omega = \chi\circ \mathrm{Ver}$, where $\mathrm{Ver} : \mathrm{Gal}(k^{ab}/k) \longrightarrow \mathrm{Gal}(F^{ab}/F)$ is the transfer map), then

$$(12)\qquad \frac{\Omega(\chi\cdot(\Psi\circ N_{F/k}))}{\Omega(\omega\cdot\Psi^n)} = \frac{\Omega(\Psi\circ N_{F/k})}{\Omega(\Psi^n)} = \Delta(F/k,\beta) .$$

Let us mention in passing that the proof of (12) also shows that the quotients

$$\frac{\Omega(\omega\Psi)}{\Omega(\Psi)}$$

may always be expressed by Gauss sums.

§ 5. The rationality conjecture for Hecke L-functions

The proof of Deligne's conjecture (see end of § 2) for the critical values of L-functions of algebraic Hecke characters falls into two parts: The case where the base field k is a CM-field is treated first. From there one passes to the general case by a theorem about the behaviour of special values under extension of the base field.

(I) Let us briefly describe the CM-case:

Historically, the main idea for the CM-case goes back to Eisenstein. But it was Damerell who, in his thesis [Da], published the first comprehensive account of algebraicity results for critical values of Hecke L-functions <u>of imaginary quadratic fields</u>. He also announced finer rationality theorems in that case, but never published them. (The case of imaginary quadratic k was later settled completely in [GS] and [GS'].) In the Fall of 1974, André Weil gave an exposition of work of Eisenstein and Kronecker including, among other things, Damerell's theorem as an application. This course at the IAS - which was later on developed into the book [WEK] - inspired G. Shimura to generalize Damerell's algebraicity results to critical values of Hecke L-functions of arbitrary CM-fields: [Sh 3] . (At that point, he still needed a technical assumption on the infinity-type of the Hecke character.)

To explain the starting point of this method of proof, recall our example in § 1: the L-value there appeared (up to a factor of $\frac{1}{4}$) as an <u>Eisenstein series</u> :

$$\sum_{a,b\in\mathbb{Z}}' \frac{1}{(a+bi)^{4\nu}} ,$$

relative to the lattice $\mathbb{Z}+\mathbb{Z}i$. Now, sometimes the relation between L-value and Eisenstein series is not quite as straightforward - e.g., if, in § 1, we were to study the values $L(\Psi\mathbb{N}^a,0)$ for integers $a \neq 0$ such that $s = 0$ is critical for $\Psi\mathbb{N}^a$, then we would have to transform the Eisenstein series by certain (non holomorphic) differential operators. But except for such operators it remains true that, in any pair of critical values of an Hecke L-function of a CM-field k which are symmetric with respect to the functional equation, there is a value which can be written as a linear combination of Eisenstein series (viz., Hilbert modular forms with respect to the maximal totally real subfield of k), relative to lattices in k .

When k is imaginary quadratic, the algebraicity properties of the Eisenstein series can be derived directly from explicit polynomial relations among them (see, e.g., Weil's treatment of Damerell's theorem in [WEK]). But in general the proof of their algebraicity depends on a theory of canonical models for the Hilbert modular group (as in [Sh 3]) or, equivalently, on an algebraic theory of Hilbert modular forms.

This latter approach was used by Katz in [K1], [K2]. Just like Shimura, Katz did not stop to look at more precise rationality theorems about the special values he had determined up to an algebraic number. In fact, Katz' main concern was with integrality properties and p-adic interpolation.

When Deligne formulated his conjecture in 1977 he felt the need to check that, up to a factor in $\overline{\mathbb{Q}}^*$, it predicted Shimura's theorem. This turned out to be a confusing problem, for the following reason. Shimura expresses the L-values in terms of periods of abelian varieties constructed from lattices in k, which therefore have <u>complex multiplication by</u> k, and are defined over some number field E'. On the other hand, the L-function in question is that of a Hecke character of the field k, with values in some number field E. The motive of such a character arises from abelian varieties <u>defined over</u> k, <u>with complex multiplication by</u> E (or some field closely related to E). This double role of k as field of definition and of coefficients was dealt with by Deligne - up to factors in $\overline{\mathbb{Q}}^*$ - by an ad hoc dualization, see [D1], 8.19. (Its refinement for more precise rationality statements remained the most serious obstacle in the attempt to prove Deligne's conjecture made in [Sch 1].)

Don Blasius managed to solve this problem by writing down an analogue of Deligne's dualization on the level of motives over k, resp. E: his "reflex motive". Thus he was able to prove

<u>Theorem 1</u>: <u>Let</u> k <u>be a CM-field, and</u> Ψ <u>a Hecke character of</u> k, <u>with values in some CM-field</u> E. <u>If</u> $s = 0$ <u>is critical for</u> Ψ, <u>then</u>

$$\frac{L(\Psi,0)}{\Omega(\Psi)} \in E \hookrightarrow E \otimes \mathbb{C} .$$

(Note that any algebraic Hecke character of any number field takes values in a CM-field.)

As Blasius' paper [B] is about to be available we shall not enter into describing the technique of his proof in detail. Suffice it to say that, apart from the "reflex motive" mentioned above, he needs, of course, a very careful analysis of the behaviour of the Eisenstein series under $\mathrm{Gal}(\overline{\mathbb{Q}}/\mathbb{Q})$ (i.e., Shimura's reciprocity law in CM-points), and also the explicit description - due to Tate and Deligne - of the action of $\mathrm{Gal}(\overline{\mathbb{Q}}/\mathbb{Q})$ on abelian varieties of CM-type: see [LCM], chapter 7.

(II) We shall now describe a little bit more in detail the second part of the proof of Deligne's conjecture for Hecke L-functions. It relies on a generalization of [Ha], § 3, from GL_2 to GL_n, and might not be published completely before some time.

Consider the following situation: Let k be a totally imaginary number field, and F/k a finite extension of degree $n \geq 2$. Let Ψ be an algebraic Hecke character with values in a number field E, of infinity-type β. Assume $s = 0$ is critical for Ψ. Let $\chi : F^*_{\mathbb{A}}/F^* \longrightarrow E^*$ be a character of finite order, and put $\omega = \chi|_{k^*_{\mathbb{A}}}$, like in § 4.7 above. Recall the array

$$\Delta(F/k,\beta) = (\Delta(F/k, \tau\circ\beta))_{\tau\in T}$$

defined in § 4.6, formula (10).

<u>Theorem 2</u>:

$$\Delta(F/k,\beta)\ \frac{L_F(\chi\cdot(\Psi\circ N_{F/k}),0)}{L_k(\omega\cdot\Psi^n,0)} \in E \hookrightarrow E\otimes\mathbb{C}\ .$$

<u>Remarks</u>: (i) As the Euler product for $L(\Psi,s)$ converges for $\mathrm{Re}(s) > \frac{w}{2} + 1$, and $s = 0$ is critical for Ψ, it is well-known that the denominator in the theorem is not zero.

(ii) Here is how theorems 1 and 2 imply Deligne's conjecture for all critical values of all Hecke L-functions: Given any totally imaginary number field F , and any Hecke character φ of F , with values in a number field $E_\circ$, of infinity-type $\beta_\circ$, the homogeneity condition (5) of § 2 forces $\beta_\circ$ to factor through the <u>maximal CM-field</u> k <u>contained in</u> F :

$$\beta_\circ = \beta \circ N_{F/k} ,$$

for some algebraic homomorphism

$$\beta : k^* \longrightarrow E_\circ'^* .$$

Choose a Hecke character Ψ of k with infinity-type β , write $\varphi = \chi \cdot (\Psi \circ N_{F/k})$, for some finite order character χ of F , and choose $E \supset E_\circ$ big enough to contain the values of Ψ as well as those of χ . Define $\omega = \chi|_{k^*_{\mathbb{A}}}$. Put $n = [F : k]$. By theorem 1,

$$\frac{L(\omega \cdot \Psi^n, 0)}{\Omega(\omega \cdot \Psi^n)} \in E .$$

But we know the behaviour of the periods Ω under twisting and base extension: see end of § 4. Theorem 2 therefore implies that

$$\frac{L(\varphi, 0)}{\Omega(\varphi)} = \frac{L(\chi \cdot (\Psi \circ N_{F/k}), 0)}{\Omega(\chi \cdot (\Psi \circ N_{F/k}))} \in E \longrightarrow E \otimes \mathbb{C} .$$

Finally, E may now be replaced by $E_\circ$ because Deligne's conjecture is invariant under finite extension of the field of coefficients: [D1], 2.10.

This gives Deligne's conjecture for Hecke L-functions of totally imaginary number fields. These are the only fields with honest regard Hecke L-functions. But it should be said, for the sake of completeness, that Deligne's conjecture for Hecke (=Dirichlet) L-functions of <u>totally real fields</u> follows from results of Siegel's (cf. [D1], 6.7) and, in the case of number fields which are neither totally real nor totally imaginary, no Hecke (=Dirichlet) L-function has any critical value.

The remainder of this section is devoted to sketching the proof of theorem 2. Let us set up some notation.

We consider the following algebraic groups over k :

$$G_\circ/k = GL_n/k .$$

$$T_\circ/k = \text{standard maximal torus}$$

$$B_\circ/k = \text{standard Borel subgroup of upper triangular matrices,}$$

and the two maximal parabolic subgroups

$$P_\circ/k = \left\{ \begin{pmatrix} t & * \\ 0 & p \end{pmatrix} \;\middle|\; p \in GL_{n-1} , \; t \in GL_1 \right\}$$

$$Q_\circ/k = \left\{ \begin{pmatrix} q & * \\ 0 & t \end{pmatrix} \;\middle|\; q \in GL_{n-1} , \; t \in GL_1 \right\} .$$

Dropping the subscript zero will mean taking the restriction of scalars to $\mathbb{Q}$. So,

$$G/\mathbb{Q} = R_{k/\mathbb{Q}}(G_\circ/k) ,$$

and so on.

We introduce the two characters

$$\gamma_P : g = \begin{pmatrix} t & * \\ 0 & p \end{pmatrix} \longmapsto \frac{t^n}{\det(g)} ,$$

and

$$\gamma_Q : g = \begin{pmatrix} q & * \\ 0 & t \end{pmatrix} \longmapsto \frac{\det(g)}{t^n}$$

which we view as characters on the torus extending to $P_\circ$ (resp. $Q_\circ$). The representations of $G_\circ/k$ with highest weight $\nu\gamma_P$ (resp. $\mu\gamma_Q$) are the ν-th (resp. μ-th) symmetric power of the standard representation of $G_\circ/k$ on k^n (resp. its dual $(k^n)^\vee$) .
Coming back to the situation of theorem 2, define a homomorphism

$$\Phi : P(\mathbb{Q}_{\mathbb{A},f}) = P_\circ(k_{\mathbb{A},f}) \longrightarrow E^*$$

by

$$\Phi : \begin{pmatrix} \underline{t}_f & * \\ 0 & \underline{p}_f \end{pmatrix} = \underline{g}_f \longmapsto \Psi_1(\underline{t}_f)\,\Psi\,(\det(\underline{g}_f)) \ .$$

We require that the central character of Φ be our ω . This means that Ψ_1 is determined by

$$\Phi \begin{pmatrix} \underline{t}_f & & \\ & \ddots & \\ & & \underline{t}_f \end{pmatrix} = \omega(\underline{t}_f) = \Psi_1(\underline{t}_f)\,\Psi\,(\underline{t}_f)^n \ .$$

We may view Φ as an "algebraic Hecke character" on $P/\mathbb{Q}$, and it has an infinity-type

$$\text{type } (\Phi) = \gamma \in \text{Hom}(P/\mathbb{Q}\ ,\ R_{E/\mathbb{Q}}(\mathbb{G}_m)) \ .$$

Hence we get an array of types, indexed by $\tau \in T$, with components

$$\text{type}(\tau\circ\Phi) = \tau\circ\gamma \in \text{Hom}(P,\mathbb{G}_m) \ .$$

Recall that

$$\text{Hom}(P,\mathbb{G}_m) = \bigoplus_{\sigma\in\Sigma} \text{Hom}(P_\circ,\mathbb{G}_m) \ ,$$

and that the type β of Ψ is given by the integers $n(\sigma,\tau)$ - see § 2.

It is then easy to check that

$$\tau\circ\gamma = (n(\sigma,\tau)\cdot\gamma_P)_{\sigma\in\Sigma} ,$$

for every $\tau\in T$.

Given τ , define an array of dominant weights

$$\Lambda(\tau) = (\lambda(\sigma,\tau))_{\sigma\in\Sigma}$$

by the rule

$$\lambda(\sigma,\tau) = \begin{cases} (-n(\sigma,\tau)-1)\gamma_Q & \text{if } n(\sigma,\tau) < 0 \\ n(\sigma,\tau)\gamma_P & \text{if } n(\sigma,\tau) \geqq 0 . \end{cases}$$

This affords a representation

$$\rho : G\times_{\mathbb{Q}}\bar{\mathbb{Q}} = \prod_{\sigma\in\Sigma}(GL_n/k) \longrightarrow GL(M(\Lambda(\tau))) ,$$

where $M(\Lambda(\tau)) = \bigotimes_{\sigma\in\Sigma} M(\lambda(\sigma,\tau))$, $M(\lambda(\sigma,\tau))$ being the representation with highest weight $\lambda(\sigma,\tau)$. The system $\{M(\Lambda(\tau))\}_{\tau\in T}$ is a $\mathbb{Q}$-rational system of representations in the sense of [Ha], 2.4 - i.e., the representations are conjugate under $\mathrm{Gal}(\bar{\mathbb{Q}}/\mathbb{Q})$.

As in [Ha], we study the cohomology of congruence subgroups of $GL_n(\mathcal{O})$ with coefficients in these modules: Form the quotients

$$S_K = G(\mathbb{Q})\backslash G(\mathbb{Q}_{\mathbb{A}})/K_\infty\cdot K_f ,$$

where $K_\infty = \prod_{v|\infty} U(n)Z_\infty$ is a standard maximal compact subgroup, times the centre of $G(\mathbb{R}) = G_\infty$, and where K_f is open compact in $G(\mathbb{Q}_{\mathbb{A},f})$. The modules $M(\Lambda(\tau))$ provide coefficient systems $\widetilde{M(\Lambda(\tau))}$ on S_K , and we consider the $\bar{\mathbb{Q}}$ - $G(\mathbb{Q}_{\mathbb{A},f})$ - module

$$H^\bullet(\tilde{S},\widetilde{M(\Lambda(\tau))}) := \varinjlim_{K_f} H^\bullet(S_K, \widetilde{M(\Lambda(\tau))}) .$$

The embedding of S_K into its Borel-Serre compactification $\overline{S_K}$ is a homotopy equivalence. The boundary $\partial\overline{S_K}$ of this compactification has a stratification, with strata corresponding to the conjugacy classes of parabolic subgroups of $G/\mathbb{Q}$. The stratum of lowest dimension, $\partial_B \overline{S_K}$, corresponds to the conjugacy class of Borel-subgroups. The coefficient system can be extended to the boundary, and the limit

$$H^\bullet(\partial_B \widetilde{S}, \widetilde{M(\Lambda(\tau))}) = \varinjlim_{K_f} H^\bullet(\partial_B \overline{S_K}, \widetilde{M(\Lambda(\tau))})$$

is again a $G(\mathbb{Q}_{\mathbb{A},f})$ -module. The diagram

$$S_K \xhookrightarrow{\ i\ } \overline{S}_K \longleftarrow \partial_B \overline{S}_K$$

induces a $G(\mathbb{Q}_{\mathbb{A},f})$ -module homomorphism

$$r_B : H^\bullet(\widetilde{S}, \widetilde{M(\Lambda(\tau))}) \longrightarrow H^\bullet(\partial_B \widetilde{S}, \widetilde{M(\Lambda(\tau))}) .$$

Just as in [Ha] ,II, the right hand side turns out to be a direct sum of modules, induced from an algebraic Hecke character

$$\begin{array}{ccc} \eta : B(\mathbb{Q}_{\mathbb{A},f}) & \longrightarrow & \overline{\mathbb{Q}}^* \\ \searrow & & \nearrow \\ & T(\mathbb{Q}_{\mathbb{A},f}) & \end{array}$$

on $B(\mathbb{Q}_{\mathbb{A},f})$, up to $G(\mathbb{Q}_{\mathbb{A},f})$. The types of these characters are determined by Kostant's theorem, [Ko] ; cf. [Ha], II, for $n = 2$. In particular, it is easily checked that the following induced module (for Φ as above, and $\tau \in T$) is contained in the cohomology of $\partial_B\widetilde{S}$:

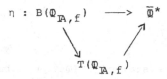

$$V_{\tau\circ\Phi} = \mathrm{Ind}_{B(\mathbb{Q}_{\mathbb{A},f})}^{G(\mathbb{Q}_{\mathbb{A},f})} \tau\circ\Phi =$$

$$= \left\{ h : G(\mathbb{Q}_{\mathbb{A},f}) \longrightarrow \overline{\mathbb{Q}} \;\middle|\; \begin{array}{l} h \text{ is } C_\infty \text{ , and} \\ h(\underline{b}_f \underline{g}_f) = (\tau\circ\Phi)(\underline{b}_f).h(\underline{g}_f) \text{ ,} \\ \text{for all } \underline{b}_f \in B(\mathbb{Q}_{\mathbb{A},f}) \text{ and} \\ \underline{g}_f \in G(\mathbb{Q}_{\mathbb{A},f}) \end{array} \right\}$$

(Here, "C_∞" means right invariance under a suitably small open compact subgroup in $G(\mathbb{Q}_{\mathbb{A},f})$.)

More precisely, we have

$$V_{\tau\circ\Phi} \xhookrightarrow{i_\tau} H^{(n-1)d_\circ}(\partial_B \widetilde{S}, \widetilde{M(\Lambda(\tau))}) ,$$

where $d_\circ = \frac{1}{2}[k : \mathbb{Q}]$, and the system of maps $\{i_\tau\}_{\tau\in T}$ is $\mathbb{Q}$-rational with respect to the two obvious $\mathbb{Q}$-structures on the systems on both sides.

Consider the non-trivial submodule

$$J_{\tau\circ\Phi} = \mathrm{Ind}_{P(\mathbb{Q}_{\mathbb{A},f})}^{G(\mathbb{Q}_{\mathbb{A},f})} \tau\circ\Phi \subset V_{\tau\circ\Phi} .$$

Obviously, $\{J_{\tau\circ\Phi}\}_{\tau\in T}$ is a $\mathbb{Q}$-rational system of $G(\mathbb{Q}_{\mathbb{A},f})$-submodules of $H^{(n-1)d_\circ}(\partial_B \widetilde{S}, \widetilde{M(\Lambda(\tau))})$. The first essential step of the proof is to construct a $\mathbb{Q}$-rational "section" of r_B ,

$$\mathrm{Eis}_\tau : J_{\tau\circ\Phi} \longrightarrow H^{(n-1)d_\circ}(\widetilde{S}, \widetilde{M(\Lambda(\tau))}) ,$$

for all $\tau\in T$. Thus, $r_B \circ \mathrm{Eis}_\tau = \mathrm{Id}$ on $J_{\tau\circ\Phi}$. This section is constructed first over $\mathbb{C}$ by means of residual Eisenstein series or, in other words, non cuspidal Eisenstein series attached to $P/\mathbb{Q}$. To prove that $\{\mathrm{Eis}_\tau\}_{\tau\in T}$ is defined over $\mathbb{Q}$ one has to use a multiplicity one argument, like in [Ha], III. But here this is more complicated. One has to use the spectral sequence which computes the cohomology of the boundary in terms of the cohomology of the strata. Then the cohomology has to be related to automorphic forms, and one has to appeal to results of Jacquet-Shalika on multiplicity one, and of Jacquet on the discrete non cuspidal spectrum.

Once we have the modules

$$\mathrm{Eis}_{\tau}\,(J_{\tau\circ\Phi}) \subset H^{(n-1)d_\circ}(\tilde{S}, \widetilde{M(\Lambda(\tau))})$$

we can proceed more or less in the same way as in [Ha], V: We construct an embedding

$$i_H : F^* \longrightarrow GL_n(k) \quad ,$$

H being the torus with $H(\mathbb{Q}) = i_H(F^*)$. Using this torus we can construct homology classes (compact modular symbols)

$$\mathfrak{Z}(i_H, \tau \circ \chi, \underline{g}) \in H_{(n-1)d_\circ}(\tilde{S}, \widetilde{M(\Lambda(\tau))})$$

depending on a point $\underline{g} \in G(\mathbb{Q}_{\mathbb{A}})$ and on a finite order character

$$\chi : F^*_{\mathbb{A}}/F^* \longrightarrow E^*$$

whose restriction to $k^*_{\mathbb{A}}$ should be ω .

As in [Ha], V, we get an intertwining operator

$$\mathrm{Int}(\mathfrak{Z}(i_H,\chi)) : J_{\tau\circ\Phi} \longrightarrow \mathrm{Ind}_{H(\mathbb{Q}_{\mathbb{A},f})}^{G(\mathbb{Q}_{\mathbb{A},f})}\, \tau\circ\chi$$

by evaluating $\mathrm{Eis}_{\tau}(J_{\tau\circ\Phi})$ on $\mathfrak{Z}(i_H,\chi,\underline{g})$. There is another intertwining operator

$$\mathrm{Int}^{\mathrm{loc}} : J_{\tau\circ\Phi} \longrightarrow \mathrm{Ind}_{H(\mathbb{Q}_{\mathbb{A},f})}^{G(\mathbb{Q}_{\mathbb{A},f})}\, \tau\circ\chi \quad ,$$

constructed as a product of local intertwining operators. Both operators are $\mathbb{Q}$-rational, and for some $x \in E^*$ we find that, for all $\tau \in T$,

$$\mathrm{Int}(\mathfrak{Z}(i_H, \tau\circ\chi)) = \tau(x)\,\Delta(F/k, \tau\circ\beta)\,\frac{L_F(\tau\circ(\chi\cdot(\Psi\circ N_{F/k})),0)}{L_k(\tau\circ(\omega\cdot\Psi^n),0)}\,\mathrm{Int}^{\mathrm{loc}}\,.$$

This implies theorem 2. - The factor $x \in E^*$ can actually be given more explicitly.

§ 6. A formula of Lerch

The fact that a Hecke character determines its motive up to isomorphism produces a period relation whenever two different geometric constructions of a motive for the same character can be given. We have seen a first example of this principle in formula (8) of § 4. The periods $p(\sigma,\tau)$ occuring in this formula comprise those for which Shimura [Sh2] has proved various monomial period relations (up to an algebraic number). These monomial relations were reproven, by means of motives over $\overline{\mathbb{Q}}$, by Deligne, [D2]. They can be refined using the above principle. But we leave aside here this application,as well as some others, referring the reader to [Sch]. Instead, let us concentrate on a typical case involving G. Anderson's motives for Jacobi-sum Hecke characters.

Let $K = \mathbb{Q}(\sqrt{-D})$ be an imaginary quadratic field of discriminant $-D$. Assume for simplicity that $D > 4$. Recall the construction of the simplest Jacobi-sum Hecke character of K, in the sense of [W III], [1974 d]: K is contained in $\mathbb{Q}(\mu_D)$, the field of D-th roots of 1. Write $n = [\mathbb{Q}(\mu_D) : K] = \frac{\varphi(D)}{2}$. For a prime ideal $\mathfrak{P}$ of $\mathbb{Q}(\mu_D)$ not dividing D, put

$$G(\mathfrak{P}) = -\sum_{x \in \mathbb{Z}[\mu_D]/\mathfrak{P}} \chi_{D,\mathfrak{P}}(x)\cdot\lambda(x)\quad,$$

with "$\chi_{D,\mathfrak{P}}(x) \equiv x^{(\mathbb{N}\mathfrak{P}-1)/D} \pmod{\mathfrak{P}}$" the $D^{\underline{th}}$-power residue symbol mod $\mathfrak{P}$, and $\lambda(x) = \exp(2\pi i\cdot \mathrm{tr}_{(\mathbb{Z}[\mu_D]/\mathfrak{P})/\mathbb{F}_p}(x))$.

Then extend the function of prime ideals $\mathfrak{p}$ of K with $\mathfrak{p} \nmid D$:

$$J(\mathfrak{p}) = \prod_{\mathfrak{P} | \mathfrak{p}} G(\mathfrak{P}) ,$$

multiplicatively to all ideals of K prime to D . Elementary properties of Gauss sums show that J takes values in K . By a theorem of Stickelberger and an explicit version of the analytic class number formula for K one finds that, <u>if</u> J is an algebraic Hecke character, <u>then</u> its infinity-type

$$\beta : K^* \longrightarrow K^*$$

is given by

$$x \longmapsto x^{(n+h)/2} \, \bar{x}^{(n-h)/2} ,$$

where h is the class number of K . (Note that n and h have the same parity, by genus theory.) In other words, if J is a Hecke character, then

(13) $$J . \mathbb{N}^{-(n+h)/2} = \mu \cdot \Psi^h ,$$

for some character μ of $K^*_{\mathbb{A}}$ of finite order and some Hecke character Ψ of K of weight -1 .

That J is in fact a Hecke character, i.e., is well-behaved at the places dividing D , if viewed on idèles, was proved by Weil (loc. cit). But it can also be deduced, e.g., from the following construction of a motive for J which was given by Greg Anderson, [A1],[A2].

Anderson finds a motive defined over K , with coefficients in K , whose ℓ-adic representations are given by J , in H^{n-2} of the zero-set Z of the function

$$\begin{aligned} \mathbb{Q}(\mu_D) &\longrightarrow K \\ x &\longmapsto tr_{\mathbb{Q}(\mu_D)/K}(x^D) \end{aligned} ,$$

viewed as a projective variety in $\mathbb{P}_K(\mathbb{Q}(\mu_D))$, the projective space of the K-vector space $\mathbb{Q}(\mu_D)$. Note that

$$Z\ x_K\mathbb{Q}(\mu_D) = \left\{ x_1^D + \ldots + x_n^D = 0 \right\} \subset \mathbb{P}^{n-1} .$$

Anderson's construction is of course motivated by the well-known fact that Fermat-hypersurfaces contain motives (carved out by the action of their large automorphism groups) attached to Jacobi-sum Hecke characters of cyclotomic fields: see [DMOS], pp. 79 - 96. For details of Anderson's more general construction, we refer to his preprints, or to [Sch].

At any rate, thanks to Anderson's work, we have at our disposal a motive $M(J)$ for the character J , which lies in the category of motives obtained from abelian varieties. (This last fact is proved by Shioda-induction: [DMOS], p. 217). Thus, by (13), the periods of the motive

$$M(J\,\mathbb{N}^{-(n+h)/2}) = M(J) \otimes_K K((n+h)/2)$$

will be the same as those of any motive constructed for the character $\mu \cdot \psi^h$.

The period calculations on Fermat-hypersurfaces always reduce eventually to Beta-integrals. For $M(J)$ one essentially gets the product

$$\prod_{\chi(a)=-1} \Gamma\left(<\frac{a}{D}>\right)^{-1} .$$

Here, $\chi(p) = (\frac{-D}{p})$ is the Dirichlet character of the quadratic field K , and the product is taken over those $a \in (\mathbb{Z}/D\,\mathbb{Z})^*$ for which $\chi(a) = -1$. $<\frac{a}{D}>$ is the representative of the class $\frac{a}{D}$ mod $\mathbb{Z}$ which lies between 0 and 1 .

A motive for $\mu\cdot\Psi^h$ can be built up from elliptic curves with complex multiplication by K . - Assume for simplicity that $\Psi\circ N_{H/K}$ takes values in K^* , for H the Hilbert class field of K . Choose any elliptic curve A/H such that $H_1(A)$ is a motive for $\Psi\circ N_{H/K}$, and call $B = R_{H/K}A$ its restriction of scalars to K . Calling E the field of values of Ψ , $H_1(B)^{\otimes_E h}$ can be shown to be a motive for Ψ^h (viewed as taking values in E): cf. [GS], § 4. Using formulas derived in [GS], § 9, the periods of this motive can be computed in terms of the periods Ω_σ of the conjugates A^σ/H of our elliptic curve A , for

$$\sigma \in \mathrm{Gal}(H/K) = C\ell(K) .$$

Straightening out the twists by the norm and the finite order character (cf. § 4.6 and 4.7), one finally obtains the following relation, up to a factor of K^* :

$$(14) \qquad y \cdot \prod_{\sigma\in C\ell(K)} \left(\sqrt{\frac{D}{2\pi}}\cdot\Omega_\sigma\right) \sim \prod_{\chi(a)=-1} \frac{\sqrt{\pi}}{\Gamma(<\frac{a}{D}>)} ,$$

where y generates the abelian extension of K belonging to μ . Multiplying (14) with its complex conjugate, we get

$$(15) \qquad z \cdot \prod_{\sigma\in C\ell(K)} \left(\frac{D}{2\pi}\,\Omega_\sigma\,\overline{\Omega}_\sigma\right) = \prod_{a\in(\mathbb{Z}/D\mathbb{Z})^*} \Gamma(<\frac{a}{D}>)^{\chi(a)} ,$$

for some z with $z^4\in\mathbb{Q}^*$. Except for the different interpretation of z and the Ω_σ , this is the exponential of an identity proved analytically by Lerch in [Le], p.303. The first geometric proof of (15), up to a factor in $\overline{\mathbb{Q}}^*$, was given by Gross in [Gr], a paper which in turn inspired Deligne's proof of the theorem about absolute Hodge cycles on abelian varieties - which again is essential in proving uniqueness of the motive for an algebraic Hecke character.

REFERENCES

[A1] G. Anderson, The motivic interpretation of Jacobi sum Hecke characters; preprint.

[A2] G. Anderson, Cyclotomy and an extension of the Taniyama group , preprint.

[B] Don Blasius, On the critical values of Hecke L-series; preprint

[Da] R.M. Damerell, L-functions of elliptic curves with complex multiplication. I, Acta Arithm. 17 (1970), 287 - 301; II, Acta Arithmetica 19 (1971), 311 - 317.

[D1] P. Deligne, Valeurs de fonctions L et périodes d'intégrales; Proc. Symp. Pure Math. 33 (1979), part 2; 313 - 346.

[D2] P. Deligne (texte rédigé par J.L. Brylinski), Cycles de Hodge absolus et périodes des intégrales des variétés abéliennes; Soc. Math. France, Mémoire n° 2 (2ème sér.) 1980, p. 23 - 33.

[DMOS] P. Deligne, J. Milne, A. Ogus, K. Shih, Hodge Cycles, Motives and Shimura Varieties; Springer Lect. Notes Math. 900 (1982).

[Ei] G. Eisenstein, Über die Irreductibilität und einige andere Eigenschaften der Gleichung, von welcher die Theilung der ganzen Lemniscate abhängt, - and the sequels to this paper -; Math. Werke II, 536 - 619.

[GS] C. Goldstein, N. Schappacher, Séries d'Eisenstein et fonctions L de courbes elliptiques à multiplication complexe; J.r. ang. Math. 327 (1981), 184 - 218.

[GS'] C. Goldstein, N. Schappacher, Conjecture de Deligne et Γ-hypothèse de Lichtenbaum sur les corps quadratiques imaginaires. CRAS Paris, t. 296 (25 Avril 1983), Sér. I, 615-618.

[Gr] B.H. Gross, On the periods of abelian integrals and a formula of Chowla and Selberg; Inventiones Math. 45 (1978), 193 - 211.

[Ha] G. Harder, Eisenstein cohomology of arithmetic groups - The case GL_2 ; preprint Bonn 1984.

[Hu] A. Hurwitz, Über die Entwicklungskoeffizienten der lemniskatischen Funktionen; first communication in: Nachr. k. Ges. Wiss. Göttingen, Math. Phys. Kl. 1897, 273 - 276 = Math. Werke II, n° LXVI, 338 - 341. Published in extenso: Math. Ann. 51 (1899), 196 - 226 = Math. Werke II, n° LXVII, 342 - 373.

[K1] N. Katz, p-adic interpolation of real analytic Eisenstein series; Ann. Math. 104 (1976), 459 - 571.

[K2] N. Katz, p-adic L-functions for CM-fields; Inventiones math. 49 (1978); 199 - 297.

[Ko] B. Kostant, Lie algebra cohomology and the generalized Borel-Weil theorem; Ann. Math. 74 (1961), 329 - 387.

[LCM] S. Lang, Complex Multiplication; Springer: Grundlehren 255, 1983.

[La] R.P. Langlands, Automorphic Representations, Shimura Varieties, and Motives. Ein Märchen; Proc. Symp. Pure Math. 33 (1979), part 2; 205 - 246.

[Le] M. Lerch, Sur quelques formules relatives au nombre des classes; Bull. Sc. Mathém. (2) 21 (1897), prem. partie, 290 - 304.

[Sch1] N. Schappacher, Propriétés de rationalité de valeurs spéciales de fonctions L attachées aux corps CM; in: Séminaire de théorie de nombres, Paris 1981-82, Birkhäuser (PM 38), 1983; 267 - 282.

[Sch] N. Schappacher, On the periods of Hecke characters; in preparation.

[Sℓ] J-P. Serre, Abelian ℓ-adic representations and elliptic curves; Benjamin 1968.

[Sh1] G. Shimura, On the zeta-function of an abelian variety with complex multiplication; Ann. Math. 94 (1971), 504 - 533.

[Sh2] G. Shimura, Automorphic forms and the periods of abelian varieties; J. Math. Soc. Japan 31 (1979), 561 - 592.

[Sh3] G. Shimura, On some arithmetic properties of modular forms of one and several variables; Ann. Math. 102 (1975), 491-515.

[WIII] A. Weil, Oeuvres Scientifiques - Collected Papers, vol. III. Springer 1980.

[WEK] A. Weil, Elliptic functions according to Eisenstein and Kronecker. Springer 1976.

AN INTRODUCTION TO INFINITESIMAL VARIATIONS OF HODGE STRUCTURES

Joe Harris
Mathematics Department, Brown University
Providence, RI 02912, U.S.A.

The purpose of this note is to give a simple introduction to the notion of infinitesimal variation of Hodge structure. This is an object first defined and used in [1] (though the underlying ideas had been in the air for a while) and more recently the subject of an excellent monograph by Peters and Steenbrink [2]. Unfortunately, this theory, which in fact should make life easier for mathematicians trying to apply Hodge theory to geometry, gives at first the impression of being complicated and technical. It is my hope here to avoid this impression by presenting the basic ideas of the theory in as simple a fashion as possible.

We begin by recalling the basic set-up of Hodge theory. The goal of this theory is to associate to an m-dimensional complex manifold X (for simplicity we will take X a submanifold of $\mathbb{P}^N$) a linear-algebraic invariant, as follows. To begin with, we can for each n associate to X its $n^{\underline{th}}$ topological cohomology group modulo torsion $H_{\mathbb{Z}} = H^n(X,\mathbb{Z})/\text{tors}$, or its complexification the $n^{\underline{th}}$ deRham cohomology group $H_{\mathbb{C}} = H_{\mathbb{Z}} \otimes \mathbb{C} = H^n_{DR}(X)$. We can also associate the cup product in cohomology; or rather, since we are only dealing with one group at a time, the bilinear pairing

$$Q : H_{\mathbb{Z}} \times H_{\mathbb{Z}} \longrightarrow \mathbb{Z}$$

defined by

$$Q(\alpha,\beta) = \int_X \alpha \cup \beta \cup \omega^{m-n}$$

where ω is the restriction to X of the generator of $H^2(\mathbb{P}^N,\mathbb{Z})$. Of course, these are invariants of the underlying differentiable manifold of X, and do not reflect its complex structure. What does determine the complex structure of X is the decomposition of the complexified tangent spaces to X into holomorphic and anti-holomorphic parts; or, equivalently, the decomposition of the space $A^n(X)$ of differential forms of degree n on X by type:

$$A^n(X) = \bigoplus_{p+q=n} A^{p,q}$$

Naturally, this data is too cumbersome to carry around, but here we are in luck: by the Hodge theorem, this decomposition descends to the level of cohomology. Precisely, if we let $H^{p,q} = H^{p,q}(X) \subset H^n_{DR}(X)$ be the subspace of classes representable by forms of type (p,q), we get a decomposition

$$H^n_{DR}(X) = H_{\mathbb{C}} = \bigoplus_{p+q=n} H^{p,q}$$

satisfying the obvious relations

$$\overline{H^{p,q}} = H^{q,p}$$

and

$$Q(H^{p,q}, H^{p',q'}) \equiv 0 \quad \text{unless} \quad p+p' = q+q' = n .$$

The package of data introduced so far -- a lattice $H_{\mathbb{Z}}$ with integral bilinear form Q and decomposition $H_{\mathbb{Z}} \otimes \mathbb{C} = \oplus H^{p,q}$ satisfying these relations -- we call a Hodge structure of weight n associated to X. It is an object that is on one hand essentially finite, and that on the other hand we may hope will reflect the geometry of X.

Now, whenever we associate to a geometric object a (presumably simpler) invariant, two questions arise: to what extent does the invariant actually determine the original object; and to what extent can we read off directly from the invariant answers to naive questions about the geometry of the object. In the present circumstances, the first question translates into the Torelli problem, which asks when the members of a given family of varieties (e.g., curves of genus g, hypersurfaces of degree d in $\mathbb{P}^n$) are determined by their Hodge structures; or the "generic" or "birational" Torelli problem, which asks when this is true for a general member of the family. The Torelli problem has been answered in a number of cases (e.g., for curves of genus g it was proved classically by Torelli; and the generic Torelli for hypersurfaces was proved recently by Donagi); it remains very much an open question in general.

The most famous example of a question in Hodge theory along the lines of the second sort above is of course the Hodge conjecture. It is not hard to see that if $Y \subset X$ is an analytic subvariety of codimension k, its fundamental class must lie in the subspace $H^{k,k} \subset H^{2k}(X,\mathbb{C})$. The Hodge conjecture asks whether the converse is true: that is, whether a class $\gamma \in H^{k,k} \cap H_{\mathbb{Z}}$ is necessarily a rational linear combination of classes of subvarieties.

The simplest case of Hodge theory is its application to curves, and here by any standards it is successful. To the Hodge structure

$(H_{\mathbb{Z}}, Q, H^{1,0} \oplus H^{0,1})$ of a curve we associate the projection Λ of $H_{\mathbb{Z}}$ to $H^{0,1}$ (traditionally represented by the period matrix: we choose a basis for $H_{\mathbb{Z}}$ normalized with respect to Q and write out the (0,1)-components of these vectors in a $g\times 2g$ matrix Ω) and then the complex torus $H^{0,1}/\Lambda = J(C)$, called the Jacobian of C. This in turn gives rise to a host of subvarieties of $J(C)$ and theta-functions that reflect and elucidate the geometry of C.

For higher-dimensional varieties, the application of Hodge theory has been less successful, for which there are perhaps two reasons. The first of these is the apparent absence of any reasonably natural geometric and/or analytic object associated to a Hodge structure in general. Looking at the case of curves, one sees that it is exactly through the geometry of the Jacobian, and the analysis of the theta-function, that Hodge theory is useful. Unfortunately, no analogous objects have been found in general.

The second factor is simply this: that only in a very few cases can one ever hope to determine explicitly the Hodge structure of a given variety. To be specific, consider the case of a smooth hypersurface $X \subset \mathbb{P}^{n+1}$ given by a homogeneous polynomial $F(Z) = 0$ of degree d. By the Lefschetz theorem, all the cohomology of X below the middle dimension (and hence above it as well) is at most one-dimensional, so we focus on $H^n(X)$. We can immediately identify one of the Hodge groups: $H^{n,0}(X)$, the space of holomorphic n-forms on X, may be realized as Poincaré residues of $(n+1)$-forms on $\mathbb{P}^{n+1}$ with poles along X; explicitly,

$$\omega = \operatorname{Res}\left(\frac{G(Z_0,\dots,Z_{n+1})\, d\left(\frac{Z_1}{Z_2}\right)\wedge\dots\wedge d\left(\frac{Z_{n+1}}{Z_0}\right)\cdot Z_0^{n+1}}{F(Z_0,\dots,Z_n)}\right)$$

$$= \frac{G(Z_0,\dots,Z_{n+1})\, d\left(\frac{Z_1}{Z_0}\right)\wedge\dots\wedge d\left(\frac{Z_i}{Z_0}\right)\wedge\dots\wedge d\left(\frac{Z_{n+1}}{Z_0}\right)\cdot Z_0^{n}}{\frac{\partial F}{\partial Z_i}(Z_0,\dots,Z_{n+1})}$$

for $G(Z)$ a homogeneous polynomial of degree $d-n-1$. Thus

$$H^{n,0} = S_{d-n-1}$$

where S is the graded ring $\mathbb{C}[Z_0,\dots,Z_{n+1}]$. Similarly, the other Hodge groups of X may be realized as residues of forms on $\mathbb{P}^{n+1}$ with higher-order poles on X (actually, we get in this way just the <u>primitive</u> cohomology $H^n_{pr}(X)$, which here means the classes orthogonal to ω). We obtain an identification

$$H^{n-k,k}_{pr}(X) = (S/J)_{(k+1)d-n-1}$$

where $J \subset S$ is the Jacobian ideal of X, that is, the homogeneous ideal generated by the partial derivatives of X.

We have thus found the vector space decomposition $H^n_{pr}(X) = \oplus H^{n-k,k}_{pr}(X)$. The problem is, it is impossible in general to identify in these terms the lattice $H_{\mathbb{Z}}$ of integral classes. Indeed, this has been done only in the presence of a large automorphism group acting on X, e.g., for Fermat hypersurfaces. Thus, for example, if one is given a particular hypersurface of even dimension $n = 2k$, it is impossible to determine in general $H^{k,k}(X) \cap H^n(X,\mathbb{Z})$, or when two such X have the same Hodge structure. Simply put, we cannot find the lattice; but without the lattice we have no invariants.

One solution of this difficulty appears at first to be moving in the wrong direction, toward increased difficulty. One considers not just a variety X, but a family of varieties $\{X_b\}_{b\in B}$ parametrized by a variety B, of which $X = X_0$ is a member; we assume $0 \in B$ is a smooth point. Locally around X_0, then, we can identify the lattices $H^n(X_b,\mathbb{Z})/\mathrm{tors}$ with a single lattice $H_{\mathbb{Z}}$ and the vector spaces $H^n(X_b,\mathbb{C})$ with $H_{\mathbb{C}}$ correspondingly. We then consider the spaces $H^{n-k,k}(X_b)$ -- or the associated

$$F^k = \bigoplus_{\ell=0}^{k} H^{n-\ell,\ell}(X_b) \text{ --}$$

as variable subspaces of $H_{\mathbb{C}}$. The basic facts then are:

i) The map ϕ_k from B (or a neighborhood of $0 \in B$) to the Grassmannian sending b to $F^k(X_b) \subset H_{\mathbb{C}}$ is holomorphic; and

ii) In terms of the identification of the tangent space to the Grassmannian at $\Lambda \subset H$ with $\mathrm{Hom}(\Lambda,H/\Lambda)$, the image under $\delta_k = d\phi_k$ of any tangent vector to B at 0 carries F_k into F_{k+1}/F_k. We thus arrive at a collection of maps $\delta_k : T_0B \longrightarrow \mathrm{Hom}(H^{n-k,k}(X),H^{n-k-1,k+1}(X))$. By equality of mixed partials, they satisfy the relations

$$(*) \qquad \delta_{k+1}(v)\circ\delta_k(w) = \delta_{k+1}(w)\circ\delta_k(v) \qquad \forall \ v,w \in T$$

and since the spaces $F_k(X_b)$ satisfy the relation $Q(F_k,F_{n-k-1}) \equiv 0$ for all b, we have

(**) $$Q(\delta_k(v)(\alpha),\beta)+Q(\alpha,\delta_{n-k-1}(v)(\beta)) = 0$$

$$\forall\, \alpha \in H^{n-k,k}(X)\ ,\ \beta \in H^{k+1,n-k-1}(X)\ ,\ v \in T\ .$$

We now define an <u>infinitesimal variation of Hodge structure</u> (IVHS) to be just this collection of data: that is, a quintuple $(H_{\mathbb{Z}},Q,H^{p,q},T,\delta_q)$ in which $(H_{\mathbb{Z}},Q,H^{p,q})$ is a Hodge structure, T a vector space, and

$$\delta_q : T \longrightarrow \mathrm{Hom}(H^{p,q},H^{p-1,q+1})$$

maps satisfying (*) and (**) above. By what we have just said, to every member $X = X_0$ of a family of varieties $\{X_b\}$ we have associated such an object.

Two key observations here are the following:

i) The infinitesimal variation of Hodge structure associated to a family is in general computable; or at least as computable as the Hodge structures associated to the members. For example, going back to our example of hypersurfaces, if we let $X \subset \mathbb{P}^{n+1}$ be smooth with equation $F(Z) = 0$, the tangent space at X to the family of hypersurfaces of degree d up to projective isomorphism is just the space S_d of homogeneous polynomials of degree d, modulo the Jacobian ideal. (A variation of X in $\mathbb{P}^{n+1}$ is given by $F+\varepsilon G$ for $G \in S_d/\mathbb{C}F$; if $G = \Sigma a_{ij}X_i \frac{\partial F}{\partial X_j}$ this corresponds to first order to the motion of X along the 1-parameter group e^{tA} of automorphisms of $\mathbb{P}^{n+1}$). Thus $T = (S/J)_d$; and the maps

$$\delta_k : (S/J)_d \longrightarrow \mathrm{Hom}((S/J)_{(k+1)d-n-1},(S/J)_{(k+2)d-n-1})$$

turn out to be nothing but polynomial multiplication.

It should be noted here that this in itself has some nice consequences: for example, while we are as indicated earlier unable to determine $H_{k,k}(X) \cap H^{2k}(X,\mathbb{Z})$ for any given hypersurface in $\mathbb{P}^{n+1}$, $n = 2k$, the fact that for $d \geq n+1$ the map

$$(S/J)_d \times (S/J)_{kd-n-1} \longrightarrow (S/J)_{(k+1)d-n-1}$$

is surjective immediately implies that for <u>general</u> X, $H^{k,k}_{pr}(X) \cap H^{2k}(X,\mathbb{Z}) = 0$, and so $H^{k,k}(X) \cap H^{2k}(X,\mathbb{Z}) = \mathbb{Z}$. Thus on a general hypersurface every algebraic subvariety is homologous to a rational multiple of a complete intersection. In particular in case $n = 2$ this yields the famous

Theorem (Noether; Lefschetz): a surface $S \subset \mathbb{P}^3$ of degree $d \geq 4$, having general moduli, contains no curves other than complete intersections $S \cap T$ with other surfaces.

2) The second key point is this: that even without the lattice $H_{\mathbb{Z}}$, an infinitesimal variation of Hodge structure will in general possess non-trivial invariants, and will give rise to geometric objects. These of course come from the maps δ_k which, being trilinear objects, have lots of accessible invariants (e.g. their associated determinantal varieties).

To illustrate the use of this, consider the generic Torelli theorem for hypersurfaces. The application of IVHS to this problem is based on the following trick: for any map $f : X \longrightarrow Y$ of varieties, the condition that f is birational onto its image, i.e. that

$$\text{for general } p \in X , \not\exists\, q \in X : q \neq p , f(q) = f(p)$$

is in fact equivalent to the a priori weaker statement

$$\text{for general } p \in X , \not\exists\, q \in X :$$

$$q \neq p , f(q) = f(p) \text{ and } \operatorname{Im}(f_*)_q = \operatorname{Im}(f_*)_p .$$

In our present circumstances, this equivalence means that

A general hypersurface of degree d in $\mathbb{P}^{n+1}$ is determined by its Hodge structure

$\Longleftrightarrow$

A general hypersurface of degree d in $\mathbb{P}^{n+1}$ is determined by its infinitesimal variation of Hodge structure.

Thus, to prove the generic Torelli theorem for hypersurfaces, Donagi shows that from the data of the vector spaces

$$(S/J)_{(k+1)d-n-1} , (S/J)_d$$

and the multiplication maps

$$\delta_{k-1} : (S/J)_d \times (S/J)_{kd-n-1} \longrightarrow (S/J)_{(k+1)d-n-1}$$

$$Q : (S/J)_{(k+1)d-n-1} \times (S/J)_{(n-k+1)d-n-1} \longrightarrow (S/J)_{(n+2)d-n-1} \cong \mathbb{C}$$

one can reconstruct the entire ring S/J, and from this the hypersurface X. This suffices to establish the theorem; again, it should be observed that at no point in the argument is the lattice $H_{\mathbb{Z}}$ mentioned.

Donagi's argument is a beautiful one, but this is not the place to reproduce it. Let me instead conclude by giving a similar and easier example of the use of IVHS: to prove the generic Torelli theorem for curves of genus $g \geq 5$.

Of course, the Torelli theorem has been proved many times over, in as strong a form as one could wish. One common characteristic of the proofs, however, is that they all make essential use of the geometry of the Jacobian and its subvarieties. A natural question if one is studying higher-dimensional Torelli theorems is: does there exist a proof of the Torelli for curves that avoids the use of the Jacobian? The answer to this is unknown to me; however, using IVHS we can give a very short proof of the generic Torelli in genus $g \geq 5$ as follows.

The tangent space, at a curve C, to the family of all curves is dual to the space $H^0(C,K^2)$ of quadratic differentials on C. The IVHS associated to C in this family thus consists of the Hodge structure of C, together with a map

$$\delta : H^0(C,K^2) \longrightarrow \mathrm{Hom}(H^{1,0},H^{0,1}) .$$

Here the relation (*) above is trivial; while the relation (**) says that in terms of the identification of $H^{0,1}$ with $(H^{1,0})^*$ given by Q, the image of δ lies in the subspace $\mathrm{Sym}^2(H^{1,0})^* \subset \mathrm{Hom}(H^{1,0},(H^{1,0})^*)$, i.e.

$$\delta : H^0(C,K^2)^* \longrightarrow \mathrm{Sym}^2(H^0(C,K)^* .$$

The transpose of δ is now easy to identify: it is the map

$${}^t\delta : \mathrm{Sym}^2 H^0(C,K) \longrightarrow H^0(C,K^2)$$

that simply takes a quadratic polynomial $P(\omega_1,\ldots,\omega_g)$ in the holomorphic differentials on C and evaluates it as a quadratic differential on C. In particular, the kernel of ${}^t\delta$ is just the vector space of quadratic polynomials vanishing on the image of the canonical curve $C \subset \mathbb{P}H^0(C,K)^* = \mathbb{P}^{g-1}$; since it is well known that a general canonical curve of genus $g \geq 5$ is the intersection of the quadrics containing it, we can recover the curve C. Explicitly, in terms of the infinitesimal variation of Hodge structure $(H_{\mathbb{Z}}, Q, H^{1,0} \oplus H^{0,1}, T, S)$ associated to C, we have

$$C = \mathbb{P}\{ Z \in H^{0,1} : Q(Z,\lambda(Z)) = 0 \quad \text{for all} \quad \lambda : H^{0,1} \longrightarrow H^{1,0}$$
$$\text{such that} \quad \mathrm{trace}(\lambda \circ \delta(v)) = 0 \quad \text{for all} \quad v \in T \}$$

and this suffices to establish generic Torelli.

References

[1] R. Donagi, J. Carlson, M. Green, P. Griffiths and J. Harris, Compositio Math 50(1983).

[2] C. Peters and J. Steenbrink, Infinitesimal variations of Hodge structure and the ϕ generic Torelli problem for projective hypersurfaces, in Birkhauser Progress in Mathematics series number 39. (1984)

NEW DIMENSIONS IN GEOMETRY

Yu. I. Manin

Steklov Mathematical Institutc
Moscow USSR

Introduction

Twenty-five years ago André Weil published a short paper entitled "De la métaphysique aux mathématiques" [37]. The mathematicians of the XVIII century, he says, used to speak of the "methaphysics of the calculus" or the "metaphysics of the theory of equations". By this they meant certain dim analogies which were difficult to grasp and to make precise but which nevertheless were essential for research and discovery.

The inimitable Weil style requires a quotation.

"Rien n'est plus fécond, tous les mathématiciens le savent, que ces obscures analogies, ces troubles reflets d'une théorie à une autre, ces furtives caresses, ces brouilleries inexplicables; rien aussi ne donne plus de plaisir au chercheur. Un jour vient où l'illusion se dissipe, le pressentiment se change en certitude; les théories jumelles rêvèlent leur source commune avant de disparaître; comme l'enseigne la Gitá on atteint à la connaissance et à l'indifférence en même temps. La métaphysique est devenu mathématique, prête à former la matière d'un traité dont la beauté froide ne saurait plus nous émouvoir".

I think it is timely to submit to the 25th Arbeitstagung certain variations on this theme. The analogies I want to speak of are of the following nature.

The archetypal m-dimensional geometric object is the space $\mathbb{R}^m$ which is, after Descartes, represented by the polynomial ring $\mathbb{R}[x_1,\dots,x_m]$.

Consider instead the ring $\mathbb{Z}[x_1,\dots,x_m;\xi_1,\dots,\xi_n]$, where $\mathbb{Z}$ denotes the integers and ξ_i are "odd" variables anticommuting among themselves and commuting with the "even" variables x_K . It is convenient to associate with this ring a certain geometric object of dimension $1+m+n$, or better still $(1;m|n)$, where 1 refers to the "arithmetic dimension" $\mathbb{Z}$, m to the ordinary geometric dimensions $(x_1,\ldots,x_m)$ and n to the new "odd dimensions" represented by the coordinates ξ_i.

Before the advent of ringed spaces in the fifties it would have been difficult to say precisely what we mean when we speak about this geometric object. Nowadays we simply define it as an "affine superscheme" Spec $\mathbb{Z}[x_i,\xi_K]$, an object of the category of topological spaces locally ringed by a sheaf of $\mathbb{Z}_2$-graded supercommutative rings (cf. n°4 below) I have tried to draw the "three-space-2000", whose plain x-axis is supplemented by the set of primes and by the "black arrow", corresponding to the odd dimension.

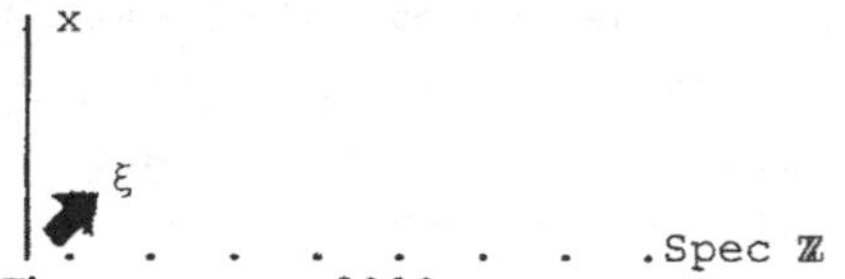

Three-space-2000

The message of the picture is intended to be the following methaphysics underlying certain recent developments in geometry: "<u>all three types of geometric dimensions are on an equal footing</u>".

Actually the similarity of Spec $\mathbb{Z}$ to Spec $k[x]$, or in general of algebraic number fields to algebraic function fields, is a well known heuristic principle which led to the most remarkable discoveries in the diophantine geometry of this century. This similarity was in fact the subject matter of the Weil paper I just quoted. Weil likens the three theories, those of Riemann surfaces, algebraic numbers and algebraic curves over finite fields, to a trilingual inscription with parallel texts. The texts have a common theme but not identical. Also they have been partly destroyed, each in different places, and we are to decipher the enigmatic parts and to reconstruct the missing fragments.

In this talk I shall be concerned with only one aspect of this similarity, reflected in the idea that one may compactify a projective scheme over $\mathbb{Z}$ by adding to it a fancy infinite closed fibre. In the remarkable papers [1], [2] S. Arakelov has shown convincingly that in this way the arithmetic dimension acquires truly geometric global properties, not just by itself, but in its close interaction with the "functional" coordinates. G. Faltings [10], [11] has pushed through Arakelov's idea much further and beyond doubt (for me) the existence of a general arithmetic geometry, or A-geometry. This A-geometry is expected to contain the analogues of all main results of conventional algebraic geometry. The leading idea for the construction of the arithmetic compactifications seems to be as follows:

Kähler-Einstein geometry = ∞-adic arithmetic

I have tried in sections 1-3 of this talk to bring together our scattered knowledge on this subject.

Starting with section 4 the odd dimensions enter the game. The algebraic geometers are well accustomed to envisage the spectrum of the dual numbers Spec $\mathbb{R}[\varepsilon]$, $\varepsilon^2 = 0$, as the infinitesimal arrow and will hardly object to a similar visualization of Spec $\mathbb{R}[\xi]$. Still, there is an essential difference between these two cases. The even arrow Spec $\mathbb{R}[\varepsilon]$ is not a manifold but only an infinitesimal part of a manifold. This can be seen e.g. in the fact that $\Omega^1[\varepsilon]/\mathbb{R}$ is not $\mathbb{R}[\varepsilon]$-free, since from $\varepsilon^2 = 0$ it follows that $\varepsilon d\varepsilon = 0$. By contrast, the odd arrow Spec $\mathbb{R}[\xi]$ is an honest manifold from this point of view, since the $\mathbb{Z}_2$-graded Leibniz formula for, say, the even differential $d\xi$, is valid automatically, $\xi \cdot d\xi + d\xi \cdot \xi = 0$ and one easily sees that $\Omega^1\mathbb{R}[\xi]/\mathbb{R}$ is $\mathbb{R}[\xi]$-free.

In spite of the elementary nature of this example it shows why the odd nilpotents in the structure sheaf may deserve the name of coordinates. But of course this is only a beginning.

The most remarkable result of supergeometry up to now is probably the extension of the Killing-Cartan classification to the finite dimensional simple Lie supergroups made in [15], [16]. The Lie supergroups acting on supermanifolds mix the even and the odd coordinates, which is one reason more to consider them on an equal footing.

In sections 5 and 6 we state some recent results of A. Vaintrob, J. Skornyakov, A. Voronov, I. Penkov and the author on the geometry of supermanifolds. They refer to the Kodaira deformation theory and the construction of the Schubert supercells and show that in this respect also supergeometry is a natural extension of the pure even geometry.

The following radical idea seems more fascinating:

the even geometry is a collective effect in the ∞-dimensional odd geometry

There is a very simple algebraic model showing how this might happen. The homomorphism of the formal series in ∞ variables

(1) $$\mathbb{R}[[x_1,x_2,\ldots]] \to \mathbb{R}[[\ldots\xi_{-1},\xi_0,\xi_1,\ldots]] ,$$

$$x_i \to \sum_{n=-\infty}^{\infty} \xi_n\xi_{n+i}$$

is injective.

A considerably more refined version of this construction has recently emerged in the work on representations of Kac-Moody algebras [17],[7]. This result establishes that the two realizations of $\mathfrak{gl}(\infty)$ in the differential operator algebras $\mathcal{D}_{ev} = \mathrm{Diff}^{\infty}(\mathbb{R}[x_1,x_2,\ldots])$ and $\mathcal{D}_{odd} = \mathrm{Diff}(\mathbb{R}[[\ldots\xi_{-1},\xi_0,\xi_1,\ldots]])$ are explicitly isomorphic:

(2) $$\xi_i\frac{\partial}{\partial\xi_j} - \delta_{ij} \leftrightarrow Z_{ij}(x,\frac{\partial}{\partial x}) .$$

Here Z_{ij} are defined from the formal series

$$\Sigma Z_{ij}p^iq^{-j} = qp^{-1}(1-qp^{-1})[\exp(\sum_1^{\infty} x_i(p^i - q^i))\exp(\sum_1^{\infty}\frac{\partial}{\partial x_i}\frac{q^{-i}-p^{-i}}{i})-1].$$

The isomorphism (2) is established by comparison of two natural representations, that of $\mathcal{D}_{ev}$ on $\mathbb{R}[x_1,x_2,\ldots]$ and that of $\mathcal{D}_{odd}$ on $F = \mathbb{R}[\xi_i,\frac{\partial}{\partial\xi_j}]/I$, where I is the left ideal generated by ξ_i, $i<0$, and $\frac{\partial}{\partial\xi_j}$, $j\geq 0$. The generator 1 mod I of the cyclic $\mathbb{R}[\xi_i,\frac{\partial}{\partial\xi_j}]$-module F can be conveniently represented as the infinite wedge-product $\bigwedge_{n=-\infty}^{-1}\xi_i$ and the total module F as the span of half-infinite monomials $\bigwedge_{i\in J}\xi_i$, $J\subset\mathbb{Z}$, card $\mathbb{Z}\setminus J<\infty$. The isomorphism

(3) $$F \tilde{\to} \mathbb{R}[x_1,x_2,\ldots]$$

may then be considered as the development of the simplistic idea (1).

The investigation of geometry with odd coordinates was started by physicists and is continued mainly in the physically motivated work [12],[13],[34]. In particular, the mathematical foundations of supergeometry were laid by F.A. Berezin [5] who early understood the role and the necessity of this extension of our geometric intuition. Of course the general philosophy of algebraic geometry is of great help.

Odd functions serve for modelling the internal degrees of freedom of the fundamental matter fields, leptons and quarks. Their quanta have spin $\frac{1}{2}$ and obey the Fermi-Dirac statistics. On the other hand the quanta of gauge fields (photons, gluons, $W^{\pm}$, Z, ...) have spin 1 and are bosons. The map (1) is a toy model of the bosonic collective excitations in the condensate of pairs of fermions. The formulas (2) and (3) also were essentially known to specialists in dual strings theory.

The idea that fermionic coordinates are primary with respect to the bosonic ones has been repeatedly advertised in various disguises. It is still awaiting the precise mathematical theory. It may well prove true that our four space-time coordinates $(x_0 = ct, x_1, x_2, x_3)$ are only the phenomenologically effective entities convenient for the description of the low energy world in which our biological life can exist only, but not really fundamental ones.

Meanwhile physicists are discussing grand unification schemes and supergravity theories which account for all fundamental interactions (or some of them) united in a Lagrangian invariant with respect to a Lie supergroup or covariant with respect to the general coordinate transform in a superspace.

Section 6 of this talk describes the geometry of simple supergravity from a new viewpoint which presents superspace as a "curved flag space" keeping a part of its Schubert cells.

The geometry of supergravity being essentially different from the simple-minded super-riemannian geometry, one is led to believe that the substitute of the Kählerian structure in supergeometry must be rather sophisticated. Therefore I do not venture here to make any guesses about the $\mathbb{Z}$-geometry with odd coordinates.

Comparing our present understanding of the arithmetic dimension with that of the odd ones we discover that the destroyed texts are reconstructed in different parts of the parallel texts. Trying to guess more, we can ask two questions.

a) <u>Is it possible to compactify a supermanifold with respect to the odd dimensions ?</u>

We seemingly need a construction of such a compactification if we want to have a cohomology theory in which the Schubert supercells would have nontrivial (i.e. depending essentially on the odd part) cohomology classes.

D. Leites has conjectured that in an appropriate category an "odd projective space" might exist, that is the quotient of $\operatorname{Spec} k[\xi_1,\ldots,\xi_n] \setminus \operatorname{Spec} k$ modulo the multiplicative group action $(t,(\xi_i)) \mapsto (t\xi_i)$. Of course, in the ordinary sense it is empty.

b) <u>Does there exist a group, mixing the arithmetic dimension with the (even) geometric ones ?</u>

There is no such group naively, but a "category of representations of this group" may well exist. There may exist also certain correspondence rings (or their representations) between $\operatorname{Spec} \mathbb{Z}$ and x. A recent work by Mazur and Wiles [27] shows that the p-adic Kubota-Leopoldt ζ-function divides a certain modular p-adic ζ-function defined in characteristic p. Such things usually happen if a correspondence exists.

Finally, I would like to acknowledge my gratitude to many friends whose ideas helped to consolidate certain beliefs expressed here. I am particularly grateful to I.R. Shafarevitch who taught the arithmetic-geometry analogies to his students for three decades, to A.A. Beilinson who has generously shared his geometric insight with the author.

1. A-manifolds and A-divisors

1. A-manifolds. Let K be a finite algebraic number field, R its ring of integers, $S = S_f \cup S_\infty$ the set of finite and infinite places of K. If $v \in S$, K_v denotes the completion of K with respect to the valuation $|\,|_v : K \to \mathbb{R}^*$. We put $|a|_v = |a|$ if $K_v = \mathbb{R}$, $|a|_v = |a|^2$, if $K_v = \mathbb{C}$. Then $\prod_v |a|_v = 1$ for all $a \in K$. Moreover, $R = \{a \in K \mid |a|_v \leq 1$ for all $v \in S_f\}$.

We shall call the following data an A-manifold:

$$X = (X_f ; \omega_v, v \in S_\infty) \tag{1}$$

Here X_f is a scheme of finite type, proper, surjective and flat over Spec R, with smooth irreducible generic fiber. Furthermore, ω_v is a Kählerian form on the complex variety $X_v = (X_f \underset{\mathbb{R}}{\otimes} K_v)(\mathbb{C})$, and $\overline{\omega}_v = \omega_v$ if $K_v = \mathbb{R}$; if $K_v = \mathbb{C}$, then the forms corresponding to the two embeddings $K_v \to \mathbb{C}$ should be conjugate.

We shall denote by $\mathrm{vol}_v = \omega_v^{\dim X_v}$ the corresponding volume forms.

The simplest example of an A-manifold is the A-curve $X_f = \mathrm{Spec}\, R$ endowed with the volumes of all points $v \in S_\infty$; ω_v do not exist in this case.

I want to stress the preliminary nature of the definition (1). First of all, one should not restrict oneself to the relatively proper schemes. If X_v is not proper, the ω_v presumably may have logarithmic growth at infinity, cf. [8]. Furthermore, a very special role is played by the Kähler-Einstein forms ω_v, see n°5 below.

2. Invertible A-sheaves. An invertible A-sheaf on the A-manifold (1) is the data

$$L = (L_f ; h_v , v \in S_\infty) . \tag{2}$$

Here L_f is an invertible sheaf on X_f, h_v - a Hermitian metric on $L_v = L_f \underset{R}{\otimes} K_v$ with the evident reality conditions and the following property:

(3) <u>the curvature form</u> F_v <u>of the hermitian connection corresponding to</u> h_v <u>is</u> ω_v<u>-harmonic.</u>

We recall that if s is a local holomorphic section of L_v, then $F_v = \bar{\partial}\partial \log h_v(s,s)$ in the domain of s. For an A-curve the condition (3) is empty.

It is evident how to define the tensor product $L \otimes L'$ of two invertible A-sheaves. The group of isomorphism classes of invertible A-sheaves is denoted by $\mathrm{Pic}_A X$. The identity is the class of the structure sheaf

$$\mathcal{O}_X = (\mathcal{O}_{X_f}; h_v \mid h_v(1,1) = 1 \quad \text{for all} \quad v \in S_\infty)$$

Later on we shall use the following fact: for a fixed ω_v, h_v is defined by the condition (3) up to a multiplicative constant.

3. <u>Sections.</u> Let L be the invertible sheaf (2). Set $H^i(X,L) = H^i(X_f,L_f)$. These cohomology groups are the R-moduli of finite type, and the ordinary Riemann-Roch-Grothendieck theorem for schemes [35] tells much about their structure. An essentially new object in A-geometry is the Euler A-characteristic $\chi_A(L)$. If a canonical map $\rho: X \to$"A-point" were to exist, one would define $\chi_A(L)$ as $R\rho_*(L)$. This being otherwise, only certain ad hoc definitions of $\chi_A(L)$ in a few particular cases are known, which are reviewed in n°2. The general idea is that in case $H^i(X,L) = 0, i > 0$, one must define $\chi_A(L)$ as the covolume of the image

$$H^\circ(X,L) \to \bigoplus_{v \in S_\infty} H^\circ(X_v,L_v) = H^\circ_\infty(X,L)$$

relative to a certain volume form on H°_∞. The general definition of this volume form is still lacking. Following Faltings [10], one may conjecture that to construct it one can use a canonical metric on the bundle on $\mathrm{Pic}_0 X_v$ with fiber $\bigotimes_i \det H^i(X_v,L_v)^{(-1)^i}$ and to supplement this by inductive reasoning on the Néron-Severi group.

A correctly defined $\chi_A(L)$ should be calculable via an A-Riemann-Roch theorem so that we shall need divisors of sections of L and, more generally, A-characteristic classes.

4. <u>A-divisors.</u> We shall mean by an A-divisor on X the following data:

$$D = (D_f; r_v \mid r_v \in \mathbb{R}, v \in S_\infty).$$

where D_f is a Cartier divisor on X_f. The following symbolic notation is more convenient

$$D = D_f + \sum_{v \in S_\infty} r_v X(v) \quad .$$

The A-divisors $X(v)$ (not to be confused with X_v) are called the "closed fibers" of X at infinity. The A-divisors form a group $\mathrm{Div}_A X$.

By the A-divisor of a section $s \in H^0(X, L)$ we shall mean the following element of $\mathrm{Div}_A X$:

(4) $$\mathrm{div}\, s = \mathrm{div}_f s - \sum_{v \in S_\infty} \left(\int_{X_v} \log |s|_v \cdot \mathrm{vol}_v\right) X(v),$$

$$|s|_v^{[\mathbb{C}:K_v]} = h_v(s,s).$$

Here $\mathrm{div}_f s$ is the Cartier divisor of s. If a rational function g on X_f is a quotient of two sections of L, it is natural to define its principal A-divisor by the formula

(5) $$\mathrm{div}\, g = \mathrm{div}_f g - \sum_{v \in S_\infty} \left(\int_{X_v} \log |g|_v \,\mathrm{vol}_v\right) X(v),$$

which does not depend on L. Finally, the same formula (4) may be used to define that A-divisor of a meromorphic section of L.

Now we can easily introduce the A-sheaves $\mathcal{O}(D)$ where D is an arbitrary A-divisor, together with the canonical section whose A-divisor is D. First, for $D = D_f$ we set:

(6) $$\mathcal{O}(D_f) = (\mathcal{O}_{X_f}(D_f); h_v)$$

where h_v is the unique metric on $\mathcal{O}_{X_f}(D_f) \otimes K_v$ satisfying equation (3) and normalized by

(7) $$\int_{X_v} \log |1_{D_f}|_v \cdot \mathrm{vol}_v = 0, \quad v \in S_\infty,$$

where 1_{D_f} is the meromorphic section of $\mathcal{O}_{X_f}(D_f)$ whose Cartier divisor is D_f. Using (4) and (7) we get $\operatorname{div} 1_D = \operatorname{div}_f 1_D = D_f$ which justifies (6).

Furthermore, we set

(8) $$\mathcal{O}(\sum_{v \in S_\infty} r_v X(v)) = (\mathcal{O}_{X_f}; h_v \quad ||1||_v = \exp(-r_v(\int_{X_v} \mathrm{vol}_v)^{-1})).$$

Again using (4) we obtain

$$\operatorname{div} 1 = \sum_{v \in S} r_v X(v).$$

as is to be expected.

As in the geometric case we can construct the exact sequence

$$\begin{array}{ccccccc} 0 \to \operatorname{Div}^\circ_A X \to & \operatorname{Div}_A X & \to & \operatorname{Pic}_A X & \to 0, \\ & \cup & & \cup & \\ & D & \to & \text{class of } \mathcal{O}(D) & \end{array}$$

where $\operatorname{Div}^\circ_A X$ is the group of principal A-divisors.

5. <u>Green's functions.</u> It is clear from the previous definitions that the essential information about the Archimedean part of the A-divisors is encoded in the functions

(9) $$G_v(D_v, x) = |1_{D_v}|_v(x) \quad , \quad x \in X_v$$

On the compact Kählerian manifold (X_v, ω_v) they are uniquely defined by the following conditions:

a) $G_v(D_v,x)$ is real analytic for $x \notin \operatorname{supp.} D_v$. The function $G_v(D_v,x)/|g(x)|_v$, where g_v is a local equation of D_v, is extendable to $\operatorname{supp} D_v$.

b) The (1,1)-form $\bar\partial\partial \log G_v(D_v,x)$ is ω_v-harmonic outside of D_v. The corresponding current is a linear combination of a harmonic form and the δ_{D_v}.

c) $\int_{X_v} \log G_v(D_v,x)\cdot \mathrm{vol}_v = 0$

Furthermore,

d) $G_v(D_v' + D_v'',x) = G_v(D_v',x)\cdot G_v(D_v'',x)$.

e) $G_v(\mathrm{div}\, g_v,x) = c_g \cdot |g_v(x)|_v$ for each meromorphic function g_v where c_v is defined by c).

It is explained in the last chapter of Lang's book [22] how to calculate Green's functions on abelian varieties and algebraic curves using theta-functions and differentials of the third kind respectively. The Kählerian metric involved is flat in the first case and induced by the flat metric of the Jacobian in the second one. The same metrics are used in the Arakelov-Faltings-Riemann-Roch theorem on A-surfaces which we shall state in n°2.

Since the function (9) is not constant except in trivial cases, the closed fibers $X(v)$, $v \in S_\infty$, should be imagined as "infinitely degenerate". To make it more credible note that if for $v \in S_f$ the closed fiber is degenerate then a meromorphic function or a section of an invertible sheaf can have different orders at different components of $X(v)$. In other words, instead of $|1_{D_f}|_v$ one should consider in this case $|1_{D_f}|_{v_i}$ where the valuations v_i correspond to the components $X(v)_i$ of $X(v)$. Finally, one can unify these numbers into a function $|1_{D_f}|(x)$, $x \in X(R_v) = X_v(K_v)$ setting $|1_{D_f}|(x) = |1_{D_f}|_{v_i}$ if the section x intersects $X(v)_i$.

This analogy suggests refining the definition of the divisor supported by $X(v)$, $v \in S_\infty$. Conjecturally, instead of a constant r_v one should consider a volume form ρ_v and delete the integrals from (4) and (5). A comparison with the Mumford-Schottky curves may serve to clarify the situation. Meanwhile we shall use the coarse definitions.

6. <u>The intersection index.</u> Let $\sigma: \mathrm{Spec}\, R \to X_f$ be a section of the structural morphism $\pi: X_f \to \mathrm{Spec}\, R$. We shall consider the image of σ as the closed A-curve Y lying in X. We define the intersection index of Y with an A-divisor D such that $\mathrm{supp}\, Y \cap \mathrm{supp}\, D$ is disjoint from the generic fiber of π:

$$(10)\qquad \langle Y,D\rangle = \sum_{v\in S_f} (Y,D_f)_v \log q_v + \sum_{v\in S_\infty} \langle Y,D\rangle_v .$$

Here (Y,D_f) for $v \in S_f$ denotes the sum of the local intersection indices of Y and D in the closed points of X_f over v, q_v is the order of the residue field. Furthermore, for $v \in S_\infty$ we set

$$\langle Y,D_f\rangle_v = -\log G_v(D_{f,v},y_v) ,$$

$$\langle Y, \sum_{v\in S_\infty} r_v X(v)\rangle = \sum_{v\in S_\infty} r_v\Big(\int_{X_v} \mathrm{vol}_v\Big)^{-1} .$$

An equivalent definition is obtained if one puts (10) and (4) together. Denote by $\sigma^*(1_D)$ the section of the A-sheaf $\sigma^*(\mathcal{O}(D))$ on Y, induced by $\mathcal{O}(D)$. Then

$$(11)\qquad \langle Y,D\rangle = -\log \prod_v |\sigma^*(1_D)|_v$$

From the product formula one sees that the right hand side of (11) remains unchanged if one takes a different non zero section of $\sigma^*(\mathcal{O}(D))$ instead of $\sigma^*(1_D)$. This justifies the following general definition of degree of an invertible A-sheaf L on the A-curve $Y = \operatorname{Spec} R$:

$$(12)\qquad \deg L = -\log \prod_v |s|_v$$

One can take any non zero meromorphic section s of L in (12).

2. The Riemann-Roch theorems.

1. <u>The geometric Riemann-Roch theorems.</u> We shall recall first the simplest Riemann-Roch-Hirzebruch theorem for projective manifolds over a field. Let X be a d-dimensional manifold, L an invertible sheaf on it. Set $\chi(L) = \sum_i (-1)^i \dim H^i(L)$ and denote by $c_1(L)$, $td_i(X)$ the Chern and Todd classes respectively. Then

$$(1)\qquad \chi(L) = \Big\langle \sum_{i=0}^{d} \frac{1}{i!} c_1(L)^i \, td_{d-i}(X) \Big\rangle$$

where $\langle\ \rangle$ in the right hand side of (1) means the intersection index calculated in the Chow ring or in a cohomology ring with characteristic zero coefficients. In particular

(2) $$\chi(L) = \sum_{i=0}^{d} \frac{n^i}{i!} \langle c_1(L)^i \, td_{d-i}(X) \rangle$$

In this section we shall describe three particular cases of a would be Riemann-Roch theorem for A-manifolds, for the projective space, A-curve and A-surface respectively, the last case being by far the deepest one. As we have said already, the first problem is to define $\chi_A(L)$.

2. Projective A-space. Let us consider the A-manifold, for simplicity over $\mathbb{Z}$, $\mathbb{P}^d = (\mathbb{P}^d_{\mathbb{Z}}, \omega)$, where ω is a Kählerian form on $\mathbb{P}^d(\mathbb{C})$. We shall realize $\mathbb{P}^d_{\mathbb{Z}}$ as $\mathrm{Proj}\, S(T_{\mathbb{Z}})$, where $T_{\mathbb{Z}}$ is a $\mathbb{Z}$-free module of the rank $d+1$, and we set $T = \mathbb{R} \otimes T_{\mathbb{Z}}$. There is a canonical hermitian metric on $\mathcal{O}_f(n)$ whose curvature from is a multiple of ω . We shall denote by $\mathcal{O}(n)$ the corresponding A-sheaf. Since

$$H^i(\mathbb{P}^d, \mathcal{O}(n)) = S^n(T_{\mathbb{Z}}) \;, \; H^i(\mathbb{P}^d, \mathcal{O}(n)) = 0,$$

for $i > 0$, $n \geq 0$, we must choose a volume form $w_n \in \Lambda_{\mathbb{R}}^{\binom{n+d}{d}}(S^n(T))$ and then define

$$\chi_A(\mathcal{O}(n)) = \log \left|\frac{w_n}{v_n}\right| \quad ,$$

where $v_n \in \Lambda_{\mathbb{Z}}^{\binom{n+d}{d}}(S^n(T_{\mathbb{Z}}))$ is one of the generators of this cyclic group.

The simplest imaginable choice of w_n is the following one. Consider the isomorphism

$$\varphi_n : \Lambda_{\mathbb{Z}}^{\binom{n+d}{d}}(S^n(T_{\mathbb{Z}})) \to [\Lambda_{\mathbb{Z}}^{d+1}(T_{\mathbb{Z}})]^{\otimes \frac{n}{d+1}\binom{n+d}{d}}$$

which maps v_n onto $v_1^{\otimes \frac{n}{d+1}\binom{n+d}{d}}$ (if $\frac{n}{d+1}\binom{n+d}{d} \notin \mathbb{Z}$ one can still correctly define $\varphi_n^{\otimes(d+1)}$, which suffices for our needs). Now choose somehow w_1 and set

$$w_n = (\varphi_n \otimes id_{\mathbb{R}})^{-1}\left(w_1^{\otimes \frac{n}{d+1}\binom{n+d}{d}}\right) .$$

Then

$$\chi_A(\mathcal{O}(n)) = \log \left|\frac{w_1}{v_1}\right|^{\otimes \frac{n}{d+1}\binom{n+d}{d}} = \frac{n}{d+1}\binom{n+d}{d} \chi_A(\mathcal{O}(1)) .$$

In view of (8), n°1, the tensor multiplication of an A-sheaf $\mathcal{L}$ by $\mathcal{O}(\frac{r}{n}\,\mathbb{P}(\infty))$ multiplies the metric on $\mathcal{L}_\infty$ by $\exp(-\frac{r}{n \operatorname{vol} \mathbb{P}_\infty})$. Assuming the corresponding change of w_1, we get

$$\chi_A(\mathcal{O}(1) \otimes \mathcal{O}(\tfrac{r}{n}\,\mathbb{P}(\infty))) = \chi_A(\mathcal{O}(1)) + \frac{r(d+1)}{n \operatorname{vol} \mathbb{P}_\infty}$$

and finally

$$(3) \qquad \chi_A(\mathcal{O}(n) \otimes \mathcal{O}(r\mathbb{P}(\infty))) = \frac{n}{d+1}\binom{n+d}{d}\chi_A(\mathcal{O}(1)) + \binom{n+d}{d}\frac{r}{\operatorname{vol} \mathbb{P}_\infty}$$

Comparing (3) with (1) and (2) we see that $\mathbb{P}^d$ looks like a $(d+1)$-dimensional geometric manifold. We can also guess the Todd A-classes $td^A_{d+1-i}(\mathbb{P}^d)$.

3. A-curve. Let $X = \operatorname{Spec} R$, $n = [R:\mathbb{Z}] = r_1 + 2r_2$, $r_1 = \operatorname{card}\{v \in S_\infty \mid K_v = \mathbb{R}\}$. Denote by $\mathcal{L} = (\mathcal{L}_f; h_v)$ an invertible A-sheaf on X, $L = H^0(X, \mathcal{L}_f)$. According to n°.1.3, we have

$$\chi_A(\mathcal{L}) = \operatorname{vol}((\bigoplus_{v \in S_\infty} L \otimes K_v)/L).$$

We choose the volume form on $\bigoplus_{v \in S_\infty} L \otimes K_v$ implicit in this definition following A. Weil [36] and Szpiro [33] in the following way:

$$w = \frac{2^n}{2^{r_1}\pi^{r_2}} \prod_{v \in S_\infty} w_v ,$$

where w_v is the volume form corresponding to the Euclidean metric on L_v defined by h_v. With this choice, the following statements, closely parallel to the case of curves over finite fields, are valid.

The Riemann-Roch theorem:

$$\chi_A(\mathcal{L}) = \deg \mathcal{L} + \chi_A(\mathcal{O}_X)$$

where $\deg \mathcal{L}$ is defined by (12), n°1.

The Euler number of the structure sheaf:

$$\chi_A(\mathcal{O}_X) = r_2 \log \frac{\pi}{2} - \frac{1}{2} \log |\Delta_K| ,$$

where Δ_K is the discriminant of K.

Furthermore, set $H_A^\circ(L) = \{s \in L \mid |s|_v \leq 1 \text{ for all } v\}$. Then $H_A^\circ(O) = \{0\} \cup \{\text{roots of unity in } R\}$, which is the analog of the constant field. From the Minkowski theorem one easily deduces that $\chi_A(L) \geq 0$ implies $H_A^\circ(L) \neq 0$.

4. A-surface. An A-surface $X = (X_f, \omega_v)$, according to Arakelov and Faltings, is a semistable family of curves $X_f \xrightarrow{\pi} \operatorname{Spec} R$ with smooth irreducible generic fiber of genus $g > 0$ and the following metrics at infinity:

$$\omega_v = \mathrm{vol}_v = -\frac{1}{2\pi i}\sum_{k=1}^{g} \nu_{k,v} \wedge \bar{\nu}_{k,v}$$

Here $(\nu_{1,v}, \ldots, \nu_{g,v})$ is a base of the differentials of the first kind on X_v orthonormal with respect to the scalar product

$$\langle \nu, \nu' \rangle = -\frac{1}{2\pi i}\int_{X_v(\mathbb{C})} \nu \wedge \bar{\nu}'.$$

We denote by Ω_f the relative dualizing sheaf of π. Then $\Omega_v = \Omega^1 X_v$. The canonical A-sheaf $\Omega = (\Omega_f, h_v)$ is unambigously defined by the following prescription which normalizes h_v. For an arbitrary point $x \in X_v$ the residue map $\mathrm{res}_x : \Omega^1 X_v \otimes \mathcal{O}_{X_v}(x) \to \mathbb{C}$ is an isometry of the geometric fiber of the former sheaf and of $\mathbb{C}$.

5. The Euler characteristic. Faltings [10] defines $\chi_A(L)$ for an invertible A-sheaf L on X in the following way.

The decisive step is the definition of canonical metrics on the spaces $\det_{K_v} H^\circ(L_v) \otimes \det_{K_v}^{-1} H^1(L_v) \otimes \bar{K}_v$, $v \in S_\infty$, in the case $H^\circ(L_v) = H^1(L_v) = 0$. This being done, Faltings uses this case as the induction base with respect to the ordinary degree of L on the generic fiber. To this end he represents L in the form $L_0(D)$, where $H^\circ(L_{0,v}) = H^1(L_{0,v}) = 0$ and D is a horizontal A-divisor which can be taken as a sum of sections after a base extension. One can then simultaneously define $\chi_A(L_0(D))$ and prove the Riemann-Roch formula if only one establishes the independence of this construction on the choice of the isomorphism $L = L_0(D)$ which is highly non-unique.

This independence is valid for the following particular choice of metrics on all $\mathcal{L}_0$'s simultaneously. Set $\mathrm{Pic}_{g-1} X_v = M_v$ and denote by $\mathcal{E}_v$ the universal sheaf over $X_v \times M_v$. Let $\pi_2 : X_v \times M_v \to M_v$ by the projection map. We can construct the invertible sheaf $\det R\pi_{2*}\mathcal{E}_v$ on M_v. Its geometric fiber at a point $y \in M_v$ corresponding to the sheaf $\mathcal{L}_v(y) = \mathcal{E}_v \mid (X_v \times \{y\})$ can be canonically identified with

$$\det H^\circ(\mathcal{L}_v(y)) \otimes \det^{-1} H^1(\mathcal{L}_v(y)) \otimes \overline{K}_v .$$

Over the set $U = \{y \in M_v \mid H^\circ(\mathcal{L}_v(y)) = H^1(\mathcal{L}_v(y)) = 0\}$ the sheaf $\det R\pi_{2*}\mathcal{E}$ has a canonical unit section. On the other hand, $M_v \setminus U$ is the theta-divisor, and an easy consideration shows that under the suitable identification $\det R\pi_{2*}\mathcal{E} = \mathcal{O}_{M_v}(-\theta)$ the unit section goes into 1. Therefore, a choice of an A-structure on $\mathcal{O}_M(-\theta)$ normalizes all $\chi_A(\mathcal{L}_0)$. Faltings proves that the θ-polarization induces precisely the A-structure suitable for the inductive argument.

We can now state the Riemann-Roch.

6. Theorem. a) $\chi_A(\mathcal{O}(D)) = \frac{1}{2} \langle D, D-K \rangle + \chi_A(\Omega)$, where $\Omega = \mathcal{O}(K)$, $\langle\ \rangle$ is the intersection index defined in n°1.6.

b) $$\chi_A(\Omega) = \frac{1}{12}(\langle K,K \rangle + \delta) \ , \ \delta = \sum_{v \in S_\infty} \delta_v(X_v) ,$$

where $\delta_v(X_v) = \log$ card (singular points of $X(v)$) for $v \in S_f$; for $v \in S_\infty$, δ_v is a real analytic function on the moduli space of Riemann surfaces which measures the distance of X_v to the boundary. ∎

Of course, in the geometric case, the Noether formula b) follows from the Riemann-Roch-Grothendieck theorem applied to the morphism π.

The structure of $\delta_v(X_v)$ for $v \in S_\infty$ vaguely agrees with our philosophy about $X(v)$ as a degenerate fiber.

Faltings proves the Noether formula by an argument using the moduli space of X instead of Pic_{g-1} of the first part.

The governing idea always is to use some canonical A-structures on the moduli spaces and their tautological sheaves, to apply the ordinary

Riemann-Roch-Grothendieck to the finite part and then to "compactify" this information by the Kählerian geometry.

Hence we need the A-geometry of arbitrary dimension anyway, even to deal with A-surfaces only. In the next section we shall discuss what is to be done to put this program on a firm foundation.

3. Prospects and problems of A-geometry

1. The problem of the definition of the fundamental categories.

In the definition of A-manifolds given in 1.1. no conditions on the Kählerian forms ω_v were imposed. However, the Arakelov and Faltings theorems are proved for distinguished Kählerian structures. We shall give the tentative definitions in a more general context.

Let $\mathcal{E}_v$ be a locally free sheaf on a compact Kählerian manifold (X_v,ω_v) and h_v a Hermitian metric on $\mathcal{E}_v$. We choose holomorphic local coordinates (z^α) on X_v and a base of local holomorphic sections (s_i) of $\mathcal{E}_v$ and set $h_{ij} = h_v(s_i,s_j)$. The curvature tensor of the canonical connection associated with h_v is

$$F_{ij\alpha\beta} = -\frac{\partial^2 h_{ij}}{\partial z^\alpha \partial \bar{z}^\beta} + h^{ab}\frac{\partial h_{ib}}{\partial z^\alpha}\frac{\partial h_{ij}}{\partial \bar{z}^\beta},$$

where $(h^{ab}) = (h_{ij})^{-1}$. Set $\omega_v = \sqrt{-1}\, g_{\alpha\beta}dz^\alpha \wedge d\bar{z}^\beta$ and $(g^{\gamma\delta}) = (g_{\alpha\beta})^{-1}$. Then $(\mathcal{E}_v,h_v)$ is called a Hermite-Einstein sheaf if

$$g^{\alpha\beta}F_{ij\alpha\beta} = \lambda h_{ij},$$

where λ is a constant. (It can be explicity calculated: setting $n = \dim X_v$ we get $\lambda = (2\pi n \int_{X_v} c_1(\mathcal{E}_v)\omega_v^{n-1}) / (\mathrm{rk}\, \mathcal{E} \int_{X_v} \omega_v^n))$.

On the holomorphic tangent sheaf $\mathcal{T}X_v$ there is the hermitian metric $g_v = 2g_{\alpha\beta}\, dz^\alpha d\bar{z}^\beta$. The manifold (X_v,ω_v) is called the Hermite-Einstein manifold if $(\mathcal{T}X_v,g_v)$ is a Hermite Einstein sheaf.

The existence and uniqueness problems for Hermite-Einstein structures on a sheaf E_v were considered by Kobayashi [19] and Donaldson [9]. Kobayashi has shown that the existence of such structure implies the semistability of E_v and that any semistable Hermite-Einstein sheaf (E_v, h_v) is a direct sum $\oplus (E_v^{(i)}, h_v^{(i)})$ with stable $E_v(i)$. (Stability here means that the function $\mu(F) = \deg_{\omega_r} F/\mathrm{rk}\, F$, where $\deg_{\omega_v} F = \int_{X_v} c_1(F)\omega_v^{n-1}$, is monotonous on subsheaves $F \subset E$).

On the other hand, Donaldson proved that on projective algebraic surfaces (X_v, ω_v) any stable sheaf has a unique Hermite-Einstein metric (up to a multiplicative constant). The same is true for algebraic curves and, conjecturally, for all projective manifolds.

Deep existence and uniqueness properties of Kähler-Einstein metrics on manifolds were obtained by Yau [38] and Aubin [3]. From Yau's results it follows in particular that for $c_1(X_v) = 0$ each cohomology class of Kählerian metrics contains a unique Kähler-Einstein metric. Aubin has established the existence on X_v of a unique Kähler-Einstein metric with the constant $\lambda = -1$ under condition that $c_1(X_v)$ contains a form with negative definite metric.

Our limited understanding of A-geometry suggests the special role of those A-manifolds for which (X_v, ω_v) are Kähler-Einstein. This condition appears to be a reasonable analog of the minimality of X_f over Spec R. Furthermore, on a given A-manifold, the following definition of a locally free A-sheaf seems plausible enough: it is the data $E = (E_f; h_v, v \in S_\infty)$ for which E_f is a locally free sheaf on X_f and (E_v, h_v) are Hermite-Einstein sheaves on (X_v, ω_v). For $\mathrm{rk} E_f = 1$ this is our initial definition.

The category-theoretic aspects of these definitions need clarification.

Since the Hermite-Einstein property may possibly be relevant only for locally free and semistable sheaves, to define a substitute for coherent A-sheaves one probably is bound to consider something like "perfect complexes of locally free A-sheaves", as in [35]. Unfortunately, the differential geometry of complexes of sheaves in a derived category is not sufficiently developed.

The complexes of A-sheaves must have a torsion invariant χ_A. For

$v \in S_\infty$ the corresponding component of χ_A under certain conditions should be given by the Quillen construction [30].

I do not know how to define morphisms of A-manifolds. The problem seems to be related to the hyperbolicity theory by Kobayashi [18]. In fact, it is based on the study of morphisms $D \to X_v$ and $X_v \to D$, where $D = \{z \in \mathbb{C}, |z| \leq 1\}$ is the analog of Spec $\mathbb{Z}_p$. Therefore it can be considered as the counterpart of the theory of $\mathbb{Z}_p$-models of $\mathbb{Q}_p$-manifolds.

2. <u>The problem of canonical A-structures on moduli spaces and of A-moduli spaces.</u> The Arakelov and Faltings work shows the existence of distinguished A-structures on the moduli spaces of curves and of invertible sheaves on a curve, by which one arrives at good statements of the principal results. (For the moduli space of curves having a boundary this statement should be considered as heuristic, granting the existence of good definitions).

One should study from this view point the moduli spaces of stable sheaves on a curve, with rank and degree relatively prime. The first unsolved problem is to generalize the Riemann-Roch-Arakelov-Faltings theorem to the A-sheaves of arbitrary rank, where the second Chern A-class $c_2^A(E)$ should emerge,an intersecting new invariant.

When a category of A-spaces is properly defined one would naturally hope for existence of moduli-objects in this category. Of course, the first problems here are again connected with the situation "at infinity", i.e. the Kählerian geometry. In this respect a recent work of Koiso [20] deserves to be mentioned. Koiso shows in particular that the base space of normal and stable family of Kähler-Einstein structures carries a canonical Kähler structure. Unfortunately in most cases it is unknown whether it in addition satisfies the Einstein equation.

3. <u>The problem of intersection theory of A-manifolds.</u> In the important paper [4] A. Beilinson defined regulators for K-theory and introduced the general technique for construction of intersection theory on A-manifolds. We shall briefly describe here a part of his results, stressing the role of K-theory as a cohomology theory.

Let X_f be a regular projective scheme flat over $\mathbb{Z}$, $\dim X_f = d + 1$.

We fix two cycles z_i of pure codimension ℓ_i in X_f. Assume that $\ell_0 + \ell_1 = d + 1$ and $\mathrm{supp}\,(z_0 \wedge z_1) \cap X_{f,\mathbb{Q}} = \emptyset$. We shall describe Beilinson's construction of the A-intersection index $\langle z_0, z_1 \rangle$. For simplicity, we shall assume that both cycles have zero cohomology classes on the generic fiber. In this case we can forget ω_∞ since the intersection index will not depend on this metric. We get

$$\langle z_0, z_1 \rangle = \langle z_0, z_1 \rangle_\infty + \sum_{p \in S_f} \log p \cdot \langle z_0, z_1 \rangle_p$$

and define the fiberwise indices $\langle\ \rangle_v$ with the help of several cohomology theories.

To calculate $\langle\ \rangle_\infty$ we shall use the Deligne-Beilinson cohomology $H^j_{\mathcal{D}}$. For a smooth compact complex manifold X_∞ and a coefficient ring $B \subset \mathbb{R}$, $B(j) = (2\pi\sqrt{-1})^j B$, this cohomology is defined as follows:

$$H^K_{\mathcal{D}}(X_\infty, B(j)) = R^K\Gamma(X_\infty, B(j)_{\mathcal{D}}),$$

$$B(j)_{\mathcal{D}} = \mathrm{Cone}\,(F^j \oplus B(j) \overset{\alpha}{\to} \Omega^\cdot)[-1],$$

where $F^j = \Omega^{\geq j}$ (the truncated complex of holomorphic forms), $\alpha = \alpha_1 - \alpha_2$, α_i the natural injections. In the noncompact case, the forms with logarithmic singularities are used. The now standard homological methods permit us to define the $\mathcal{D}$-cohomology of simplicial schemes, the relative $\mathcal{D}$-cohomology, to define the classes of algebraic cycles and to prove the Poincaré duality theorem.

Now we return to the situation described earlier and set $U_i = X_\infty - \mathrm{supp}\, z_{i,\infty}$. If the classes of $z_{i,\infty}$ in $H^{2\ell_i}_{\mathcal{D}}(X_\infty, \mathbb{R}(\ell_i))$ vanish, the Mayer-Vietoris sequence shows that the classes $c\ell_{\mathcal{D}} z_i \in H^{2\ell_i}_{\mathcal{D}}(X, U_i, \mathbb{R}(\ell_i))$ are of the form $\partial\zeta_i$ where $\zeta_i \in H_{\mathcal{D}}^{2\ell_i - 1}(U_i, \mathbb{R}(\ell_i))$. We can construct the class

$$\zeta_0 \cup \zeta_1 \in H^{2d}_{\mathcal{D}}(U_0 \cap U_1)\mathbb{R}\,(d + 1))$$

and its image

$$(z_0 \cap z_1)_\infty = \partial(\zeta_0 \cup \zeta_1) \in H_{\mathcal{D}}^{2d+1}(X, \mathbb{R}(d + 1))$$

Let $\pi : X_\infty \to \mathrm{Spec}\, \mathbb{R}$ be the structure morphism. The final formula for the A-intersection index at the arithmetical infinity is

$$\langle z_0, z_1\rangle_\infty = \pi_*(z_0 \cap z_1)_\infty \in H^1_{\mathcal{D}}(\mathrm{Spec}\,\mathbb{R}, \mathbb{R}(1)) = \mathbb{C}/\mathbb{R}(1) = \mathbb{R}\ .$$

To define $\langle z_1, z_2\rangle_p$ in a similar way, Beilinson introduces the K-cohomology:

$$H^j_K(X_f, \mathbb{Q}(i)) = K^{(i)}_{2i-j}(X_f) \otimes \mathbb{Q} = \{a \in K_{2i-j}(X_f) \otimes \mathbb{Q} \,|\, \Psi^p(a) = p^i a\},$$

where Ψ^p are the Adams operations. In this case also the relative version and the formalism of the cyclic classes can be defined. Setting $U_i = X_f \setminus \mathrm{supp}\, z_i$ as earlier we can now construct the intersection class

$$(z_0 \cap z_1)_f \in H^{2d+2}_K(X_f, U_0 \cup U_1; \mathbb{Q}(d+1))\ .$$

Set $S' = \mathrm{supp}\, \pi_f(X_f \setminus (U_0 \cup U_1))$, where $\pi_f : X_f \to \mathrm{Spec}\, \mathbb{Z}$ is the structure morphism. This is a finite set of primes. We have $\langle z_0, z_1\rangle_p = 0$ for for $p \notin S'$, and for $p \in S'$ the index $\langle z_0, z_1\rangle_p$ is a sort of direct image $\pi_f(z_0 \cap z_1)_f$ localized at p.

4. <u>The problem of the Euler A-characteristic and of the Riemann-Roch A-theorem</u>. I cannot add much to what has been said earlier. Two remarks may be in order.

First, granting that the definition of χ_A in a general situation can be done in terms of the analytic torsion of the Dalbeault complexes, we shall need the relative analytic torsion to treat the general Riemann-Roch-Grothendieck case.

Second, independently of the conjectural general theory, very interesting and directly accessible problems of A-geometry may be found, e.g., in the theory of flag manifolds $G/P_{\mathbb{Z}}$. A recent work by Bombieri-Vaaler [6] is an example. It suggests in particular that the classical Minkowski "geometry of numbers" should be interpreted in A-geometry as a theory of characteristic classes at the arithmetic infinity.

4. Superspace

1. <u>Examples of superspaces</u>. A smooth or analytic manifold can be described

by a family of local coordinate systems and transition functions. Before introducing a formal definition of superspace, we shall give several examples of supermanifolds with the help of local coordinates.

a) The $m|n$ -dimensional affine superspace. It has global coordinates $(x_1,\dots,x_m;\xi_1,\dots,\xi_n)$, where x_i commute among themselves and with ξ_j and ξ_j anticommute. In the category of superschemes over a commutative ring A the ring of functions on the relative affine $m|n$-space is the Grassmann algebra with generators ξ_j over polynomial ring $A[x_1,\dots,x_m]$. In the category of C^∞-supermanifolds the ring of functions is $C^\infty(x_1,\dots,x_m)\,[\xi_1,\dots,\xi_n]$.

b) The $m|n$-dimensional projective superspace. It is defined by the atlas U_i, $i=0,\dots,m$, each U_i being a $m|n$-dimensional affine space. It is convenient to introduce a homogeneous coordinate system $(X_0,X_1,\dots,X_m;Z_1,\dots,Z_n)$ and to relate the coordinates $(x^i_j,\ j\neq i\ |\ \xi^i_1,\dots,\xi^i_r)$ by setting $x^i_j = X_j/X_i$, $\xi^i_j = Z_j/X_i$.

c) The supergrassmannian of the $d_0|d_1$-dimensional linear superspaces in the $(d_0+c_0|d_1+c_1)$-dimensional linear superspace. We shall describe it by the following standard atlas. Consider matrices of the form $(d_0+d_1)\times(d_0+c_0+d_1+c_1)$ divided into four blocks such that the format of the upper left block is $d_0\times(d_0+c_0)$. For each subset I of columns containing d_0 columns of the left part and d_0 columns of the right part consider the matrix

$$(1)\qquad Z_I = \begin{array}{c|c|c|c|c|l} \multicolumn{1}{c}{c_0} & \multicolumn{1}{c}{d_0} & \multicolumn{1}{c}{d_1} & \multicolumn{1}{c}{c_1} & \\ \hline x_I & \begin{matrix}1 & 0\\ 0 & 1\end{matrix} & 0 & \xi_I & d_0 \\ \hline \xi_I & 0 & \begin{matrix}1 & 0\\ 0 & 1\end{matrix} & x_I & d_1 \\ \hline \end{array}$$

$$\underbrace{\qquad\qquad\qquad\qquad}_{I}$$

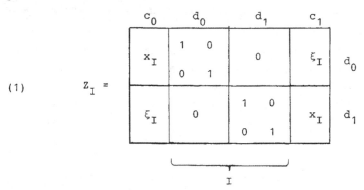

The columns I in Z_I form the identity matrix. All the remaining places are filled by the independent even and odd variables x_I^{ab}, ξ_I^{cd}, even places being in the upper left and lower right blocks. These variables (x_I, ξ_I) are the coordinates of the local chart U_I. Denote by B_{IJ} the submatrix of Z_I formed by the columns with indices in J. Then the transition rules are $Z_J = B_{IJ}^{-1} Z_I$.

Setting in this prescription $d_0|d_1 = 1|0$, $d_0 + c_0|d_1 + c_1 = m+1|n$, we get the projective superspace. On the other hand, setting $d_1 = c_1 = 0$, we get an ordinary grassmannian.

Proceeding in a more systematic way, we shall start with several basic notions of superalgebra and then define superspaces by means of a structure sheaf.

2. Superalgebra. The algebraic composition laws relevant in geometry are naturally divided into additive and multiplicative ones. All additive groups in superalgebra are endowed with a $\mathbb{Z}_2$-gradation and all multiplications are compatible with it. We use notation $A = A_0 \oplus A_1$ and $\tilde{a} = \varepsilon$ in case $a \in A_\varepsilon$, then $\widetilde{ab} = \tilde{a} + \tilde{b}$. The elements of A_0 are called even ones and those of A_1 odd ones. The characteristic feature of the superalgebra is the appearance of certain signs ± 1 in all definitions, axioms and polynomial identities of fundamental structures.

We shall give a representative list of examples.

Let $A = A_0 \oplus A_1$ be an associative ring. The supercommutator of homogeneous elements $a, b \in A$ is defined by the formula $[a,b] = ab - (-1)^{\tilde{a}\tilde{b}} ba$. The ring A is called supercommutative iff $[a,b] = 0$ for all a, b. If 2 is invertible (which we shall always assume), $a^2 = \frac{1}{2}[a,a] = 0$ for all $a \in A_1$. The supercommutators in general satisfy two identities

$$[a,b] = -(-1)^{\tilde{a}\tilde{b}} [b,a] ,$$

$$[a,[b,c]] + (-1)^{\tilde{a}(\tilde{b}+\tilde{c})} [b,[c,a]] + (-1)^{\tilde{c}(\tilde{a}+\tilde{b})} [c,[a,b]] = 0 .$$

These identities (together with superbilinearity) are taken as the definition of Lie superalgebras. The ring morphisms, by definition, respect the gradation.

Let A be a supercommutative ring. The notions of ($\mathbb{Z}_2$-graded, of course) left, right and bimodule S over A coincide, just as in the commutative case, left and right multiplications being connected by the formula $as = (-1)^{\tilde{a}\tilde{s}} sa$. A new feature is the parity-change functor: $(\Pi S)_0 = S_1, (\Pi S)_1 = S_0$, right multiplication by A coincide on S and ΠS . An A-module S is called free of rank $p|q$ iff it is isomorphic to $A^{p|q} = A^p \oplus (\Pi A)^q$. The tensor algebra of A-modules differs from the ordinary one by the introduction of Z sign into certain canonical isomorphisms, e.g. $\varphi : S \otimes T \tilde{\to} T \otimes S$ is defined by $\varphi(s \otimes t) = (-1)^{\tilde{s}\tilde{t}} t \otimes s$. There are internal Hom's in the category of A-modules consisting of ordinary morphisms and also of odd ones, with the linearity rule $f(as) = (-1)^{\tilde{a}} af(s)$.

The morphisms between the free A-modules can be given by matrices.

One must not forget that the passage from the left (row) coordinates to the right (column) coordinates of an even element implies sign change in odd coordinates etc. The matrices are often written in the standard format, like (1) where the even-even places are kept in the upper left block. The group GL represents the functor of the invertible matrices corresponding to even morphisms.

F. Berezin has invented the superdeterminant, or Berezinian, $\mathrm{Ber} : GL(p|q,A) \to A_1^*$. It is a rational function of the elements of the matrix which in the standard format is given by the formula

$$\mathrm{Ber} \begin{pmatrix} B_1 & B_2 \\ B_3 & B_4 \end{pmatrix} = \det (B_1 - B_2 B_4^{-1} B_3) \det B_4^{-1}.$$

The kernel of Ber is denoted SL. Since Ber is rational, supergrassmannians fail to have Plücker coordinates, as we shall see later.

The Berezinian of a free module Ber S is defined as a free module of rank $1|0$ or $0|1$ (depending on the parity of q in $\mathrm{rk} S = p(q)$ freely generated by any element of the form $D(s_1,\ldots,s_{p+q})$ where (s_i) is a free base of S, with the relations $D(f(s_1),\ldots,f(s_{p+q})) = \mathrm{Ber} f \cdot D(s_1,\ldots,s_{p+q})$. This notion is a specific substitute for the maximal exterior power in commutative algebra.

The bilinear forms on a free A-module T with the symmetry conditions are divided into four main types: OSp (even symmetric), SpO (even alternate), ΠO (odd symmetric), ΠSp (odd alternate). The form b^{π} on ΠT is defined by the formula $b^{\pi}(\Pi t_1, \Pi t_2) = (-1)^{t} b(t_1, t_2)$. This construction preserves parity but changes the symmetry of the form so that usually it suffices to consider only types OSp and ΠSp.

The automorphism functor of a non-degenerate form defines the supergroups of the corresponding type (we shall consider below only the split ones). Besides SL, OSp and ΠSp, there is in superalgebra one more series Q of supergroups of classical type: the centralizer group of an odd involution $p : T \to T$, $\tilde{p} = 1$, $p^2 = id$. The Lie superalgebras of these groups, slightly diminished if necessary (to kill a center etc.) constitute the classical part of the Kac classification [15] of simple finite dimensional Lie superalgebras. There are also two exceptional types (one having parameters) and Catzdan-type superalgebras of formal vector fields which happen to be finite-dimensional when defined on Grassmann algebras.

A superderivation $X: A \to A$ verifies the Leibniz formula $X(ab) = (Xa)b + (-1)^{\tilde{X}\tilde{a}} aXb$. There are two natural modules of relative differentials of a commutative A-algebra B: $\Omega^1_{ev}B/A$ and $\Omega^1_{odd}B/A$, classifying even and odd differentials respectively. Later on we use mainly $\Omega^1_{odd}B/A$ since the corresponding de Rham complex is supercommutative while for $\Omega^1_{ev}B/A$ it is super anticommutative.

3. Supergeometry. The most general known notion of a "space with even and odd coordinates" is that of superspace. Superspace is a pair $(M, \mathcal{O}_M)$, where M is a toplogical space, $\mathcal{O}_M$ a sheaf of local supercommutative rings on it. Morphisms of superspaces are morphisms of locally ringed spaces compatible with the gradations of structure sheaves.

All objects of the main geometric categories,-differentiable and analytic manifolds, analytic spaces, schemes, - are trivially superspaces, with $\mathcal{O}_M = \mathcal{O}_{M,0}$. Such superspaces we shall call purely even ones.

In the general case we set $J_M = \mathcal{O}_M \cdot \mathcal{O}_{M,1}$, $Gr_i\, \mathcal{O}_M = J^i_M / J^{i+1}_M$. Furthermore, $M_{rd} = (M, Gr_0 \mathcal{O}_M)$, $GrM = (M, \bigoplus_{i=0}^{\infty} Gr_i \mathcal{O}_M)$.

The structure sheaf GrM has a natural $\mathbb{Z}$-gradation. To consider GrM as a superspace we reduce it modulo 2.

With the help of these constructions we can define the most simple and important class of superspaces. We shall call a superspace $(M,\mathcal{O}_M)$ a <u>supermanifold</u>, analytic or algebraic, iff a) M_{rd} is a pure even manifold of the respective class; b) the sheaf $\mathcal{O}_M$ is locally isomorphic to the sheaf $Gr\mathcal{O}_M$, which is in turn isomorphic to the Grassmann algebra of the locally free (over $Gr_0 M$) sheaf J_M/J_M^2 of finite rank. (Note that this Grassmann algebra should be called symmetric in the superalgebra since J_M/J_M^2 is of pure odd rank).

One proves then that C^∞ and analytic supermanifolds can be described by local charts $(x_1,\dots,x_m;\ \xi_1,\dots,\xi_n)$. The sheaf $\mathcal{O}_M$ locally consists of the expressions $\sum_\alpha f_\alpha(x)\xi^\alpha$ where f_α are even functions of the corresponding type. An essential feature of supergeometric constructions is their invariance with respect to the coordinate changes mixing even and odd functions. Set, e.g., $y = x + \xi_1\xi_2, \eta_1 = (1+x^2)\xi_1,\ \eta_2 = \xi_2$. Then the local function $f(x)$ goes into $f(y - \frac{1}{1+x^2}\ \eta_1\eta_2) = f(y) - \frac{f'(y)}{1+x^2}\ \eta_1\eta_2$. The appearance of derivatives in the coordinate change formulas plays an essential role in the Langrangian formalism of supersymmetric field theoretic models. It also shows that in continuous supergeometry a natural structure sheaf ought to contain certain distributions. The peculiarities of continuous supergeometry were not studied for this reason.

The most important superspaces which are not necessarily supermanifolds can be easily defined in the analytic category. They are the superspaces $(M,\mathcal{O}_M)$ such that $(M,\mathcal{O}_{M,0})$ is an analytic space and $\mathcal{O}_{M,1}$ is $\mathcal{O}_{M,0}$-coherent. In the same way one defines superschemes.

The notion of a locally free sheaf of $\mathcal{O}_M$-modules is a natural substitute for vector bundles. For supermanifolds over a field the tangent sheaf $\mathcal{T}M$ and the cotangent sheaf $\Omega^1 M = \Omega^1_{odd} M$ are defined in the usual way, using superderivations over the ground field. The generalization to the relative case is selfevident. The rank of $\mathcal{T}M$ is called the dimension of the supermanifold.

Now the reader will easily transcribe the descriptions given in n°4.1 into the definitions of the superspaces in the algebraic, analytic or C^∞ categories. Notice that the grassmannian is endowed with the tautological sheaf, which is generated by the rows of Z_I over U_I. Over the projective superspace it is denoted $\mathcal{O}(-1)$.

4. Methods of construction of superspaces. a) Let $(M,\mathcal{O}_M)$ be a pure even locally ringed space, $\mathcal{E}$ a locally free sheaf of modules of rank $0|q$ over $\mathcal{O}_M$. Then $(M,\mathcal{O}_{M,s} = S(\mathcal{E}))$ is a superspace, which is called split. In the C^∞ category every supermanifold is split, i.e. can be obtained in this way from a manifold and a vector bundle on it. In the analytic and algebraic categories this is not true anymore, e.g. the Grassmannians are not analytically split unless the tautological bundle is of pure even or pure odd rank. (cf. below).

b) As in pure even analytic and algebraic geometry, very important superspaces are defined by their functor of points. We have already mentioned the algebraic supergroups GL. SL , OSp, ΠSp and in the next section we shall work with the flag superspaces $F(d_1,\ldots,d_K;T)$, where $d_1 < d_2 < \ldots < d_K$ are the dimensions of the components of a flag in the linear superspace $\mathcal{T}$. We expect that the main theorems on the representability of various functors and moduli problems admit their counterparts in supergeometry although the systematic work has barely begun. We shall state two results proved by A. Vaintrob which show the existence of a local deformation theory of Kodaira-Spencer type. The basic definitions are readily stated in the context of analytic superspaces. The infinitesimal deformations are represented by the ring $\mathbb{C}[x,\xi] \mid (x^2,x\xi)$.

Let M be a compact supermanifold, $\mathcal{T}M$ its tangent sheaf.

5. Theorem. a) Let $\dim H^1(M,\mathcal{T}M) = a|b$. If $H^2(M,\mathcal{T}M) = 0$, then in the category of analytic supermanifolds there exists a local deformation of M over $B = \mathbb{C}^{a|b}$ such that the Kodaira-Spencer map $\rho : T_0B \to H^1(M,\mathcal{T}M)$ is an isomorphism.

b) Any deformation of M over a supermanifold with surjective Kodaira-Spencer map is complete; in particular, it is versal, if ρ is isomorphic. ∎

Let us give an example. Let M be a compact analytic supermanifold of dimension $1|1$. It is completely defined by the Riemann surface $M_0 = M_{rd}$ and the invertible sheaf $\Pi J_M = \mathcal{L}$ on it. Assume that genus $M = g > 1$, $\deg L = 0$ and L is not isomorphic to $\mathcal{O}_{M_\theta}$. In this case $\mathrm{rk}\, H^1(M, TM) = 4g-3|4g-4$. The even part $4g-3$ of this dimension corresponds to the classical manifold Z of deformations of the pair (M_0, L) which is fibered by Jacobians over the coarse moduli space of curves. Theorem 5 shows that outside of the zero section this manifold Z is naturally extended to the supermanifold of odd dimension $4g-4$. This structure deserves further study.

6. Theorem. Let M be a closed compact subsupermanifold of the complex supermanifold M'. Let N be a normal sheaf to M.

a) If $\dim H^\circ(M, N) = a|b$ and $H^1(M, N) = 0$, there exists a versal local deformation of M in M' over $B = \mathbb{C}^{a|b}$.

b) A deformation of M in M' over B is complete iff the Kodaira-Spencer map $\rho: T_0 B \to H^0(M, N)$ is surjective.

7. Example. Let us return to the definition of a supergrassmannian and illustrate certain of our constructions. The supergrassmannian $G = G(1|1; \mathbb{C}^{2|2})$ is covered by four $2|2$-dimensional affine superspaces. The corresponding Z_I-matrices are

$$\begin{pmatrix} x_1 & 1 & \xi_1 & 0 \\ \eta_1 & 0 & y_1 & 1 \end{pmatrix}, \quad \begin{pmatrix} x_2 & 1 & 0 & \xi_2 \\ \eta_2 & 0 & 1 & y_2 \end{pmatrix},$$

$$\begin{pmatrix} 1 & x_3 & \xi_3 & 0 \\ 0 & \eta_3 & y_3 & 1 \end{pmatrix}, \quad \begin{pmatrix} 1 & x_4 & 0 & \xi_4 \\ 0 & \eta_4 & 1 & y_4 \end{pmatrix}$$

Using the prescription in the beginning of this section, one calculates the transition functions, e.g.

$$\begin{pmatrix} x_1 & \xi_1 \\ \eta_1 & y_1 \end{pmatrix} = \begin{pmatrix} x_4 & \xi_4 \\ \eta_4 & y_4 \end{pmatrix}^{-1} =$$

$$= \begin{pmatrix} x_4^{-1} + x_4^{-2} y_4^{-1} \xi_4 \eta_4 \; , & -x_4^{-1} y_4^{-1} \xi_4 \\ -x_4^{-1} y_4^{-1} \eta_4 \; , & y_4^{-1} - x_4^{-1} y_4^{-2} \xi_4 \eta_4 \end{pmatrix}$$

It follows immediately, that $G_{rd} = P^1 \times P^1$. A calculation shows also that

$$J_G^2 \simeq \Omega^2(P^1 \times P^1)$$

so that the obstruction ω to the splitting of G lies in the group

$$H^1(T(P^1 \times P^1) \otimes \Omega^2(P^1 \times P^1)) = H^1(P^1, \Omega^1 P^1)^2 = \mathbb{C} \oplus \mathbb{C} .$$

One can check directly, using the Čech cocycle in the standard atlas, that this obstruction is (1,1). Hence G is not split. (Notice that the projective superspace is split: $P^{m|n} = (P^m, S(\prod \mathcal{O}^n_{pm}(-1)))$.

Moreover, G is not a projective supermanifold. In fact, the image of the Picard group $H^1(G, \mathcal{O}^*_{G,0}) \to \mathrm{Pic}\ (P^1 \times P^1)$ consists of the classes of sheaves $\mathcal{O}(a,-a)$, $a \in \mathbb{Z}$, since the obstruction to extending $\mathcal{O}(a,b)$ from G_{rd} to G is essentially $(a+b)\omega$. Therefore, any supergrassmannian $G(a|b;\mathbb{C}^{m|n})$ with $0 < a < m$, $0 < b < n$ is non-projective since G admits a closed embedding in such a Grassmannian.

This example shows that the use of projective technique in the algebraic supergeometry is restricted, and one is obliged to generalize those methods of algebraic and analytic geometry which do not rely upon the existence of ample invertible sheaves.

For example, P. Deligne conjectured that the dualizing sheaf on a smooth complex supermanifold X in Ber $\Omega^1_{ev} X$. This was proved by I. Penkov [29] who has demonstrated that in this case working with $\mathcal{D}$-modules on a supermanifold permits one to effectively reduce the situation to the pure even one.

5. Schubert supercells.

1. Basic notions. Let G be a semisimple algebraic group, $B \subset G$ its Borel subgroup. The G-orbits of $G/B \times G/B$ form a finite stratification on this manifold whose strata Y_w are numbered by elements w of the Weyl group $W = W(G)$ (Bruhat decomposition). We shall call locally closed submanifolds $X_w(b) = (\{b\} \times G/B) \cap Y_w \subset G/B$ the Schubert cells. In the same way, using G-orbits of $G/B \times G/P$, one defines the Schubert cells for a parabolic subgroup $P \subset G$. The geometry of the Schubert cells plays an important role in many developments of characteristic classes theory and representation theory.

In this section we shall define Schubert supercells for complete flag superspaces of classical type and explain how some classical results generalize in this context.

Let T be a linear superspace of dimension $m|n$ over a field. We shall consider the following algebraic supergroups G given together with their fundamental representation T: a) $G = SL(T)$; b) $G = OSp(T)$, the automorphism group of a nondegenerate even symmetric form $b: T \to T^*$; c) $G = \Pi Sp(T)$, the automorphism group of a nondegenerate odd alternate form $b: T \to T^*$; d) $G = Q(T)$, the automorphism group of an odd involution $p: T \to T$, $p^2 = id$. In the cases ΠSp and Q we have $m = n$.

The counterpart of the classical manifold G/B is the supermanifold F of complete flags in T, which are in addition invariant with respect to b or p in the cases $G = OSp$, ΠSp, Q (the exact definitions are given below). Several differences between this situation and the classical one are worth mentioning.

First, the stabilizers of complete flags B in general are not maximal solvable subsupergroups. However, they play the same role as the classical Borel subgroups both in the theory of highest weight [16] and in the theory of the Schubert supercells. Second, not all subgroups B are pairwise conjugate, and the flag manifolds F consist of several components. Third, the stratification we want to construct is not purely set-theoretic. In fact it will be a decomposition of $F \times F$ into a union of locally closed subsuperschemes. The flag realization is suitable for this construction.

The first unsolved problem is to define a cohomology theory in which the classes of the Schubert cells would be free generators. As we conjectured in the introduction, this may require a sort of compactification along the odd dimensions.

We shall now give some details.

2. <u>The connected components of flag supermanifolds</u>. These connected components are naturally numbered by the sets ${}^{G}I$ which are defined as follows. Set

$$^{SL}I = \{(\delta_1, \ldots, \delta_{m+n}) \mid \delta_i = 1|0 \text{ or } 0|1, \sum_{i=1}^{m+n} \delta_i = m|n\}.$$

Furthermore,

$$^{OSp}I = \{(\delta_1, \ldots, \delta_{m+n}) \in {}^{SL}I \mid \delta_1 = \delta_{m+n+1-i}, \; i = 1, \ldots, m+n\},$$

$$^{\Pi Sp}I = \{(\delta_1, \ldots, \delta_{2m}) \in {}^{SL}I \mid \delta_i = \delta^c_{2m+1-i}, \; i = 1, \ldots, 2m\},$$

where $(p|q)^c = q|p$. Finally, $^{Q}I = \{(\overbrace{1|1, \ldots, 1|1}^{m})\}$ (the one element set). We say that the flag $f: 0 = S_0 \subset S_1 \subset \ldots \subset S_{m+n} = T$ is of type $I \in {}^{G}I$, if $\delta_i(I) = \mathrm{rk}\, S_i/S_{i+1}$. For groups $G = OSp, \Pi Sp, Q$ the flag f is called G-stable if the following conditions are fulfilled:

$$b(S_i) = S^{\perp}_{m+n-i} \text{ for } OSp, \Pi Sp \; ; \; p(S_i) = S_i \text{ for } Q .$$

The functor ${}^{G}F_I$ on the category of superschemes over ground field associates with a superscheme S the set of flags of type $I \in {}^{G}I$ in the sheaf $\mathcal{O}_S \otimes T$; for $G \neq SL$ the flags should be G-stable. (A flag is a filtration of $\mathcal{O}_S \otimes T$ by subsheaves S_i such that all injections $S_i \subset S_j$ locally split; G-stability is defined with respect to $\mathrm{id}_{\mathcal{O}_S} \otimes b$ or $\mathrm{id}_{\mathcal{O}_S} \otimes p$).

3. Theorem. Functor ${}^{G}F_{I}$ is representable by a supermanifold which is irreducible except for $G = \mathcal{O}Sp(2r|2s)$, $r \geq 1$, in which case ${}^{G}F_{I}$ consists of two ismorphic components. Furthermore,

$$\dim {}^{SL}F_{I} = \left(\frac{m(m-1)}{2} + \frac{n(n-1)}{2} \mid mn\right) ;$$

$$\dim {}^{0Sp}F_{I} = \begin{cases} (r^2 + s^2 \mid (2r+1)s) & \text{for } m = 2r+1,\ n = 2s > 0; \\ (r^2 - r + s^2 \mid 2rs) & \text{for } m = 2r,\ n = 2s > 0 \end{cases}$$

$$\dim {}^{Q}F_{I} = \left(\frac{m(m-1)}{2} \mid \frac{m(m-1)}{2}\right) ;$$

$$\dim {}^{\Pi Sp}F_{I} = \left(rs + \frac{r(r-1)}{2} + \frac{s(s+1)}{2} \quad rs + \frac{r(r+1)}{2} + \frac{s(s-1)}{2}\right) ,$$

where $(r|s) = \sum_{i=1}^{m} \delta_i(I)$. ■

Of course, the functors of noncomplete flags also are representable and the morphisms of projection onto a subflag are representable by morphisms of supermanifolds. It is convenient to prove theorem 3 by induction on the length of a flag, starting with relative Grassmannians as in section 4. The reader shall find most of details in [24].

Now we shall set ${}^{G}F = \coprod_{I \in G_I} {}^{G}F_{I}$ and denote by $S_i \subset \mathcal{O}_F \otimes T$ the components of the tautological flag on ${}^{F}G$. There are two natural flags on ${}^{G}F \times {}^{G}F$, $\{p_1{}^*(S_i)\}$ and $\{p_2{}^*(S_j)\}$, where $p_{1,2}$ are the projections. In the classical theory every G-orbit consists of those points of ${}^{G}F \times {}^{G}F$, over which the type of relative position of flags $\{p_1{}^*(S_i)\}$ and $\{p_2{}^*(S_j)\}$ is fixed. We can imitate this definition in supergeometry taking functor of points instead of geometric points.

The type of relative position of complete flags (S_i) and (S_j') in T is, by definition, the matrix $d_{ij} = \mathrm{rk}(S_i + S_j')$.

Let us introduce the Weyl groups G_W , acting on ${}^{G}I$:

$${}^{SL}W = S_{m+n} \; ; \; {}^{Q}W = S_m \; ; \; {}^{G}W = \{g \in S_{m+n} \mid g({}^{G}I) \subset {}^{G}I\} \quad \text{for } G = 0Sp, \Pi Sp.$$

The reader will notice that our ${}^{G}W$ is in general different from the Weyl group of G_{rd}, e.g. the latter is $S_m \times S_n$ for SL and not S_{m+n}. As we shall see in a moment, in the theory of Schubert supercells it is this big group, which contains the odd reflections, which is the right one.

4. Lemma. The types of relative positions of complete G-stable flags are in (1,1)-correspondence with the triples (I,J,w), where $I,J \in {}^{G}I$, $w \in {}^{G}W$, $J = w(I)$ ∎.

The proof is purely combinatorial.

5. Bruhat subsets. We set now for $w \in {}^{G}W$, $I,J = {}^{G}I$:

$$|Y_{w,IJ}| = \{x \in {}^{G}F_I \times {}^{G}F_J \mid rk(p_1{}^*S_i(x) + p_2{}^*S_j(x)) = d_{ij,w}\}$$

where $(d_{ij,w})$ is the type of relative position corresponding to the triple (I,J,w) in view of Lemma 4 (if $J \neq w(I)$ we set $|Y_{w,IJ}| = \emptyset$). Furthermore, put

$$|Y_w| = \coprod_{I,J} |Y_{w,IJ}| \subset {}^{G}F \times {}^{G}F \quad .$$

6. Theorem. Each set $|Y_w|$ carries a canonical structure of the locally closed subsuperscheme $Y_w \subset {}^{G}F \times {}^{G}F$, such that the decomposition $\coprod_w Y_w$ is the flattening stratification for the family of sheaves $S_{ij} = p_1{}^* S_i + p_2{}^*S_j$ on ${}^{G}F \times {}^{G}F$ ∎.

We recall that by definition of a flattening stratification this condition means that each morphism $q: X \to {}^{G}F \times {}^{G}F$ for which all the sheaves $q^*(S_{ij})$ are flat uniquely decomposes as $X \to \coprod_w Y_w \to {}^{G}F \times {}^{G}F$. The proof of the existence of the flattening stratification is the same as in the pure even case.

7. Superlength. In the classical theory the dimension of a Schubert cell associated with $w \in W$ equals the minimal length of a decomposition of w into a product of basic reflections. To state the counterpart of this fact in supergeometry we need several definitions.

We shall call the following elements of ${}^{G}W$ the basic reflections:

$$\sigma_i = (i,i+1) \quad \text{for} \quad G = SL, Q\ ;$$

$$\sigma_i = (i,i+1)\ (m+n+1-i,\ m+n-i)\ ,\ i+1 \leq [\tfrac{m+n}{2}]\ ;$$

$$\tau_\ell = (\ell, m+n+1-\ell)\ ,\quad \ell = [\tfrac{m+n}{2}] \quad \text{for}\ G = OSp,\ \Pi Sp.$$

The <u>superlength</u> ${}^G\ell(w)$, $w \in {}^GW$ will be defined inductively . This is a vector of superdimensions $({}^G\ell_{IJ}(w) \mid I,J \in {}^GI)$ such that

a) For a basic reflection $\sigma \in {}^GW$ we have ${}^G\ell_{IJ}(\sigma) = 0$ if $J \neq \sigma(I)$; the other possibilities are contained in the table:

σ \ G	SL	OSp	ΠSp	Q
σ_i	$1\mid 0$ $(I=J)$; $0\mid 1$ $(I\neq J)$;	$1\mid 0$ $(I=J)$; $0\mid 1$ $(I\neq J)$;	$1\mid 0$ $(I=J)$; $0\mid 1$ $(I\neq J)$,	$1\mid 1$
τ_ℓ	-	$1\mid 0$ $(\delta_\ell(I) = 1\mid 0,$ $m\mid n = 2r+1\mid 2s)$; $1\mid 1$ $(\delta_\ell(I) = 0\mid 1,$ $m\mid n = 2r+1\mid 2s)$; $0\mid 0$ $(\delta_\ell(I) = 1\mid 0,$ $m\mid n = 2r\mid 2s)$; $1\mid 0$ $(\delta_\ell(I) = 0\mid 1,$ $m\mid n = 2r\mid 2s)$.	$0\mid 1$ $(\delta_\ell(I)=1\mid 0)$ $0\mid 0$ $(\delta_\ell(r)=0\mid 1)$	-

b) Let $w = \sigma^K \ldots \sigma^1$ be an irreducible decomposition of w as a product of basic reflections. Set $I_i = \sigma^i \ldots \sigma^1(I)$ and

$${}^G\ell_{IJ}(w) = \sum_{i=0}^{K-1} {}^G\ell_{I_i,I_{i+1}}(\sigma^{i+1})\ ,\ \text{if}\quad J = w(I);$$

$${}^G\ell_{IJ}(w) = 0\ ,\ \text{if}\quad J \neq w(I).$$

8. <u>Theorem</u>. The projection map $Y_{w,IJ} \to {}^GF_I$ is surjective, and locally over ${}^GF_I$ the Bruhat manifold $Y_{w,IJ}$ is a relative affine superspace

of dimension ${}^{G}\ell_{IJ}(w)$. ∎

In other words, the dimension of a Schubert cell $Y_w(b)$ coincides with the superlength of w.

From the geometric proof of the theorem some purely combinatorial facts follow. For example, ${}^{G}\ell_{IJ}(w)$ does not depend on the choice of an irreducible decomposition of w; furthermore, ${}^{G}\ell_{IJ}(w) = {}^{G}\ell_{IJ}(w^{-1})$ for $G \neq \Pi Sp$, finally for $G = \Pi Sp$

$$ {}^{G}\ell_{IJ}(w) + \dim {}^{G}F_I = {}^{G}\ell_{JI}(w^{-1}) + \dim {}^{G}F_J, $$

if $J = w(I)$.

6. Geometry of supergravity

1. Minkowski space and Schubert cells. The objective of this section is to describe a model of simple supergravity from the view point which was introduced in [26] where the kinematic constraints of supergravity were interpreted as the integrability conditions for a curved version of a flag superspace.

To explain the essence of our approach let us recall the usual exposition of general relativity. The space-time without gravitational field is the Minkowski space of special relativity $\mathbb{R}^4$ with a metric which in an inertial frame takes the form $dx^2 - \sum_{i=1}^{3} dx_i^2$. The gravitation field reflects itself in the curvature of space-time which becomes a smooth four-manifold M^4 with the pseudoriemannian metric $g_{ab}dx_a dx_b$. The dynamics is governed by the Lagrangian (action density) $R\,\mathrm{vol}_g$ where R is the scalar curvature, vol_g the volume form of g.

The models of supergravity in superspace studied in many recent works [12],[13],[34] also start with certain geometric structures on a differentiable supermanifold $M^{m|n}$ which are then used to define a (super) Lagrangian which is a section of the sheaf $\mathrm{Ber}\, M = \mathrm{Ber}\, \Omega^1_{ev} M$. There are physically meaningful cases with $m \neq 4$, e.g. the case $m|n = 11|32$ is now considered as the most fundamental one.

What is still very much unclear, is the question what exactly is the geometry to start with, i.e. the kinematics of supergravity. The naive suggestion to use a supermetric was quickly seen inadequate. The most universal known method is the Cartan approach. One starts with an affine connection and then painstakingly guesses the so called constraints and the action density. The constraints are the differential equations which must imply no equations of motion. The physical interpretation and quantization of constrained fields is a difficult task and one faces the problem of solving constraints and expressing everything in terms of free fields. This approach was successful more than once but the poor command of underlying geometry hinders the work considerably.

Our approach essentially interprets the constraints as integrability conditions ensuring the existence of certain families of submanifolds in $M^{m|n}$, the geometry of these families being a curved geometry of Schubert supercells.

Let us first describe from this viewpoint the simplest example, the Plücker-Klein-Penrose model of Minkowski space.

Let T be a four dimensional complex space (Penrose's twistor space). Let $G = G(2;T)$ be the Grassmannian of planes in T, $S = S_\ell$ the tautological sheaf on G, $S_r = (T \otimes \mathcal{O}_G/S_\ell)^*$. There is a canonical isomorphism $\Omega^1 G = S_\ell \otimes S_r$, and the subsheaf $\Lambda^2 S_\ell \otimes \Lambda^2 S_r \subset S^2(\Omega^1 G)$ can be interpreted as the holomorphic conformal metric on G. Choose a big cell $U \subset G$. The complement $G \setminus U$ is a singular divisor, the light-cone at infinity, and there are sections $\varepsilon_\ell, \varepsilon_r$ of the sheaves $\Lambda^2 S_\ell, \Lambda^2 S_r$ on U having a pole of first order at this divisor. The complex metric $\varepsilon_\ell \otimes \varepsilon_r \in \Gamma(U, S^2\Omega^1 G)$ is well defined up to multiplicative constant. Now introduce a real structure ρ on $T \oplus T^*$ interchanging T and T^*. The involution ρ acts on $\mathbb{C}$-points of $G(2,T)$ since $G(2,T)$ canonically identifies with $G(2;T^*)$. Let ρ be compatible with $(U, \varepsilon_\ell, \varepsilon_r)$ in the sense that $U^\rho = U$, $\varepsilon_\ell{}^\rho = \varepsilon_r$. The following statements can be directly verified.

a) The real (i.e. ρ-invariant) points of the big cell U form the space $\mathbb{R}^4$. The restriction of $\varepsilon_\ell \otimes \varepsilon_r$ to it is a Minkowski metric.

b) The real three-dimensional Schubert manifolds in the Grassmannian

G intersected with $U(\mathbb{R})$ form the system of light cones of this metric.

c) There are no real two-dimensional Schubert cells in G. In $U(\mathbb{C})$ they define two connected families of complex planes. These families play an essential role in the theory of Yang-Mills fields. In fact, the integrability of a connection along one family means that this connection is an (anti) self dual solution of the Yang-Mills equation.

In a curved space-time of general relativity null geodesics and light cones still exist and, moreover, define the corresponding metric up to a conformal factor. To break the conformal invariance one may choose metrics $\varepsilon_\ell, \varepsilon_r$ for two-component Weyl spinors, left and right.

We describe supergravity along these lines. In nn° 2,3 a flag model of Minkowski superspace is introduced, In n°4 we explain that in a curved superspace two families of $0|2N$ - dimensional Schubert supercells should be preserved. Finally, in nn°5,6 we define the dynamics by means of an action density, expressed through the Ogievetsky-Sokachev prepotential [28].

2. Minkowski superspace. Fix an integer $N \geq 1$ and a linear complex superspace T of dimension $4|4N$. Set $M = F(2|0,\ 2|N;T)$, i.e. a S-point of M is a flag $S^{2|0} \subset S^{2|N}$ in $\mathcal{O}_S \otimes T$. Moreover, define the left and right superspaces as Grassmannians

$$M_\ell = G(2|0;T),\ M_r = G(2|N;T) = G(2|0;T^*).$$

Denote by $S^{2|0} \subset S^{2|N}$ the tautological flag in $\mathcal{O}_M \otimes T$, by $\tilde{S}^{2|0} \subset \tilde{S}^{2|N}$ the orthogonal flag in $\mathcal{O}_M \otimes T^*$. Set $F_\ell = S^{2|N}/S^{2|0}$, $F_r = \tilde{S}^{2|N}/\tilde{S}^{2|0}$. Let $\pi_{\ell,r}: M \to M_{\ell,r}$ be the canonical maps. Let $T_\ell M = TM/M_r$, $T_r M = TM/M_\ell$ (recall that we work in the category of complex superspaces). Since M over $M_{\ell,r}$ is a relative Grassmannian, a standard argument gives canonical isomorphisms

$$T_\ell M = (S^{2|0})^* \otimes F_\ell \quad ,\ T_r M = F_\ell^* \otimes (\tilde{S}^{2|0})^*.$$

Combining this with the map $F_\ell \otimes F_\ell \to \mathcal{O}_M$ we get a natural map

(1) $a\colon T_\ell M \otimes T_r M \to (S^{2|0})^* \otimes (\tilde{S}^{2|0})^*.$

On the other hand, the relative tangent sheaves $T_{\ell,r}M$ are integrable distributions, i.e. locally free subsheaves of Lie superalgebras in TM, of rank $0|2N$. Every point of M is contained in two closed sub-supermanifolds of dimension $0|2N$ tangent to $T_\ell M$ and $T_r M$ respectively. They are the Schubert cells we are interested in. The supercommutator between $T_\ell M$ and $T_r M$ defines the Frobenius map

(2) $b\colon T_\ell M \otimes T_r M \to TM/(T_\ell M + T_r M) = T_0 M$

The following statements contain the essential geometric features of the picture we want to keep in the curved case.

3. Proposition. a) The sum $T_\ell M + T_r M$ in TM is a direct subsheaf in TM of rank $0|4N$.

b) There is a well defined isomorphism $T_0 M = (S^{2|0})^* \otimes (\tilde{S}^{2|0})^*$ making the maps (1) and (2) to coincide. ■

Finally, as in n°1, we must introduce a real structure ρ on $T \oplus T^*$ (in superalgebra $(ab)^\rho = (-1)^{\tilde{a}\tilde{b}} a^\rho b^\rho$; cf. [24] for further details). We shall assume that $T^\rho = T^*$, in this case $(T_\ell M)^\rho = T_r M$, $(S^{2|0})^\rho = \tilde{S}^{2|0}$. One can check that over a ρ-stable big cell in M some natural sections of $T_\ell M$, $T_r M$, $T_0 M$ generate the Poincare superalgebra introduced by physicists (see e.g. [34]).

4. Curved superspace. A complex supermanifold $M^{4|4N}$ with the following structures will be called superspace of N-extended supergravity.

a) Two integrable distributions $T_\ell M, T_r M \subset TM$ or rank $0|2N$ whose sum is direct.

b) Two locally free sheaves S_ℓ, S_r of rank $2|0$, two locally free sheaves $F_\ell ; F_r = F_\ell^*$ of rank $0|N$ and structure isomorphisms $T_\ell M = S_\ell^* \otimes F_\ell$, $T_r M = F_r \otimes S_r^*$.

c) A real structure ρ on M such that its real points in M_{rd} form a four-manifold, and extensions of this real structure to $S_\ell \oplus S_r$,

$F_\ell \oplus F_r$ interchanging left subsheaves with right ones.

d) Volume forms $v_{\ell,r} \in \Gamma(M_{\ell,r}, \text{Ber } M_{\ell,r})$ such that $v_\ell^\rho = v_r$. A choice of these volume forms corresponds to the choice of spinor metrics $\varepsilon_\ell, \varepsilon_r$ in n°1.

This data is subjected to one axiom. Set $T_0M = TM/(T_\ell M \oplus T_r M)$. Then the Frobenius map $\varphi: T_\ell M \otimes T_r M \to T_0 M$ coincides with the natural map $S_\ell^* \otimes F_\ell \otimes F_r \otimes S_r^* \to S_\ell^* \otimes S_r^*$ under the appropriate identification $T_0 M = S_\ell^* \otimes S_r^*$ as in Proposition 3 b).

5. Lagrangian. Let M be a superspace of N-extended supergravity. Using the data above one can construct a canonical isomorphism:

$$\text{Ber } M = [\pi_\ell^* \text{ Ber } M_\ell \otimes \pi_r^* \text{ Ber } M_r]^{\frac{2-N}{4-N}} \quad (N \neq 4) \tag{3}$$

(From this point on we define Ber M as $\text{Ber}^*(\Omega^1_{odd} M)$).

Hence the volume forms v_ℓ, v_r make it possible to define a section (for $N \neq 4$)

$$w = (\pi_\ell^* v_\ell \otimes \pi_r^* v_r)^{\frac{N-2}{N-4}} \in \Gamma(M, \text{Ber } M) \tag{4}$$

In this way we get for $N = 1$ the correct action of simple supergravity. In the case $N = 2$ the action is certainly wrong since it gives trivial equations of motion. It seems that considering N as a formal parameter and taking the left (or right) part of the coefficient of the Taylor expansion of (4) at $N = 2$ we get an action suggested by E. Sokachev. Anyway, for $N > 1$ one must take into account new constraints which might take the form of integrability of more Schubert cells.

It is also certain that the other types of flag supermanifolds and their curved versions are necessary for a fuller understanding of supergravity and super Yang-Mills equations. For example, in a recent paper by A. Galperin, E. Ivanov, S. Kalytsyn, V. Ogievetsky and E. Sokachev the manifold $F(2|0,2|1,2|2;T)$ implicitly appears which in the curved version can be defined as the projectivized bundle $P(F_\ell) = P(F_r) \to M$.

In the same vein, for the largest physically acceptable case $N = 8$, the $11|32$-dimensional flag supermanifold $F(2|0,2|1,2|8,T)$ or its curved version $P(F_\ell)\to M$ seems to be the space considered in the context of the so called dimensional reduction, or the generalized Kaluza-Klein model.

6. Prepotential. To conclude, we give some coordinate calculations which make it possible to identify our geometric picture with that of the article [28]. Set $N = 1$ and choose in M_ℓ a local coordinate system $(x_\ell{}^a, \theta_\ell^\alpha)$. Assume that the following properties are true: functions $(x_\ell^a)_{rd}$ are ρ-stable and functions $(\dot{x}^a = \frac{1}{2}(x_\ell^a + x_r^a),$ $\theta_\ell^\alpha, \theta_r^{\dot\alpha})$ are local coordinates on M, where $\theta_r^{\dot\alpha} = (\theta_\ell^\alpha)^\rho$, $x_r^a = (x_\ell^a)^\rho)$. Such coordinates (x_ℓ, θ_ℓ) on M_ℓ , (x_r, θ_r) on M_r and $(x, \theta_\ell, \theta_r)$ on M will be called distinguished ones.

Now we set

$$(5) \qquad H^a = \frac{1}{2i}(x_\ell{}^a - x_r{}^a).$$

These four real nilpotent superfunctions on M are called the Ogievetsky-Sokachev prepotential. Working locally and identifying $F_{\ell,r}$ with ΠO_M we can say that the prepotential completely defines the geometry of the superspace, except for the forms $v_{\ell,r}$ which must be given separately :

$$(6) \qquad v_{\ell,r} = \Phi^3_{\ell,r}\, D^*(d\theta^\alpha_{\ell,r}\; d\theta^a_{\ell,r})$$

Some calculations (cf. [26] for details) show that the action (3) can be expressed through (5) and (6) by means of the Wess-Zumino formula

$$(7) \qquad w = \frac{1}{8}\,\mathrm{Ber}(E^A_B)\; D^*(d\theta^\alpha_\ell, d\theta_r^{\dot\beta}, dr^a) \;,$$

where E^A_B is the transition matrix between the frames

$$\left(\frac{\partial}{\partial x^a}, \frac{\partial}{\partial\theta_\ell{}^\alpha}, \frac{\partial}{\partial\theta_r{}^{\dot\alpha}}\right) = (\partial_a, \partial_\alpha, \partial_{\dot\alpha}) \quad\text{and}\quad (\tfrac{i}{2}[\tilde\Delta_\alpha, \tilde\Delta_{\dot\alpha}], \tilde\Delta_\alpha, \tilde\Delta_{\dot\alpha}).$$

This last frame can be defined in three steps.

Step 1. $\Delta_\alpha = \partial_\alpha + X^a_\alpha \partial_a$ and $\Delta_{\dot\alpha} = -\partial_{\dot\alpha} - X^a_{\dot\alpha}\partial_a$ are defined as local bases for $T_\ell M$ and $T_r M$ respectively. From this one finds the coefficients

$$X^a_\alpha = i\left[(1 - i\frac{\partial H}{\partial x})^{-1}\right]^a_b H^b; \; X^a_{\dot\alpha} = -i\left[(1 + i\frac{H}{x})^{-1}\right]^a_b \partial_{\dot\alpha} H^b$$

Step 2. The volume forms $v_{\ell,r}$ define the spinor metrics $E_{\ell,r} \in \mathrm{Ber}\, S_{\ell,r}$:

$$E_\ell = (\pi_\ell{}^* v_\ell)^{1/3} \otimes (\pi_r{}^* v_r)^{-2/3}, \; E_r = (\pi_\ell{}^* v_\ell)^{-2/3} \otimes (\pi_r{}^* v_r)^{1/3}.$$

Step 3. The multiplier F, defining $\tilde\Delta_\alpha = F\Delta_\alpha$ and $\tilde\Delta_{\dot\alpha} = F^\rho \Delta_{\dot\alpha}$, is constructed in such a way, that $D^*(\tilde\Delta_\alpha) = E_\ell$, $D^*(\tilde\Delta_{\dot\alpha}) = E_r$.

The structure frame $(\frac{i}{2}[\tilde\Delta_\alpha, \tilde\Delta_{\dot\alpha}], \tilde\Delta_\alpha, \tilde\Delta_{\dot\alpha})$ can be used to describe the geometry of simple supergravity Cartan style. In this approach it appears as the final product rather than the starting point.

REFERENCES

[1] S.J. Arakelov. An intersection theory for divisors on an arithmetic surface. - Izv. Akad. Nauk SSSR, ser. mat. 1974, 38:6, pp. 1179-1192 (in russian).

[2] S.J. Arakelov. Theory of intersections on arithmetic surface. - Proc. Int. Congr. Vancouver, 1974, vol. 1, pp. 405-408.

[3] T. Aubin. Equations du type de Monge-Ampère sur les varietés kähleriennes compactes. - C.R. Acad. Sci. Paris, 1976, 283, pp. 119-121.

[4] A.A. Beilinson. Higher regulators and the values of L-functions. - Sovremenny problemy matematiki, vol. 24, Moscow, VINITI 1984 (in russian), pp 181-238.

[5] F.A. Berezin. Introduction to algebra and calculus with anticommuting variables. - Moscow University Press, 1983 (in russian).

[6] E. Bombieri, J. Vaaler. On Siegel's lemma. - Inv. math. 1983, vol. 73:1, pp. 11-32.

[7] E. Date, M. Jimbo, M. Kashiwara, T. Miwa. Solitons, τ-functions and euclidean Lie algebras. - in: Mathématique et Physique, Séminaire ENS 1979-82, Birkhäuser, Boston 1983, pp. 261-280.

[8] P. Deligne. Preuve des conjectures de Tate et Shafarevitch (d'apres G. Faltings).- Sem. Bourbaki, Nov. 1983, N 616.

[9] S. Donaldson. Stable holomorphic bundles and curvature. - Oxford, preprint 1983.

[10] G. Faltings. Calculus on arithmetic surfaces.Ann.Math. 119 (1984), pp. 387-424.

[11] G. Faltings. Endlichkeitssätze für abelesche Varietäten über Zahlkörpern. - Inv. Math., 1983, 73, pp. 349-366.

[12] S.J. Gates, M.T. Grisaru, M. Rocek, W. Siegel. Superspace or one thousand and less lessons in supersymmetry. - London, Benjamin, 1983.

[13] Geometry ideas in physics. Ed. by Yu.I. Manin (in russian). - Moscow, Mir Publishers, 1983.

[14] K. Iwasawa. Analogies between number fields and function fields. - Preprint, 1965, Lecture given at the Yeshira Science Conference.

[15] V.G. Kac. Lie superalgebras. - Advances in Math., 1977, 26:1, pp. 8-96.

[16] V.G. Kac. Representations of classical Lie superalgebras.- in: Lecture Notes in Math., N 676, pp. 597-627, Springer 1978.

[17] V.G. Kac, D.A. Kazhdan. Structure of representations with highest weight of infinite dimensional Lie algebras.-Advances in Math., 34, (1979), pp. 97-108.

[18] Sh. Kobayashi. Hyperbolic manifolds and holomorphic mappings.- New York, M. Dekker, 1970.

[19] Sh. Kobayashi. Curvature and stability of vector bundles. - Proc. Jap. Ac. Sci., 1982, 58:4, pp. 158-162.

[20] N. Koiso. Einstein metrics and complex strutures.- Inv. Math., 1983, 73:1, pp. 71-106.

[21] B. Kostant. Graded manifolds, graded Lie theory and prequantization. - in: Lecture Notes in Math., N 570, pp. 177-306, Springer 1977.

[22] S. Lang. Fundamentals of Diophantine geometry.- Berlin, Springer, 1983.

[23] D.A. Leites. Information to theory of supermanifolds. (in russian). Uspekhi, 1980, 35:1, pp. 3-57.

[24] Yu. I. Manin. Gauge fields and complex geometry. (in russian). Moscow, Nauka, 1984.

[25] Yu. I. Manin. Flag superspaces and supersymmetric Yang-Mills equations.- in: Arithmetic and geometry. Papers dedicated to I.R. Shafarevich on the occasion of his sixtieth birthday. vol. 2, pp. 175-198, Birkhäuser, 1983.

[26] Yu. I. Manin. The geometry of supergravity and the Schubert supercells. (in russian).- Zapiski nachn. sem. LOMI, Leningrad, 1984, vol. 133, pp. 160-176.

[27] B. Mazur, A. Wiles. Analogies between function fields and number fields. - Amer. Journ. of Math., 1983, 104,pp. 507-521.

[28] V.I Ogievetsky, E.S. Sokatchev. The gravitational axial superfield and the formalism of the differential geometry. (in russian). - Yad. fiz., 1980,31:3, pp. 821-839.

[29] I.B. Penkov. $\mathcal{D}$-modules on supermanifolds. -Inv. Math., 1983, 71:3, pp. 501-512.

[30] D. Quillen. Arithmetic surfaces and analytic torsion.- Bonn, Arbeitstagung, 1983.

[31] I.R. Shafarevich. Algebraic number fields.-Proc. Int. Congress of Math.,Stockholm, 1962.

[32] I.R. Shafarevich. Minimal models of two dimensional schemes.- Tata Institute of Fund. Res., 37, Bombay 1965.

[33] L. Szpiro. La conjecture de Mordell (d'apres G.Faltings).- Sém Bourbaki, Nov. 1983, N 619.

[34] Superspace and supergravity (ed. by S.W. Hawking and M. Rocek). -Cambridge Univ. Press, 1981.

[35] Théorie des intersections et théorème de Riemann-Roch.- Lecture Notes in Math., N 225, Springer 1971.

[36] A.Weil. Sur l'analogie entre les corps de nombers algébriques et les corps de fonctions algébriques.-in: Collected Papers, vol. 1, pp. 236-240. New York, Springer 1980.

[37] A. Weil. De la métaphysique aux mathématiques.-in: Collected Papers, vol. 2, pp. 408-412. New York, Springer 1980.

[38] S.T. Yau. On Calabi's conjecture and some new results in algebraic geometry. - Proc. Nat. Acad. Sci. USA, 1977, 74, pp. 1798-1799.

COMMENTARY ON THE ARTICLE OF MANIN

Michael Atiyah
Mathematical Institute
Oxford OX1 3LB

§1. Mathematics and Physics

Manin's stimulating contribution to the 25th Arbeitstagung which, in his absence, I attempted to present, provided me with an opportunity of adding some further reflections of my own. This commentary, which is therefore a very personal response to Manin's article, consists of very general and speculative remarks about large areas of contemporary mathematics. Such speculations are, for good reason, rarely put down on paper but the record of the 25th Arbeitstagung provides a rather singular occasion where ideas of this type may not be out of place.

In recent years there has been a remarkable resurgence of the traditional links between mathematics and physics. A number of striking ideas and problems from theoretical physics have penetrated into various branches of mathematics, including areas such as algebraic geometry and number theory which are rarely disturbed by such outside influences. Perhaps a few specific examples will illustrate the point. The Kadomtsev-Petiashvilli equation, which arose in plasma physics, has been shown to be extremely relevant to the classical Schottky problem about the characterization of Jacobian varieties of algebraic curves (this was explained in the Arbeitstagung lecture of van der Geer). Witten's analytical approach to the Morse inequalities, based on the physicist's use of stationary-phase approximation, has led Deligne and others to imitate his ideas in number theory with great success. The Yang-Mills equations and their 'instanton' solutions have been brilliantly exploited by Donaldson to solve outstanding problems on 4-manifolds.

All these examples connect physics with various branches of geometry, and it is therefore natural that Manin should have attempted an overview of geometry in the widest sense. The picture he describes is best indicated by the following schematic diagram:

$$\underbrace{\text{Arithmetic} \xleftarrow{\text{Algebra}} \text{Geometry} \xrightarrow{\text{Analysis}} \text{Physics}}_{\displaystyle \text{Topology}}$$

Topology

Homology   K-theory

As this suggests, ideas from topology, notably homology and K-theory, provide a common language and underpinning for the whole structure. The bridge between geometry and arithmetic was greatly expanded and developed during the Grothendieck era with the introduction of 'schemes'. The bridge between geometry and physics begins essentially with Einstein's theory of gravitation but has become much stronger with the recent development of gauge theories of elementary particles.

The picture just envisaged is restricted, on the physics side, to classical physics. However, one should be more ambitious and try to fit quantum physics into the picture also. I will have more to say on this aspect later.

§2. Arithmetic manifolds

An algebraic curve defined by equations with integer coefficients can be viewed as a scheme over Spec Z. It is the analogue of a surface mapped onto a curve, the 'fibre' over a prime p being the curve reduced mod p. Such an arithmetic surface can be 'compactified' by adding the Riemann surface of the curve over the 'prime' at ∞. The Arakelov-Faltings theory is then concerned with extending as much as possible of the usual theory of surfaces to this arithmetic case.

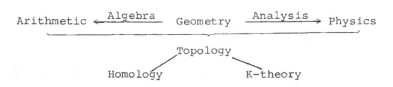

For this purpose it turns out that one needs to introduce or find canonical metrics on various objects associated to the Riemann surface. For example, given a line-bundle $\mathcal{L}$ on the Riemann surface one has the one-dimensional complex vector space

$$\det H^0(\mathcal{L})/\det H^1(\mathcal{L})$$

(where det denotes the highest exterior power), and one wants a natural metric on this space.

This particular problem which was solved in one way by Faltings has been examined in a wider context (e.g. replacing $\mathcal{L}$ by a vector bundle) by Quillen. He has shown that a natural definition arises by using the regularized determinants of Laplace type operators which were introduced into differential geometry by Ray and Singer [3]. Such operator determinants are extensively used by physicists in quantum field theory, and this link between geometry and physics is currently the scene of many investigations. In any case it provides a clear link with quantum and not purely classical physics.

On the Riemann surface itself there are two natural metrics (for $g \geq 2$), one being the Poincaré metric and the other being the metric induced by the holomorphic differentials. In higher dimensions the analogues of the Poincaré metric are the Kähler-Einstein metrics. Similarly for stable vector bundles there are distinguished metrics, and Manin proposes they should be used for a higher dimensional theory of arithmetic manifolds. It is interesting to note that all of these metrics arose in a physical context.

Thus the geometry of Kähler manifolds, and in particular the study of operator determinants on such manifolds, appears as a natural meeting point for arithmetic and physics. In this context it is perhaps worth pointing out that a number of differential-geometric invariants constructed from operator determinants have already been identified with quantities arising in number theory. Thus Ray and

Singer [4] made a connection with the Selberg zeta-function of a Riemann surface while Millson [2] did something similar in higher odd dimensions. Also values of certain L-functions of totally real number fields have been related in [1] to the eta-invariant (essentially the logarithm of a certain operator determinant) : the eta-invariant is also what appears in [2].

§3. Fermions

There is a basic distinction in physics between two types of particles, namely bosons and fermions. Bosons involve commuting variables and so are easily understood on a geometric level, but fermions involve anti-commuting variables and so are more mysterious geometrically. On a purely algebraic level of course there is no mystery: polynomial and exterior algebras are both well understood and extensively studied. However, the development of gauge theories in physics where geometric insight and interpretation greatly assist the purely formal algebraic aspect has naturally led to the attempt to develop a 'super-geometry' in which both sets of variables are incorporated.

The development of super-manifolds as outlined by Manin appears to be an elegant extension of classical geometric ideas and it should throw light on the algebraic computation of physicists who build 'super-symmetric' theories. Nevertheless the theory still appears to lack some essential ingredients and Manin asks whether the fermionic coordinates can somehow be 'compactified' so as to make them topologically more interesting. In this context I would like to make a tentative suggestion concerning the right geometric way to interpret fermionic variables.

Consider first a smooth manifold M, its de Rham complex $\Omega^*(M)$ and in particular the ring $\Omega^0(X)$ of smooth functions. For a super-symmetric analogue, suppose now that M is a closed sub-manifold of another manifold N, and let $A = \Omega_M^*(N)$ denote the complex of currents on N which are supported on M and smooth in the M-directions (recall that a current is just a differential form with distributional coefficients). Locally an element of A can be expressed in the form

$$\Sigma\, f_{\alpha\beta\gamma}(x)\,\partial_y^\gamma\delta\; dx^\alpha \wedge dy^\beta$$

where $x = (x_1 \ldots x_m)$ are coordinates on M, $y = (y_1,\ldots,y_r)$ are normal coordinates to M in N, δ is the Dirac δ-function of M, α, β are skew-symmetric multi-indices and γ is a symmetric multi-index (so that ∂_y^γ represents derivatives in y). If we take α,γ to be empty and β to be a single index we get a subspace $R\delta$ of A where R is the super-ring of the super-manifold given by M and its normal bundle in N. On the other hand A itself should be viewed as the super de Rham complex of this super-manifold.

The advantage of this point of view is that approximating the δ-function by suitable smooth functions (e.g. Gaussians) we can try to interpret fermions as bosons on N which are very sharply peaked along M. More precisely the fermions should appear as 'leading terms' of such sharply peaked bosons. Geometrically this might correspond to putting a metric on N which is very sharply curved along M, so that M is an 'edge of regression' in the language of classical differential geometry.

I am trying to suggest that super-geometry should be some kind of limit of ordinary geometry and not an entirely different kind of entity constructed simply by formal analogy.

§4. The quantum level

Quantum theory is characterized by infinite-dimensionality and by non-commutativity. When trying to understand the possible geometric counterpart of some aspect of quantum-theory this must be borne in mind.

As I have already mentioned the study of linear elliptic operators provides one bridge between geometry and quantum field theory. For example ideas from supersymmetric field theories have cast new light on the index theorem.

In a different direction it is I think not inappropriate to consider Connes' non-commutative differential geometry (see the survey talk by Connes in this volume) as a version of quantized geometry. Recall that Connes studies situations such as the ergodic action of a discrete group on a manifold where the geometric quotient does not exist in any way as a reasonable space. However, a non-commutative algebra exists with which various geometric constructions can still be made.

In the lecture of Lang he explained a conjecture of Vojta based on an interesting analogy between arithmetic surfaces and Nevanlinna theory. It is perhaps interesting in this connection that John Roe in his Oxford D.Phil. thesis shows how the Nevanlinna theory fits into Connes' framework. Analysing this situation might shed light on the analogy between Connes' theory and questions in Arithmetic.

If one asks for the analogue of quantum theory in Arithmetic one can hardly avoid considering the whole Langlands programme. Adelic groups are obvious analogues of gauge groups and Hilbert space representations are the basic objects of the theory. This analogy deserves closer scrutiny, particularly in view of the fact that non-abelian dualities, generalizing class-field theory on the one hand and electric-magnetic Maxwell duality on the other, seem to be a main objective in both number theory and physics. Perhaps our classical diagram should be enlarged to a quantum diagram in the following way:

Langlands	←	Connes	→	Quantum Field Theory	Quantum
|		|		|	
Arithmetic	←	Geometry	→	Physics	Classical

References

1. M.F. Atiyah, H. Donnelly and I.M. Singer, Eta invariants, signature defects of cusps and values of L-functions, Ann. of Math. 118 (1983), 131-177.
2. J. Millson, Closed geodesics and the η-invariant, Ann of Math. 108 (1978), 1-39.
3. D.B. Ray and I.M. Singer, R-torsion and the Laplacian on Riemannian manifolds, Advances in Math. 7 (1971), 145-210.
4. D.B. Ray and I.M. Singer, Analytic torsion for complex manifolds, Ann. of Math. 98 (1973), 154-177.

The Mandelbrot Set in a Model for Phase Transitions *)

Heinz-Otto Peitgen and Peter H. Richter **)

INTRODUCTION

According to D. Ruelle [18] "... the main problem of equilibrium statistical mechanics is to understand the nature of phases and phase transitions ...". A remarkable observation of B. Derrida, L. De Seze and C. Itzykson [4] has put these problems of theoretical physics into a new perspective: For a very particular model (the hierarchical q-state Potts model on a hierarchical lattice) they indicated that the Julia set of the corresponding renormalization group transformation is the zero set of the partition function in the classical theory of C. N. Yang and T. D. Lee [22]. The Yang-Lee theory describes a physical phase as a domain of analyticity for the free energy, viewed as a function of complex temperature. The boundaries of these domains are given by the zeroes of the partition function. Carrying on these ideas we show a connection with a discovery of B. Mandelbrot [13]. More precisely, in a discussion of the morphology of the above zero sets we discover a structure which is related to the Mandelbrot set (see [15]) attached to the one-parameter family $\mathbb{C} \ni z \to z^2 + c$, $c \in \mathbb{C}$ a fixed constant. For this we exploit recent results of D. Sullivan [21] which classify the stable regions of rational maps on $\overline{\mathbb{C}} = \mathbb{C} \cup \{\infty\}$. Though the physical meaning of the hierarchical Potts model is certainly very questionable it seems that the classical (see G. Julia [12] and P. Fatou [8]) and recent (see A. Douady and J. Hubbard [5,6,7], D. Sullivan [21], M. Herman [11]) theory of complex dynamical systems may produce a major step towards a deeper understanding of the nature of phase transitions. Besides the hierarchical Potts model we have analyzed 1- and 2-dimensional Ising models with and without an external magnetic field and have found that the theory of Julia sets and

---

*) This paper surveys the recent interaction between the theory of phase transitions in statistical mechanics and the theory of complex dynamical systems.

**) Forschungsschwerpunkt Dynamische Systeme, Universität Bremen D-2800 Bremen 33

their typical fractal properties play a very significant role in the interaction between the Yang-Lee theory and the renormalization group approach. None of these and the findings in [15] would have been possible without the aid of extensive computer graphical studies and experiments.

PRELIMINARIES AND NOTATION

The hierarchical Potts model is associated with a very particular and physically unrealistic lattice construction which we introduce schematically in fig. 1.

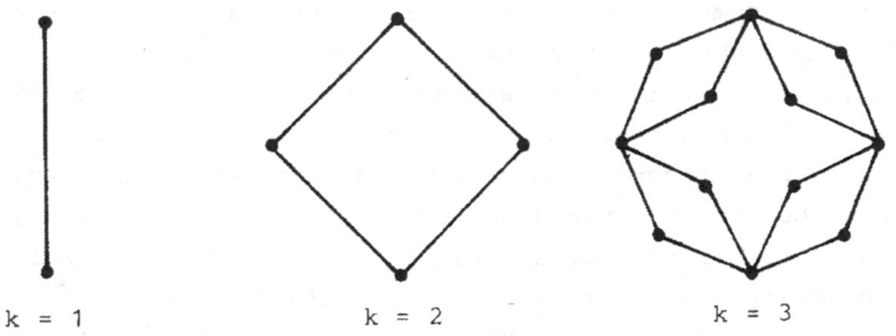

Figure 1. The diamond hierarchical lattice with $n = n(k) = 4 + 2(4^{k-1} - 4)/3$ atoms (dotts) and 4^{k-1} bonds (line segments) for $k > 1$.

For this particular lattice and nearest neighbor coupling an explicit form of the renormalization group transformation is known and that is why it is valuable here. On each lattice site i we assume a spin with $q \in \mathbb{N}$ possible states

$$\sigma_i = 1, \ldots, q .$$

The partition function $Z_k(T)$ is the sum of Boltzmann factors extended over all configurations

$$\{\sigma : \{1,\dots,n\} \to \{1,\dots,q\}\} , \quad n = \#\text{ of lattice points,}$$

(1) $$Z_k(T) = \sum_{\sigma} \exp\left(-\frac{1}{k_B T} E(\sigma)\right) ,$$

where $E(\sigma)$ is the potential energy of the configuration σ. Assuming that the interaction of different lattice sites is restricted to nearest neighbors only, i.e. only across a bond indicated by a line segment in figure 1, the energy across such a bond for a fixed configuration σ is:

(2) $$E(i,j) = \begin{cases} -U , & \text{if } \sigma_i = \sigma_j \\ 0 , & \text{else.} \end{cases}$$

Hence,

(3) $$E(\sigma) = \sum_{\text{bonds}} E(i,j) .$$

For convenience we introduce new variables

(4) $$x = \exp(U/k_B \cdot T)$$

so that $Z_n(x)$ becomes a polynomial in x with integer coefficients. The coupling constant U is characteristic for the material, $U > 0$ for ferromagnetic, and $U < 0$ for antiferromagnetic coupling. From Z_k one derives the free energy per atom

(5) $$f_n = -\frac{k_B T}{n} \ln Z_k , \quad n = n(k) .$$

Thus, zeroes of Z_k correspond to logarithmic singularities of f_n and are reasonable candidates for phase transitions. Note, however, that $Z_k(x) \neq 0$ for any finite lattice with $n = n(k)$ points and for all $x > 0$, which is the physically meaningful temperature range.

THE YANG-LEE MODEL OF PHASE TRANSITIONS

In essence the idea of C. N. Yang and T. D. Lee [22], which had a substantial impact on the forthcoming attempts to solve phase transition problems, is as follows:

Let

(6) $$N_k = \{x \in \mathbb{C} \ , \ Z_k\ (x) = 0\} \ ,$$

i.e. one embeds the partition function in a complex temperature plane. To make boundary effects negligable one has to pass to the *thermodynamic limit*, i.e. one lets $n \to \infty$. It is not obvious, of course, that such a limit makes sense and exists. If, however, the potential energy E admits an appropriate growth condition and the range of the interaction is sufficiently small, which is trivially satisfied in our case, then (see [18]) the limit exists and we denote by N_∞ the zero-set of the partition function Z_∞ in the thermodynamic limit. Now Yang and Lee postulated that N_∞ would distinguish a unique point $x_c > 0$,

(7) $$N_\infty \cap \mathbb{R}_+ = \{x_c\} \ ,$$

so that T_c , $x_c = \exp(U/k_B \cdot T_c)$, is the phase transition point (see fig. 2).

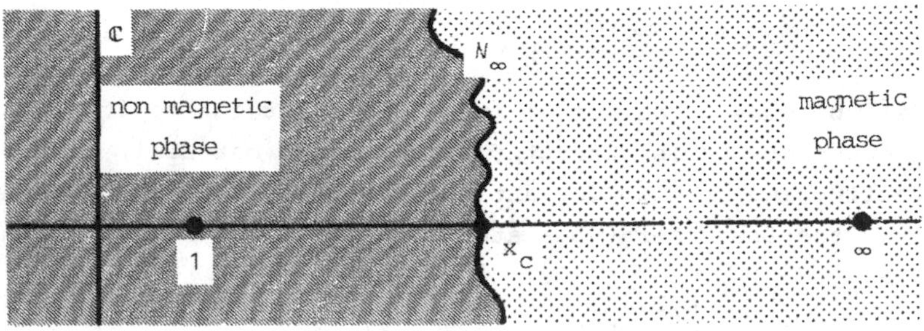

Figure 2. Note that $T = \infty$ corresponds to $x = 1$.

Thus to find and characterize T_c it remains to find x_c and interprete N_∞ in the neighborhood of x_c . For example the critical index α , which characterizes the singularity of the *specific heat*,

(8) $$C \sim |\ T - T_c\ |^{-\alpha}$$

can be obtained from the density of the zeroes in the thermodynamic limit near x_c (see [9]).

THE RENORMALIZATION GROUP APPROACH

In general the partition function Z_k is not only a function of temperature x but also of other variables like for example an external magnetic field H. In essence the idea of the renormalization group approach is to relate

$$Z_{k-1} \text{ with } Z_k \text{ , i.e.}$$

$$(9) \quad \begin{cases} Z_k(x,H,\ldots) = Z_{k-1}(\overline{x},\overline{H},\ldots) \cdot \varphi(x,H,\ldots) \\ (\overline{x},\overline{H},\ldots) = R(x,H,\ldots) \text{ .} \end{cases}$$

Thus, up to a trivial factor φ the partition function of step k is obtained by that of step $k-1$ modulo an appropriate adaption of the variables $(x,H,\ldots)$. This determines a map R, the *renormalization transformation*.

In our specific hierarchial q-states Potts model Z_k is only a function of x, the temperature variable. However, Z_k depends on the material constant q. An elementary calculation shows that (see [4], [15])

$$(10) \quad \begin{cases} Z_k(x) = Z_{k-1}(\overline{x}) \cdot \varphi(x) \text{ , } \quad k \geq 2 \\ R(x) = \overline{x} = \left(\dfrac{x^2+q-1}{2x+q-2}\right)^2 \\ Z_1(x) := q(x+q-1) \\ \varphi(x) = (2x+q-2)^{2 \cdot 4^{k-2}} \end{cases}$$

Thus, the renormalization transformation is a rational map of degree 4. Actually, as we let q vary in $\mathbb{C}$ we obtain a 1-parameter family $R = R_q$. For any q we have that

$$(11) \quad \begin{cases} R_q(1) = 1 \text{ and } R_q'(1) = 0 \\ R_q(\infty) = \infty \text{ and } R_q'(\infty) = 0 \text{ ,} \end{cases}$$

i.e. 1 and ∞ are superstable attractors. Their basins of attraction are defined by ($\overline{\mathbb{C}} = \mathbb{C} \cup \{\infty\}$)

$$(12)\qquad \begin{cases} A_q(1) = \{x \in \overline{\mathbb{C}} : R_q^n(x) \to 1 \text{ as } n \to \infty\} \\ A_q(\infty) = \{x \in \overline{\mathbb{C}} : R_q^n(x) \to \infty \text{ as } n \to \infty\} \ . \end{cases}$$

As a consequence of the classical theory of G. Julia [12] and P. Fatou [8] on the interation of rational functions in $\mathbb{C}$ we have that

$$(13)\qquad \partial A_q(1) = J_q = \partial A_q(\infty)$$

is the Julia set of R_q .

JULIA SETS AND PHASE TRANSITIONS

We are now in a position to discuss the Yang-Lee model in terms of the renormalization approach from the point of view of the theory of Julia sets. We begin by listing a few interesting conjectures and problems:

CONJECTURE 1.1.

$$N_\infty = J_q \ , \text{ i.e.}$$

$A_q(1)$ (resp. $A_q(\infty)$) corresponds to the non-magnetic (resp. magnetic) phase.

To discuss this crucial conjecture the following immediate observation from (10) is of importance:

Note that $Z_{k-1}(R_q(x)) = \eta(x)/\varphi(x)$ for some $\eta(x)$ and therefore

$$(14)\qquad N_k = \{x \in \overline{\mathbb{C}} : R_q^{k-1}(x) = 1 - q\} \ .$$

Moreover, the free energy in the thermodynamic limit f_∞ satisfies the functional equation (15) as a consequence of (5) and (10):

$$(15)\qquad \begin{cases} f_\infty(x) = \frac{1}{4} f_\infty(R_q(x)) + g(x) \ , \quad \text{with} \\ g(x) = \frac{1}{2}\ln(2x+q-2) \ . \end{cases}$$

PROBLEM 1.2.

(a) In what sense is $N_k \to N_\infty$ as $k \to \infty$?

(b) For which $q \in \mathbb{C}$ is $N_\infty = J_q$?

(c) For which $q \in \mathbb{C}$ does R_q admit further attractors, other than 1 and ∞ ?

If R_q admits a further attractor other than 1 and ∞ then its corresponding basin of attraction may characterize a third magnetic phase such as for example the *antiferromagnetic phase*.

In view of (14) and (15) conjecture (1.1) means that the singularities of f_∞, the phase transitions points, are given by points from J_q, and this is intimately related to an understanding of the forward and backward orbit

$$\{R_q^i(1-q)\}_{i \in Z} \tag{16}$$

for $q \in \mathbb{C}$. Thus, the question which remains is: Is $(1-q) \in J_q$ or in which component of $\overline{\mathbb{C}} \setminus J_q$ is it?

THE CLASSIFICATION OF STABLE REGIONS

This leads us directly into one of the most celebrated recent results in the theory of complex dynamical systems: The classification of stable regions of D. Sullivan [14,21]. Let f be an analytic endomorphism of $\overline{\mathbb{C}}$. A point $x \in \overline{\mathbb{C}}$ is *stable* for f if on some neighborhood of x the family of iterates $f, f^2, f^3, \ldots$ is an equicontinuous family of mappings of that neighborhood into $\mathbb{C}$. Note that when x is not stable, i.e *unstable*, for any neighborhood the union of images of iterates must cover $\overline{\mathbb{C}}$ except for two points at most. The set of unstable points for f is the Julia set J of f. It is the closure of the expanding periodic points. The open set of stable points $\overline{\mathbb{C}} \setminus J$ consists of countably many connected components, the *stable regions* of f, which are transformed among themselves by f. The following three theorems of D. Sullivan [21] and P. Fatou [8] are crucial for conjecture 1.1 and problem 1.2. Let f be a rational mapping with $d = \deg(f) \geq 2$.

THEOREM 1.3. (Sullivan)

Each stable region is eventually cyclic.
(For any component $C \subset \overline{\mathbb{C}} \setminus J$ there is $n \in \mathbb{N}$ such that $D = f^n(C)$ is cyclic, i.e. $f^k(D) = D$ for some $k \in \mathbb{N}$.)

THEOREM 1.4. (Sullivan)

The cycles of stable regions D are classified into five types:

(a) An *attractive basin* D arises from an attractive periodic cycle γ with non zero derivative of modulus less than one,
$\gamma = \{z, f(z), \ldots, f^{n-1}(z)\}$, $f^n(z)=z$, $0 < |(f^n)'(z)| < 1$,
and D consists of components of

$$\bigcup_{x\in\gamma} \{y : \lim_{n\to\infty} \text{distance } (f^n(y), f^n(x))=0\}$$

containing points of γ .

(b) A *parabolic basin* D arises from a non-hyperbolic periodic cycle γ with derivative a root of unity,
$\gamma = \{z, f(z), \ldots, f^{n-1}(z)\}$, $f^n(z)=z$, $((f^n)'(z))^m=1$,
γ is contained in the boundary of D , and each compact in D converges to γ under forward iteration of f .

(c) A *superattractive basin* D is defined just like an attractive basin but now $(f^n)'(z) = 0$.

(d) A *Siegel disk* D is a stable region which is cyclic and on which the appropriate power of f is analytically conjugate to an irrational ratotion of the standard unit disk.

(C.L. Siegel [19] proved these occur near a non-hyperbolic fixed point if the argument α of its derivative satisfies the following diophantine condition: there exists $c > 0$ and $\nu > 2$ such that

$$| \alpha - p/q | > c / q^{\nu}$$

for all relatively prime integers p and q .)

(e) A *Herman ring* D is a stable region similar to a Siegel disk. Now we have a periodic cycle of annuli and a power of f which restricted to any of these annuli is analytically equivalent to an irrational rotation of the standard annulus.

(For appropriate θ and a M. Herman [10] found such regions for the map:

$$x \longmapsto \frac{e^{i\theta}}{x} \left(\frac{x-a}{1-\bar{a}x} \right)^2 \quad)$$

The fate of critical points $\{c : f'(c) = 0\}$ is crucial in connection with theorem (1.4).

THEOREM 1.5. (Fatou)

(a) If D is an attractive or parabolic basin then D contains at least one critical point of f .

(b) If D is a Siegel disk or Herman ring then ∂D is contained in the ω-limit sets of critical points.

Thus f can have only finitely many cyclic stable regions. But it is still an open problem whether $2d-2$ ($d \geq 2$ the degree of f) is a sharp upper bound. Another open problem is whether a Siegel disk always has a critical point on its boundary. M. Herman [11] in a remarkable paper proved this conjecture recently for $f(z) = z^2 + \lambda$.

Note that theorem 1.5 and theorem 1.4 provide an excellent basis for computer experiments. For the detection and characterization of all cyclic stable regions of a map f one simply has to follow the forward orbits of all critical points. The following example illustrates the strength of these results:

EXAMPLE 1.6. $f(x) = \left(\frac{x-2}{x} \right)^2$, $J = \bar{\mathbb{C}}$.

The critical points are: 2,0 . Observe that $2 \mapsto 0 \mapsto \infty \mapsto 1 \mapsto 1$ and $f'(1) = -4$. Thus $\bar{\mathbb{C}} \setminus J = \emptyset$, because none of the cases (a), (b) in theorem 1.5 is possible.

THE CRITICAL POINTS OF THE RENORMALIZATION MAP R_q AND A MORPHOLOGY OF N_∞

Our map

$$R_q(x) = \left(\frac{x^2+q-1}{2x+q-2} \right)^2$$

has the six critical points:

$$1 \ , \quad \infty \ , \quad 1-q \ , \quad \pm\sqrt{1-q} \ , \quad (2-q)/2 \ .$$

Since 1 and ∞ are attractive fixed points and since $(2-q)/2 \mapsto \infty$, $\pm\sqrt{1-q} \mapsto 0$ it suffices to examine the orbits of $1-q$ and 0 only. We do this in the spirit of B. Mandelbrot's history making experiment:

Let

$$\text{(17)} \qquad \begin{cases} A_1 := \{q \in \mathbb{C} \ : \ R_q^n(1-q) \to 1 \ , \ n \to \infty\} \\ A_\infty := \{q \in \mathbb{C} \ : \ R_q^n(1-q) \to \infty \ , \ n \to \infty\} \\ M_R := \mathbb{C} \setminus (A_1 \cup A_\infty) \end{cases}$$

Figures 3,4 and 5 show A_1, A_∞ and M_R. Figure 6 shows a blow up of a detail of figure 5. Surprisingly it displays a structure which looks like a copy of the original Mandelbrot set [13]. I.e. it is exactly similar to the bifurcation set of the quadratic family $x \to x^2+c$, $c \in \mathbb{C}$. It is obvious that any q such that $|q| \gg 1$ is in A_∞, thus A_1 and M_R are bounded. Experimentally it turned out that the fate of the two crucial orbits of $(1-q)$ and 0 were related, i.e. whenever

$$\text{(18)} \qquad \begin{cases} R_q^n(1-q) \to 1 \quad \text{then} \quad R_q^n(0) \to \infty \ , \ \text{as } n \to \infty \\ R_q^n(1-q) \to \infty \quad \text{then} \quad R_q^n(0) \to 1 \ , \ \text{as } n \to \infty \ . \end{cases}$$

Indeed, this is an immediate consequence of the commutative diagram

$$\text{(19)} \qquad \begin{array}{ccc} \overline{\mathbb{C}} & \xrightarrow{\ R_q\ } & \overline{\mathbb{C}} \\ \Psi_q \downarrow & & \downarrow \Psi_q \\ \overline{\mathbb{C}} & \xrightarrow[\ S_q \circ S_q\ ]{} & \overline{\mathbb{C}} \end{array}$$

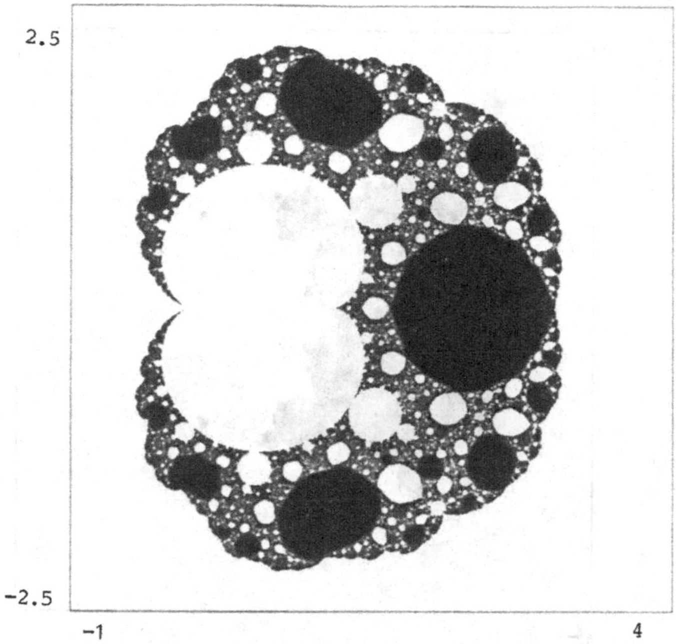

Figure 3. A_1 in black

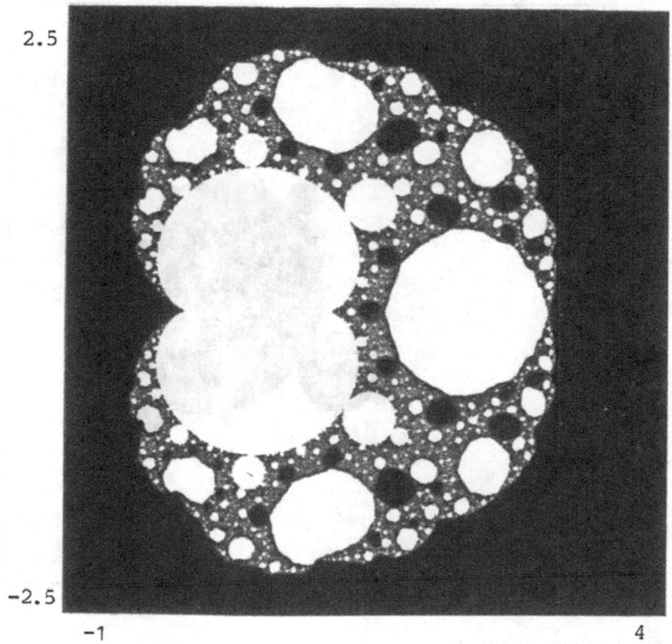

Figure 4. A_∞ in black

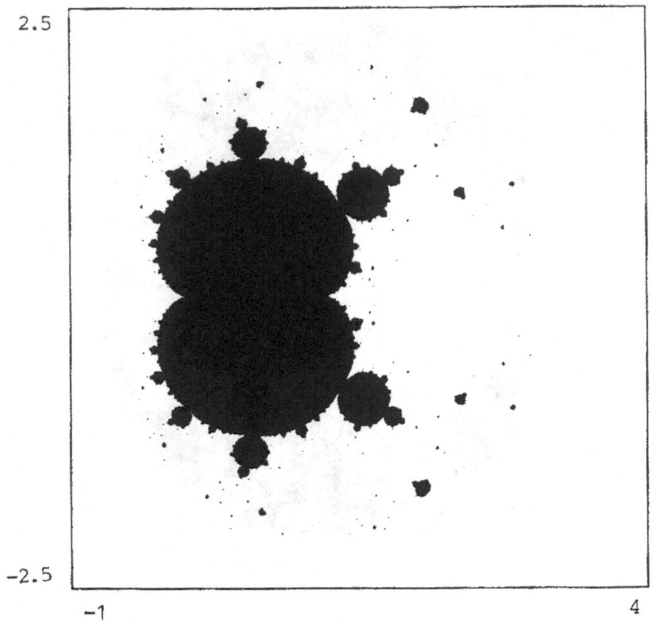

Figure 5. M_R in black

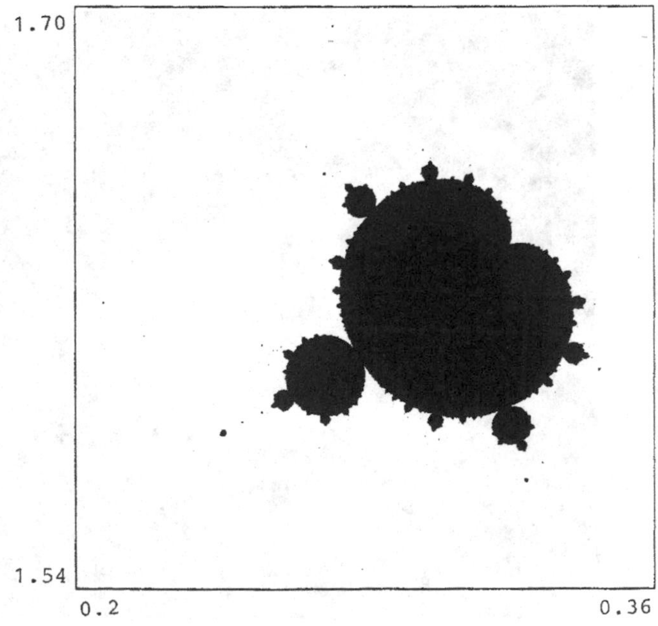

Figure 6. Detail in M_R

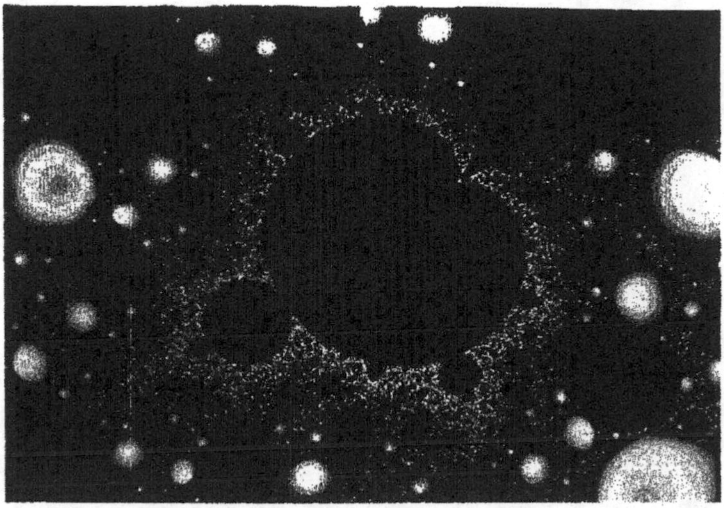

Detail of M_R (see figure 6) in black surrounded by A_1 in yellow and A_∞ in green.

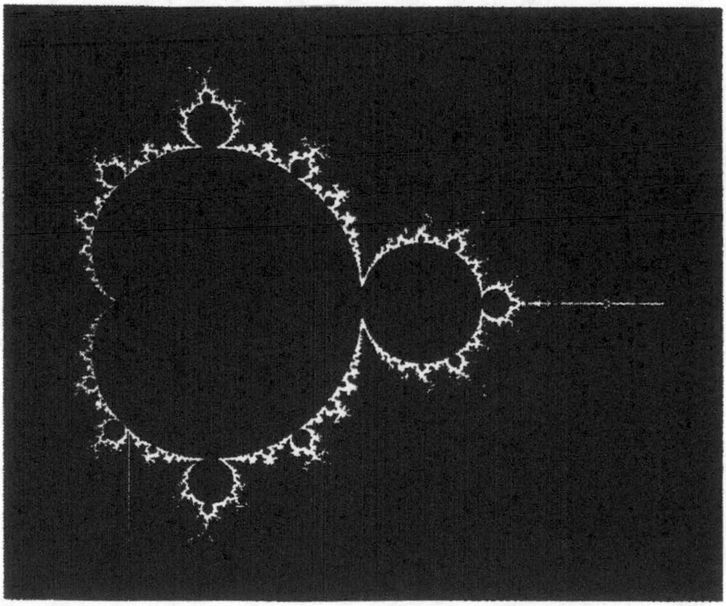

The Mandelbrot set M in black together with its electrostatic potential given by the Douady-Hubbard conformal homeomorphism $\mathbb{C}\setminus D \to \mathbb{C}\setminus M$.

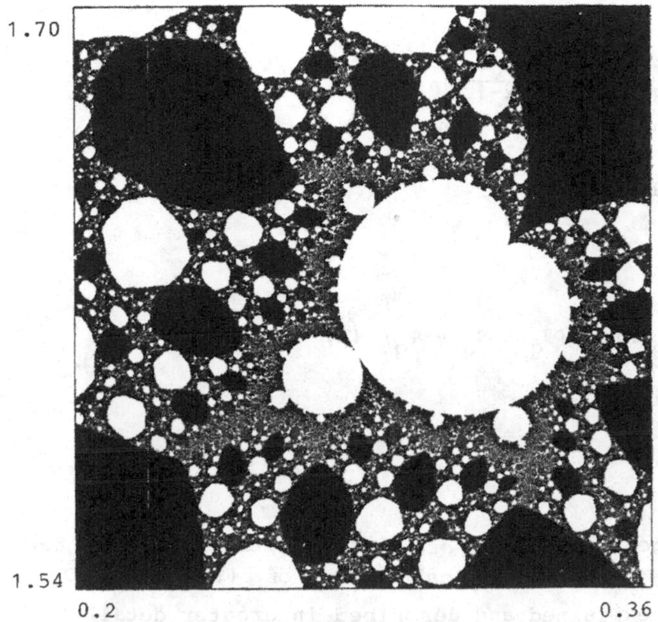

A_∞ in black

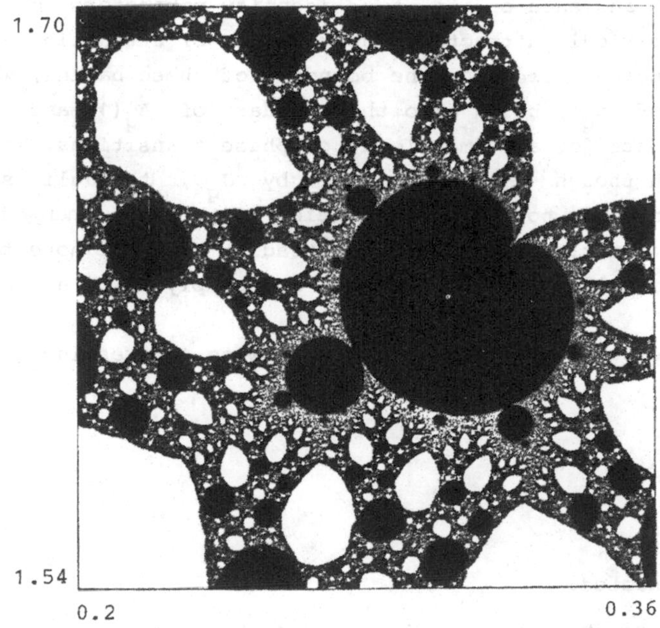

Figure 6. (continued) Detail of M_R A_1 and M_R in black

where

(20)
$$\begin{cases} \Psi_q(x) = \dfrac{x+q-1}{x-1} \\ \text{and} \\ S_q(x) = \dfrac{x^2+q-1}{x^2-1} \end{cases} .$$

This means that

(21)
$$\begin{cases} R_q(x) = (\Psi_q \circ S_q \circ \Psi_q)^2(x) = D_q^2(x) , \\ \text{with} \\ \qquad D_q(x) = \left(\dfrac{x+q-1}{x-1} \right)^2 \end{cases} .$$

Thus, Ψ_q exchanges the *hot phase* (x=1) with the *cold phase* (x=∞) and the two crucial critical orbits of (1-q) and 0 .

Figures 3-6 are explained and described in greater detail in [15]. In particular problem 1.2 (c) is answered. Roughly speaking the *main body* of M_R and each of its *buds* as well as the *main body* of the detail in figure 6 and each of its *buds* identify parameters q for which there is a periodic attractor. Their basins of attraction establish a *third magnetic phase* and the boundary of these basins, which is the Julia set of R_q , being also the boundary of $A_q(1)$ and $A_q(\infty)$, is a candidate for a formal locus of phase transitions. Note, however, that even though N_∞ may be given by J_q , the Julia set of R_q , its points may not be singularities of the free energy f_∞ in the thermodynamic limit. This seems to contradict (5), but note that in the thermodynamic limit the free energy may simply allow an analytic continuation.

In summary our experiments leed to the following interesting conjectures:

CONJECTURE 1.7.

(1) M_R is connected.

(2) The subset of M_R shown in figure 6 is homeomorphic (quasi conformally) to the Mandelbrot set M , where

$$\begin{cases} M = \{c\in\mathbb{C} : f_c^n(0) \not\to \infty , \quad \text{as} \quad n \to \infty\} \\ f_c(x) = x^2 + c . \end{cases}$$

(3) $N_\infty = J_q$ for any $q\in(\overline{\mathbb{C}} \setminus M_q) \cup \mathring{M}_q$.

Note that according to [5] the Mandelbrot set M is connected. Actually, Douady and Hubbard showed that $\mathbb{C} \setminus M$ and $\mathbb{C} \setminus D$, $D = \{x\in\mathbb{C} : |x| < 1\}$, are homeomorphic subject to a conformal mapping. Sullivan [21] gave an alternative proof which may apply also to our case. To indicate the idea we briefly survey another remarkable result of J. Curry, L. Garnett and D. Sullivan [3]:

NEWTON'S METHOD AND THE MANDELBROT SET

Consider the one-parameter family of rational maps

$$(22) \qquad \begin{cases} g_\lambda(x) = x - p_\lambda(x)/p'_\lambda(x) , \quad \text{where} \\ p_\lambda(x) = x^3 + (\lambda-1)x - \lambda . \end{cases}$$

Note that Newton's method for any cubic is equivalent by a linear change of variables to at least one of the g_λ 's. The 4 critical points of g_λ are the 3 roots of p_λ and the distinguished point 0 , which in view of theorem 1.5 is the only non-trivial critical point. The black regions in the complex λ-plane in figures 7, 8 and 9 were determined by the condition of the forward orbit of 0 converging to the root 1 of $p_\lambda(x)$. Let

$$(23) \qquad M_g = \{\lambda\in\overline{\mathbb{C}} : g_\lambda^n(0) \not\to \text{root of } p_\lambda , \text{ as } n \to \infty\} .$$

Then Sullivan [21] argues that the components in $\overline{\mathbb{C}} \setminus M_g$ correspond to quasi-conformal conjugacy classes which are analytically just punctured disks. Hence, M_g is connected. The subset of M_g shown in figure 9 is actually homeomorphic to the Mandelbrot set M , as A. Douady and J. H. Hubbard show in [6] . Arguments similar to those in

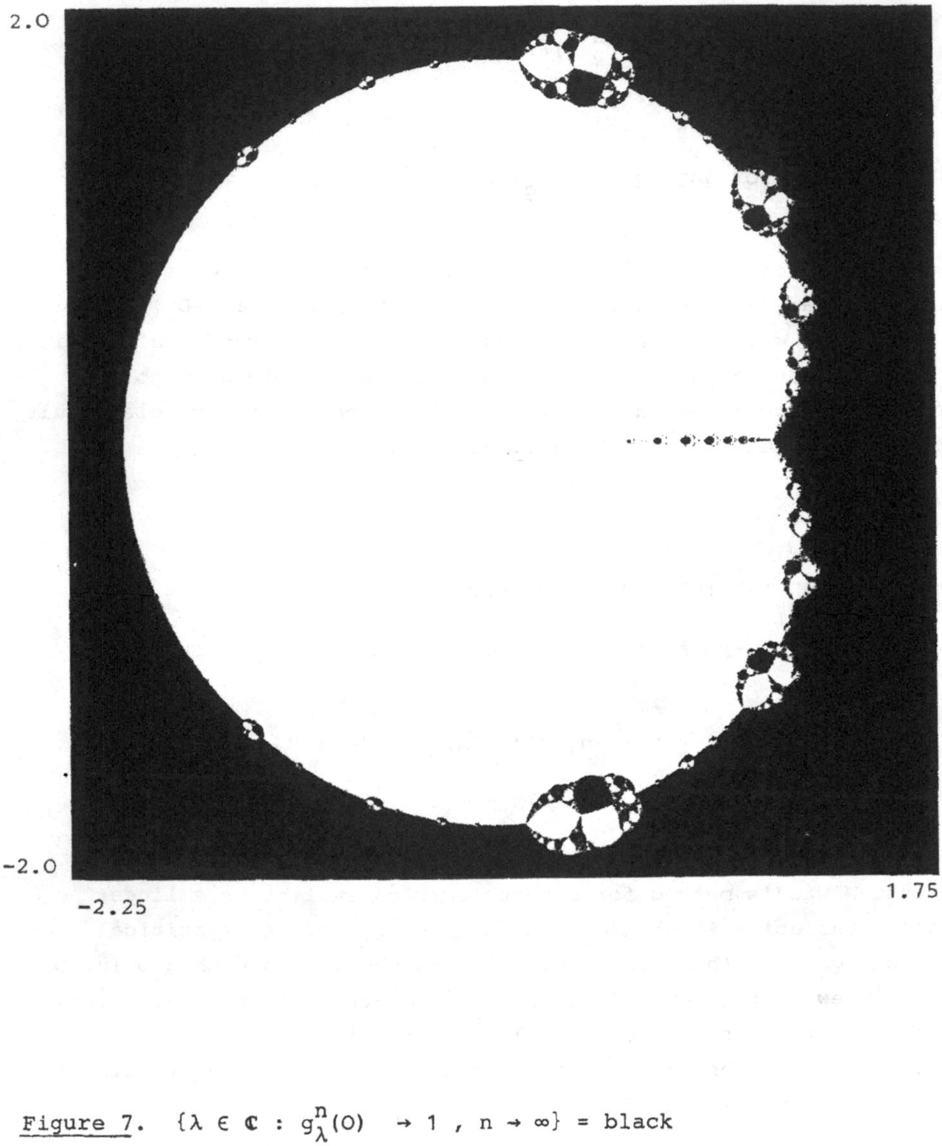

<u>Figure 7</u>. $\{\lambda \in \mathbb{C} : g_\lambda^n(0) \to 1 , n \to \infty\}$ = black

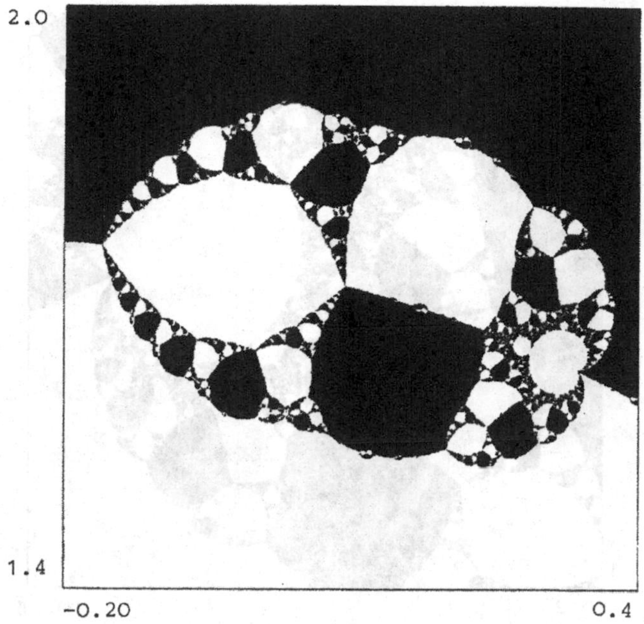

Figure 8. (a) Detail of figure 7.

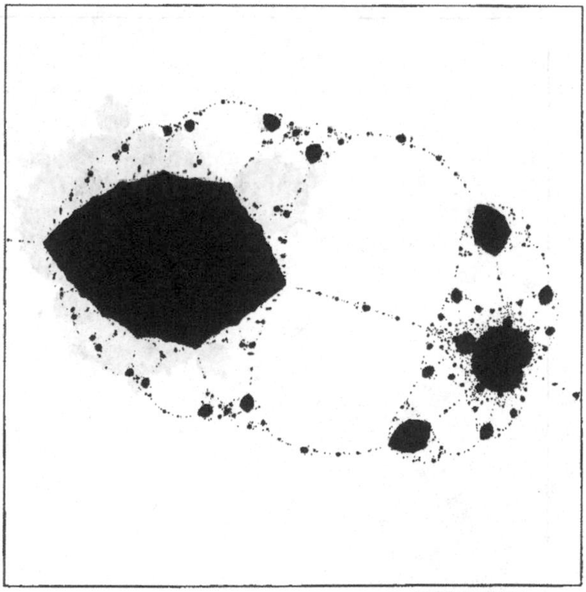

(b) $\{\lambda \in \mathbb{C} : g_\lambda^n(0) \to -\frac{1}{2} - \sqrt{\frac{1}{4} - \lambda},\ n \to \infty\}$

$\cup \{\lambda \in \mathbb{C} : g_\lambda^n(0) \not\to \text{root of } p_\lambda,\ n \to \infty\}$

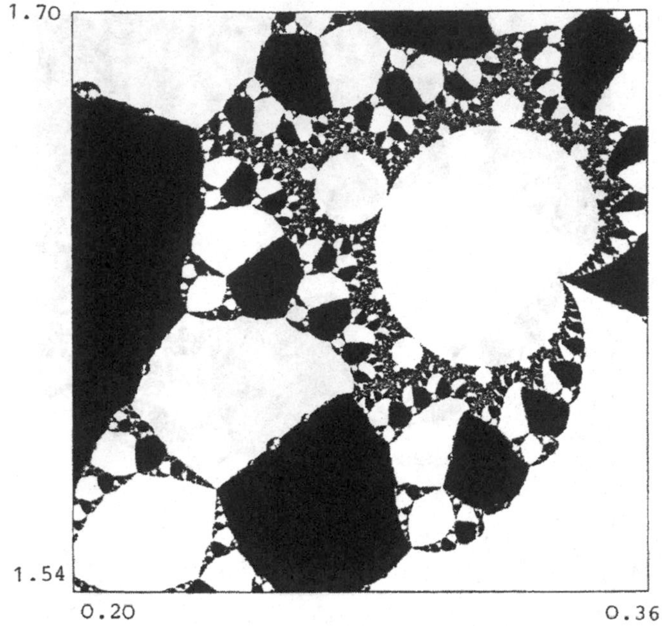

Figure 9. (a) Detail of figure 8a.

(b) The Mandelbrot-like set in

$\{\lambda \in \mathbb{C} : g_\lambda^n(0) \not\to \text{root of } p_\lambda , n \to \infty\}$

[6] and [21] should suffice to establish conjecture 1.7 (1), (2).

We add in passing that figure 9 gives some insight into a completely different set of questions: Given a polynomial, describe the set of initial values in $\mathbb{R}$ for which Newton's method converges towards a root. It is known, that for a polynomial with real coefficients and real roots this set is $\mathbb{R}$ except for a set of Lebesgue measure zero (see [1,20]). Now figure 9 teaches us that this remarkable result does not extend to $\mathbb{C}$, because for any λ in the Mandelbrot-like set (see figure 9) Newton's method allows a periodic attractor with an open basin of attraction.

Conjecture 1.7(3) is meant to contribute to problem 1.2 (a) and (b). Note that if one knew that

$$(24) \qquad \mathring{M}_R \overset{?}{=} \text{hyp}\,(M_R)\ , \quad (= \text{hyperbolic part of } M_R)$$

i.e. for any $q \in \mathring{M}_R$ the orbit of $(1-q)$ converges towards a periodic attractor of R_q, then conjecture 1.7 (3) could be established from classical theory. Note, however, that an identity corresponding to (24) is not even known for the much more fundamental Mandelbrot set M. On the other hand it is known that if M were locally connected then $\mathring{M} = \text{hyp}\,(M)$ (see [7]). For a good visual impression of the difficulties with regard to the last questions we refer to the pictures and experiments in [16].

SOME JULIA SETS FOR R_q

Finally we discuss some Julia sets of R_q for the physically meaningful choices $q = 2,3,4$; see figure 10. Firstly, one shows that

$$2 \in A_1\ , \quad 3 \in A_1\ , \quad 4 \in A_\infty\ .$$

Furthermore, for $q = 4$ one has that $A_q^*(1) = A_q(1)$ and $A_q^*(\infty) = A_q(\infty)$, where A^* denotes the immediate basin of attraction, i.e. the component which contains the attractor. Hence, it follows from [2] that the Julia set J_q, $q = 4$, is a Jordan curve, which, due to the symmetry with respect to conjugation, must intersect $\mathbb{R}_+$ in a unique point x_c, the ferromagnetic transition point.

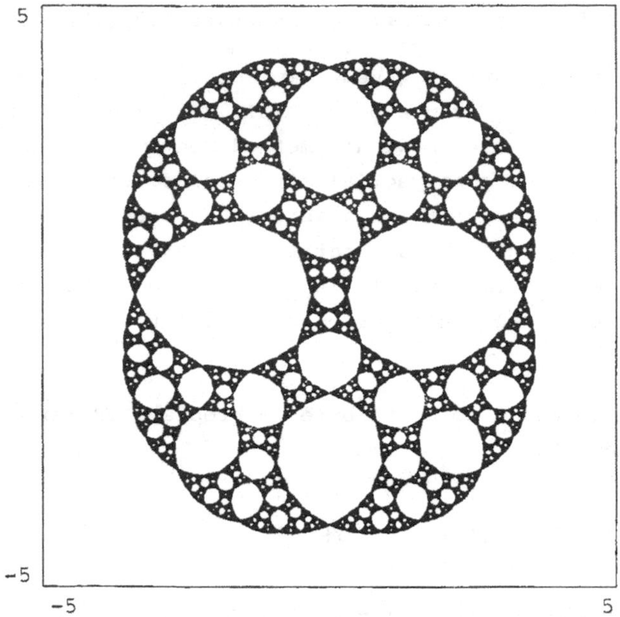

q = 2.0

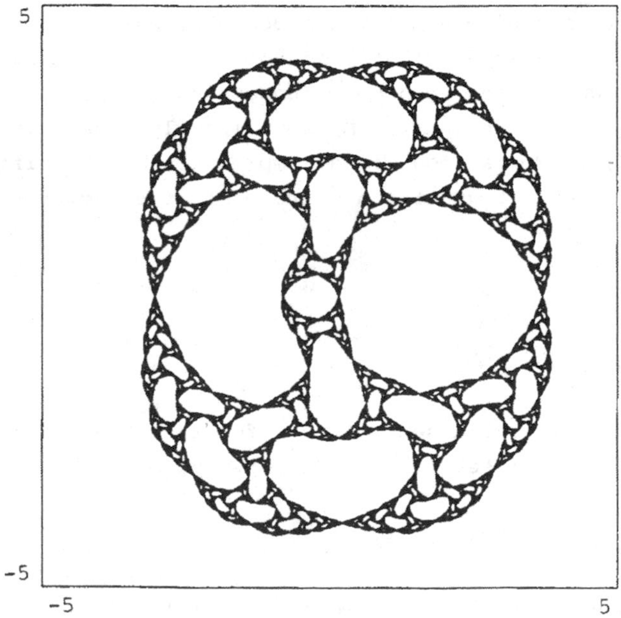

q = 2.5

Figure 10. The Julia set J_q of R_q for six values of q .

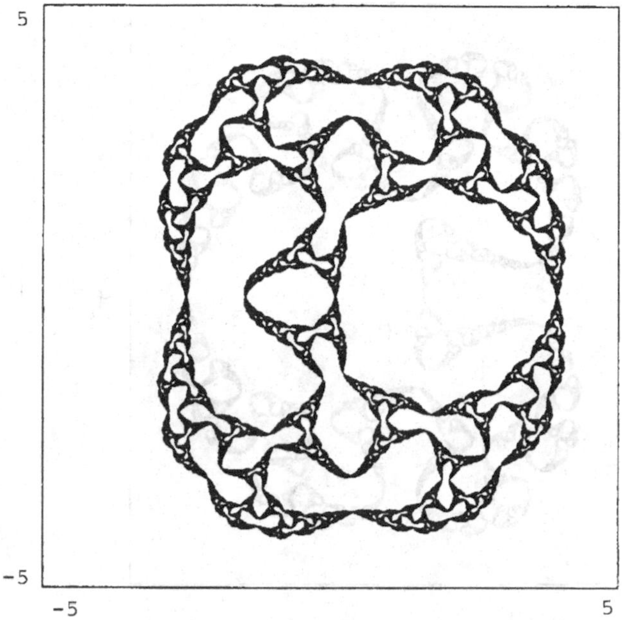

q = 2.9

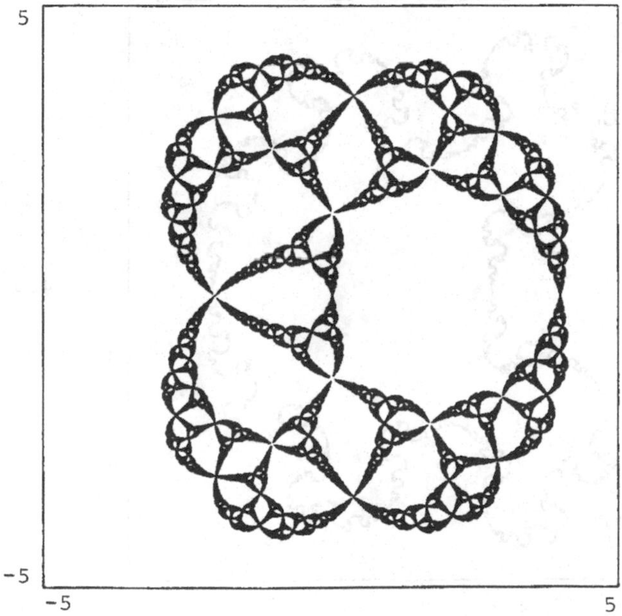

q = 3.0

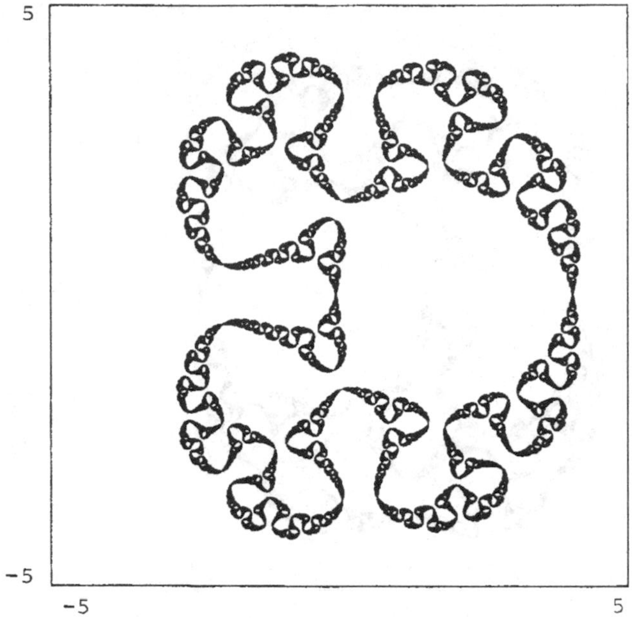

q = 3.1

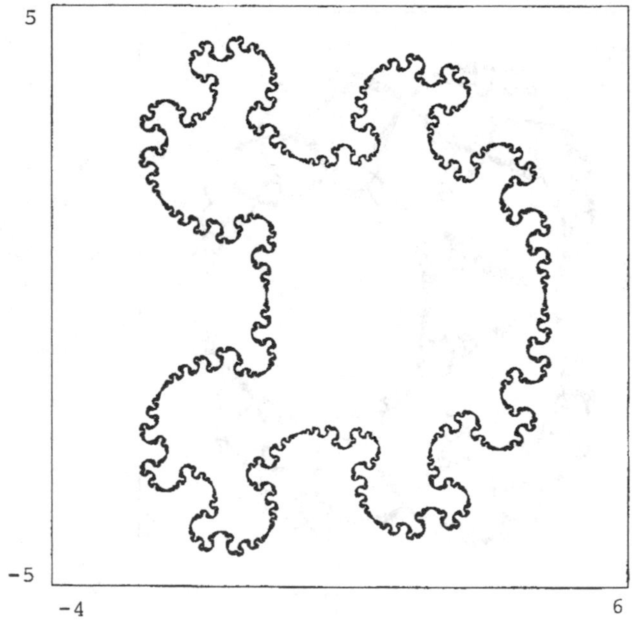

q = 4.0

Remarkably, also the Julia sets for $q < 4$ in figure 10 distinguish a unique phase transition on $\mathbb{R}_+$.

Acknowledgement. The color plates were obtained by D. Saupe and the authors on an AED 767 while figures 3-10 were obtained by H.W. Ramke and the authors on a laser printer. All pictures were produced in our "Graphiklabor Dynamische Systeme - Universität Bremen".

REFERENCES

1. B. Barna: Über die Divergenzpunkte des Newtonschen Verfahrens zur Bestimmung von Wurzeln Algebraischer Gleichungen. II, Publicatione Mathematicae, Debrecen, 4, 384-397 (1956).

2. H. Brolin: Invariant sets under iteration of rational functions, Arkiv för Math., 6, 103-144 (1966).

3. J. Curry, L. Garnett, D. Sullivan: On the iteration of rational functions: Computer experiments with Newton's method, Commun. Math. Phys., 91, 267-277 (1983).

4. B. Derrida, L. De Seze, C. Itzykson: Fractal structure of zeroes in hierarchical models, J. Statist. Phys. 33, 559 (1983).

5. A. Douady, J. H. Hubbard: Iteration des polynomes quadratiques complexes, CRAS Paris, 294, 123-126 (1982).

6. A. Douady, J. H. Hubbard: On the dynamics of polynomial - like mappings, preprint, 1984.

7. A. Douady, J. H. Hubbard: Etude dynamique des polynomes complexes, Publications Mathematiques D'Orsay, 1984.

8. P. Fatou: Sur les équations fonctionnelles, Bull. Soc. Math. Fr., 47, 161-271 (1919), 48, 33-94, 208-314 (1920).

9. S. Grossmann: Analytic Properties of Thermodynamic Functions and Phase Transitions, in: Festkörperprobleme IX, Ed. O. Madelung Vieweg 1969.

10. M. Herman: Examples de fractions rationnelles ayant une orbite dense sur la sphère de Riemann, to appear in Bull. Soc. Math. Fr.

11. M. Herman: Are there critical points on the boundaries of singular domains? Report 14, Institut Mittag-Leffler, 1984.

12. G. Julia: Sur l'iteration des fonctions rationnelles, Journal de Math. Pure et Appl., 8, 47-245 (1918).

13. B. Mandelbrot: Fractal aspects of the iteration of $z \to \lambda z(1-z)$ for complex λ, z, Annals N.Y. Acad. Sciences, 357, 249-259 (1980).

14. R. Mane, P. Sad, D. Sullivan: On the dynamics of rational maps, Ann. scient. Ec. Norm. sup., 16, 193-217 (1983).

15. H.-O. Peitgen, M. Prüfer, P. H. Richter: Phase transitions and Julia sets, Report 118, FS "Dynamische Systeme", Universität Bremen, Mai 1984, Proceedings 5th Meeting of the UNESCO Working Group on Systems Theory 'Lotka - Volterra Approach in Dynamic Systems', Wartburg, Eisenach, 1984.

16. H.-O. Peitgen, P. H. Richter: Die unendliche Reise, GEO June 1984, pp. 100-124, Gruner + Jahr, Hamburg.

17. H.-O. Peitgen, D. Saupe, F. v. Haeseler: Cayley's problem and Julia sets, The Mathematical Intelligencer, 6, (2), 11-20 (1984).

18. D. Ruelle: Thermodynamic Formalism, Encyclopedia of Mathematics and its Applications, Volume 5, 1978, Addison-Wesley Publishing Company, Reading, Massachusetts.

19. C. L. Siegel: Iteration of analytic functions, Ann. Math., 43, 607-612 (1942).

20. S. Smale: The fundamental theorem of algebra and complexity theory, Bull. Amer. Math. Soc., 4, 1-36 (1981).

21. D. Sullivan: Quasi conformal homeomorphisms and dynamics I, II, III, preprint 1982-1983.

22. C. N. Yang, T. D. Lee: Statistical theory of equations of state and phase transitions.
I. Theory of Condensation, Phys. Rev. 87, 404-409 (1952),
II. Lattice Gas and Ising Model, Phys. Rev. 87, 410-419 (1952).

RECENT DEVELOPMENTS IN REPRESENTATION THEORY

Wilfried Schmid*
Department of Mathematics
Harvard University
Cambridge, MA 02138

For the purposes of this lecture, 'representation theory' means representation theory of Lie groups, and more specifically, of semisimple Lie groups. I am interpreting my assignment to give a survey rather loosely: while I shall touch upon various major advances in the subject, I am concentrating on a single development. Both order and emphasis of my presentation are motivated by expository considerations, and do not reflect my view of the relative importance of various topics.

Initially G shall denote a locally compact topological group which is unimodular – i.e., left and right Haar measure coincide – and $H \subset G$ a closed unimodular subgroup. The quotient space G/H then carries a G-invariant measure, so G acts unitarily on the Hilbert space $L^2(G/H)$. In essence, the fundamental problem of harmonic analysis is to decompose $L^2(G/H)$ into a direct 'sum' of irreducibles. The quotation marks allude to the fact that the decomposition typically involves the continuous analogue of a sum, namely a direct integral, as happens already for non-compact Abelian groups. If G is of type I – loosely speaking, if the unitary representations of G behave reasonably – the abstract Plancherel theorem [12] asserts the existence of such a decomposition. This existence theorem raises as many questions as it answers: to make the decomposition useful, one wants to know it explicitly and, most importantly, one wants to understand the structure of the irreducible summands. In principle, any irreducible unitary representation of G can occur as a constituent of $L^2(G/H)$, for some $H \subset G$. The Plancherel problem thus leads naturally to the study of the irreducible unitary representations.

To what extent these problems can be solved depends on one's knowledge of the structure of the group G and on the nature of the subgroup H. Lie groups, p-adic groups, and algebraic groups over finite fields constitute the most interesting and best understood large classes of

*Supported in part by NSF grant DMS 8317436.

groups. Although the formal similarities are both striking and instructive, the technical aspects of the representation theory for these three classes diverge — hence the limitation to the case of Lie groups.

Semisimple groups play a distinguished role among all Lie groups, since they come up frequently in physical, geometric, and number-theoretic problems. The special emphasis on semisimple groups can also be justified on other grounds: one of the aims of Mackey's theory of induced representations is to reduce the harmonic analysis on general Lie groups to that on semisimple groups; recently Duflo [13] has worked out the reduction step quite concretely, at least for algebraic groups of type I.

From the point of view of harmonic analysis, irreducible unitary representations are the main objects of interest. Nevertheless, there are important reasons for being less restrictive: non-unitary representations not only occur naturally in their own right, for example as solution spaces of linear differential equations invariant under the action of a semisimple group, but they arise even in the context of unitary representations — a hint of this phenomenon will become visible below. Once one leaves the class of unitary representations, one should not insist on irreducibility; various known constructions produce irreducible representations not directly, but as quotients or subspaces of certain larger representations.

After these preliminaries, I let G denote a semisimple Lie group, connected, with finite center, and K a maximal compact subgroup of G. The choice of K does not matter, since any two maximal compact subgroups are conjugate. By a representation of G, I shall mean a continuous representation on a complete, locally convex Hausdorff space, of finite length — every chain of closed, G-invariant subspaces breaks off after finitely many steps — and 'admissible' in the sense of Harish-Chandra: any irreducible K-module occurs only finitely often when the representation is restricted to K. This latter assumption is automatically satisfied by unitary representations, and consequently G is of type I [19]. No examples are known of Banach representations, of finite length, which fail to be admissible.

To study finite dimesional representations of G, one routinely passes to the associated infinitesimal representations of the Lie algebra. Infinite dimensional representations are generally not differentiable in the naive sense, so the notion of infinitesimal representation requires some care. A vector v in the representation space V_π of a representation π is said to be K-finite if its K-translates span a finite dimensional subspace. By definition, $v \in V_\pi$ is a differentiable vector if the assignment $g \longrightarrow \pi(g)v$ maps G into V_π in a C^∞ fashion. Differentiable and K-finite vectors can be constructed readily, by averaging the translates of arbitrary vectors against compactly supported C^∞ or K-finite functions [17]. One may conclude that the K-finite vectors make up a dense subspace $V \subset V_\pi$, which consist entirely of differentiable vectors — at this point the standing assumption of admissibility plays a crucial role. In particular, the complexified Lie algebra $\mathfrak{g}$ of G acts on V by differentiation. The subgroup K also acts, by translation, but G does not. Partly for trivial reasons, and partly as consequence of the original hypotheses on the representation π, the $\mathfrak{g}$- and K-module V satisfies the following conditions:

a) as K-module, V is a direct sum of finite dimensional irreducibles, each occuring only finitely often;

(1) b) the actions of $\mathfrak{g}$ and K are compatible;

c) V is finitely generated over the universal enveloping algebra $U(\mathfrak{g})$.

Here b) simply means that the $\mathfrak{g}$-action, restricted to the complexified Lie algebra $\mathfrak{k}$ of K, coincides with the derivative of the K-action. This definition of the infinitesimal representation, which was introduced by Harish-Chandra [19], has the very desirable feature of associating algebraically irreducible $\mathfrak{g}$-modules to topologically irreducible representations of G; by contrast, $\mathfrak{g}$ acts in a highly reducible fashion on the spaces of all differentiable or analytic vectors of an infinite dimensional representation π.

A simultaneous $\mathfrak{g}$- and K-module V with the properties (1a-c) is called a Harish-Chandra module. All Harish-Chandra modules can be lifted to representations of G [10,41], not uniquely, but the range of possible topologies is now well-understood [45,54]. If V arises from an irreducible unitary representation π, it inherits an inner product which makes the action of $\mathfrak{g}$ skew-hermitian. An irreducible Harish-Chandra module admits at most one such inner product, up to a positive factor; if it does, the completion becomes the representation space of a unitary representation of G [19,39]. In other words, there is a one-to-one correspondence between irreducible unitary representations of G and Harish-Chandra modules which carry an inner product of the appropriate type. The problem of describing the irreducible unitary representations thus separates naturally into two sub-problems: the description of all irreducible Harish-Chandra modules, and secondly, the determination of those which are "unitarizable". Of the two, the latter seems considerably more difficult, and has not yet been solved, except in special cases — more on this below.

The irreducible Harish-Chandra modules of a general semisimple Lie group were classified by Langlands [33] and Vogan [48]; one of the ingredients of Langlands' classification is due to Knapp-Zuckerman [31]. To describe the classification in geometric terms, I introduce the flag variety of $\mathbf{g}$,

(2) $$X \;=\; \text{set of Borel subalgebras of } \mathbf{g} \;.$$

It is a complex projective variety and a homogeneous space for the complex Lie group

(3) $$G_{\mathbf{C}} \;=\; \text{identity component of } \operatorname{Aut}(\mathbf{g}) \;.$$

In the case of the prototypical example $G = Sl(n,\mathbf{R})$, X can be identified with the variety of all "flags" in $\mathbf{C}^n$, i.e. chains of subspaces $0 \subset V_1 \subset V_2 \subset \ldots \subset V_n = \mathbf{C}^n$, with $\dim V_k = k$: every Borel subalgebra of the complexified Lie algebra $\mathbf{g} = \mathbf{sl}(n,\mathbf{C})$ stabilizes a unique flag $\{V_n\}$. The group G acts on the flag variety via the adjoint homomorphism. There are finitely many G-orbits — for $G = Sl(n,\mathbf{R})$, for example, these are characterized by the position of flags relative to the real structure $\mathbf{R}^n \subset \mathbf{C}^n$. Now let $D \subset X$ be a G-orbit, and $L \longrightarrow D$ a homogeneous line bundle — a line bundle with a G-action compatible with that on the base D. A cohomological construction, which I shall describe next, if only in rough outline, associates a family of Harish-Chandra modules to the pair (D,L).

At one extreme, if G contains a compact Cartan subgroup[1], and if D is an open orbit, the homogeneous line bundles over D are parametrized by a lattice. As an open subset of X, D has the structure of complex manifold. Every homogeneous line bundle $L \longrightarrow D$ can be turned into a holomorphic line bundle, so that G acts as a group of holomorphic bundle maps. Thus G acts also on the sheaf cohomology groups of L. The differentiated action of the Lie algebra $\mathbf{g}$ turns

(4) $$H^*(D,\mathbf{O}(L))_{(K)} \;=\; \{\, \omega \in H^*(D,\mathbf{O}(L)) \;\mid\; \omega \text{ is } K\text{-finite} \,\} \,,$$

into Harish-Chandra modules. Whenever the line bundle L is negative in the appropriate sense — for example, if L extends to a line bundle over the projective variety X whose inverse is ample — the cohomology appears in only one degree and is irreducible:

(5) $$H^p(D,\mathbf{O}(L)) \;=\; 0 \qquad \text{if } p \neq s \,,$$
$$H^s(D,\mathbf{O}(L))_{(K)} \text{ is a non-zero, irreducible Harish-Chandra module}$$

[43]; here s denotes the largest dimension of compact subvarieties of

---

[1]equivalently, a torus which is a maximal Abelian subgroup.

D. These modules can be mapped $(\mathfrak{g},K)$-equivariantly into $L^2(G) \cap C^\infty(G)$, and are consequently unitarizable. The unitary structure is visible also in terms of the geometric realization: the L^2-cohomology of L injects into the Dolbeault cohomology, its image is dense, has a natural Hilbert space structure, and contains all K-finite cohomology classes [3,44]. The resulting unitary representations make up the discrete series, which was originally constructed by Harish-Chandra via character theory [21].

The opposite extreme, of a totally real G-orbit $D \subset X$, occurs only when $\mathfrak{g}$ contains Borel subalgebras defined over $\mathbb{R}$, as is the case for $G = Sl(n,\mathbb{R})$. In this situation D is necessarily compact, and the cohomological construction collapses to that of the single Harish-Chandra module

(6) $\quad C^\infty(D,L)_{(K)} \;=\;$ space of K-finite, C^∞ sections of L .

The module (6) need not be irreducible, but it has a unique irreducible quotient, provided L satisfies a suitable negativity condition. Harish-Chandra modules of this type are induced from a Borel subgroup of G ; they belong to the principal series.

The construction for a general G-orbit D combines elements of "complex induction", as in (4), and ordinary induction, as in (6). It can be viewed as a cohomological form of geometric quantization. There is completely parallel, algebraic version of the construction, due to Zuckerman, which offers certain technical advantages. It is this version that has been studied and used extensively [49]. Subject to certain hypotheses on the pair (L,D), Zuckerman's "derived functor construction" — equivalently, the geometric construction — produces cohomology in only one degree, a Harish-Chandra module that arises also by induction from a discrete series module of a subgroup of G. Under more stringent assumptions, the Harish-Chandra module corresponding to (L,D) has a unique irreducible quotient [38]. Every irreducible Harish-Chandra module can be realized as such a quotient in a distinguished manner — this, in effect, is Langlands' classification [33]. The problem of understanding the irreducible Harish-Chandra modules does not end here, however. The irreducible quotient may be all of the original module, or may be much smaller. In principle, the Kazhdan-Lusztig conjectures for Harish-Chandra modules, proved by Vogan [51] in the generic case, provide this type of information, but not as explicitly or concretely as one might wish.

I now turn to a different, more recent construction of Harish-Chandra modules, that of Beilinson-Bernstein [6]; similar ideas, in the context of Verma modules, can be found also in the work of Brylinski and

Kashiwara [9]. Some preliminary remarks are necessary. The flag variety X may be thought of as a quotient $G_{\mathbb{C}}/B$; here B is a particular Borel subgroup of $G_{\mathbb{C}}$, i.e. the normalizer of a Borel subalgebra $\mathfrak{b} \subset \mathfrak{g}$. The differentials of the algebraic characters of B constitute a lattice Λ. Each $\chi \in \Lambda$ — or more precisely, the corresponding character e^{χ} of B — associates a $G_{\mathbb{C}}$-homogeneous, holomorphic line bundle $L_{\chi} \to X$ to the principal bundle $B \to G_{\mathbb{C}} \to X$. Its cohomology groups are finite dimensional $G_{\mathbb{C}}$-modules, which are described by the Borel-Weil-Bott theorem [8]. In particular,

$$(7) \qquad \begin{array}{l} H^*(X,\mathcal{O}(L_\chi)) \ \text{ vanishes except in one degree } \ p = p(\chi)\ , \\ H^p(X,\mathcal{O}(L_\chi)) \ \text{ is irreducible}\ . \end{array}$$

The center $Z(\mathfrak{g})$ of the universal enveloping algebra $U(\mathfrak{g})$ acts on the cohomology by scalars (Schur's lemma!), so

$$(8) \qquad I_\chi = \text{annihilator of } H^*(X,\mathcal{O}(L_\chi)) \text{ in } Z(\mathfrak{g})$$

is a maximal ideal in $Z(\mathfrak{g})$. As a result of its construction, the line bundle L_χ carries an algebraic structure, and it makes sense to define

$$(9) \qquad \begin{array}{l} \mathcal{D}_\chi = \text{sheaf of linear differential operators, with} \\ \text{algebraic coefficients, acting on the sections of } L_\chi\ ; \end{array}$$

the notion of sheaf is taken with respect to the Zariski topology, as befits the algebraic setting. To picture $\mathcal{D}_\chi$, one should note that it is locally isomorphic to the sheaf of scalar differential operators on X. The Lie algebra $\mathfrak{g}$ operates on sections of L_χ by infinitesimal translation. This operation extends to a homomorphism from $U(\mathfrak{g})$ into $\Gamma\mathcal{D}_\chi$ (= algebra of global sections of $\mathcal{D}_\chi$), which in turn drops to an isomorphism

$$(10) \qquad \Gamma\mathcal{D}_\chi \cong U_\chi =_{\text{def}} U(\mathfrak{g})/I_\chi U(\mathfrak{g})$$

[6]. This is the point of departure of the Beilinson-Bernstein construction.

The passage from $U(\mathfrak{g})$ to the sheaf $\mathcal{D}_\chi$ has a counterpart on the level of $U(\mathfrak{g})$-modules: a pair of functors between

$$(11) \qquad \mathcal{M}(U_\chi) = \text{category of } U_\chi\text{-modules}$$

— equivalently, the category of $U(\mathfrak{g})$-modules on which the center $Z(\mathfrak{g})$ acts as it does on the cohomology groups (7) — and

$$(12) \qquad \mathcal{M}(\mathcal{D}_\chi) = \text{category of quasi-coherent sheaves of } \mathcal{D}_\chi\text{-modules}\ .$$

Quasi-coherence means simply that the sheaves admit local presentations in terms of generators and relations, though not necessarily finite presentations. In one direction, the global section functor

(13) $$\Gamma : M(D_\chi) \longrightarrow M(U_\chi)$$

maps sheaves of D_χ-modules to modules over $\Gamma D_\chi \cong U_\chi$. Extension of scalars from the algebra of global sections U_χ to the stalks of D_χ determines a functor in the opposite direction,

(14) $$\Delta : M(U_\chi) \longrightarrow M(D_\chi), \quad \Delta V = D_\chi \otimes_{U_\chi} V ;$$

the sheaves ΔV are quasi-coherent because every $V \in M(U_\chi)$ can be described by generators and relations.

Those parameters $\chi \in \Lambda$ which correspond to ample line bundles L_χ span an open cone $C \subset \mathbb{R}\otimes_{\mathbb{Z}}\Lambda$. One calls χ dominant if it lies in the closure of C, dominant nonsingular if χ lies in C itself. The inverse of the canonical bundle is ample, and is therefore indexed by a particular dominant nonsingular quantity, customarily denoted by 2ρ. With these conventions it possible to state the following remarkable analogue of Cartan's theorems A and B:

<u>(15) Theorem (Beilinson-Bernstein [6])</u> A) The global sections of any quasi-coherent sheaf of D_χ-modules generate its stalks, provided $\chi + \rho$ is dominant and nonsingular. B) If $\chi + \rho$ is dominant, the sheaf cohomology groups $H^p(X,V)$ vanish, for every $V \in M(D_\chi)$ and $p > 0$.

As a direct consequence, Beilinson-Bernstein deduce:

<u>(16) Corollary</u> In the situation of a dominant nonsingular $\chi + \rho$, the functor Γ defines an equivalence of categories $M(U_\chi) \cong M(D_\chi)$, with inverse Δ.

Perhaps surprisingly, the equivalence of categories implies properties of general U_χ-modules that were previously unknown. The most fruitful applications, however, occur in the context of certain smaller categories, in particular the category $\mathcal{O}$ of Bernstein-Gelfand-Gelfand [7] and the category of Harish-Chandra modules.

Irreducible modules in either of these categories are annihilated by maximal ideals in $Z(\mathfrak{g})$, but not every maximal ideal is of the form (8), with $\chi \in \Lambda$. According to Harish-Chandra [20], the correspondence $\chi \longrightarrow I_\chi$ extends naturally to a surjective map from the vector space $\mathbb{C}\otimes_{\mathbb{Z}}\Lambda$ onto the set of all maximal ideals; any two of the ideals I_χ, I_ψ, for $\chi, \psi \in \mathbb{C}\otimes_{\mathbb{Z}}\Lambda$, coincide precisely when $\chi + \rho$ and $\psi + \rho$ belong to the same orbit of the Weyl group W — a finite group which acts linearly on $\mathbb{C}\otimes_{\mathbb{Z}}\Lambda$. I want to emphasize one consequence of Harish-Chandra's description of the maximal ideal space:

(17) every maximal ideal in $Z(\mathfrak{g})$ can be realized as I_χ , with $\chi \in \mathbb{C}\otimes_{\mathbb{Z}}\Lambda$ having the property that $\mathrm{Re}(\chi + \rho)$ is dominant.

Although the bundle L_χ ceases to exist as soon as the parameter χ leaves the lattice Λ , there are "phantom line bundles" attached to all $\chi \in \mathbb{C}\otimes_{\mathbb{Z}}\Lambda$, locally defined line bundles to which the action of $\mathfrak{g}$ on X lifts. The construction (9) of the sheaf of algebras $\mathcal{D}_\chi$ continues to make sense in this wider setting, as do the isomorpism (10), the categories $\mathcal{M}(\mathcal{D}_\chi)$, $\mathcal{M}(U_\chi)$, and the functors Γ, Δ . Most importantly, theorem (15) and its corollary remain valid [6], with one minor adjustment: the phrase "$\chi + \rho$ is dominant" should be replaced by "$\mathrm{Re}(\chi + \rho)$ is dominant". Different values of χ may correspond to identical maximal ideals I_χ and quotients $U_\chi = U(\mathfrak{g})/I_\chi$, but an appropriate choice of χ will bring any such quotient within the scope of part B of the theorem — this follows from (17). The same χ makes part A and the corollary apply at least generically, for parameters outside a finite number of hyperplanes. The equivalence of categories breaks down in the remaining, singular cases only because certain sheaves fail to have global sections.

The maximal compact subgroup $K \subset G$ possesses a complexification, a complex algebraic group $K_{\mathbb{C}}$, defined over $\mathbb{R}$, which contains K as the group of real points. If $V \in \mathcal{M}(U_\chi)$ is a Harish-Chandra module, the K-action induces an algebraic $K_{\mathbb{C}}$-action on the sheaf ΔV . The differential of this action agrees with the multiplication action of $\mathfrak{k}$, viewed as Lie subalgebra of $\Gamma\mathcal{D}_\chi \cong U_\chi$ — in short, $K_{\mathbb{C}}$ and $\mathfrak{k}$ act compatibly. For the purpose of the preceeding discussion, the finiteness condition (1a) in the definition of Harish-Chandra module becomes irrelevant. It is necessary only that K act locally finitely, i.e. the K-translates of any $v \in V$ must span a finite dimensional subspace. The passage from locally finite K-actions on U_χ-modules to algebraic $K_{\mathbb{C}}$-actions on sheaves of $\mathcal{D}_\chi$-modules can be reversed; in other words, both Γ and Δ restrict to functors between

(18)

$\mathcal{M}(U_\chi, K)$ = category of U_χ-modules with a compatible, locally finite K-action , and

$\mathcal{M}(\mathcal{D}_\chi, K_{\mathbb{C}})$ = category of sheaves of quasi-coherent $\mathcal{D}_\chi$-modules with a compatible, algebraic $K_{\mathbb{C}}$-action

Whenever $\chi + \rho$ is nonsingular and $\mathrm{Re}(\chi + \rho)$ is dominant, the equivalence of categories (16) identifies these two subcategories,

(19) $\Gamma : M(D_\chi, K_C) \cong M(U_\chi, K)$.

A theorem of Harish-Chandra asserts, in effect, that finitely generated modules in the category $M(U_\chi, K)$ satisfy the finiteness condition (1a) [19], hence

(20) the irreducible objects in the categories $M(U_\chi, K)$, $\chi \in C \otimes_Z \Lambda$, exhaust the class of irreducible Harish-Chandra modules .

In particular, the identification (19) relates irreducible Harish-Chandra modules to irreducible sheaves $V \in M(D_\chi, K_C)$.

Geometric considerations suggest how to find such sheaves. The support of any $V \in M(D_\chi, K_C)$ is invariant under the translation action of K_C on X , via the adjoint homomorphism. If V is irreducible, the support must be an irreducible variety — necessarily the closure of an orbit, since

(21) K_C acts on X with finitely many orbits

[36]. Now let $Y \subset X$ be a K_C-orbit, $\bar{Y}$ its closure. The operation of pushforward yields irreducible sheaves with support in $\bar{Y}$, as I shall explain next.

Ordinarily the D-module pushforward of a sheaf exists only as an object of the derived category. In the situation at hand it can be described quite explicitely. The analogy with the C^∞ case is instructive. Linear differential operators on a C^∞ manifold M cannot be applied naturally to the C^∞ functions on a closed submanifold $N \subset M$. However, after the choice of a smooth measure, functions in $C^\infty(N)$ may be viewed as distributions on M , and the sheaf of differential operators D_M — here in the C^∞ sense — does act on these. The D-module pushforward of the sheaf of smooth measures on N is the sheaf generated by that action; in other words, the sheaf of distributions on M , with support in N , which are smooth along N . Formally, measures and distributions must be treated as sections not of the trivial bundle, but the top exterior power of the cotangent bundle. For this reason the pushforward from N to M involves a twist by the quotient of the two determinant bundles, i.e., a twist by the top exterior power of the conormal bundle. The preceeding discussion can be expressed in terms of local coordinates, and then makes sense equally in the algebraic setting.

Back to the K_C-orbit $Y \subset X$! Under an appropriate integrality condition on the parameter χ , the 'phantom line bundle' corresponding to χ extends to the orbit as a true K_C-equivariant line bundle, possibly in several different ways. I let $L_{Y,\chi}$ denote a particular such extension, tensored by the top exterior power of the normal bundle. Its sheaf of sections $\mathcal{O}_Y(L_{Y,\chi})$ is a module for a twisted sheaf of differential operators $\mathcal{D}_{Y,\chi}$ on Y. The complement of the boundary ∂Y in X contains Y as a smooth, closed subvariety. Because of the shift built into the definition, the D-module pushforward of $\mathcal{O}_Y(L_{Y,\chi})$ from Y to $X - \partial Y$ is a sheaf of modules over the sheaf $\mathcal{D}_\chi$, restricted to $X - \partial Y$. It becomes a sheaf of $\mathcal{D}_\chi$-modules over all of X when pushed forward once more — naively, simply as a sheaf — from the open subset $X - \partial Y$ to X. The resulting sheaf, which I denote by $\mathcal{V}_{Y,\chi}$, belongs to the category $M(\mathcal{D}_\chi, K_C)$, since K_C operates at each step of its construction. A basic result of Kashiwara, on sheaves of D-modules supported by smooth subvarieties, implies

(22)
a) the sheaf of $\mathcal{D}_\chi$-modules $\mathcal{V}_{Y,\chi}$ has a finite composition series and contains a unique irreducible subsheaf ;

b) every irreducible sheaf in the category $M(\mathcal{D}_\chi, K_C)$ arises in this manner, for some K_C-orbit Y and line bundle $L_{Y,\chi}$.

Under the hypotheses of the equivalence of categories, this statement translates immediately into a classification of the irreducible Harish-Chandra modules which are annihilated by the maximal ideal $I_\chi \subset Z(\mathfrak{g})$:

(23) $\Gamma\mathcal{V}_{Y,\chi}$ has a unique irreducible submodule; the assignment of that module to the datum of the orbit Y and line bundle $L_{Y,\chi}$ establishes a bijection between such pairs $(Y, L_{Y,\chi})$ and irreducible Harish-Chandra modules in $M(U_\chi, K)$.

When $\chi + \rho$ is singular, the situation becomes more complicated, as it does also from the point of view of the Langlands classification. Irreducible Harish-Chandra modules in $M(U_\chi, K)$ can still be realized as submodules of some $\Gamma\mathcal{V}_{Y,\chi}$, but not always as an only irreducible submodule, nor in a unique manner.

The reducibility or irreduciblity of $\mathcal{V}_{Y,\chi}$, in the category of sheaves of $\mathcal{D}_\chi$-modules, is a local phenomenon. All stalks at points of the complement of $\bar{Y}$ vanish, and a small calculation shows those over points of Y to be automatically irreducible. If a non-trivial quotient of $\mathcal{V}_{Y,\chi}$ exists, it also belongs to the category $M(\mathcal{D}_\chi, K_C)$ and has support in the boundary. In particular, the sheaf $\mathcal{V}_{Y,\chi}$ cannot possibly reduce unless there is a non-empty boundary: subject to the usual positivity condition on χ,

(24) the Harish-Chandra modules $\Gamma\mathcal{V}_{Y,\chi}$ associated to closed K_C-orbits are irreducible .

Non-trivial quotients of $\mathcal{V}_{Y,\chi}$ do exist whenever the line bundle $L_{Y,\chi}$ extends, $\mathfrak{g}$-equivariantly, across some K_C-orbit in ∂Y. Matsuki [37] and Springer [47] have worked out the closure relations between $K_{\mathbb{C}}$-orbits; their results make it possible to interpret this geometric irreducibility criterion quite explicitely.

The proof of the original Kazhdan-Lusztig conjectures was the first triumph of D-modules in representation theory. Irreducible modules in the category O arise from orbits of a Borel subgroup $B \subset G_{\mathbb{C}}$, via the same process of pushforward, taking sections, and passing to the unique irreducible submodule. Kazhdan and Lusztig [29] had already related their conjectured composition multiplicities for Verma modules to the intersection cohomology of closures of B-orbits, i.e., of Schubert varieties. Both Brylinski-Kashiwara [9] and Beilinson-Bernstein [6] saw the connection with the theory of D-modules; they independently established the conjectures, by relating the intersection cohomology to the composition multiplicities of the appropriate sheaves. This second step carries over, essentially unchanged, to the setting of Harish-Chandra modules. The paper [35] of Lusztig and Vogan contains the analogue of the first ingredient, namely the combinatorics of the intersection cohomology of closures of $K_{\mathbb{C}}$-orbits. Vogan [51], finally, deduces multiplicity formulas for the Langlands classification, which he had conjectured earlier [50]. I should point out that the original Kazhdan-Lusztig conjectures cover only U_χ-modules with $\chi \in \Lambda$, as does the known version for Harish-Chandra modules; Vogan's conjectures, by contrast, apply to the general case.

At first glance, the Beilinson-Bernstein construction appears far removed from the construction of Harish-Chandra modules in terms of line bundles on G-orbits. The former leads quickly to geometric reducibility criteria, as we just saw, and opens a path towards the Kazhdan-Lusztig conjectures for Harish-Chandra modules. It also has points of contact with Vogan's classification via K-types [48]; indeed, it probably implies the results of [48]. The G-orbit construction, on the other hand, is closely tied to Langlands' classification, which in turn relates it to certain analytic invariants of Harish-Chandra modules: the asymptotic behavior of matrix coefficients, for example, and the global character [10,11,24]. Since the two constructions complement each other, the possible connections between them merit attention.

Results of Matsuki [36], on orbits in flag varieties, provide an important clue. For each G-orbit $D \subset X$, there exists a unique $K_{\mathbb{C}}$-orbit Y, such that K acts transitively on the intersection $D \cap Y$; conversely every $K_{\mathbb{C}}$-orbit intersects a unique G-orbit in this manner. The correspondence $D \longleftrightarrow Y$ reverses the relative sizes of orbits, as measured by their dimensions — I shall therefore call D "dual" to the orbit Y. Once the parameter $\varkappa$ has been fixed, the duality $D \longleftrightarrow Y$ extends to the line bundles which enter the two constructions: a homogeneous line bundle $L \longrightarrow D$ is dual to $L_{Y,\varkappa} \longrightarrow Y$ if the tensor product $L \otimes L_{Y,\varkappa}$ restricts to a trivial K-homogeneous vector bundle over $D \cap Y$.

It is instructive to examine the special case of $Sl(2,\mathbb{R})$, or equivalently, its conjugate $SU(1,1)$. The diagonal matrices in $G = SU(1,1)$ constitute a maximal compact subgroup $K \cong U(1)$. Both G and $K_{\mathbb{C}} \cong \mathbb{C}^*$ act on the flag variety $X \cong \mathbb{CP}^1 \cong \mathbb{C} \cup \{\infty\}$, as groups of Möbius transformations. The duality relates the three $K_{\mathbb{C}}$-orbits $\{0\}$, $\{\infty\}$, $\mathbb{C}^*$, in the given order, to the G-orbits Δ (= unit disc), Δ' (= complement of the closure of Δ), S^1. A homogeneous line bundle over Δ is determined by a character of the isotropy subgroup at 0, i.e. a character e^λ of K. Dually, a $K_{\mathbb{C}}$-homogeneous line bundle over the one-point space $\{0\}$ has a single fibre, on which $K_{\mathbb{C}}$ acts by an algebraic character $e^\varkappa$. The differentials λ, $\varkappa$ may be viewed as linear functions on $\mathfrak{k} \cong \mathbb{C}$, whose values on $\mathbb{Z} \subset \mathbb{C}$ are integral multiples of $2\pi i$; here the duality reduces to $\lambda \longleftrightarrow \varkappa = -\lambda$. The situation for the orbits Δ', $\{\infty\}$, is entirely analogous. At points $z \in S^1$, the isotropy subgroup $G_z \subset G$ has two connected components. Its characters are parametrized by pairs (λ, ζ), consisting of a complex number λ and a character ζ of the center $\{\pm 1\} \subset G$, which meets both connected components of G_z. The corresponding G-equivariant line bundle over S^1 extends holomorphically at least to the $K_{\mathbb{C}}$-orbit $\mathbb{C}^*$, as does the dual, or inverse line bundle. If the line bundle is to extend even across $\{0\}$ or $\{\infty\}$, the pair (λ, ζ) must lift to a character of the complexification of G_z — this happens precisely when $\lambda/2\pi i$ is integral and ζ trivial or non-trivial, depending on the parity of $\lambda/2\pi i$.

To a $K_{\mathbb{C}}$-homogeneous line bundle $L_\varkappa \longrightarrow \{0\}$, the Beilinson-Bernstein construction assigns the Harish-Chandra module of 'holomorphic distributions' supported at 0, with values in the bundle $L_\varkappa$ — in other words, the $U(\mathfrak{g})$-submodule generated by 'evaluation at 0' in the algebraic dual of the stalk $\mathcal{O}_{\{0\}}(L_\varkappa^* \otimes T_X^*)$; the formal duality between functions and differentials accounts for the appearance of the cotangent bundle T_X^*. By its very definition, this module is dual, in the sense of Harish-Chandra modules, to $H^0(\Delta, \mathcal{O}(L_\varkappa^* \otimes T_X^*))_{(K)}$, the module associated to the G-orbit Δ and the line bundle $L_\varkappa^* \otimes T_X^*$. For non-negative values of the integral parameter $\varkappa/2\pi i$, the resulting Harish-Chandra modules are irreducible and belong to the discrete series. They become reducible if $\varkappa/2\pi i < -1$; in this situation the equivalence of categories (19) no longer applies. The preceeding discussion carries over, word-for-word, to the pair of orbits $\{\infty\}$, Δ'. As for the orbits S^1 and $\mathbb{C}^*$, the two constructions start with the choice of a G-homogeneous line bundle $L_{\lambda,\zeta} \to S^1$. Its extension to $\mathbb{C}^*$, which I denote by the same symbol, comes equipped with an action of $\mathfrak{g}$ and an algebraic structure. Since $\mathbb{C}^*$ is open in X, the pushforward construction attaches the space of algebraic sections $H^0(\mathbb{C}^*,\mathcal{O}(L_{\lambda,\zeta}))$ to the datum of the $K_{\mathbb{C}}$-orbit $\mathbb{C}^*$ and line bundle $L_{\lambda,\zeta}$. Integration over S^1 pairs this Harish-Chandra module nondegenerately with $C^\infty(S^1,L^*_{\lambda,\zeta}\otimes T^*_X)_{(K)}$, the module corresponding to the G-orbit S^1 and line bundle $L^*_{\lambda,\zeta}\otimes T^*_X$. The hypothesis of the equivalence of categories translates into the inequality $\operatorname{Re}\lambda > -1$. On the Beilinson-Bernstein side, this implies the existence of a unique irreducible submodule: the entire module generically, when $L_{\lambda,\zeta}$ cannot be continued across 0 and ∞, otherwise the finite dimensional submodule consisting of sections regular at the two punctures. The realization of the dual module $C^\infty(S^1,L^*_{\lambda,\zeta}\otimes T^*_X)_{(K)}$ exhibits both Harish-Chandra modules as members of the principal series.

One phenomenon that does not show up in the case of $G = SU(1,1)$ is the occurence of higher cohomology. For general groups, without any positivity assumption on the parameter χ, the sheaves $\mathcal{V}_{Y,\chi}$ can have non-zero cohomology groups above degree zero, but these are still Harish-Chandra modules. Zuckerman's derived functor construction also produces a family of Harish-Chandra modules $I^p(D,L)$, for each G-orbit D and G-equivariant line bundle $L \to D$, indexed by an integer $p \geq 0$. The example of $SU(1,1)$ suggest a duality between the two constructions, and indeed this is the case. I fix pairs of data $(Y,\mathcal{V}_{Y,\chi})$, (D,L), which are dual in the sense described above, and define $s = \dim_{\mathbb{R}}(Y\cap D) - \dim_{\mathbb{C}} Y$, $d = \dim_{\mathbb{C}} X$. Then

(25) there exists a natural, nondegenerate pairing between the Harish-Chandra modules $H^p(Y,\mathcal{V}_{Y,\chi})$ and $I^{s-p}(D,L\otimes\Lambda^d T^*_X)$,

for all $p \in \mathbb{Z}$, with no restriction on χ (Hecht-Miličić-Schmid-Wolf [23]). In both constructions homogeneous vector bundles can be substituted for line bundles. The duality carries over to this wider setting, and then becomes compatible with the coboundary operators. Earlier, partial results in the direction of (25) appear in Vogan's proof of the Kazhdan-Lusztig conjectures for Harish-Chandra modules [51]; there Vogan identifies certain Beilinson-Bernstein modules with induced modules, by explicit calculation.

The duality does not directly relate the Beilinson-Bernstein classification to that of Langlands: in the language of geometric quantization, the latter uses partially real polarizations, whereas the former works with arbitrary, mixed polarizations. This problem can be dealt with on the level of Euler characteristics, and the known vanishing theorems for the two constructions are sufficiently complementary to permit a comparison after all. In particular, it is possible to carry techniques and results back and forth between the two constructions [23].

By definition, the discrete series is the family of irreducible unitary representations which occur discretely in $L^2(G)$. It was remarked earlier that G has a non-empty discrete series if it contains a compact Cartan subgroup; these representations then correspond to open G-orbits, and their unitary structures are related to the geometric realization. Open G-orbits are dual to closed $K_{\mathbb{C}}$-orbits, so the observation (24) "explains" the irreducibility statement (5). The discrete series lies at one extreme of the various non-degenerate series of irreducible unitary representations — the other series consist of representations unitarily induced from discrete series representations of proper subgroups. These are precisely the representations which occur in the decomposition of $L^2(G)$ [22]. Roughly speaking, they are parametrized by hermitian line bundles over G-orbits. Here, too, the inner products have geometric meaning [56]. As for the rest of the unitary dual, the picture remains murky, though substantial progress has been made during the past few years. I shall limit myself to some brief remarks, since more detailed summaries can be found in the articles [30,52] of Knapp-Speh and Vogan.

A unitarizable Harish-Chandra module V is necessarily conjugate isomorphic to its own dual, a property which translates readily into a condition on the Harish-Chandra character, or on the position of V in the Langlands classification. If the condition holds, V admits a non-trivial $\mathfrak{g}$-invariant hermitian form — only one, up to scalar multiple, provided V is irreducible. The real difficulty lies in deciding whether the hermitian form has a definite sign. For a one parameter family V_t of irreducible Harish-Chandra modules of this type, the form stays definite if it is definite anywhere: not until the family reduces at some $t = t_0$ can the hermitian form become indefinite; even the composition factors at the first reduction point are unitarizable. In the case of $Sl(2,\mathbb{R})$, such deformation techniques generate the complementary series and the trivial representation — in other words, all of the unitary dual outside the discrete series and the unitary principal series [5]. Examples of Knapp and Speh [30] suggest that the analogous phenemenon for general groups can become quite complicated.

Neither induction nor deformation techniques account for isolated points in the unitary dual. Typically isolated unitary representations do exist, beyond those of the discrete series, but with certain formal similarities to the discrete series. Zuckerman's derived functor construction, and the Beilinson-Bernstein construction as well, extends to orbits in generalized flag vartieties, i.e., quotients G_C/P by parabolic subgroups $P \subset G_C$. The G-invariant hermitian line bundles over an open G-orbit $D \subset G_C/P$ are parametrized by the character group of the center of the isotropy subgroup $G_z \subset G$ at some reference point $z \in D$. Whenever that center is compact, the derived functor construction produces a discrete family of irreducible Harish-Chandra modules. According to a conjecture of Zuckerman, which was recently proved by Vogan [53], these modules are unitarizable. Vogan actually proves more; in geometric language, the cohomology of G-invariant vector bundles, modeled on irreducible unitary representations of the isotropy group G_z, vanishes in all but one degree and can be made unitary, again under an appropriate negativity assumption on the bundles. The proof consists of an algebraic reduction to the case of the non-degenerate series: Vogan introduces a notion of signature for g-invariant hermitian forms on Harish-Chandra modules, formal sums of irreducible characters of K with integral coefficients; he then calculates these signatures for the derived functor modules, in terms of the K-multiplicities of induced modules. Because of the origin of Zuckerman's conjecture, one might hope for a geometric proof. Earlier attempts in this direction were only marginally successful, but give a hint of a possible strategy [42].

A complete description of the unitary dual exists for groups of low dimension, for groups of real rank one [4,27,32], and the family SO(n,2) [1]. Vogan has just announced a classification also for the special linear groups over **R** , **C** , **H** — a big step, since there is no bound on real rank. In effect, the methods of unitary induction, degeneration, and Vogan's proof of the Zuckerman conjecture generate all irreducible unitary representations of the special linear groups. One feature that makes these groups more tractable is a hereditary property of their parabolic subgroups: all simple factors of the Levi component are again of type Sl_n. In the general case, conjectures of Arthur [2] and Vogan [52] predict the unitarity of certain highly singular representations. There are also results about particular types of unitary representations [14, 15,28], but a definite common pattern has yet to emerge.

I close my lecture by returning to its starting point, the decomposition of $L^2(G/H)$. The solution of this problem for $H = \{e\}$ — the explicit Plancherel formula [22] — was aim and crowning achievement of Harish-Chandra's work on real groups. A discussion of his proof would lead too far afield. However, I should mention a recent 'elementary', though not simple, argument of Herb and Wolf [26]. It is based on Herb's formulas for the discrete series characters [25], and emulates Harish-

Chandra's original proof in the case of $Sl(2,\mathbb{R})$, by integration by parts [18].

The decomposition problem has been studied systematically for two classes of subgroups H, besides the identity group: arithmetically defined subgroups, and symmetric subgroups, i.e. groups of fixed points of involutive automorphisms. The symmetric case contains the case of the trivial group, since G can be identified with $G \times G/\text{diagonal}$. Oshima and Matsuki [40], building on a remarkable idea of Flensted-Jensen [16], have determined the discrete summands of $L^2(G/H)$, for any symmetric $H \subset G$; these representations are parametrized by homogeneous line bundles over certain orbits in generalized flag varieties. Oshima has also described a notion of induction in the context of symmetric quotients. Presumably $L^2(G/H)$ is made up of representations which are induced in this sense, from discrete summands belonging to smaller quotients, but the explicit decomposition remains to be worked out. The case of arithmetically defined subgroups is the most interesting from many points of view, and the most difficult. Again the discrete summands constitute the "atoms" of the theory, as was shown by Langlands [34] — the Eisenstein integral takes the place of induction. There is an extensive literature on the discrete spectrum, too extensive to be summarized here, yet a full understanding does not seem within reach.

References

[1] E. Angelopolos: Sur les représentations de $\bar{S}\bar{O}_0(p,2)$. C.R. Acad. Sci. Paris 292 (1981), 469-471

[2] J. Arthur: On some problems suggested by the trace formula. In Lie Group Representations II. Springer Lecture Notes in Mathematics 1041 (1984), pp. 1-49

[3] M. F. Atiyah and W. Schmid: A geometric construction of the discrete series for semisimple Lie groups. Inventiones Math. 42 (1977), 1-62

[4] M. W. Baldoni-Silva: The unitary dual of Sp(n,1), n≥2. Duke Math. J. 48 (1981), 549-584

[5] W. Bargman: Irreducible unitary representations of the Lorentz group. Ann. of Math. 48 (1947), 568-640

[6] A. Beilinson and J. Bernstein: Localisation de $\mathfrak{g}$-modules. C.R. Acad. Sci. Paris, 292 (1981), 15-18

[7] J. Bernstein, I. M. Gelfand and S. I. Gelfand: Differential operators on the base affine space and a study of $\mathfrak{g}$-modules. In Lie Groups and their Representations. Akadémiai Kiadó, Budapest 1975

[8] R. Bott: Homogeneous vector bundles. Ann. of Math. 66 (1957), 203-248

[9] J.-L. Brylinski and M. Kashiwara: Kazhdan-Lusztig conjectures and holonomic systems. Inventiones Math. 64 (1981), 387-410

[10] W. Casselman: Jacquet modules for real reductive groups. In Proceedings of the International Congress of Mathematicians. Helsinki 1978, pp. 557-563

[11] W. Casselman and D. Miličić: Asymptotic behavior of matrix coefficients of admissible representations. Duke Math. J. 49 (1982), 869-930

[12] J. Dixmier: Les C*-algèbres et Leur Représentations. Gauthier-Villars, Paris 1964

[13] M. Duflo: Construction de représentations unitaires d'un groupe de Lie. Preprint

[14] T. J. Enright, R. Howe and N. R. Wallach: A classification of unitary highest weight modules. In Representation Theory of Reductive Groups. Progress in Mathematics 40 (1983), pp. 97-144

[15] T. J. Enright, R. Parthasarathy, N. R. Wallach and J. A. Wolf: Unitary derived functor modules with small spectrum. Preprint

[16] M. Flensted-Jensen: Discrete series for semisimple symmetric spaces. Ann. of Math. 111 (1980), 253-311

[17] L. Gårding: Note on continuous representations of Lie groups. Proc. Nat. Acad. Sci. USA 33 (1947), 331-332

[18] Harish-Chandra: Plancherel formula for the 2×2 real unimodular group. Proc. Nat. Acad. Sci. USA 38 (1952), 337-342

[19] Harish-Chandra: Representations of semisimple Lie groups I. Trans. Amer. Math. Soc. 75 (1953), 185-243

[20] Harish-Chandra: The characters of semisimple Lie groups. Trans. Amer. Math. Soc. 83 (1956), 98-163

[21] Harish-Chandra: Discrete series for semisimple Lie groups II. Acta Math. 116 (1966), 1-111

[22] Harish-Chandra: Harmonic analysis on real reductive groups III. Ann. of Math. 104 (1976), 117-201

[23] H. Hecht, D. Miličić, W. Schmid and J. Wolf: Localization of Harish-Chandra modules and the derived functor construction. To appear

[24] H. Hecht and W. Schmid: Characters, asymptotics, and n-homology of Harish-Chandra modules. Acta Math. 151 (1983), 49-151

[25] R. Herb: Discrete series characters and Fourier inversion on semisimple real Lie groups. Trans. Amer. Math. Soc. 277 (1983), 241-261

[26] R. Herb and J. Wolf: The Plancherel theorem for general semisimple Lie groups. Preprint

[27] T. Hirai: On irreducible representations of the Lorentz group of n-th order. Proc. Japan. Acad. 38 (1962), 83-87

[28] R. Howe: Small unitary representations of classical groups. To appear in the Proceedings of the Mackey Conference, Berkeley 1984

[29] D. Kazhdan and G. Lusztig: Representations of Coxeter groups and Hecke algebras. Inventiones Math. 53 (1979) 165-184

[30] A. Knapp and B. Speh: Status of classification of irreducible unitary representations. In Harmonic Analysis, Proceedings, 1981. Springer Lecture Notes in Mathematics 908 (1982), pp. 1-38

[31] A. Knapp and G. Zuckerman: Classification of irreducible tempered representations of semisimple groups. Ann. of Math. 116 (1982), 389-501

[32] H. Kraljević: Representations of the universal covering group of the group SU(n,1). Glasnik Mat. 8 (1973), 23-72

[33] R. Langlands: On the classification of irreducible representations of real algebraic groups. Mimeographed notes, Institute for Advanced Study 1973

[34] R. Langlands: On the Functional Equations Satified by Eisenstein series. Springer Lecture Notes in Mathematics 544 (1976)

[35] G. Lusztig and D. Vogan: Singularities of closures of K orbits on flag manifolds. Inventiones Math. 71 (1983)

[36] T. Matsuki: The orbits of affine symmetric spaces under the action of minimal parabolic subgroups. J. Math. Soc. Japan 31 (1979), 331-357

[37] T. Matsuki: Closure relation for K-orbits on complex flag manifolds. Preprint

[38] D. Miličić: Asymptotic behavior of matrix coefficients of the discrete series. Duke Math. J. 44 (1977), 59-88

[39] E. Nelson: Analytic vectors. Ann. of Math. 70 (1959), 572-615

[40] T. Oshima and T. Matsuki: A description of discrete series for semisimple symmetric spaces. To appear in Adv. Studies in Math.

[41] S. J. Prichepionok: A natural topology for linear representations of semisimple Lie algebras. Soviet Math. Dokl. 17 (1976), 1564-66

[42] J. Rawnsley, W. Schmid and J. A. Wolf: Singular unitary representations and indefinite harmonic theory. J. Funct. Anal. 51 (1983), 1-114

[43] W. Schmid: Homogeneous complex manifolds and representations of semisimple Lie groups. Thesis, UC Berkeley 1967

[44] W. Schmid: L^2-cohomology and the discrete series. Ann. of Math. 102 (1975), 535-564

[45] W. Schmid: Boundary value problems for group invariant differential equations. To appear in Proceedings of the Cartan Symposium, Lyon 1984

[46] B. Speh and D. Vogan: Reducibility of generalized principal series representations. Acta Math. 145 (1980), 227-299

[47] T. A. Springer: Some results on algebraic groups with involutions. Preprint

[48] D. Vogan: The algebraic structure of the representations of semisimple Lie groups I. Ann. of Math. 109 (1979), 1-60

[49] D. Vogan: Representations of Real Reductive Lie Groups. Birkhäuser, Boston 1981

[50] D. Vogan: Irreducible characters of semisimple Lie groups II. The Kazhdan-Lusztig conjectures. Duke Math. J. 46 (1979), 61-108

[51] D. Vogan: Irreducible characters of semisimple Lie groups III. Proof of Kazhdan-Lusztig conjecture in the integral case. Inventiones Math. 71 (1983), 381-417

[52] D. Vogan: Understanding the unitary dual. In Lie Group Representations I. Springer Lecture Notes in Mathematics 1024 (1983), pp. 264-288

[53] D. Vogan: Unitarizability of certain series of representations. Ann. of Math. 120 (1984), 141-187

[54] N. Wallach: Asymptotic expansions of generalized matrix entries of representations of real reductive groups. In Lie Group Representations I. Springer Lecture Notes in Mathematics 1024 (1983), pp. 287-369

[55] N. Wallach: On the unitarizability of derived functor modules. Preprint

[56] J. A. Wolf: Unitary Representations on Partially Holomorphic Cohomology Spaces. Amer. Math. Soc. Memoir 138 (1974)

LOOP GROUPS

G.B. Segal,
St. Catherine's College,
Oxford.

§1 General remarks

In this talk a loop group LG will mean the group of smooth maps from the circle S^1 to a compact Lie group G. One reason for studying such groups is that they are the simplest examples of infinite dimensional Lie groups. Thus LG has a Lie algebra $L\mathfrak{g}$ - the loops in the Lie algebra $\mathfrak{g}$ of G - and the exponential map $L\mathfrak{g} \to LG$ is a local diffeomorphism. Furthermore LG has a complexification $LG_{\mathbb{C}}$, the loops in the complexification of G. Neither of these properties is to be expected of infinite dimensional groups: neither holds, for example, for the group of diffeomorphisms of the circle [17].

From this point of view the group Map(X;G) of smooth maps $X \to G$, where X is an arbitrary compact manifold, seems almost as simple as LG. Such groups are of great importance in quantum theory, where they occur as "gauge groups" and "current groups"; the manifold X is physical space. Thus loop groups arise in quantum field theory in two-dimensional space-time. In fact it is not much of an exaggeration to say that the mathematics of two-dimensional quantum field theory is almost the same thing as the representation theory of loop groups.

If dim(X) > 1, however, surprisingly little is known about the group Map(X;G). Essentially only one irreducible representation of it is known - the representation of Vershik, Gelfand and Graev [9] - and that representation does not seem relevant to quantum field theory. For loop groups, in contrast, there is a rich and extensively developed theory. They first became popular because of their connection with the intriguing combinatorial identities of Macdonald [16]. They are the groups whose Lie algebras are the "affine algebras" of Kac-Moody - roughly speaking, the algebras associated to positive-semidefinite Cartan matrices. From that point of view the groups have been discussed in Tits's talk In this talk I shall keep away from the Lie algebra theory, of which there is an excellent exposition in the recent book of Kac [11], and instead shall attempt to survey what is known about the global geometry and analysis connected with the groups.

From any point of view the crucial property of loop groups is the existence of the one-parameter group of automorphisms which simply rotates the loops. It permits one to speak of representations of LG of <u>positive energy</u>. A representation of LG on a topological vector space H has positive energy if there is given a positive action of the circle group $\mathbb{T}$ on H which intertwines with the action of LG so at to provide a representation of the semidirect product $\mathbb{T} \tilde{\times} LG$, where $\mathbb{T}$ acts on LG by rotation. An action of $\mathbb{T}$ on H is <u>positive</u> if $e^{i\theta} \in \mathbb{T}$ acts as $e^{iA\theta}$, where A is an operator with positive spectrum. It turns out that representations of LG of positive energy are necessarily projective (cf. (4.3) below).

The theory of the positive energy representations of LG (or, more accurately, of $\mathbb{T} \tilde{\times} LG$) is strikingly simple, and in strikingly close analogy with the representation theory of compact groups.(*) Thus the irreducible representations

(i) are all unitary,

(ii) all extend to holomorphic representations of $LG_{\mathbb{C}}$, and

(iii) form a countable discrete set, parametrized by the points of a positive cone in the lattice of characters of a torus.

None of these properties holds, for example, for the representations of $SL_2(\mathbb{R})$.

The positive energy condition is strongly motivated by quantum field theory: the circle action on H corresponds to the time evolution on the Hilbert space H of states. It would be very interesting if one could formulate an analogous condition for more general groups Map(X;G). Certainly in quantum field theory one might expect such a gauge group to act on a state space on which time evolution was defined by a positive Hamiltonian operator, and

(*) We are thinking of continuous representations on arbitrary complete locally convex topological vector spaces. But we do not distinguish between representations on H and $\tilde{H}$ if there is an injective intertwining operator $H \to \tilde{H}$ with dense image.

the gauge transformations should intertwine in some perhaps complicated way with the time evolution. But there has been no progress on this front, and the attempt may well be misconceived. (Cf. §3 below.)

To conclude these introductory remarks I should say that the material I am going to present is all essentially well-known, and has been worked out independently by many people in slightly different contexts. As representative treatments of various aspects of the subject from standpoints somewhat different from mine let me refer to Garland [8], Lepowsky [15], Kac and Peterson [12], Goodman and Wallach [10], Frenkel [5]. More details of my own approach can be found in [18], [19] and [20].

§2 The fundamental homogeneous space X

In the study of LG the homogeneous space $X = LG/G$ (where G is identified with the constant loops in LG) plays a central role. One can think of X as the space ΩG of <u>based</u> loops in G; but we prefer to regard it as a homogeneous space of LG. I shall list its most important properties.

(i) X is a complex manifold, and in fact a homogeneous space of the complex group $LG_{\mathbb{C}}$:

$$X = LG/G \cong LG_{\mathbb{C}}/L^{+}G_{\mathbb{C}} . \tag{2.1}$$

Here $L^{+}G_{\mathbb{C}}$ is the group of smooth maps $\gamma : S^1 \to G_{\mathbb{C}}$ which are the boundary values of holomorphic maps

$$\gamma : \{z \in \mathbb{C} : |z| < 1\} \to G_{\mathbb{C}} .$$

The isomorphism (2.1) is equivalent to the assertion that any loop γ in $LG_{\mathbb{C}}$ can be factorized

$$\gamma = \gamma_u \cdot \gamma_+$$

with $\gamma_u \in LG$ and $\gamma_+ \in L^{+}G_{\mathbb{C}}$. This is analogous to the factorization of an element of $GL_n(\mathbb{C})$ as (unitary) × (upper triangular).

(ii) For each invariant inner product < , > on the Lie algebra $\mathfrak{g}$ of G there is an invariant closed 2-form ω on X which makes it a symplectic manifold, and even fits together with the complex structure to make a Kähler manifold. The tangent space to X at its base-point is $L\mathfrak{g}/\mathfrak{g}$, and ω is given there by

$$\omega(\xi,\eta) = \frac{1}{2\pi}\int_0^{2\pi} \langle\xi'(\theta), \eta(\theta)\rangle d\theta \; . \tag{2.2}$$

(iii) The <u>energy</u> function $\mathcal{E} : X \to \mathbb{R}_+$ defined by

$$\mathcal{E}(\gamma) = \frac{1}{4\pi}\int_0^{2\pi} \|\gamma'(\theta)\|^2 d\theta$$

is the Hamiltonian function corresponding in terms of the symplectic structure to the circle-action on X which rotates loops. The critical points of $\mathcal{E}$ are the loops which are homomorphisms $\mathbb{T} \to G$. Downwards gradient trajectories of $\mathcal{E}$ emanate from every point of X, and travel to critical points of $\mathcal{E}$. The gradient flow of $\mathcal{E}$ and the Hamiltonian circle action fit together to define a holomorphic action on X of the multiplicative semigroup $\mathbb{C}^{\times}_{\leqslant 1} = \{z \in \mathbb{C} : 0 < |z| \leqslant 1\}$.

The connected components $C_{[\lambda]}$ of the critical set of $\mathcal{E}$ are the conjugacy classes of homomorphisms $\lambda : \mathbb{T} \to G$. They correspond to the orbits of the Weyl group W on the lattice $\pi_1(T)$, where T is a maximal torus of G. The gradient flow of $\mathcal{E}$ stratifies the manifold X into locally closed complex submanifolds $X_{[\lambda]}$, where $X_{[\lambda]}$ consists of the points which flow to $C_{[\lambda]}$. Each stratum $X_{[\lambda]}$ is of finite codimension.

<u>Proposition (2.3)</u>. The stratification coincides with the decomposition of X into orbits of $L^-G_{\mathbb{C}}$; i.e. $X_{[\lambda]} = L^-G_{\mathbb{C}}.\lambda$.

Here $L^-G_{\mathbb{C}}$ is the group of loops in $G_{\mathbb{C}}$ which are boundary values of holomorphic maps $D_\infty \to G_{\mathbb{C}}$, where $D_\infty = \{z \in S^2 : |z| > 1\}$. Proposition (2.3) is the classical Birkhoff factorization theorem: a loop γ in $G_{\mathbb{C}}$ can be factorized

$$\gamma = \gamma_-.\lambda.\gamma_+ \; ,$$

with $\gamma_\pm \in L^\pm G_{\mathbb{C}}$, and $\lambda : S^1 \to G$ a homomorphism. This is the analogue of factorizing an element of $GL_n(\mathbb{C})$ as

(lower triangular)×(permutation matrix)×(upper triangular).

There is one dense open stratum X_0 in X. It is contractible, and can be identified with the nilpotent group $L_0^-G_{\mathbb{C}} = \{\gamma \in L^-G_{\mathbb{C}} : \gamma(\infty) = 1\}$.

(iv) The complex structure of X can be characterized in another way, pointed out by Atiyah [1]. To give a holomorphic map $Z \to X$, where Z is an arbitrary complex manifold, is the same as to give a holomorphic principal $G_{\mathbb{C}}$-bundle on $Z \times S^2$ together with a trivialization over $Z \times D_\infty$. If Z is compact it follows that the space of based maps $Z \to X$ in a given homotopy class is finite dimensional; for the moduli space of $G_{\mathbb{C}}$-bundles of a given topological type is finite dimensional. This is a rather striking fact, showing that X, although a rational variety, is quite unlike, say, an infinite dimensional complex projective space: for in X the set of points which can be joined to the base-point by holomorphic curves of a given degree is only finite dimensional.

§3 The Grassmannian embedding of X

Let us choose a finite dimensional unitary representation V of compact group G, and let H denote the Hilbert space $L^2(S^1;V)$. Evidently $LG_{\mathbb{C}}$ acts on H, and we have a homomorphism $i : LG_{\mathbb{C}} \to GL(H)$ an embedding if V is faithful.

To make a more refined statement we write $H = H_+ \oplus H_-$, where H_+ (resp. H_-) consists of the functions of the form $\sum_{n\geq 0} v_n e^{in\theta}$ (resp. $\sum_{n<0} v_n e^{in\theta}$) with $v_n \in V$. The restricted general linear group $GL_{res}(H)$ is defined as the subgroup of $GL(H)$ consisting of elements

$$\begin{pmatrix} a & b \\ c & d \end{pmatrix} \qquad (3.1)$$

whose off-diagonal blocks b,c (with respect to the decomposition $H_+ \oplus H_-$) are Hilbert-Schmidt. The blocks a and d are then automatically Fredholm.

Proposition (3.2). $i(LG_{\mathbb{C}}) \subset GL_{res}(H)$.

The set of closed subspaces of H obtained from H_+ by the action of $GL_{res} = GL_{res}(H)$ will be called the <u>Grassmannian</u> Gr(H). It is naturally a Hilbert manifold, and has the homotopy type of the space known to topologists as $\mathbb{Z} \times BU$. The homomorphism

$$i : LG_{\mathbb{C}} \to GL_{res}(H) \tag{3.3}$$

induces a smooth map (again an embedding if V is faithful)

$$i : X \to Gr(H) .$$

This map is closely connected with the Bott periodicity theorem. In fact Bott's theorem asserts that when $G = U_n$ and $V = \mathbb{C}^n$ the map is a homotopy equivalence up to dimension 2n-2.

It should be remarked that i(X) is far from being a <u>closed</u> submanifold of Gr(H). Indeed it is so highly curved that its closure is not a submanifold of Gr(H).

There is a holomorphic line bundle Det on Gr(H) whose fibre at $W \subset H$ can be thought of as the renormalized "top exterior power" of W. Because of the renormalization needed to define it it is not a homogeneous bundle under GL_{res}, but its group of holomorphic automorphisms is a central extension $\widetilde{GL}_{res}$ of GL_{res} by $\mathbb{C}^\times$. The homomorphism (3.3) then gives us a central extension of $LG_{\mathbb{C}}$ by $\mathbb{C}^\times$; up to finite-sheeted coverings, <u>all</u> extensions of $LG_{\mathbb{C}}$ by $\mathbb{C}^\times$ are obtained in this way. (The Lie algebra cocycle of the extension is given by (2.2), where $\langle\ ,\ \rangle$ is the trace form of V.)

The line bundle Det has no holomorphic sections, but its dual Det* has an infinite dimensional space of sections Γ on which $\widetilde{GL}_{res}$ acts irreducibly. Just as the space of sections of the dual of the determinant bundle on the Grassmannian Gr(E) of a finite dimensional vector space E can be identified with the exterior algebra $\Lambda(E^*)$ we find

<u>Proposition 3.4</u>. $\Gamma \cong \Lambda(H_+ \oplus \overline{H}_-)$.

This space is very familiar in quantum field theory as the "fermionic Fock space" got by quantizing a classical state space H (e.g. the space of solutions of the Dirac equation) in which H_+ and H_- are the states of positive and negative energy.

From the point of view of loop groups the importance of Γ is that when $G = U_n$ and $V = \mathbb{C}^n$ the projective action of LU_n on Γ via (3.3) is the "basic" irreducible representation of LU_n (cf. §4 below). It even extends from LU_n to LO_{2n}, for Γ is most correctly regarded as the spin representation of the restricted orthogonal group of the real Hilbert space underlying H.

Let us briefly consider generalizing the foregoing discussion to the group Map(X;G), where X is a compact odd-dimensional Riemannian manifold. If H is the space of spinor fields on X then $\mathrm{Map}(X;U_n)$ acts naturally on $H \otimes \mathbb{C}^n$. We can decompose

$$H \otimes \mathbb{C}^n = (H_+ \otimes \mathbb{C}^n) \oplus (H_- \otimes \mathbb{C}^n) \;,$$

where $H_\pm$ are the positive and negative eigenspaces of the Dirac operator. We get an embedding

$$\mathrm{Map}(X;U_n) \to GL_{(m)}(H \otimes \mathbb{C}^n) \;, \tag{3.5}$$

where $GL_{(m)}$ denotes the group of operators of the form (3.1) in which the off-diagonal blocks belong to the Schatten ideal $\mathcal{J}_m$ with $m-1 = \dim(X)$. (Cf. [21],[4].)

The homomorphism (3.5) is very interesting: topologically it represents the index map in K-theory [4]. On the other hand no representations of $GL_{(m)}$ are known, and one even feels that representations are not the natural thing to look for, as the two-dimensional cohomology class which forces $GL_{res} = GL_{(2)}$ to have a projective rather than a genuine representation is replaced by an m-dimensional class for $GL_{(m)}$. Alternatively expressed, on the Grassmannian $Gr_{(m)}(H)$ associated with $GL_{(m)}$ there is a tautological infinite dimensional bundle with a connection. The "determinant" line bundle of this - i.e. its first Chern class - cannot be defined, but nevertheless the higher components of its Chern character do make geometric sense.

§4 The Borel-Weil theory

(i) The basic representation

To simplify the discussion we shall assume from now on that the compact group G is simply connected and simple. Then $H^2(X;\mathbb{Z}) \cong \mathbb{Z}$, and so the complex line bundles L on X are classified by an integer invariant $c_1(L)$. In fact each bundle has a unique holomorphic structure, and has non-zero holomorphic sections if and only if $c_1(L) \geqslant 0$. The space of holomorphic sections of the bundle L_1 with $c_1(L_1) = 1$ is called the basic representation of $LG_{\mathbb{C}}$: we have remarked that when $G = SU_n$ the bundle L_1 is the restriction of Det* on Gr(H). As we saw in that case, L_1 is not quite homogeneous under $LG_{\mathbb{C}}$. The holomorphic automorphisms of L_1 which cover the action of $LG_{\mathbb{C}}$ on X form a group $\widetilde{L}G_{\mathbb{C}}$ which is a central extension of $LG_{\mathbb{C}}$ by $\mathbb{C}^\times$ - in fact its universal central extension. It corresponds to the Lie algebra cocycle (2.2) for an inner product $\langle\ ,\ \rangle$ on which I shall also call "basic".

One reason for the name "basic" is provided by

Proposition (4.1). If G is a simply-laced group and Γ is the basic representation of $LG_{\mathbb{C}}$ then any irreducible representation of positive energy is a discrete summand in $\rho^*\Gamma$, where $\rho : LG_{\mathbb{C}} \to LG_{\mathbb{C}}$ is an endomorphism.

(ii) The Borel-Weil theorem

To describe all the positive energy irreducible representations of LG we must consider the larger complex homogeneous space $Y = LG/T$, where T is a maximal torus of G. This manifold Y is fibred over X with the finite dimensional complex homogeneous space G/T as fibre. Complex line bundles on Y are classified topologically by

$$H^2(Y;\mathbb{Z}) \cong \mathbb{Z} \oplus H^2(G/T;\mathbb{Z}) \cong \mathbb{Z} \oplus \hat{T} ,$$

where $\hat{T}$ is the character group of T. Once again each bundle has a unique holomorphic structure, and is homogeneous under $\widetilde{L}G_{\mathbb{C}}$. If we denote the bundle corresponding to $(n,\lambda) \in \mathbb{Z} \oplus \hat{T}$ by $L_{n,\lambda}$ then we have the following "Borel-Weil" theorem.

Proposition (4.2).

(a) The space $\Gamma(L_{n,\lambda})$ of holomorphic sections of $L_{n,\lambda}$ is either zero or an irreducible representation of $LG_{\mathbb{C}}$ of positive energy.

(b) Every projective irreducible representation of LG of positive energy arises in this way.

(c) $\Gamma(L_{n,\lambda}) \neq 0$ if and only if (n,λ) is positive in the sense that

$$0 \leqslant \lambda(h_\alpha) \leqslant n\langle h_\alpha, h_\alpha\rangle$$

for each positive coroot h_α of G, where $\langle\ ,\ \rangle$ is the basic inner product on $\mathfrak{g}$.

It should be emphasized that except for the "if" part of (c) this proposition is quite elementary, amounting to little more than the observations that (i) any representation of positive energy contains a ray invariant under $L^-G_{\mathbb{C}}$, and (ii) $L^-G_{\mathbb{C}}$ acts on Y with a dense orbit. Thus the elementary part already yields

Corollary (4.3). For positive energy representations of LG:

(a) each representation is necessarily projective,

(b) each representation extends to a holomorphic representation of $LG_{\mathbb{C}}$, and

(c) each irreducible representation is of finite type, i.e. if it is decomposed into energy levels $H = \bigoplus H_q$, where H_q is the part where the rotation $e^{i\theta} \in \mathbb{T}$ acts as $e^{iq\theta}$, then each H_q has finite dimension.

Assertion (c) holds because a holomorphic section of $L_{n,\lambda}$ is determined by its Taylor series at the base-point. That gives one an injection

$$\Gamma(L_{n,\lambda}) \to \hat{S}(T_Y^*) \quad , \tag{4.4}$$

where T_Y is the tangent space to Y at the base-point, and $\hat{S}$ denotes the completed symmetric algebra. The injection is compatible with the action of $\mathbb{T}$, and the right hand side of (4.4) is of finite type.

(iii) Unitarity

We have mentioned that all positive energy representations of LG are unitary. In fact a simple formal argument shows that each irreducible representation has a non-degenerate invariant sesquilinear form, but it is not so simple to show that it is positive definite. By (4.1) it is enough to consider the basic representation. When $G = SU_n$ or SO_{2n} the unitarity is then clear from the description (3.4) of the basic representation; and one can deal similarly with all simply laced groups by the method of §5 below. The only proof known in the general case is an inductive argument in terms of generators and relations, due to Garland [7].

It would obviously be very attractive to prove the unitarity directly by putting an invariant measure on the infinite dimensional manifold Y and using the standard L^2 inner product. That has not yet been done, though it seems to be possible. The measure will be supported not on Y but on a thickening Y*, to which the holomorphic line bundles L extend. One expects to have an LG-invariant measure on sections of $\bar{L} \otimes L$ for each positive bundle L. There is no difficulty in finding a candidate for Y*: the manifold Y is modelled on the Lie algebra $N^-\mathfrak{g}_{\mathbb{C}}$ of holomorphic maps $\xi : D_\infty \to \mathfrak{g}_{\mathbb{C}}$ (with $\xi(\infty)$ lower triangular) which extend smoothly to the boundary of D_∞; the thickening is modelled on the dual space, i.e. the holomorphic maps with distributional boundary values on S^1. (*)

(iv) The Kac character formula and the Bernstein-Gelfand-Gelfand resolution

Because each irreducible representation of $\mathbb{T} \tilde{\times} LG$ is of finite type it makes sense to speak of its formal character, i.e. of its decomposition under the torus $\mathbb{T} \times T$. This is given by the Kac character formula, an exact analogue of the classical Weyl character formula.

Thinking of $Y = LG/T$ as $\mathbb{T} \tilde{\times} LG/\mathbb{T} \times T$, we observe that the torus $\mathbb{T} \times T$ acts on Y with a discrete set of fixed points. This set is the affine Weyl group $W_{aff} = N(\mathbb{T} \times T)/(\mathbb{T} \times T)$. If one ignores the infinite dimensionality of Y and writes down formally

(*) An interesting family of measures on Y is constructed in [5], but it does not include the measure needed to prove unitarity.

the Lefschetz fixed-point formula of Atiyah-Bott [2] for the character of the torus action on the holomorphic sections of a positive line bundle L on Y then one obtains the Kac formula, at least if one assumes that the cohomology groups $H^q(Y; \mathcal{O}(L))$ vanish for $q > 0$. (Here $\mathcal{O}(L)$ is the sheaf of holomorphic sections of L.) Unfortunately it does not seem possible at present to prove the formula this way.

One can do better by using more information about the geometry of the space Y. It possesses a stratification just like that of X described in §2. The strata $\{\Sigma_w\}$ are complex affine spaces of finite codimension, and are indexed by the elements w of the group W_{aff}: indeed Σ_w is the orbit of w under $N^-G_{\mathbb{C}} = \{\gamma \in L^-G_{\mathbb{C}} : \gamma(\infty) \text{ is lower triangular}\}$.

Let Y_p denote the union of the strata of complex codimension p. The cohomology groups $H^*(Y; \mathcal{O}(L))$ are those of the cochain complex $K^\cdot$ formed by the sections of a flabby resolution of $\mathcal{O}(L)$. Filtering $K^\cdot$ by defining $K_p^\cdot$ as the subcomplex of sections with support in $\overline{Y}_p$ gives us a spectral sequence converging to $H^*(Y; \mathcal{O}(L))$ with $E_1^{pq} = H^{p+q}(K_p^\cdot/K_{p+1}^\cdot)$. Because Y_p is affine and has an open neighbourhood U_p isomorphic to $Y_p \times \mathbb{C}^p$ the spectral sequence collapses, and its E_1-term reduces to

$$E_1^{p0} = H^p_{Y_p}(U_p; \mathcal{O}(L)) ,$$

$$E_1^{pq} = 0 \qquad \text{if} \qquad q \neq 0 .$$

In other words $H^*(Y; \mathcal{O}(L))$ can be calculated from the cochain complex $\{H^p_{Y_p}(U_p; \mathcal{O}(L))\}$. Here $H^p_{Y_p}(U_p; \mathcal{O}(L))$ means the cohomology of the sheaf $\mathcal{O}(L)|U_p$ with supports in Y_p. It is simply the space of holomorphic sections of the bundle on Y_p whose fibre at y is

$$L_y \otimes H^p_{\{0\}}(N_y; \mathcal{O}) ,$$

where $N_y \cong \mathbb{C}^p$ is the normal space to Y_p at y; furthermore, $H^p_{\{0\}}(N_y; \mathcal{O})$ is the dual of the space of holomorphic p-forms on N_y. Thus as a representation of $\mathbb{T} \times T$

$$E_1^{p0} \cong \bigoplus_w S(T_w^* \oplus N_w) \otimes \det(N_w) \otimes L_w ,$$

where w runs through the elements of W_{aff} of codimension p, and T_w and N_w are the tangent and normal spaces to Σ_w at w. If we know that $H^q(Y; \mathcal{O}(L)) = 0$ for $q > 0$ then we can read off the Kac formula.

The cochain complex $E_1^{\cdot 0}$ is the dual of the Bernstein-Gelfand-Gelfand resolution, described in the finite dimensional case in [3] (cf. also [13]). Its exactness can be proved by standard algebraic arguments, and one can deduce the vanishing of the higher cohomology groups $H^q(Y; \mathcal{O}(L))$. But it would be attractive to reverse the argument by proving the vanishing theorem analytically.

§5 "Blips" or "vertex operators"

The Borel-Weil construction of representations is quite inexplicit. I shall conclude with a very brief description of an interesting explicit construction of the basic representation of LG, for simply laced G, which was independently extracted from the physics literature in [14], [6] and [19].

The idea is to start with a standard irreducible projective representation H of LT, and to extend the action from LT to LG. The abelian group LT is essentially a vector space, and for H we take its "Heisenberg" representation. To make the Lie algebra $L\mathfrak{g}_{\mathbb{C}}$ act on H amounts to defining, for each basis element ξ_i of $\mathfrak{g}_{\mathbb{C}}$, an "operator-valued distribution" B_i on S^1: for then an element $\Sigma f_i \xi_i$ of $L\mathfrak{g}_{\mathbb{C}}$ will act on H by

$$\sum_i \int_{S^1} f_i(\theta) B_i(\theta) d\theta \ .$$

We must construct B_i for each basis element of $\mathfrak{g}_{\mathbb{C}}/\mathfrak{t}_{\mathbb{C}}$. These are indexed by the <u>roots</u> of G, and the remarkable fact about simply-laced groups (i.e. those for which all the roots have the same length) is that the roots correspond precisely to the set of all homomorphisms $\alpha : \mathbb{T} \to T$ of minimal length. Now for each $\theta \in S^1$ and each small positive ε let us consider the blip-like element $B_{\alpha,\theta,\varepsilon}$ of LT such that

$$B_{\alpha,\theta,\varepsilon}(\theta') = 1 \quad \text{if} \quad |\theta' - \theta| > \varepsilon \ ,$$

while on the interval $(\theta - \varepsilon, \theta + \varepsilon)$ of the circle $B_{\alpha,\theta,\varepsilon}$ describes the loop α in T. When $B_{\alpha,\theta,\varepsilon}$ is regarded as an operator on H it turns out that the renormalized limit

$$\lim_{\varepsilon \to 0} \varepsilon^{-1} B_{\alpha,\theta,\varepsilon}$$

exists in an appropriate sense, and is the desired $B_\alpha(\theta)$. Such operators have been called "vertex operators" in the physics literature.

Extending the representation from theLie algebra to LG presents no problems.

REFERENCES

[1] M.F. Atiyah, Instantons in two and four dimensions. To appear

[2] M.F. Atiyah and R. Bott, A. Lefschetz fixed point formula for elliptic complexes: II. Applications. Ann. of Math., 88 (1968), 451-491

[3] I.N. Bernstein, I.M. Gelfand, and S.I. Gelfand, Differential operators on the base affine space and a study of $\mathfrak{g}$-modules. In Lie groups and their representations. Summer School of the Bolyai Janos Math. Soc., ed. I.M. Gelfand. Wiley, New York, 1975.

[4] A Connes, Non commutative differential geometry. Chapter I, the Chern character in K homology. IHES preprint, 1982.

[5] I.G. Frenkel, Orbital theory for affine Lie algebras. To appear.

[6] I.G. Frenkel and V.G. Kac, Basic representations of affine Lie algebras and dual resonance models. Invent. Math., 62 (1980), 23-66.

[7] H. Garland, The arithmetic theory of loop algebras. J. Algebra, 53 (1978), 480-551.

[8] H. Garland, The arithmetic theory of loop groups. Publ. Math. IHES, 52 (1980), 5-136.

[9] I.M. Gelfand, M.I. Graev, and A.M. Vershik, Representations of the group of smooth mappings of a manifold into a compact Lie group. Compositio Math., 35 (1977), 299-334.

[10] R. Goodman and N. Wallach, Structure and unitary cocycle representations of loop groups and the group of diffeomorphisms of the circle. To appear.

[11] V.G. Kac, Infinite dimensional Lie algebras. Birkhauser, 1983.

[12] V.G. Kac and D.H. Peterson, Spin and wedge representations of infinite dimensional Lie algebras and groups. Proc. Nat. Acad. Sci. USA, 78 (1981), 3308-3312.

[13] G. Kempf, The Grothendieck-Cousin complex of an induced representation. Advanced in Math., 29 (1978), 310-396.

[14] J. Lepowsky, Construction of the affine algebra $A_1^{(1)}$. Comm. Math. Phys., 62 (1978), 43-53.

[15] J. Lepowsky, Generalized Verma modules, loop space cohomology, and Macdonald type identities. Ann. Sci. Ec. Norm. Sup., 12 (1979), 169-234.

[16] I.G. Macdonald, Affine Root Systems and the Dedekind η-function. Invent. Math., 15 (1972), 91-143.

[17] J. Milnor, On infinite dimensional Lie groups. To appear.

[18] A.N. Pressley and G.B. Segal, Loop groups. Oxford Univ. Press, to appear.

[19] G.B. Segal, Unitary representations of some infinite dimensional groups. Comm. Mlath. Phys., 80 (1981), 301-342.

[20] G.B. Segal and G. Wilson, Loop groups and equations of KdV type. Publ. Math. IHES, to appear.

[21] B. Simon, Trace ideals and their applications. London Math. Soc. Lecture Notes No.35, Cambridge Univ. Press, 1979.

Some Recent Results in Complex Manifold Theory Related to Vanishing Theorems for the Semipositive Case

Yum-Tong Siu
Department of Mathematics
Harvard University
Cambridge, MA 02138, U.S.A.

To put this survey in the proper perspective, let me first make some rather general remarks. To study complex manifolds or in general complex spaces, one works with holomorphic objects like holomorphic maps, holomorphic functions, holomorphic vector bundles and their holomorphic sections. One has to construct such objects. For example, to prove that a complex manifold is biholomorphic to $\mathbb{C}^n$, one tries to produce n suitable holomorphic functions. To prove that a complex manifold is biholomorphic to $\mathbb{P}_n$, one tries to produce n+1 good holomorphic sections of a suitable holomorphic line bundle. To prove that two complex manifolds are biholomorphic, one tries to produce a biholomorphic map. How does one produce such holomorphic objects? So far we have mainly the following methods:

(1) The method of constructing harmonic objects first and then getting holomorphic objects from them. An example is the use of the Dirichlet principle to construct harmonic functions on open Riemann surfaces and then obtaining holomorphic functions from them. Examples of the construction of harmonic objects are the results of Eells-Sampson [10] and Sachs-Uhlenbeck [40] on the existence of harmonic maps. However, unlike the one-dimensional case, in the higher-dimensional case the gap between a harmonic object and a holomorphic object is very wide and, except for some special cases, is impossible to bridge.

(2) The method of using the vanishing theorem of Kodaira to construct holomorphic sections of high powers of positive line bundles [24].

(3) Grauert's bumping technique to construct holomorphic functions on strongly pseudoconvex domains [13].

(4) The method of using L^2 estimates of $\bar{\partial}$ to construct holomorphic functions on strongly pseudoconvex domains (Morrey [30], Andreotti-Vesentini [1], Kohn [25], Hörmander [20]).

These methods produce holomorphic objects from scratch so to speak. There are also other methods like the use of Theorems A and B of Cartan-Serre to construct holomorphic objects, but one has to have a Stein manifold (i.e. a complex submanifold of $\mathbb{C}^n$) or a Stein space to apply Theorems A and B and on such manifolds previously existing global holomorphic functions are essential for the construction.

Let me briefly explain the notions of positive line bundles and strongly pseudoconvex domains and how they are related. A holomorphic line bundle with a Hermitian metric along its fibers is said to be positive if the curvature form associated to the Hermitian metric is a positive-definite quadratic form. A relatively compact domain with smooth boundary in a complex manifold is said to be strongly pseudoconvex if it is defined near its boundary by $r < 0$ for some smooth function r with nonzero gradient such that the complex Hessian of r as a Hermitian form is positive-definite. If L is a Hermitian holomorphic line bundle over a compact complex manifold, then the set Ω of all vectors of the dual bundle L^* of L whose lengths are less than 1 is a strongly pseudoconvex domain in L^* if and only if L is positive. Grauert [14] observed that a holomorphic function on Ω gives rise to holomorphic sections of powers of L because its kth coefficient in the power series expansion along the fibers of L^* is a section of the kth power of L. So producing holomorphic sections of powers of a positive line bundle is a special case of producing holomorphic functions of a strongly pseudoconvex domain.

In the above methods of producing holomorphic objects some positive-definite quadratic form is used, be it the curvature form in the case of a positive line bundle or the complex Hessian of the defining function in the case of a strongly pseudoconvex domain. In the method of using harmonic objects to construct holomorphic objects no positive-definite quadratic form is used. However, in the higher-dimensional case there is a wide gap between harmonic and holomorphic objects and methods known up to now [45, 46, 47, 51, 52] to bridge the gap require positive-definiteness of a certain quadratic form coming from the curvature tensor. This survey talk discusses the situation when the quadratic forms used in producing holomorphic objects are only positive semidefinite instead of strictly positive-definite. In certain cases we may even allow certain benign negativity. One may wonder why one should bother to study the semidefinite case. There are a number of reasons. Let me give two here. One is that some situations are naturally semidefinite, like the seminegativity of the sectional curvature for a bounded symmetric domain. Another is that when limits of holomorphic objects are used in proofs (like in the continuity method), the limit of strictly positive definite objects can only be assumed first to be semidefinite though in the final result it may turn out to be strictly positive definite. The semidefinite case is by far much more complicated than the definite case.

In this talk we will survey some recent results concerning vanishing theorems for the semidefinite case and their applications. More specifically we will discuss the following three topics:

(i) The construction of holomorphic sections for line bundles with curvature form not strictly positive or even with bengin negativity somewhere. An application is a proof of the Grauert-Riemenschneider conjecture characterizing Moishezon manifolds by semipositive line bundles [49, 50].

(ii) The strong rigidity of compact Kähler manifolds with seminegative curvature, in particular the results of Jost-Yau [22] and Mok [29] on the strong rigidity of irreducible compact quotients of polydiscs.

(iii) Subelliptic estimates of Kohn's school [26, 6] and their applications to vanishing theorems for semipositive bundles.

I. Producing Sections for Semipositive Bundles

We want to discuss how one can produce holomorphic sections for a Hermitian line bundle whose curvature form is only semipositive or may even be negative somewhere. The original motivation for this kind of study is to prove the so-called Grauert-Riemenschneider conjecture[15, p.277]. Kodaira[24] characterized projective algebraic manifolds by the existence of a Hermitian holomorphic line bundle whose curvature form is positive definite. The conjecture of Grauert-Riemenschneider attempts to generalize Kodaira's result to the case of Moishezon manifolds. A Moishzon manifold is a compact complex manifold with the property that the transcendence degree of its meromorphic function field equals its complex dimension. Moishezon showed [28] that such manifolds are precisely those which can be transformed into a projective algebraic manifold by proper modification. The concept of a Moishezon space is similarly defined.

The conjecture of Grauert-Riemenschneider asserts that a compact complex space is Moishezon if there exists on it a torsion-free coherent analytic sheaf of rank one with a Hermitian metric whose curvature form is positive definite on an open dense subset. Here a Hermitian metric for a sheaf is defined by going to the linear space associated to the sheaf and the curvature form is defined only on the set of points where the sheaf is locally free and the space is regular. The difficulty with the proof of the conjecture is how to prove the following special case.

<u>Conjecture of Grauert-Riemenschneider</u>. Let M be a compact complex manifold which admits a Hermitian holomorphic line bundle L whose curvature form is positive definite on an open dense subset G of M. Then M is Moishezon.

Since the conjecture of Grauert-Riemenschneider was introduced, a number of other characterizations of Moishezon spaces have been obtained [38,57,53,12,35] which circumvent the difficulty of proving the Grauert-Riemenschneider conjecture by stating the characterizations in such a way that a proof can be obtained by using blow-ups, Kodaira's vanishing and embedding theorems, or L^2 estimates of $\bar{\partial}$ for complete Kähler manifolds. If the manifold M is assumed to be Kähler, then Riemenschneider [39] observed that Kodaira's original proof of his vanishing and embedding theorems together with the identity theorem for solutions of second-order elliptic partial differential equations [2] already yields right away the conjecture of Grauert-Riemenschneider. If the set of points where the curvature form of L is not positive definite is of complex dimension zero [38] or one

[44] or if some additional assumptions are imposed on the eigenvalues of the curvature form of L [47], the conjecture of Grauert-Riemenschneider can rather easily be proved. Recently Peternell [33] used degenerate Kähler metrics to obtain some partial results about the Grauert-Riemenschneider conjecture. However, all the above results fail to deal with the fundamental question of how to produce in general holomorphic sections for a line bundle not strictly positive definite.

Recently a new method of obtaining holomorphic sections for nonstrictly positive line bundles was introduced [49]. There it was used to give a proof of the conjecture of Grauert-Riemenschneider in the special case where M-G is of measure zero in M. It was later refined to give a proof of the general case and a stronger version of the conjecture of Grauert-Riemenschneider [50]. The method imitates the familiar technique in analytic number theory of using the Schwarz lemma to prove the identical vanishing of a function by estimating its order and making it vanish to high order at a sufficient number of points. Such a technique applied to the holomorphic sections of a holomorphic line bundle was used by Serre [41] and also later by Siegel [43] to obtain an alternative proof of Thimm's theorem [54] that the transcendence degree of the meromorphic function field of a compact complex manifold cannot exceed its complex dimension. In [49, 50] the technique was applied to harmonic forms with coefficients in a holomorphic line bundle and its use was coupled with the theorem of Hirzebruch-Riemann-Roch [19, 3].

We give a more precise brief description of the method of [49,50]. To make the description easier to understand, we first impose the condition that M-G is of measure zero in M. By the theorem of Hirzebruch-Riemann-Roch (which for the case of a general compact complex manifold is a consequence of the index theorem of Atiyah-Singer [3]), $\Sigma_{q=0}^{n}(-1)^q \dim H^q(M,L^k) \geq ck^n$ for some positive constant c when k is sufficiently large, where n is the complex dimension of M. To prove that L^k admits enough holomorphic sections to give sufficiently many meromorphic functions to make M Moishezon, it suffices to show that $\dim H^0(M,L^k) \geq ck^n/2$ for k sufficiently large. Thus the problem is reduced to proving that for any given positive number ε and for $q \geq 1$ one has $\dim H^q(M, L^k) \leq \varepsilon k^n$ for k sufficiently large. Give M a Hermitian metric and represent elements of $H^q(M,L^k)$ by L^k-valued harmonic forms. By using the L^2 estimates of $\bar\partial$ one obtains a linear map from the space of harmonic forms to the space of cocycles. Take a lattice of points with distances $k^{-1/2}$ apart in a small neighborhood W of M-G. Then one uses the usual technique of Bochner-Kodaira for the case of a compact Hermitian (not necessarily Kähler) manifold [16, p.429, (7.14)] and uses the Schwarz lemma to show that any cocycle coming from a harmonic form via the linear map and vanishing at all the lattice points to an appropriate fixed order must vanish identically, otherwise its norm is so small that the $\bar\partial$-closed form constructed from it by using a partition of unity would have a norm smaller than that of the harmonic form in its cohomology class, contradicting the minimality of the norm of a harmonic form in its cohomology class. It follows that dim $H^q(M,L^k)$ is dominated by a fixed constant times the number of lattice points (which is comparable to the volume of W times k^n), otherwise there is a nonzero combination of cocycles coming from a basis of harmonic forms via the linear map and having the required vanishing orders. Since M-G is of measure zero in M, we can make the volume of W as small as we please and therefore can choose ε smaller than any prescribed positive number after making k sufficiently large. The reason why such a lattice of points is chosen is that the pointwise square norm of a local holomorphic section of L^k is of the form $|f|^2e^{-k\phi}$, where f is holormorphic function and ϕ is a plurisubharmonic function corresponding to the Hermitian metric of L. The factor $e^{-k\phi}$ is an obstacle to applying the Schwarz lemma. To overcome this obstacle, one chooses a local trivialization of L so that ϕ as well as $d\phi$ vanishes at a point. The on the ball of radius $k^{-1/2}$ centered at that point, $e^{-k\phi}$ is bounded below from zero and from above by constants independent of k. The reason why one uses cocycles instead of dealing directly with harmonic forms is that the Schwarz lemma is a consequence of the log plurisubharmonic property of the absolute value of holomorphic functions and there is no corresponding Schwarz lemma for harmonic forms.

The method outlined above can be refined in the following way so that it works in the general case where G is only assumed to be nonempty. Let R be the set of points of M where the smallest eigenvalue of the curvature form $\partial\bar\partial\phi$ of L does not exceed some positive number λ. For every point O in R one can choose a coordinate polydisc D with coordinates $z_1,\dots,z_n$ centered at O and can choose a global trivialization of L over D such that for some constant $C > 0$

$$|\phi(P_1) - \phi(P_2)| \leq C\left(\lambda|z_1(P_1) - z_1(P_2)|^2 + \sum_{i=2}^{n} |z_i(P_1) - z_i(P_2|^2\right)$$

for P_1, P_2 in D. Moreover, both C and the polyradius of D can be chosen to be the same for all points O of R. Cover R by a finite number of such coordinate polydiscs so that for some constant m depending only on n no more than m of them intersect. Then one chooses the lattice points so that they are $(\lambda k)^{-1/2}$ apart along the z_1 direction but are $k^{-1/2}$ apart along the directions of $z_2,\dots,z_n$. Now the total number of lattice points is no more than a constant times λk^n times the volume of R. By choosing λ sufficiently small, we conclude that for any given positive number ε and for $q \geq 1$ one has $\dim H^q(M,L^k) \leq \varepsilon k^n$ and therefore $\dim H^0(M,L^k)$ is no less than ck^n for some positive number c when k is sufficiently large. Thus we have the following theorem [49, 50].

Theorem 1. *Let M be a compact complex manifold and L be a Hermitian holomorphic line bundle over M whose curvature form is everywhere semipositive and is strictly positive at some point. Then M is a Moishezon manifold.*

By the result of Grauert-Riemenschneider, one has as a corollary the vanishing of $H^q(M,LK_M)$ for $q \geq 1$, where K_M is the canonical line bundle of M.

In conjunction with the characterization of Moishezon manifolds, I would like to mention the recent result of Peternell [34] that a 3-dimensional Moishezon manifold is projective algebraic if in it no positive integral linear combination of irreducible curves is homologous to zero. Together with Hironaka's example [18 and17, p.443] of a 3-dimensional non-projective-algebraic Moishezon manifold Peternell's result gives us the complete picture of the difference between projective-algebraic threefolds and Moishezon threefolds.

The noncompact analog of Theorem 1 is the following conjecture which is still open.

<u>Conjecture</u>. Let Ω be a relatively compact open subset of a complex manifold such that its boundary is weakly pseudoconvex at every point and is strictly pseudoconvex at some point P. Then there exists a holomorphic function on Ω going to infinity along some sequence in Ω approaching P.

Theorem 1 corresponds to the case where Ω is the set of vectors in the dual bundle of L with length < 1.

The method used in the proof of Theorem 1 can be further refined to yield results about the existence of holomorphic sections for line bundles whose curvature form is allowed to be negative somewhere [50]. An example of such results is the following.

<u>Theorem 2. For every positive integer</u> n <u>there exists a constant</u> C_n <u>depending only on</u> n <u>with the following property: Let</u> M <u>be a compact Kähler manifold of complex dimension</u> n <u>and</u> L <u>be a Hermitian line bundle over</u> M. <u>Let</u> G <u>be an open subset of</u> M <u>and</u> a, b <u>be positive numbers such that the curvature form of</u> L <u>admits</u> a <u>as a lower bound at every point of</u> G <u>and admits</u> -b <u>as a lower bound at every point of</u> M-G. <u>Assume that</u>

$$C_n \;(1 + \log^+(b/a))^n \;(b^2/a)^n(\text{volume of } M\text{-}G) \leq c_1(L)^n$$

<u>where</u> $c_1(L)$ <u>is the first Chern class of</u> L. <u>Then</u> $\dim H^0(M, L^k)$ <u>is</u> $\geq$ $c_1(L)^n \; k^n/2(n!)$ <u>for</u> k <u>sufficiently large.</u>

When the metric of the manifold is Hermitian instead of Kähler, there is a corresponding theorem with the constant C_n depending on the torsion of the Hermitian metric. The inequality in the assumption of Theorem 2 is not natural. There should be better and more natural formulations of this kind of results.

We describe below the refinement needed to get a proof of Theorem 2. The method described above can readily yield Theorem 2 if we allow the constant C_n to depend on M, but then Theorem 2 would be far less interesting. The reason why the above method can only yield a

C_n depending on M is that in constructing a correspondence from the space of harmonic forms to the space of cocycles, besides solving the $\bar{\partial}$ equations, one has to use a partition of unity and the constants obtained in the process depend very heavily on the manifold M. To solve this problem, we make use of the estimate of the (0,1)-covariant derivative of the harmonic form from the Bochner-Kodaira formula. We locally solve with estimates the inhomogeneous $\bar{\partial}$ equations with $\bar{\partial}$ of the coefficients of harmonic form on one side so that the differences between the coefficients of the harmonic form and the solutions are holomorphic and then apply the Schwarz lemma to the differences. This way we avoid passing from the Dolbeault cohomology to the Cech cohomology and can make the constant C_n independent of M.

II. Strong Rigidity of Seminegatively Curved Compact Kähler Manifolds

A compact Kähler manifold is said to be strongly rigid if any other Kähler manifold homotopic to it is biholomorphic or antibiholomorphic to it. Strong rigidity can be regarded as the complex analog of Mostow's strong rigidity [31]. Compact Kähler manifolds M with curvature tensor negative in a suitable sense are known to be strongly rigid [45, 46, 47]. The way to obtain the strong rigidity of M is to consider a harmonic map f to M from the compact Kähler manifold N homotopic to M which is a homotopy equivalence. The existence of such a harmonic map is guaranteed by the result of Eells-Sampson[10] because of the nonpositivity of the sectional curvature of M. As a section of the tensor product of the bundle of (0,1)-forms of N and the pullback under f of the (1,0)-tangent bundle of M, $\bar{\partial}f$ is harmonic. By using the technique of Bochner-Kodaira we conclude that either ∂f or $\bar{\partial}f$ vanishes because of the curvature condition of M. The reason why we can only conclude the vanishing of either ∂f or $\bar{\partial}f$ is that the curvature term from the Bochner-Kodaira formula is homogeneous of degree two in ∂f and of degree two in $\bar{\partial}f$ because it comes from pulling back of the curvature tensor of M under f. Actually the Bochner-Kodaira technique is applied to the image of $\bar{\partial}f$ under the complexified version of the Hodge star operator. In other words we are applying the Bochner-Kodaira technique to the dual of the bundle. That is the reason why the curvature tensor of M has to be assumed negative instead of positive and also that is the reason why the Ricci tensor of N does not enter the picture. The most general formulation of this kind of results on strong rigidity is the following theorem [47].

Theorem 3. A compact Kähler manifold M of complex dimension n is strongly rigid if there exists a positive number p less than n with the following properties: (i) The bundle of (1,0)-forms on M is positive semidefinite in the sense of Nakano [32] and the bundle of (p,0)-forms on M is positive definite in the sense of Nakano [32]. (ii) At any point of M the complex tangent space of M does not contain two nontrivial orthogonal subspaces with combined dimension exceeding p such that the bisectional curvature of M in the direction of two vectors one from each subspace vanishes.

As a corollary any compact quotient of an irreducible bounded symmetric domain of complex dimension at least twc is stronglg rigid, because we have the following table giving the complex dimension and the smallest p satisfying the assumptions of Theorem 3 for each bounded symmetric domain.

Type	Complex Dimension	Smallest p
$I_{m,n}$	mn	(m-1)(n-1)+1
II_n	n(n-1)/2	(n-2)(n-3)/2 +1
III_n	n(n+1)/2	n(n-1)/2 +1
IV_n	n	2
V	16	6
VI	27	11

The values of the smallest p for the two exceptional domains were computed by Zhong [58].

This method also yields the holomorphicity or antiholomrorphicity of any harmonic map from a compact Kähler manifold into M whose rank over $\mathbb{R}$ is $\geq 2p+1$ at some point [47].

This method can be regarded as an application of the quasilinear version of Kodaira's vanishing theorem. Though strict negativity of the curvature tensor is not needed for this method, this method should be considered as corresponding to the strictly definite case rather than the semidefinite case of the vanishing theorem, because through the use of the complexified Hodge star operator the vanishing required is in codimension one rather than in dimension one.

The only case of compact quotients of bounded symmetric domains which are expected to enjoy the property of strong rigidity as suggested by Mostow's result [31] and which are not covered by the results of [47] is the case of an irreducible compact quotient of a polydisc of complex dimension at least two. This remaining case corresponds to the semidefinite case of the vanishing theorem. Jost-Yau [21] first considered this remaining case and obtained some partial results. Recently Jost-Yau [22] and Mok [29] completely solved this case. We would like to sketch a slightly more streamlined version of the proof in [29]. First we make some general observations about the application of the Bochner-Kodaira technique to the case of a compact quotient of a polydisc and discuss a simple but rather surprising theorem about the existence of holomorphic maps from compact Kähler manifolds into compact hyperbolic Riemann surfaces.

Let f be a harmonic map from a compact Kähler manifold M to a compact quotient Q of a polydisc D^n of complex dimension n. The following conclusions are immediate from the Bochner-Kodaira technique.

(i) f is pluriharmonic in the sense that the restriction of f to any local complex curve in M is harmonic.

(ii) $\partial f^i \wedge \partial \overline{f^i}$ is zero for $1 \leq i \leq n$, where f^i is the ith component of f when it is expressed in terms of local coordinates along the n component discs.

From conclusion (ii) above it follows that the pullback $f^*T_Q^{1,0}$ under f of the (1,0)-tangent bundle $T_Q^{1,0}$ of Q can be endowed with the structure of a holomorphic vector bundle in the following way. A local section is defined to be holomorphic if its covariant derivative in the (0,1) direction is identically zero. This can be done because (ii) implies that the (0,1) covariant exterior differentiation composed with itself is identically zero, which is the integrability condition for such a holomorphic vector bundle structure. The same argument can be applied to the pullback $f^*T_Q^{0,1}$ under f of the (0,1)-tangent bundle $T_Q^{0,1}$ of Q to give it a holomorphic vector bundle structure. Moreover, if every element of the fundamental group of Q maps each individual component disc of D^n to itelf, then each of these two holomorphic vector bundles are the direct sum of the n holomorphic line bundles which are the pullbacks of the line subbundles of the tangent bundle of Q defined by the directions of the individual component discs. In such a case let L_i be the line subbundles of $f^*T_Q^{1,0}$ and L_i' be the line subbundles of $f^*T_Q^{0,1}$.

Because of conclusion (i) ∂f is a holomorphic section of $f^*T_Q^{1,0} \times \Omega_M^1$ and $\partial \bar{f}$ is a holomorphic section of $f^*T_Q^{0,1} \times \Omega_M^1$, where Ω_M^1 is the bundle of holomorphic 1-forms on M. Assume that every element of the fundamental group of Q maps each individual component disc of D^n to itself. Then for each fixed $1 \leq i \leq n$,

(iii) ∂f^i is a holomorphic section of $L_i \otimes \Omega_M^1$ and $\partial \overline{f^i}$ is a holomorphic section of $L_i' \otimes \Omega_M^1$.

For any local holomorphic section s_i of the dual bundle of L_i, $s_i \partial f^i$ is a (local) holomorphic 1-form on M whose exterior derivative equals its product with some 1-form. By the theorem of Frobenius, near points where ∂f^i does not vanish we have a holomorphic family of local complex submanifolds of complex codimension one whose tangent spaces annihilate ∂f^i. Such a holomorphic foliation of codimension one defined by the kernel of ∂f^i exists also at points where ∂f^i can be divided by a local holomorphic function to give a nowhere zero holomorphic local section of $L_i \otimes \Omega_M^1$. If in addition the rank of df^i is two over $\mathbb{R}$ at the points under consideration, because of (ii) the local leaves of the holomorphic foliation agree with the fibers of the locally defined map f^i. The same consideration can be applied to $\partial \overline{f^i}$. Also because of (ii) the holomorphic foliation defined by the kernel of ∂f^i agrees with the holomorphic foliation of $\partial \overline{f^i}$ at points where both ∂f^i and $\partial \overline{f^i}$ can be divided by local holomorphic functions to give nowhere zero holomorphic local sections of $L_i \otimes \Omega_M^1$ and $L_i' \otimes \Omega_M^1$ respectively. These rather straightforward discussions lead us immediately to the following theorem [48].

<u>Theorem 4. Let M be a compact Kähler manifold and R be a compact hyperbolic Riemann surface such that there exists a continuous map f from M to R which is nonzero on the second homology. Then there exists a holomorphic map g from M into a compact hyperbolic Riemann surface S and a harmonic map h from S to R such that h∘g is homotopic to f.</u>

The Riemann surface S is constructed from the holomorphic foliation described above in the following way. By the result of Eells-Sampson we can assume without loss of generality that f is harmonic and therefore real-analytic. Let Z be the complex subvariety of complex codimension ≥ 2 in M consisting of all points where either ∂f or $\partial \bar{f}$ cannot be divided by any local holomorphic function to give a nowhere zero holomorphic local section of the tensor product of Ω^1_M with the pullback under f of the (1,0) or (0,1) tangent bundle of R. Let V be the set of points of M where the rank of df over $\mathbb{R}$ is $<$ 2. On M-Z we have a holomorphic foliation described above with the property that whenever a leaf of the foliation has a point in common with M-V, the leaf agrees with the real-codimension-two branch of the fiber of f passing through that point and therefore can be extended to a complex-analytic subvariety of codimension one in M. Because of the Kähler metric of M, by using Bishop's theorem [4] on the limit of subvarieties of bounded volume and by passing to limit, we conclude that every leaf of the holomorphic foliation can be extended to a complex-analytic subvariety of codimension one in M. Since Z is of complex codimension $\geq$ 2 in M, by using the theorem of Remmert-Stein on extending subvarieties [37] we conclude that M is covered by the holomorphic family of subvarieties consisting of the extensions of the leaves of the holomorphic foliation. The Riemann surface S is now obtained as the nonsingular model of the quotient of M whose points are the branches of the extensions of the leaves of the holomorphic foliation.

The rather surprising aspect of Theorem 4 is that from the existence of a continuous map from a compact Kähler manifold to a compact hyperbolic Riemann surface nonzero on the second homology we can conclude the existence of a nontrivial holomorphic map from the Kähler manifold to a compact hyperbolic Riemann surface. In particular by going to the respective universal covers we obtain a nontrivial bounded holomorphic function on the univeral cover of the Kähler manifold. So far there is no known general method of constructing bounded holomorphic functions on complex manifolds which are expected to admit a large number of bounded holomorphic functions, such as the universal cover of compact Kähler manifolds of negative curvature. Here to conclude the existence of a nontrivial bounded holomorphic function we do not use any curvature property of the compact Kähler manifold. Instead the existence of a continuous map to a compact hyperbolic Riemann surface is used. Since until now there is no general way of constructing nontrivial bounded holomorphic functions, this could only mean that the existence of the kind of continuous map we want is rather rare and if such a continuous map exists, its existence would be rather difficult to establish. Even for negatively curved compact Kähler manifolds in general we do not expect such continuous maps to exist. As a matter of fact, for compact quotients of a ball of complex dimension at least two the only known examples so far that admit nontrivial holomorphic maps into any compact hyperbolic Riemann surface are the ones constructed by Livné [27] by taking branched covers of certain elliptic surfaces. It is not known whether in other dimensions there are similar examples of maps between compact quotients of balls of different dimensions besides the obvious ones.

<u>Problem.</u> Suppose $1 < m < n$ are integers. Let M and N be respectively compact quotients of the balls of complex dimensions m and n.

(a) Is it true that there exists no surjective holomorphic map from N to M?

(b) Is it true that every holomorphic embedding of M in N must have a totally geodesic image?

Yau conjectured that Problem (b) should be a consequence of uniquenss results for proper holomorphic maps between balls of different dimensions. For $n \geq 3$ Webster [56] showed that the only proper holomorphic maps from the n-ball to the (n+1)-ball C^3 up to the boundary are the obvious ones. Faran [11] showed that, up to automorphisms of the two balls, there are only four proper holomorphic maps from the 2-ball to the 3-ball C^3 up to the boundary . Unfortunately until now there are no general results about proper holomorphic maps between balls of different dimensions without any known boundary regularity. In our case the proper map, though without any known boundary regularity, has the additional property that it comes from maps between compact quotients. Hopefully this additional property may be used instead of boundary regularity.

We now introduce the theorem on the strong rigidity of irreducible compact quotients of polydiscs and sketch its proof.

Theorem 5 (Jost-Yau [22] and Mok [29]). *Suppose Q is an irreducible compact quotient of an n-disc D^n with $n \geq 2$, M is a compact Kähler manifold, and f is a harmonic map from M to Q which is a homotopy equivalence. Let $\tilde{M}$ be the universal cover of M and $F: \tilde{M} \to D^n$ with components $(F^1,\ldots,F^n)$ be induced by F. Then for each $1 \leq i \leq n$, F^i is either holomorphic or antiholomorphic.*

Here an irreducible quotient means one that cannot be decomposed as a product of two lower-dimensional quotients of polydiscs. For the proof of this theorem, by replacing both M and Q by finite covers, we can assume without loss of generality that the fundamental group of Q is a product of n groups $G_1,\ldots,G_n$, each of which is a (nondiscrete) subgroup of the automorphism group of the 1-dimensional disc D. We regard ∂F^i and $\partial \overline{F^i}$ as holomorphic sections of the holomorphic vector bundles on M described above (rather as (1,0)-forms on $\tilde{M}$). A consequence of the irreducibility of Q is that for each $1 \leq i \leq n$ every orbit of G_i is dense in D. We have to show that for every $1 \leq i \leq n$ either ∂F^i or $\partial \overline{F^i}$ vanishes identically on M. Without loss of generality we assume that the assertion fails for $i = 1$ and try to get a contradiction. Since f is a homotopy equivalence, the length of ∂F^1 and the length of $\partial \overline{F^1}$ cannot agree at every point. Without loss of generality we can assume that the length of ∂F^1 is greater than the length of $\partial \overline{F^1}$ at some point. By (ii) and (iii) we can write $\partial F^1 = g\ \partial \overline{F^1}$ so that locally g is the product of a nowhere zero smooth function and a meromorphic function. Thus the pole-set V of g is a complex-analytic hypersurface in M if it is nonempty. The pole-set V cannot be empty, otherwise by considering the Laplacian of the log of the absolute value h of g we get a contradiction at a maximum point of h. Since f is a homotopy equivalence, the real rank of f on the regular points of V must be precisely $2n - 2$, otherwise the homology class represented by V would be mapped to zero by f. Let $p:\tilde{M} \to M$ be the projection of the universal cover and $q:D^n \to D$ be the projection onto the first component. Because of the holomorphic foliation discussed above the function $h \circ p$ on $\tilde{M}$ must be constant along the components of the fibers of F^1: $M \to D$. The proper closed subset $F(p^{-1}(V))$ of D^n contains an entire fiber of q whenever it contains one of its points. It follows that $q(F(p^{-1}(V)))$ is a proper closed subset of D which is invariant under the group G_1. This contradicts the density of every orbit of G_1 in D.

Mok [29] also showed that for any harmonic map from a compact Kähler manifold to an irreducible compact quotient of the n-disc ($n \geq 2$) with real rank 2n somewhere, each of the n components of the map between the universal covers induced by it is either holomorphic or antiholomorphic.

III. Vanishing Theorems Obtained by Subelliptic Estimates

So far all the vanishing theorems for bundles with curvature conditons make use of the pointwise property of the curvature form. The recent theory of subelliptic multipliers developed by Kohn, Catlin, and others [26, 5, 6] makes it possible to get vanishing thoerems based on the local property of the curvature form when the curvature form is semidefinite. Kohn developed his theory to deal with the question of boundary regularity for solutions of the $\bar{\partial}$ equation in the case of a weakly pseudoconvex boundary.

Let Ω be an open subset of $\mathbb{C}^n$ whose boundary is smooth and weakly pseudoconvex at a boundary point x_0. Let $1 \leq q \leq n$ be an integer. We say that a subelliptic estimate holds for (0,q)-forms at x_0 if there exists a neighborhood U of x_0 and constants $\varepsilon > 0$ and $C > 0$ such that

$$\|\phi\|_\varepsilon^2 \leq C(\|\bar{\partial}\phi\|^2 + \|\bar{\partial}^*\phi\|^2 + \|\phi\|^2)$$

for all smooth (0,q)-form ϕ on $U \cap \bar{\Omega}$ with compact support belonging to the domain of $\bar{\partial}^*$, where $\| \ \|$ means the L^2 norm and $\| \ \|_\varepsilon$ means the Sobolev ε-norm. In order to obtain subelliptic estimates Kohn introduced the concept of a subelliptic multiplier. A smooth function f on U is said to be a subelliptic multiplier if there exist positive ε

and C so that

$$\|f\phi\|_{\varepsilon}^{2} \leq C(\|\bar{\partial}\phi\|^{2} + \|\bar{\partial}^{*}\phi\|^{2} + \|\phi\|^{2})$$

for all ϕ. The subelliptic multipliers form an ideal I_q. Let c_{ij} $(1 \leq i,j \leq n-1)$ be the Levi form of the boundary of Ω near x in terms of an orthonormal frame field of (1,0) vectors tangential to the boundary of Ω. The starting point of Kohn's theory is the following results concerning the ideal I_q of subelliptic multipliers. For notational simplicity we describe the case q = 1 and the general case is similar.

(i) A smooth function r with nonzero gradient whose zero-set is the boundary of Ω belongs to I_1.

(ii) The determinant of the matrix $(c_{ij})_{1 \leq i,j \leq n-1}$ belongs to I_1.

(iii) Whenever $f_1,\dots,f_k$ belong to I_1, the determinant formed in the following way belongs to I_1. The ith column consists of the components of ∂f_i in terms of the frame field of (1,0) vectors tangential to the boundary of Ω. The other n-1-k columns are any n-1-k columns of the matrix $(c_{ij})_{1 \leq i,j \leq n-1}$.

(iv) I_1 equals to its real radical in the sense that if f belongs to I_1 and g is a smooth function with $|g|^m \leq |f|$ for some positive integer m, then g also belongs to I_1.

Kohn [26] showed that if the boundary of Ω is real-analytic near x_0 and contains no local complex-analytic subvariety of complex dimension q, then the constant function 1 belongs to the ideal I_q of subelliptic multipliers and as a consequence a subelliptic estimate for (0,q)-forms holds at x_0. (Diederich-Fornaess [9] contributed to the formulation of the assumptions in Kohn's result.) Recently Catlin [5,6] carried out the investigation for the case of smooth boundary and showed that a subelliptic estimate for (0,1)-forms holds at x_0 if and only if the boundary of Ω is of finite type at x_0 in the sense of D'Angelo [7,8]. (Similar statements hold for subelliptic estimates for (0,q)-forms.) D'Angelo's definition of finite type is as follows. The boundary of Ω is of type $\leq$ t at x_0 if for every holomorphic map $h = (h_1,\dots,h_n)$ from the open 1-dimensional disc D to $\mathbb{C}^n$ with h(0) =

x_0 the vanishing order of $r \cdot h$ at 0 does not exceed t times the minimum of the vanishing orders of $h_1, \ldots, h_n$ at 0. At every point x of the boundary of Ω let t(x) be the smallest number such that the boundary of Ω at x is of type $\leq t(x)$. D'Angelo showed that t(x) in general is not upper semicontinuous, but satisfies $t(x) \leq t(x_0)^{n-1}/2^{n-2}$ for x near x_0. The order ε in the subelliptic estimate at x_0 is expected to be the reciprocal of the maximum of t(x) for x near x_0. Catlin's result showed that ε cannot be bigger than the expected number but he can so far only show that subelliptic estimates hold for an ε of the order of $t(x_0)$ raised to the power $-t(x_0)^{n^2}$.

We now study how the subelliptic estimates can be used to get vanishing theorems. We follow Grauert's approach to vanishing theorems [14]. A number of vanishing theorems can be formulated from the method of subelliptic estimates. Some of them can readily be derived by other means. We illustrate here by an example of such vanishing theorems. Let M be a compact complex manifold and L be a Hermitian holomorphic line bundle over M whose curvature form is semipositive. Let V be a holomorphic vector bundle over M. Let $p: L^* \to M$ be the dual bundle of L. Let Ω be the open subset of L^* consisting of all vectors of L^* of length < 1. If subelliptic estimates for (0,q)-forms hold for the boundary of Ω at every one of its points, then one concludes that $H^q(\Omega, p^*V)$ is finite-dimensional by representing the cohomology by harmonic forms. It follows that $H^q(M, V \otimes L^k)$ vanishes for k sufficiently large, because the kth coefficient in the power series expansion in the fiber coordinate of L^* of a local holomorphic function defined near a point in the zero-section of L^* is a local section of L^k.

When the Hermitian metric of L is real-analytic, by Kohn's result subelliptic estimates for (0,q)-forms hold if the boundary of Ω contains no local q-dimensional complex-analytic subvariety. If there is such a subvariety, its projection under p is a local q-dimensional subvariety W with the property that with respect to some local trivialization of L the Hermitian metric of L is represented by a function which is constant on W. If we give M a Hermitian metric, then all covariant derivatives of the curvature form of L along the directions of W must vanish. Hence we have the following theorem.

Theorem 6. Let M be a compact complex manifold with a Hermitian metric, L a holomorphic line bundle over M with a real-analytic Hermitian metric, and V a holomorphic vector bundle over M. Let θ be the curvature form of L. Let q be a positive integer. Suppose θ is positive semidefinite and suppose at every point x of M the following is true. If E is a q-dimensional complex linear subspace of the space all (1,0)-vectors at x such that the restriction of θ to $E \times \overline{E}$ is zero (where $\overline{E}$ is the complex conjugate of E), then for some positive integer m the mth covariant derivative of θ evaluated at some $m+2$ vectors from E and $\overline{E}$ is not zero. Then $H^q(M, V \otimes L^k) = 0$ for k sufficiently large.

By using Catlin's result [6] for weakly pseudoconvex smooth boundary, one can drop the real-analytic assumption on the Hermitian metric of L. This kind of result tells us that in the case of a semipositive line bundle we can still get vanishing of the cohomology if the derivatives of the curvature form satisfy certain nondegeneracy conditions. Similar theorems can be formulated for holomorphic vector bundles and noncompact pseudoconvex manifolds. When $q = 1$, Theorem 5 can be proved by using the method of producing holomorphic sections for semipositive line bundles described above and Grauert's criterion of ampleness [14, p.347, Lemma] to show that the line bundle L must be ample. Though for lack of known examples there is no application yet for the kind of vanishing theorems derived from subelliptic estimates, hopefully in the future this approach may turn out to be fruitful.

We would like to remark that on compact projective algebraic manifolds there is another kind of vanishing theorems motivated by Seshadri's criterion of ampleness [42, p.549] and obtained by Ramanujam [36], Kawamata [23], and Viehweg [55] for line bundles satisfying conditions weaker than ampleness. An example of such a kind of vanishing theorems is the following. If L is a holomorphic line bundle over a compact projective algebraic manifold M of complex dimension n such that $c_1(L)^n > 0$ and $c_1(L|C) \geq 0$ for every complex-analytic curve C in M, then $H^q(M, L^k K_M) = 0$ for $q \geq 1$, where $c_1(\cdot)$ denotes the first Chern class and K_M denotes the canonical line bundle of M. The assumptions involved are weaker than local curvature conditions. However, such results apply only to the projective algebraic case.

References

1. A. Andreotti & E. Vesentini, Carleman estimates for the Laplace-Beltrami equation on complex manifolds, Inst. Hautes Études Sci. Publ. Math. 25 (1965), 91-130.

2. N. Aronszajn, A unique continuation theorem for solutions of elliptic partial differential equations or inequalities of second order, J. Math. Pures Appl. 36 (1957) 235-249.

3. M. F. Atiyah & I. M. Singer, The index of elliptic operators III, Ann. of Math. 87 (1968) 546-604.

4. E. Bishop, Conditions for the analyticity of certain sets, Michigan Math. J. 11 (1964) 289-304.

5. D. Catlin, Necessary conditions for subellipticity of the $\bar{\partial}$-Neumann problem, Ann. of Math. 117 (1983), 147-171.

6. D. Catlin, in preparation.

7. J. D'Angelo, Subelliptic estimates and failure of semicontinuity for orders of contact, Duke Math. J. 47 (1980), 955-957.

8. J. D'Angelo, Intersection theory and the $\bar{\partial}$-Neumann problem, Proc. Symp. Pure Math. 41 (1984), 51-58

9. K. Diederich & J. E. Fornaess, Pseudoconvex domains with real-analytic boundary, Ann. of Math. 107 (1978), 371-384.

10. J. Eells & J. H. Sampson, Harmonic mappings of Riemannian manifolds, Amer. J. Math. 86 (1964), 109-160.

11. J. Faran, Maps from the two-ball to the three-ball, Invent. Math. 68 (1982), 441-475

12. R. Frankel, A differential geometric criterion for Moišezon spaces, Math. Ann. 241 (1979), 107-112; Erratum, Math. Ann. 252 (1980), 259-260.

13. H. Grauert, On Levi's problem and the imbedding of real-analytic manifolds, Ann. of Math. 68 (1958), 460-472.

14. H. Grauert, Über Modifikation und exzeptionelle analytische Mengen, Math. Ann. 146 (1962), 331-368.

15. H. Grauert & O. Riemenschneider, Verschwindungssätze für analytische Kohomologiegruppen auf komplexen Räumen, Invent. Math. 11 (1970) 263-292.

16. P. Griffiths, The extension problem in complex analysis II: embeddings with positive normal bundle, Amer. J. Math. 88 (1966), 366-446.

17. R. Hartshorne, Algebraic Geometry, New York: Springer-Verlag 1977.

18. H. Hironaka, On the theory of birational blowing-up, Ph.D. thesis, Harvard (1960).

19. F. Hirzebruch, Topological methods in algebraic geometry. New York: Springer-Verlag 1966.

20. L. Hörmander, L^2 estimates and existence theorems of the $\bar{\partial}$ operator, Acta Math. 113 (1965), 89-152.

21. J. Jost & S.-T. Yau, Harmonic mappings and Kähler manifolds, Math. Ann. 262 (1983), 145-166.

22. J. Jost & S.-T. Yau, Strong rigidity theorem for a certain class of compact complex surfaces.

23. Y. Kawamata, A generalization of Kodaira-Ramanujam's vanishing theorem, Math. Ann. 261 (1982), 43-46.

24. K. Kodaira, On Kähler varieties of restricted type (an intrinsic characterization of algebraic varieties), Ann. of Math. 60 (1954), 28-48.

25. J. J. Kohn, Harmonic integrals on strongly pseudoconvex manifolds, I, II, Ann. of Math. 78 (1963), 112-148; 79 (1964) 450-472.

26. J. J. Kohn, Subellipticity of the $\bar{\partial}$-Neumann problem on weakly pseudoconvex domains: sufficient conditions, Acta Math. 142 (1979), 79-122.

27. R. A. Livné, On certain covers of the universal elliptic curve, Ph.D. thesis, Harvard University 1981.

28. B. Moishezon, On n-dimensional compact varieties with n algebraically independent meromorphic functions, Amer. Math. So, Translations 63 (1967), 51-177.

29. N. Mok, The holomorphic or antiholomorphic character of harmonic maps into irreducible compact quotient of polydiscs.

30. C. B. Morrey, The analytic embedding of abstract real analytic manifolds, Ann. of Math. 68 (1958), 159-201.

31. G. D. Mostow, Strong rigidity of locally symmetric spaces, Ann. of Math. Studies 78, Princeton: Princeton University Press 1973.

32. S. Nakano, On complex analytic vector bundles, J. Math. Soc. Japan 7 (1955), 1-12

33. T. Peternell, Fast-positive Geradenbundel auf kompakten komplexen Manifaltigkeiten.

34. T. Peternell, Algebraic criteria for compact complex manifolds.

35. J. H. Rabinowitz, Positivity notions for coherent sheaves over compact complex spaces, Invent. Math. 62 (1980), 79-87; Erratum, Invent. Math. 63 (1981), 355.

36. C. P. Ramanujam, Remarks on the Kodaira vanishing theorem, Ind. J. of Math. 36 (1972), 41-51; Supplement, Ind. J. of Math. 38 (1974), 121-124.

37. R. Remmert & K. Stein, Über die wesentlichen Singularitäten analytischer Menger, Math. Ann. 126 (1953), 263-306.

38. O. Rimenschneider, Characterizing Moišezon spaces by almost positive coherent analytic sheaves, Math. Z. 123 (1971), 263-284.

39. O. Riemenschneider, A generalization of Kodaira's embedding theorem, Math. Ann. 200 (1973), 99-102.

40. J. Sachs & K. Uhlenbeck, The existence of minimal immersion of 2-spheres, Ann. of Math.113 (1981), 1-24.

41. J.-P. Serre, Fonctions automorphes: quelques majorations dans le cas ou X/G est compact, Séminaire Cartan 1953-54, 2-1 to 2-9.

42. C. S. Seshadri, Quotient spaces modulo reductive algebraic groups, Ann. of Math. 95 (1972), 511-556.

43. C. L. Siegel, Meromorphe Funktionen auf kompakten Mannifaltigkeiten. Nachrichten der Akademie der Wissenschaften in Göttingen, Math.-Phys. Klasse 1955, No.4, 71-77.

44. Y.-T. Siu, The Levi problem, Proc. Symp. Pure Math. 30 (1977), 45-48.

45. Y.-T. Siu, The complex-analyticity of harmonic maps and the strong rigidity of compact Kähler manifolds, Ann. of Math. 112 (1980), 72-111.

46. Y.-T. Siu, Strong rigidity of compact quotients of exceptional bounded symmetric domains, Duke Math. J. 48 (1981), 857-871.

47. Y.-T. Siu, The complex-analyticity of harmonic maps, vanishing and Lefschetz theorems, J. Diff. Geom. 17 (1982) 55-13.

48. Y.-T. Siu, Lecture at the conference in honor of the 60th birthday of G. D. Mostow.

49. Y.-T. Siu, A vanishing theorem for semipositive line bundles over non-Kähler manifolds. J. Diff. Geom. 1984.

50. Y.-T. Siu, A method of producing holomorphic sections for line bundles with curvature of mixed sign. In preparation.

51. Y.-T. Siu, Curvature characterization of hyperquadrics, Duke Math. J. 47 (1980), 641-654.

52. Y.-T. Siu & S.-T. Yau, Compact Kähler manifolds of positive bisectional curvature, Invent. Math. 59 (1980), 189-204.

53. A. J. Sommese, Criteria for quasi-projectivity, Math. Ann. 217 (1975), 247-256.

54. W. Thimm, Über algebraische Relation zwischen meromorphen Funcktionen abgeschlossenen Räumen. Thesis, Königsberg, 1939, 44 pp.

55. E. Viehweg, Vanishing theorems, J. reine angew. Math. 335 (1982), 1-8.

56. S. Webster, On mapping an n-ball into an (n+1)-ball in complex space, Pacific J. Math. 81 (1979), 267-272.

57. R. O. Wells, Moishezon spaces and the Kodaira embedding theorem, <u>Value Distribution Theory</u>, Part A, Marcel Dekker, N.Y. 1974, pp.29-42.

58. J.-q. Zhong, The degree of strong nondegeneracy of the bisectional curvature of exceptional bounded symmetric domains, Proc. 1981 Hangzhou Conference on Several Complex Variables, Birkhauser 1984, pp. 127-139.

GROUPS AND GROUP FUNCTORS ATTACHED TO KAC-MOODY DATA

Jacques Tits
Collège de France
11 Pl Marcelin Berthelot
75231 Paris Cedex 05

1. The finite-dimensional complex semi-simple Lie algebras.

To start with, let us recall the classification, due to W. Killing and E. Cartan, of all complex semi-simple Lie algebras. (The presentation we adopt, for later purpose, is of course not that of those authors.) The isomorphism classes of such algebras are in one-to-one correspondence with the systems

$$(1.1) \qquad \mathfrak{H}\ ,\ (\alpha_i)_{1\leq i\leq \ell}\ ,\ (h_i)_{1\leq i\leq \ell}\ ,$$

where $\mathfrak{H}$ is a finite-dimensional complex vector space (a Cartan subalgebra of a representative $\mathfrak{G}$ of the isomorphism class in question), $(\alpha_i)_{1\leq i\leq \ell}$ is a basis of the dual $\mathfrak{H}^*$ of $\mathfrak{H}$ (a basis of the root system of $\mathfrak{G}$ relative to $\mathfrak{H}$) and $(h_i)_{1\leq i\leq \ell}$ is a basis of $\mathfrak{H}$ indexed by the same set $\{1,\ldots,\ell\}$ (h_i is the coroot associated with α_i), such that the matrix $\underline{\underline{A}} = (A_{ij}) = (\alpha_j(h_i))$ is a Cartan matrix, which means that the following conditions are satisfied:

(C1) the A_{ij} are integers ;

(C2) $A_{ij} = 2$ or ≤ 0 according as $i =$ or $\neq j$;

(C3) $A_{ij} = 0$ if and only if $A_{ji} = 0$;

(C4) $\underline{\underline{A}}$ is the product of a positive definite symmetric matrix and a diagonal matrix (by abuse of language, we shall simply say that $\underline{\underline{A}}$ is positive definite).

More correctly: two such data correspond to the same isomorphism class of algebras if and only if they differ only by a permutation of the indices $1,\ldots,\ell$. Following C. Chevalley, Harish-Chandra and J.-P. Serre, one can give a simple presentation of the algebra corresponding to the system (1.1): it is generated by $\mathcal{H}$ and a set of 2ℓ elements $e_1,\ldots,e_\ell$, $f_1,\ldots,f_\ell$ subject to the following relations (besides the vector space structure of $\mathcal{H}$):

$$\begin{cases} [\mathcal{H},\mathcal{H}] = \{0\} ; \\ [h,e_i] = \alpha_i(h).e_i \qquad (h \in \mathcal{H}) ; \\ [h,f_i] = -\alpha_i(h).f_i \qquad (h \in \mathcal{H}) ; \\ [e_i,f_i] = -h_i ; \\ [e_i,f_j] = 0 \quad \text{if } i \neq j ; \\ (\mathrm{ad}\, e_i)^{-A_{ij}+1}(e_j) = 0 \quad \text{if } i \neq j ; \\ (\mathrm{ad}\, f_j)^{-A_{ij}+1}(f_j) = 0 \quad \text{if } i \neq j . \end{cases}$$

If one does no longer assume that the α_i and the h_i generate $\mathcal{H}^*$ and $\mathcal{H}$ respectively, one obtains in that same way all complex <u>reductive</u> Lie algebras.At this point, the generalization is rather harmless (reductive = semi-simple × commutative), but it becomes more significant at the group level and will turn out to be quite essential in the Kac-Moody situation.

2. <u>Reductive algebraic groups and Chevalley schemes</u>.

It is well known that a complex Lie algebra determines a Lie group only up to local isomorphism. Thus, in order to characterize a reductive algebraic group, over $\underline{\mathbb{C}}$, say, an extra-information, besides the data (1.1), is needed. It is provided by a lattice Λ in $\mathcal{H}$ (i.e. a $\underline{\mathbb{Z}}$ -submodule of $\mathcal{H}$ generated by a basis of $\mathcal{H}$) such that $h_i \in \Lambda$ and $\alpha_i \in \Lambda^*$ (the $\underline{\mathbb{Z}}$ -dual of Λ), namely the lattice of rational co-characters of a maximal torus of the group one considers. To summarize: the isomorphism classes of complex reductive groups are in one-to-one correspondence (again up to permutation of the indices) with the systems

(2.1) $S = (\Lambda, (\alpha_i)_{1\leq i\leq \ell}, (h_i)_{1\leq i\leq \ell})$,

where Λ is a finitely generated free $\underline{\underline{Z}}$ -module, $\alpha_i \in \Lambda^*$, $h_i \in \Lambda$ and $\underline{\underline{A}} = (\alpha_j(h_i))$ is a Cartan matrix.

A remarkable result of C. Chevalley [Ch2] is that the same classification holds when one replaces $\underline{\underline{C}}$ by any algebraically closed field. Furthermore, to any system (2.1), Chevalley [Ch3] and Demazure [De2] associate a group-scheme over $\underline{\underline{Z}}$, hence, in particular, a group functor G_S on the category of rings. Thus, the main result of [Ch2] asserts that the reductive algebraic groups over an algebraically closed field K are precisely the groups $G_S(K)$, where S runs over the systems (2.1) described above.

Question: what happens if, in the above considerations, one drops Condition (C4) (in which case, the matrix $\underline{\underline{A}}$ is called a generalized Cartan matrix, or GCM)? This is what the Kac-Moody theory is about.

3. Kac-Moody Lie algebras.

From now on, when talking about a system (1.1), we only assume that $\alpha_i \in H^*$, $h_i \in H$ (the α_i and h_i need not generate H^* and H) and that $\underline{\underline{A}} = (\alpha_j(h_i))$ is a GCM. To such a system, the presentation (1.2) associates a Lie algebra which is infinite-dimensional whenever $\underline{\underline{A}}$ is not a Cartan matrix. The Lie algebras one obtains that way are called Kac-Moody algebras. A large part of the classical theory - root systems, linear representations etc. - extends to them, with a bonus: the study of root multiplicities (roots do have multiplicities in the general case) and of character formulas for linear representations with highest weights have a number-theoretic flavour which is not apparent in the finite-dimensional situation. For those questions, which are outside the subject of the present survey, see [Ka3] and its bibliography.

In general, Kac-Moody algebras are entirely new objects, but there is a case, besides the positive definite one, where they are still closely related to finite-dimensional simple Lie algebras, namely the "semi-definite" case: by the same abuse of language as above, we say that the matrix $\underline{\underline{A}}$ is semi-definite if it is the product of a semi-definite symmetric matrix and a diagonal matrix.

The simplest example of Kac-Moody algebras of semi-definite type is provided by the so-called loop algebras. Let $\mathfrak{G}$ be a complex semi-simple Lie algebra, $\mathfrak{H}$ a Cartan subalgebra of $\mathfrak{G}$, $(\alpha_i)_{1\le i\le \ell}$ a basis of the root system of $\mathfrak{G}$ relative to $\mathfrak{H}$, α_0 the opposite of the dominant root and h_j, for $0\le j\le \ell$, the coroot corresponding to α_j. Then, the system

$$\mathfrak{H}\ ,\ (\alpha_j)_{0\le j\le\ell}\ ,\ (h_j)_{0\le j\le\ell}$$

satisfies our conditions and the corresponding Kac-Moody algebra turns out to be the "loop algebra" $\mathfrak{G}\otimes\underline{\mathbb{C}}[z,z^{-1}]$. In this case, the GCM $\underline{\tilde{A}} = (\alpha_k(h_j))_{1\le j,k\le\ell}$ is described by the well-known extended Dynkin diagram ("graphe de Dynkin complété" in [Bo]) of $\mathfrak{G}$; we shall call it the extended Cartan matrix of $\mathfrak{G}$.

Let us modify the previous example slightly: instead of $\mathfrak{H}$, we take a direct sum $\tilde{\mathfrak{H}} = \coprod_{0\le j\le\ell} \underline{\mathbb{C}}.\tilde{h}_j$, where the $\tilde{h}_j$'s are "copies" of the h_j's, and we choose the elements $\tilde{\alpha}_j$ of $\tilde{\mathfrak{H}}^*$ in such a way that the matrix $(\tilde{\alpha}_k(\tilde{h}_j))$ be the same $\underline{\tilde{A}}$ as before. Then, $\tilde{\mathfrak{H}}$ is the extension of $\mathfrak{H}$ by a one-dimensional subspace $\mathfrak{r} = \underline{\mathbb{C}}.(\Sigma d_j\tilde{h}_j)$ (where the d_j's are nonzero coefficients such that $\Sigma d_j h_j = 0$), and it is readily seen that the Kac-Moody algebra defined by the system $(\tilde{\mathfrak{H}},(\tilde{\alpha}_j),(\tilde{h}_j))$ is a perfect central extension of $\mathfrak{G}\otimes\underline{\mathbb{C}}[z,z^{-1}]$ by the one-dimensional algebra $\mathfrak{r}$. In fact, it is the universal central extension of $\mathfrak{G}\otimes\underline{\mathbb{C}}[z,z^{-1}]$: this is a special case of the following, rather easy proposition, proved independently by Kac ([Ka3], exercise 3.14), Moody (unpublished) and the author ([Ti4]):

PROPOSITION 1. - *If the h_i's form a basis of $\mathfrak{H}$, the Kac-Moody algebra defined by* (1.2) *(for $(\alpha_j(h_i))$ an arbitrary GCM) has no nontrivial central extension.*

The existence of a nontrivial central extension of $\mathfrak{G}\otimes\underline{\mathbb{C}}[z,z^{-1}]$ by $\underline{\mathbb{C}}$ plays an important role in the applications of the Kac-Moody theory for instance to physics and to the theory of differential equations (cf. e. g. [Ve1], [SW] and the literature cited in those papers). It is worth noting that the Kac-Moody presentation provides a natural approach to that extension and a very simple proof of its universal property, which is much less evident when one uses direct (e. g. cohomological) methods (cf. [Ga], [Wi]). (NB. In the literature, the expression "Kac-Moody algebras" is frequently used to designate merely the loop algebras and/or their universal central extension; this unduly restrictive usage explains itself by the importance of those special cases for the applications.)

Here, a GCM will be called "<u>of affine type</u>" if it is semi-definite, nondefinite and indecomposable; we say that it is of <u>standard</u> (resp. <u>twisted</u>) affine type if it is (resp. is not) the extended Cartan matrix of a finite-dimensional simple Lie algebra. (In the literature, one often finds the words "affine" and "euclidean" to mean "standard affine" and "twisted affine" in our terminology.) In rank 2, there are two GCM of affine type, one standard $\begin{pmatrix} 2 & -2 \\ -2 & 2 \end{pmatrix}$ and one twisted $\begin{pmatrix} 2 & -1 \\ -4 & 2 \end{pmatrix}$ (up to permutation of the indices). When the rank is ≥ 3, the coefficients of a GCM (A_{ij}) of affine type always satisfy the relation $A_{ij}A_{ji} \leq 3$ (for $i \neq j$), so that the matrix can be represented by a Dynkin diagram in the usual way (cf. e. g. [BT3], 1.4.4 or [Bo], p. 195); then, it turns out that the diagrams representing the twisted types are obtained by reversing arrows in the diagrams representing standard types (i.e. in extended Dynkin diagrams of finite-dimensional simple Lie algebras). For instance,

$(\tilde{F}_4)$

is standard, whereas

$({}^2\tilde{E}_6)$

is twisted.

The most general Kac-Moody algebra of standard affine type is a semi-direct product of an abelian algebra by a central extension of a loop algebra. There is a similar description for the algebras of twisted affine type, in which the loop algebras must be replaced by suitable twisted forms. For instance, if $\mathfrak{G}$ is a complex Lie algebra of type E_6 and if σ denotes an involutory automorphism of the loop algebra $\mathfrak{G} \otimes \underline{\mathbb{C}}[z,z^{-1}]$ operating on the first factor by an outer automorphism and on the second by $z \longmapsto -z$, then the fixed point algebra $(\mathfrak{G} \otimes \underline{\mathbb{C}}[z,z^{-1}])^{\sigma}$ is a Kac-Moody algebra of type ${}^{2}\tilde{E}_6$ above (hence the notation !). The connection between Kac-Moody algebras of affine type and the loop algebras and their twisted analogues was first made explicit in [Ka2], but the corresponding relation at the group level had been known for some time: cf. [IM] and [BT2] (where, however, a local field - such as $\underline{\mathbb{C}}((z))$ - replaces $\underline{\mathbb{C}}[z,z^{-1}]$).

4. Associated groups: introductory remarks.

In the classical, finite-dimensional theory, the Lie algebras often appear as intermediate step in the study of Lie groups. It is therefore natural to try similarly to "integrate" Kac-Moody Lie algebras and to define "Kac-Moody groups". More precisely, let S be as in (2.1) except that, now, the matrix $(\alpha_j(h_i))$ is only assumed to be a GCM. To such a system S, one wishes to associate an "infinite-dimensional group over $\underline{\mathbb{C}}$", let us call it $G_S(\underline{\mathbb{C}})$, or, more ambitiously, a group functor G_S on the category of rings.

Before passing in quick review the methods that have been used to define such groups, let us make a preliminary comment. As may be expected, since the groups in question are "infinite-dimensional", one is led, for a given S, to define not one but several groups which are various completions of a smallest one (those completions corresponding usually to various completions of the Kac-Moody Lie algebra). Thus, the group theory can be developed at different levels (or, if one prefers, in different categories); roughly speaking, one may distinguish a minimal (or purely algebraic) level, a formal level and an analytic level, with many subdivisons.

Instead of trying to define those terms formally, I shall just illustrate them with one example. Let $\mathfrak{G}$ be a complex, quasi-simple simply connected algebraic group (Lie algebras will now play a minor role, and we are free to use gothic letters for other purposes !), Λ^* the lattice of rational characters of a maximal torus of $\mathfrak{G}$, Λ its $\underline{\mathbb{Z}}$-dual, $(\alpha_i)_{1\leq i\leq \ell}$ a basis of the root system of $\mathfrak{G}$ with respect to the torus in question, α_0 the opposite of the dominant root, h_j (for $0 \leq j \leq \ell$) the coroot corresponding to α_j and $S = (\Lambda, (\alpha_j)_{0\leq j\leq \ell}, (h_j)_{0\leq j\leq \ell})$. In § 3, we have seen that the Lie algebra associated with S (in which Λ is replaced by $\underline{\mathbb{C}} \otimes \Lambda$) is Lie $\mathfrak{G} \otimes \underline{\mathbb{C}}[z,z^{-1}]$. Clearly, the group most naturally associated with S over $\underline{\mathbb{C}}$ must - and will - be the group $\mathfrak{G}(\underline{\mathbb{C}}[z,z^{-1}])$ of all "polynomial maps" $\underline{\mathbb{C}}^{\times} \longrightarrow \mathfrak{G}$. In that special case, this is the answer to our question at the minimal level. At the formal level, we find $\mathfrak{G}(\underline{\mathbb{C}}((z)))$. Now, the points of $\mathfrak{G}(\underline{\mathbb{C}}[z,z^{-1}])$ can also be viewed as certain special loops $S^1 \longrightarrow \mathfrak{G}$ (by restricting $\underline{\mathbb{C}}^{\times} \longrightarrow \mathfrak{G}$ to the complex numbers of absolute value one) and this opens the way to a great variety of completions of $\mathfrak{G}(\underline{\mathbb{C}}(z,z^{-1}))$, leading to groups of loops $S^1 \longrightarrow \mathfrak{G}$ in various categories (L^2 , continuous, C^{∞} , etc.): this is the analytic level.

In the case of the above system S , there is no difficulty in guessing what should be the group functor G_S : at the minimal level, we shall have $G_S(R) = \mathfrak{G}(R[z,z^{-1}])$, where $\mathfrak{G}$ now denotes the Chevalley scheme corresponding to the system $(\Lambda, (\alpha_i)_{1\leq i\leq \ell}, (h_i)_{1\leq i\leq \ell})$, and the corresponding formal group will be $\mathfrak{G}(R((z)))$. (In this generality, I do not know what "analytic" should mean.) As one sees, all those groups can be described with elementary means, without reference to Kac-Moody algebras. But things change as soon as one slightly modifies the system S as in § 3 by taking for instance $\Lambda = \coprod_{j=0}^{\ell} \mathbb{Z}\tilde{h}_j$ (and keeping the GCM unchanged, as before). The corresponding group is then a central extension of the loop group (whichever category one is in) by $\underline{\mathbb{C}}^{\times}$ or, in the ring situation, by $R^{\times}$. As in the Lie algebra case, the existence of that extension comes out of the general theory quite formally, but in the loop group case, it reflects rather deep properties of those groups (cf. e. g. [SW]), and direct existence proofs are not easy. Note that if R is a finite field k , one gets a central extension of $\mathfrak{G}(k((z)))$ by $k^{\times}$ which appears in the work of C. Moore [Mo2] and H. Matsumoto [Ma3].

Here, we shall most of the time adopt either the minimal or the formal viewpoint (the analytic ones are usually deeper and more important for the applications, but unfortunately less familiar to the speaker). Let us briefly mention some contrasting features of those. The formal groups are usually simpler to handle (as are local fields compared to global ones !). This is due in particular to the fact that they contain "large" proalgebraic subgroups (cf. e. g. [BT2], § 5, and [Sℓ], Kap. 5). Also, they seem to be the right category for simplicity theorems (cf. [Mo1]; observe that if $\mathfrak{G}$ denotes a complex simple Lie group, then $\mathfrak{G}(\underline{\mathbb{C}}((z)))$ is a simple group, which is far from true for $\mathfrak{G}(\underline{\mathbb{C}}[z,z^{-1}])$. On the other hand, the minimal theory presents a certain symmetry (the symmetry between the e_i's and f_i's or, in the example of $\mathfrak{G}(\underline{\mathbb{C}}[z,z^{-1}])$, the symmetry between z and z^{-1}), which gets lost in the formal completion.

Let us mention an important aspect of that symmetry. All the groups $G = G_S(\underline{\mathbb{C}})$ we are talking about (and, in fact, the groups $G_S(K)$, for K a field), whether minimal or formal, are equipped with a BN-pair (B,N) (or Tits system: cf. [Bo]) whose Weyl group $W = N/B \cap N$ is the Coxeter group $W(\underline{A})$ defined as follows:

$$W(A) = \langle r_i \mid 1 \leq i \leq \ell \ ; \ r_i^2 = 1 \ ; \ (r_i r_j)^{c_{ij}} = 1 \ \text{if} \ i \neq j \ , \ A_{ij}A_{ji} \leq 3 \ , \ \text{and} \ c_{ij} = 2,3,4 \text{ or } 6 \ \text{according as} \ A_{ij}A_{ji} = 0,\ 1,\ 2 \text{ or } 3 \rangle$$

(cf. [MT], [Ma1], [Ti3] and also, for the affine case, [IM], [BT2] and [Ga]). In particular, G has a Bruhat decomposition $G = \bigcup_{w \in W} BwB$, leading to a "cell decomposition" of G/B : the quotients BwB/B have natural structures of finite-dimensional affine spaces. Now, in the minimal situation, the same N is the group N of another BN-pair (B^-,N), not conjugate to the previous one except in the finite-dimensional case (i.e. when $\underline{A}$ is positive-definite). Furthermore, one also has a partition $G = \bigcup_{w \in W} B^-wB$, called the Birkhoff decomposition of G (because of the special case considered in [Bi]; for the general result, cf. [Ti4]). While the cells BwB/B are finite-dimensional, the "cells" B^-wB are finite-codimensional in G, in a suitable sense, and, unlike the Bruhat decomposition, the Birkhoff decomposition always has a big cell, namely B^-B if one chooses B^- in its conjugacy class by N so that the intersection $B^- \cap B$ is minimum with respect to the inclusion (we then say that B and B^- are opposite). In the formal situation, a Birkhoff decomposition (and hence a big cell) still exists, but here, the groups B^- and B play completely asymmetric roles: B_- is <u>much</u> <u>smaller</u> than B in that, for instance, $B^-\backslash G/B^-$ is now highly uncountable (always excepting the case where $\underline{A}$ is positive-definite). We can be more explicit: if $G = \cup B^- wB$ is the Birkhoff decomposition of the <u>minimal</u> group G, and if $\hat{G}$ denotes the formal completion of G, then the Birkhoff decomposition of $\hat{G}$ is $\cup B^- w\bar{B}$, where $\bar{B}$ is the closure of B in $\hat{G}$; the group B^- is closed (and even discrete) in $\hat{G}$.

Different methods have been used to attach groups to Kac-Moody data. Roughly, one can classify them into four types, according to which techniques they are based upon, namely:

linear representations (cf. § 5 below);
generators and relations (cf. § 6);
Hilbert manifolds and line bundles;
axiomatic (cf. [Ti4]).

About the third approach, which is handled in Graeme Segal's lecture at this Arbeitstagung, let us just say that it gives a deeper geometric insight in the situation than the other methods, but that, at present, it concerns only the affine case. Also the axiomatic approach has been used only in the affine case so far: we shall briefly indicate below (§ 6 and Appendix 2) to which purpose.

5. Construction of the groups via representation theory.

One of the simplest way to prove the existence of a Lie group with a given (finite-dimensional) Lie algebra L consists in embedding L in the endomorphism algebra $\operatorname{End} V$ of a vector space V (by Ado's theorem) and then considering the group generated by $\exp L$.

If L is a Kac-Moody algebra, linear representations are infinite-dimensional and $\exp L$ is no longer defined in general. However, suppose that the linear representation $L \hookrightarrow \operatorname{End} V$ is such that the elements e_i, f_i, considered as endomorphisms of V, are <u>locally nilpotent</u> (an endomorphism φ of V is said to be locally nilpotent if, for any $v \in V$, $\varphi^n(v) = 0$ for almost all $n \in \underline{\mathbb{N}}$). Then, if the ground field K has characteristic zero, say, $\exp Ke_i$ and $\exp Kf_i$ are well-defined "one-parameter" automorphism groups of V which generate the group $G_S(K)$ one is looking for, at least if the h_i's generate Λ. Otherwise, one must also require that, as a Λ-module (remember that $\Lambda \subset L$), V is a direct sum $\coprod V_s$ of one-dimensional modules on which Λ operates through "integral characters" $\chi_s \in \Lambda^*$; then, one adds to the above generators the "one-parameter groups" $\lambda(K^\times)$, with $\lambda \in \Lambda$, where, by definition, $\lambda(k)$ operates on V_s via the multiplication by $k^{\langle\lambda,\chi_s\rangle}$. An L-module V is said to be <u>integrable</u> if it satisfies the above conditions (local nilpotency of e_i, f_i, plus the extra-requirement on Λ, which however follows from the first condition when the h_i's generate Λ).

That method for integrating L, inspired by C. Chevalley's Tohôku paper [Ch1], was first devised by R. Moody and K. Teo [MT], who used the adjoint representation of L. In that way, of course, they only get the <u>minimal adjoint group</u>. (More precisely, the group they construct is the analogue of Chevalley's <u>simple</u> group, namely the subgroup of the adjoint group generated by the $\exp Ke_i$ and $\exp Kf_i$; here, we say that the system S defines an <u>adjoint group</u> if the α_i's generate Λ^* and if $\mathbb{Q} \otimes \Lambda$ is generated as a $\mathbb{Q}$-vector space by the h_i's.) On the other hand, a suitable variation of the method described above enables them to include the case of a ground field with sufficiently large characteristics. Later on, Moody [Mo1] has applied the same ideas at the formal level, starting from a suitable completion of the Kac-Moody algebra.

In [Ma1], R. Marcuson works with highest weight modules, at the formal level. His method requires the characteristic to be zero.

In [Ga], H. Garland also uses highest weight representations. He restricts himself to the standard affine case - and makes heavy use of the relation between L and the loop algebra -, but in that special case, his results go much beyond those of Marcuson in that he essentially works over $\mathbb{Z}$ (with $\mathbb{Z}$-forms of the universal enveloping algebra of L and of the representation space), which enables him to define groups over arbitrary fields.

One drawback of the approach by means of linear representations is that it is not clear, <u>a priori</u>, how the group one associates to a given Kac-Moody algebra (over $\underline{\mathbb{C}}$, say) varies with the chosen representation. In [Ma1], this question is left open. Garland answers it by using the fact that the groups he constructs are central extensions of loop groups, and computing a cocycle which describes the extension.

V. Kac and D. Peterson [KP] obviate that inconvenient of the method by considering all integrable modules simultaneously. They start from the free product G^* of the additive groups Ke_i, Kf_i for all i . For any integrable module V , the maps $te_i \longrightarrow \exp te_i$, $tf_i \longrightarrow \exp tf_i$ extend to a representation $\exp_V : G^* \longrightarrow GL(V)$, and the group they consider is $G^*/\bigcap_V (\mathrm{Ker}\ \exp_V)$, where V runs through all integrable representations. This is the minimal group, in the sense of § 4, and corresponds to the case where the h_i's form a basis of Λ . (An other, earlier approach of that same group, but without this last restriction on the h_i's , can be found in [Ti3] : cf. § 6).

R. Goodman and N. Wallach [GW] are concerned with the standard affine case over $\underline{\mathbb{C}}$. Working within the theory of Banach Lie algebras and groups, they consider a large variety of Banach completions of the Kac-Moody algebras, and integrate them by using suitable topologizations of certain highest weight (so-called standard) modules. One of their purposes is to define the central extension of loop groups by $\underline{\mathbb{C}}^\times$ at various analytic levels. An alternative, more elementary approach to that problem (not touching, however, the main body of results of [GW]) may possibly be suggested by the remark of Appendix 1 below.

6. Generators and relations.

In a course of lectures summarized in [Ti3] (cf. also [Sℓ] and [Ma2]), I gave another construction for groups associated with Kac-Moody data. In order to sketch the main idea, let us return to the case of a finite-dimensional complex semi-simple Lie group G . Such a group is known to be the amalgamated product of the normalizer N of a maximal torus T and the parabolic subgroups $P_1,\ldots,P_\ell$ containing properly a given Borel subgroup B containing T and minimal with that property, with amalgamation of the intersections $P_i \cap P_j = B$ and $P_i \cap N$. (cf.[Ti2],13.3). Furthermore, P_i is the semi-direct product of its Levi subgroup L_i containing T by a unipotent group U_i . Thus, we have a presentation of G whose ingredients are the subgroups

N, L_i, U_i . The groups N and L_i can be reconstructed from the system S of (2.1) in a uniform way, without reference to the positivity of the matrix $\underline{\underline{A}}$: the group N is generated by $T = \mathrm{Hom}(\Lambda^*, C^\times)$ and ℓ elements m_i $(1 \leq i \leq \ell)$ submitted to the relations

(6.1) m_i normalizes T , and the automorphism of T it induces is the adjoint of the <u>reflection</u>

$\lambda \longmapsto \lambda - \langle\lambda, h_i\rangle \cdot \alpha_i$ of Λ^* ,

(6.2) $m_i^2 = \eta_i \in T = \mathrm{Hom}(\Lambda^*, \underline{\underline{C}}^\times)$, with $\eta_i(\lambda) = (-1)^{\langle\lambda, h_i\rangle}$ for $\lambda \in \Lambda^*$

and

(6.3) if $A_{ij}A_{ji} = 0$ (resp. 1;2;3), then $m_i m_j = m_j m_i$ (resp. $m_i m_j m_i = m_j m_i m_j$; $(m_i m_j)^2 = (m_j m_i)^2$; $(m_i m_j)^3 = (m_j m_i)^3$),

whereas L_i is nothing else but the reductive group of semi-simple rank one corresponding to the system (Λ, h_i, α_i) . As for the U_i's , being unipotent, they are easily described in terms of their Lie algebras Lie U_i , either by means of the Campbell-Hausdorff formula or, more conceptually, by exponentiating Lie U_i in the completion (for the natural filtration) of its universal enveloping algebra $\mathcal{U}(\mathrm{Lie}\ U_i)$.

All this can be carried over to an arbitrary system S , with the only difference that, now, Lie U_i is infinite-dimensional and no longer nilpotent but only pro-nilpotent (more precisely, the Lie U_i's are certain subalgebras of codimension 1 of the Lie algebra generated by the e_j's , and the latter has a <u>pro-nilpotent</u> <u>completion</u>). Moreover, by using a suitable $\underline{\underline{Z}}$-form of the universal enveloping algebra of the Kac-Moody algebra (generalizing the $\underline{\underline{Z}}$-form used by H. Garland in the affine case: cf. § 5), one is able to do everything over $\underline{\underline{Z}}$ and, by reduction, over an arbitrary ring R . Thus, one is led to attach to S a group functor on the category of rings, call it E_S . But this group functor E_S is <u>not</u> the "good" functor G_S one is looking for: indeed, if $\underline{\underline{A}}$ is a Cartan matrix, that is, in the positive definite case, G_S should of course be the Chevalley group-scheme corresponding to the Chevalley-Demazure data S , and one finds that the functor E_S

coincides with that scheme only over the principal ideal domain. This suggests that, in general, E_S may be the good functor when restricted to those rings. This is undoubtedly so in the affine case. Indeed, in that case, one can characterize the functor E_S restricted to principal ideal domains - call it $E^{(pid)}$ - by a system of very natural axioms which, it seems, should be satisfied by the "good" functor G_S (cf. [Ti4], 7.6 b)). Another application of those axioms is that they enable one to determine explicitly the functor $E_S^{(pid)}$ (whereas the more abstract definition by generators and relations is much less manageable), and that the result one obtains suggests (always in the affine case) what must be the functor G_S for arbitrary rings. We shall come back to that question in the next section (and in Appendix 2), but let us first conclude the present one by two remarks.

The above considerations can be developed both at the minimal and the formal level. In fact, the construction of [Ti3] depends on the choice of a certain subgroup X (subject to some simple conditions) of the multiplicative group of the completed universal enveloping algebra of the Lie algebra generated by the e_j's. Among the possible X, there is a minimal one, leading to the minimal group $G_S(\underline{\mathbb{C}})$ (and functor G_S), and a maximal one (which has been determined by O. Mathieu [Ma2]), leading to the formal group (and functor) associated with S.

The groups we have been considering are the generalizations, in the Kac-Moody framework, of the split reductive groups but, as in the finite-dimensional case, one can define non-split forms of those groups. In particular, over $\underline{\mathbb{R}}$, there is a "compact" form (which is by no means compact in the topological sense !): in the minimal set-up, it can be defined as the fixed-point group of the "anti-analytic" involution of $G_S(\underline{\mathbb{C}})$ induced by the semi-linear involution of the Kac-Moody algebra which permutes e_i and f_i and inverts the elements of Λ. (Another definition, involving hermitian forms in representation spaces, also works at the formal level: cf. [Ga], [KP]). Generalizing a result which was known in the finite-dimensional case [Ka1], V. Kac [KDP] has observed that that compact form can be defined as the amalgamated product of its rank 2 subgroups corresponding to the pairs of indices $i,j \in \{1,\ldots,\ell\}$, with amalgamation of the rank 1 subgroups (of type SU_2) corresponding to the indices (here, one must assume that the h_i's form a basis of Λ, or add a compact torus to the amalgam). As for the rank 2 groups, which are the ingredients of that definition, they are known in case they are finite-dimensional (i.e. when $\alpha_i(h_j).\alpha_j(h_i) \leq 3$) ; otherwise, they are shown to be amalgamated products of two groups of type U_2 with suitable amalgamation of a two-dimensional torus. (N.B. A result similar to the above is known to hold for finite-dimensional split groups or, more generally, for groups having a BN-pair with finite Weyl group: cf. [Ti2], 13.32, and, for earlier versions and special cases, [Cu] and [Ti1], 2.12.)

The fact that the definition by generators and relations does not provide the "good" functor G_S for rings that are not principal ideal domains probably lies in the nature of things (as K-theory suggests). A more likely way to get at the "right" G_S would consist in exhibiting a suitable $\mathbb{Z}$-form of the affine algebra of $G_S(\mathbb{C})$ (cf. § 8 below).

7. An example: groups of type ${}^2\tilde{E}_6$.

In this section, we adopt the *formal* viewpoint; to emphasize the fact, we shall use the notation $\hat{G}_S$, instead of G_S as above.

Let S be the system $(\Lambda,(\alpha_j,h_j)_{0\leq j\leq 4})$, where the matrix $(\alpha_j(h_i))$ is of type ${}^2\tilde{E}_6$ (cf. § 3), and where the α_i generate Λ^* whereas the h_i generate Λ : these properties characterize S . Our purpose is to describe the groups $\hat{G}_S(K)$ when K is a *field*. We discuss only this special example for the sake of concreteness, but similar results hold for any other twisted type (the type

is briefly examined in [Ti4], 7.4 , and general statements, concerning all affine types and arbitrary rings, will be given in Appendix 2).

From the explicit description of the Kac-Moody algebras of type ${}^2\tilde{E}_6$ given in § 3, one readily guesses what must be the group $\hat{G}_S(K)$ when K is a field *of characteristic not* 2, namely

$$\hat{G}_S(K) = \mathbb{G}(K((z))) ,$$

where $\mathbb{G}$ is a quasi-split algebraic group of type ${}^2E_6$, defined over $K((z))$ and whose splitting field is $K((\sqrt{z}))$. If $K = \mathbb{C}$, one proves this by straightforward "integration" (for arbitrary affine types, this part of the work is done in [Mo3]), and the general case ensues via the axiomatic method mentioned above (cf. § 6 and [Ti4], 7.6 b)).

Now, suppose that car $K = 2$. The above description cannot hold in that case since the extension $K((\sqrt{z}))/K((z))$ is not separable, hence is improper for the definition of a quasi-split group. But there is a circumstance which enables one again to guess the result, at least when K is perfect. Indeed, one knows that, in the finite-dimensional theory, the arrows carried by the double bonds of Dynkin diagrams "disappear" over perfect fields of characteristic 2: more precisely, reversing such an arrow corresponds to an inseparable isogeny which is bijective on rational points. Here, the diagram becomes "the same as" F_4 =  , hence the (correct) guess

$$G_S(K) = F_4(K((z))) .$$

But how can it be that a 78-dimensional (quasi-split) group of type E_6 suddenly degenerates into a 52-dimensional (split) group of type F_4 ?! The answer is simple: $F_4(K((z)))$ must be viewed as the group of rational points of a suitable 78-dimensional group defined over $K((z))$. The existence of such a group is not so surprising when one considers the isomorphism $(a,b) \longmapsto a^2+zb^2$ of $K((z))\times K((z))$ onto $K((z))$, hence of a 2-dimensional group onto a 1-dimensional group (K perfect).

To be more specific, set $L = \mathbb{F}_2((z))$, $L' = \mathbb{F}((\sqrt{z}))$ and denote by $\mathcal{F}$ a split group of type F_4 over L', by $\sigma : \mathcal{F} \longrightarrow \mathcal{F}$ a special isogeny of $\mathcal{F}$ into itself whose square is the Frobenius endomorphism (cf. [BT1], 3.3), by $R_{L'/L}$ the restriction of scalars and by $\mathbb{G}$ the image of $R_{L'/L}\sigma : R_{L'/L}\mathcal{F} \longrightarrow R_{L'/L}\mathcal{F}$. Then

the algebraic group $\mathbb{G}$ *is 78-dimensional. For any (non necessarily perfect) field* K *of characteristic* 2, *one has* $G_S(K) = \mathbb{G}(K((z)))$;

<u>if</u> K <u>is perfect, the map</u> $R_{L'/L}\sigma : R_{L'/L}\mathcal{F} \longrightarrow \mathfrak{G}$ <u>is bijective on rational points, therefore</u> $G_S(K) \cong (R_{L'/L}\mathcal{F})(K((z))) = \mathcal{F}(K((\sqrt{z}))) \cong \mathcal{F}(K((z)))$.

Let us explain briefly where the 78 dimensions of $\mathfrak{G}$ come from. The group F_4 has an open set Ω which is the product, in a suitable order, of 48 additive groups $\mathfrak{U}_a$ corresponding to the 48 roots a and a 4-dimensional torus $\mathfrak{T}$. The isogeny σ induces a bijection $a \longmapsto \sigma(a)$ of the root system into itself which maps short roots onto long roots and vice versa. The groups $R_{L'/L}\mathfrak{U}_a$ are 2-dimensional and $\dim R_{L'/L}\mathfrak{T} = 8$. Now, it is readily checked that:

if a is short, $R_{L'/L}\sigma$ maps $R_{L'/L}\mathfrak{U}_a$ isomorphically onto $R_{L'/L}\mathfrak{U}_{\sigma(a)}$;

if a is long, $R_{L'/L}\sigma$ maps $R_{L'/L}\mathfrak{U}_a$ onto a one-dimensional subgroup of $R_{L'/L}\mathfrak{U}_{\sigma(a)}$;

$R_{L'/L}\sigma$ maps $R_{L'/L}\mathfrak{T}$ onto a six-dimensional subtorus of itself.

Thus, $\dim (R_{L'/L}\sigma)(\Omega) = 2.24 + 24 + 6 = 78$.

We propose the following exercise to the interested reader: for K perfect of characteristic 2, write $SL_2(K((z)))$ as the group of rational points on $K((z))$ of an 8-dimensional algebraic group. This arises when one studies the case of the GCM $\begin{pmatrix} 2 & -4 \\ -1 & 2 \end{pmatrix}$; the 8-dimensional group in question appears as a characteristic 2 "degeneracy" of $SU_3(K((\sqrt{z})))$. Cf. [Ti4], 7.4, for more details.

8. <u>The algebro-geometric nature of the groups</u> $G_S(\underline{\underline{C}})$.

What kind of algebro-geometric objects are the functors G_S and, in particular, the groups $G_S(\underline{\underline{C}})$? Little is known for G_S is general, but something can be said about $G_S(\underline{\underline{C}})$ (here, $\underline{\underline{C}}$ could be replaced by any field of characteristic zero).

Set $G = G_S(\underline{\underline{C}})$. We have already mentioned the Bruhat decomposition $G = \bigcup BwB$, where B is a certain subgroup of G , which we may call "Borel subgroup", and w runs over a Coxeter group W . Coxeter groups, endowed as usual with a distinguished generating set $S = \{r_1,\ldots,r_\ell\}$ (cf. § 4), have a natural ordering: for $w,w' \in W$, one sets $w \geq w'$ if there exists a reduced expression $w = s_1\ldots s_n$ $(s_i \in S)$ and a subsequence $(i_1,\ldots,i_m)$ of $(1,\ldots,n)$ such that $w' = s_{i_1}\ldots s_{i_m}$. Then:

for any $w \in W$, _the subset_ $\mathrm{Schub}\, w = \bigcup_{w' \leq w} (Bw'B)/B$ _of_ G/B , _called the_ Schubert variety _corresponding to_ w , _has a natural structure of projective manifold_ (cf. [Ti4]); _thus,_ G/B _is a direct limit of projective manifolds_.

(In [Ti4], the projective structure of $\mathrm{Schub}\, w$ is defined by means of a highest weight representation of G , and is then shown not to depend on the choice of that representation. It would be desirable to have a more intrinsic definition, using for instance the big cell of the Birkhoff decomposition, as was suggested to the speaker by G. Lusztig.) The set G/B , and its description as a limit of projective varieties, does not depend on whether one adopts the minimal or the formal viewpoint (more precisely, the formal group is the completion of the minimal one for a topology for which B is an open subgroup). Also, since B contains $H = B \cap N$, the quotient G/B and the varieties $\mathrm{Schub}\, w$ depend only on the GCM $\underline{\underline{A}}$, and not on Λ ; when the choice of $\underline{\underline{A}}$ needs to be specified, we shall write $\mathrm{Schub}_{\underline{\underline{A}}} w$ instead of $\mathrm{Schub}\, w$.

If we now take the formal viewpoint, the Borel subgroup, or rather, to remain consistent with the notation of the end of § 4, the closure $\overline{B}$ of B in $\hat{G} = \hat{G}_S(\underline{\underline{C}})$, is a proalgebraic group, semi-direct extension of a torus by a prounipotent group (cf. [BT2], [Sℓ]).

Having thus described both $G/B = \hat{G}/\overline{B}$ and $\overline{B}$, we have gained some understanding of the algebro-geometric nature of $\hat{G}$ itself. But a more direct and promising picture is given by V. Kac and D. Peterson [KP1] who attach to G (the minimal group) a "coordinate ring", or rather two rings, the ring $\underline{\underline{C}}[G]$ of "strongly regular" functions, and the ring $\underline{\underline{C}}[G]_r$ of "regular" functions. The first one is generated by the coefficients of all highest weight representations (in [KP1], this is not chosen as definition of $\mathbb{C}[G]$ but is proved to be a property of the ring, defined in a different way) and provides a Peter-Weyl type theorem. That ring is not invariant by the map $i : g \mapsto g^{-1}$ (under that map, highest weight representations become lowest weight representations !); for a suitable topology, $\mathbb{C}[G]_r$ is topologically generated by $\mathbb{C}[G]$ and $i^*(\mathbb{C}[G])$. It is shown in [KP] that G is an affine (infinite-dimensional) algebraic group with coordinate ring $\mathbb{C}[G]_r$, in the sense of Shafarevitch [Sh] : this implies, in particular, that G can be identified with a subset of $\mathbb{C}^{\mathbb{N}}$ in such a way that $\mathbb{C}[G]_r$ is the restriction to G of the ring $\mathbb{C}[\mathbb{C}^{\mathbb{N}}]_r$ of regular functions on $\mathbb{C}^{\mathbb{N}}$ (i.e. the ring of functions whose restriction to $\mathbb{C}^{[0,n]}$ is polynomial for all n), and that G is the vanishing set of an ideal of $\mathbb{C}[\mathbb{C}^{\mathbb{N}}]_r$.

9. Applications.

"Kac-Moody groups" have been used in a variety of domains such as topology, differential and partial differential equations, singularity theory, etc. Those applications, a fast growing subject, are beyond both the scope of this survey and the competence of the speaker. Let me just unsystematically list a few basic references, which will give access to at least part of the literature on that topic: [SW] (cf. also the reference[5] of [SW]), [Ve1], [Ve2] (these concern applications of Kac-Moody Lie algebras, rather than groups), [RS], [Sℓ].

Most applications so far use only groups of affine type, and there may still be doubts about the usefulness of the general theory. To finish with, I would like to give an argument in favour of it. We have seen that to every GCM $\underline{A} = (\underline{A}_{ij})$ and every element of the corresponding Coxeter group $W(\underline{A})$, the theory associates a certain complex projective variety $\mathrm{Schub}_{\underline{A}} w$. If w is one of the canonical generators r_i of W , $\mathrm{Schub}_{\underline{A}} w$ is just a projective line. The next simple case is $w = r_i r_j$; then, $\mathrm{Schub}_{\underline{A}} w$ is a rational ruled surface, i.e. a surface fibered over $\mathbb{P}_1(\mathbb{C})$ with projective lines as fibers. It is well known that such a surface X is characterized up to isomorphism by a single invariant $\nu(X)$ which is a negative integer (if $\nu(X) \neq 0$, X is obtained by blowing up the vertex of a cone of degree $\nu(X)$ in a $(\nu(X)+1)$-dimensional projective space). Now, one shows that $\nu(\mathrm{Schub}_{\underline{A}}(r_i r_j)) = A_{ij}$. This gives a geometric interpretation of the matrix $\underline{\underline{A}}$. Moreover, observe that, if one accepts only to consider GCM of affine type, only the surfaces X with $\nu(X) \in [-4,0]$, among the rational ruled surfaces, have the right to be called "Schubert varieties", which seems rather unnatural ! I should think that the class of all $\mathrm{Schub}_{\underline{\underline{A}}}w$, for all $\underline{\underline{A}}$ and w , will turn out to be a very natural and interesting class of projective varieties to consider.

Appendix 1. Central extension.

For arbitrary S, the "minimal group" $G_S(\underline{\underline{C}})$ can be constructed by the methods described in §§ 5 and 6. In particular, those methods provide very simple, purely formal existence proofs for a nontrivial central extension of the "polynomial" loop groups by $\underline{\underline{C}}^\times$. The situation is quite different when one starts from loop groups defined by analytic conditions. However, the following rather trivial considerations may conceivably enable one to exploit the result known for polynomial loops in the analytic case. Here, all topological spaces are assumed to be Hausdorff.

Let $\pi : G' \longrightarrow G$ be a central group extension and let U'_-, H', U'_+ be three subgroups of G' such that $\operatorname{Ker}\pi \subset H'$, that H' normalizes $U'_\pm$ and that the product mapping $U'_- \times H' \times U'_+ \longrightarrow G'$ is injective. Thus, $\pi_+ = \pi|_{U'_+}$ and $\pi_- = \pi|_{U'_-}$ are isomorphisms of U'_+ and U'_- onto two subgroups U_+ and U_- of G. We set $H = \pi(H')$.

Now, let us embed G in a complete topological group $\hat{G}$ and suppose that, if $\overline{U}_-$ and $\overline{U}_+$ denote the closures of U_- and U_+ in $\hat{G}$, the product mapping in $\hat{G}$ defines a homeomorphism of $\overline{U}_- \times H \times \overline{U}_+$ onto a dense open subset Ω of $\hat{G}$. Let us also endow H' with a topology making it into a complete topological group, such that $\operatorname{Ker}\pi$ is closed in H' and that the canonical algebraic isomorphism $H'/\operatorname{Ker}\pi \longrightarrow H$ is an isomorphism of topological groups as well (observe that, by hypthesis, H is locally closed, hence closed in $\hat{G}$, and is therefore a complete topological group).

Set $X = \{(u,u') \in U_+ \times U_- \mid uu' \in \Omega\}$. This is a dense open subset of $U_+ \times U_-$ (endowed with the topology induced by that of $\overline{U}_+ \times \overline{U}_-$). For $(u,u') \in X$, there is a unique element $\varphi(u,u') \in H'$ such that

$$\pi_+^{-1}(u) \cdot \pi_-^{-1}(u') \in U'_- \cdot \varphi(u,u') \cdot U'_+ \quad .$$

The topology of $\hat{G}$ induces a topology on $U_\pm$ which we lift to $U'_\pm$ by means of $\pi_\pm^{-1}$, and we endow $\Omega' = U'_- H' U'_+$ with the product topology. The following proposition is easy.

PROPOSITION 2. *If the function $\varphi : X \longrightarrow H'$ is continuous, there is a unique topology on G' making G' into a topological group and Ω', topologized as above, into a dense open subset of G'. Suppose further that there is a neighborhood $\overline{X}_1$ of (1.1) in $\overline{U}_+ \times \overline{U}_-$ such that the restriction of φ to $X \cap \overline{X}_1$ extends to a continuous map $\overline{X}_1 \longrightarrow H'$. Then, the topological group G' admits a completion $\hat{G}'$, $\operatorname{Ker} \pi$ is a closed subgroup of $\hat{G}'$ and the homomorphism $\hat{G}' \longrightarrow \hat{G}$ extending π factors through an isomorphism of topological groups $\hat{G}'/\operatorname{Ker} \pi \longrightarrow \hat{G}$.*

Note that the left (or right) translates of all open subsets of Ω obviously form a basis of the topology of G' (hence the uniqueness assertion).

In the application I have in mind, G would be a "polynomial" loop group, $\hat{G}$ some other loop group, $\pi : G' \longrightarrow G$ the "natural" central extension of G by $\operatorname{Ker} \pi \cong \underline{\underline{C}}^\times$ (whose existence is easily proved by any of the methods described in §§ 5 and 6), U'_- and U'_+ the (non complete) "prounipotent radicals" of two opposite Borel subgroups of G' (cf. § 4) and H' the intersection of those Borel subgroups, a direct product of copies of $\underline{\underline{C}}^\times$ which one endows with its natural topology. The main problem, which I have not investigated, is of course to prove (in the interesting cases) that φ is continuous and extends to a neighborhood of (1,1) in $\overline{U}_+ \times \overline{U}_-$.

Appendix 2. The group functor $\hat{G}_S$ in the affine case.

In this appendix, we shall use the techniques and terminology of [BT4] to describe the formal functors $\hat{G}_S$ for all systems

$$S = (\Lambda, (\alpha_i)_{0 \leq i \leq \ell}, (h_i)_{0 \leq i \leq \ell})$$

satisfying the following conditions:

(A1) the matrix $\underline{\underline{A}} = (\alpha_j(h_i))$ is of irreducible, affine type;

(A2) the set $\{h_i \mid 0 \leq i \leq \ell\}$ generates Λ;

(A3) the set $\{\alpha_i \mid 0 \leq i \leq \ell\}$ contains a $\underline{\underline{Q}}$-basis of $\underline{\underline{Q}} \otimes \Lambda^*$.

More precisely, for any such S, we shall describe a topological group functor $\hat{G}_S$ having the following properties.

(P0) There is a Lie algebra functor Lie $\hat{G}_S$ defined as follows (compare [DG], pp. 209-210). For any ring R, set $R(\varepsilon,\varepsilon') = R[t,t']/(t^2,t'^2)$, where ε,ε' are the canonical images of t,t' in the quotient; in other words, $R(\varepsilon,\varepsilon')$ is the tensor product of two algebras $R(\varepsilon)$, $R(\varepsilon')$ of dual numbers. For $r \in R$, let $\pi : R(\varepsilon) \longrightarrow R$, $\iota : R(\varepsilon) \longrightarrow R(\varepsilon')$, $\sigma : R(\varepsilon) \longrightarrow R(\varepsilon,\varepsilon')$ and $\mu_r : R(\varepsilon) \longrightarrow R(\varepsilon)$ be the R-homomorphisms sending ε onto $0,\varepsilon',\varepsilon\varepsilon'$ and $r\varepsilon$ respectively. Then, the additive group $(\mathrm{Lie}\ \hat{G}_S)(R)$ is the kernel of the homomorphism

$$\hat{G}_S(\pi) : \hat{G}_S(R(\varepsilon)) \longrightarrow \hat{G}_S(R) ,$$

the scalar multiplication by r is induced by the automorphism $\hat{G}_S(\mu_r)$ of $\hat{G}_S(R(\varepsilon))$ and the commutator of two elements $x,y \in (\mathrm{Lie}\ \hat{G}_S)(R) \subset \hat{G}_S(R(\varepsilon))$ is the only element $[x,y]$ such that

$$\hat{G}_S(\sigma)([x,y]) = (x,\hat{G}_S(\iota)(y))$$

where $(\ ,\ )$ stands for the usual commutator in the group $\hat{G}_S(R(\varepsilon,\varepsilon'))$.

(P1) $(\mathrm{Lie}\ \hat{G}_S)(\underline{\underline{C}})$ is the Kac-Moody algebra associated to the system $(\underline{\underline{C}} \otimes \Lambda,\ (\alpha_i)_{0 \leq i \leq \ell},(h_i)_{0 \leq i \leq \ell})$ completed with respect to the natural gradation ($\deg e_i = 1$, $\deg f_i = -1$, $\deg h_i = 0$).

(P2) The group $\hat{G}_S(\underline{\underline{C}})$ coincides with the formal group over $\underline{\underline{C}}$ attached to S by any one of the construction processes described in §§ 5 and 6; in particular, it contains (a canonical image of) $\mathrm{Hom}(\Lambda^*,\underline{\underline{C}}^\times)$ and its center consists of all $\xi \in \mathrm{Hom}(\Lambda^*,\underline{\underline{C}}^\times)$ such that $\xi(\alpha_i) = 0$ for all i.

(P3) Modulo its center, $G_S(\underline{\underline{C}})$ is the subgroup of $\mathrm{Aut}((\mathrm{Lie}\ \hat{G}_S)(\underline{\underline{C}}))$ generated by all converging exp ad g, with $g \in (\mathrm{Lie}\ \hat{G}_S)(\underline{\underline{C}})$ (this turns out to be identical with the adjoint group considered by R.V. Moody [Mo1] and J.I. Morita [Mo3]; about this group, cf. also the last sentence of this appendix).

(P4) The functor $\hat{G}_S$ restricted to principal ideal domains, together with suitably defined functorial homomorphisms $\sigma_i : SL_2 \longrightarrow \hat{G}_S$, $\eta : \mathrm{Hom}(\Lambda^*,?^{\times}) \longrightarrow \hat{G}_S$ (which we leave as an exercise to determine explicitly), satisfies the axioms (i') to (iv') of [Ti4], 7.5, and is characterized by them, once $\hat{G}_S(\underline{\underline{C}})$, $\sigma_i(\underline{\underline{C}})$, $\eta(\underline{\underline{C}})$ are given.

Those properties clearly indicate that the functor $\hat{G}_S$ which we are going to define is the "right one", at least when restricted to principal ideal domains but maybe also for general rings, considering its fairly simple and natural definition (though it is conceivable that some algebro-geometric invariants of the ring, such as Pic R, should be brought into play).

Let e be an integer and let G be a quasi-split, simply connected absolutely almost simple group defined over the field $K = \underline{\underline{Q}}(Z)$, whose splitting field over K is generated by the e-th roots of Z; thus, $e = 1$, 2 or 3 and, in the latter case, G is of type D_4. Let S be a maximal split torus of G, Φ the system of roots of G with respect to S, $\underline{\underline{\Phi}}$ the system of root rays ("rayons radiciels": cf. [BT4], 1.1.2), i.e. of half-lines $\underline{\underline{R}}_+ . a$ with $a \in \Phi$, T the centralizer of S in G and U_a (for $a \in \underline{\underline{\Phi}}$) the root subgroup corresponding to a. We also denote by $G,S,T,\ldots,$ the groups of K-rational points of $G,S,T,\ldots$.

Let now $S = (\Lambda,(\alpha_i)_{0 \le i \le \ell},(h_i)_{0 \le i \le \ell})$ be defined as follows:

$\Lambda = X_*(S) = \mathrm{Hom}(\mathrm{Mult},S)$ is the group of cocharacters of S;

$(\alpha_1,\ldots,\alpha_\ell)$ is a basis of Φ and $-\alpha_0$ is the maximal root if $e = 1$ or if $e = 2$ and G is of type A_{2n}, and it is the maximal "short" root in the remaining cases; h_i is the coroot associated α_i.

Varying e and the type of G, one gets all systems S satisfying the conditions (A1) to (A3) above in this way. If G has type X, we say that S has type ${}^e\tilde{X}$. The Dynkin diagram representing the GCM $(\alpha_j(h_i))$ is given by the following table:

type of S	diagram	
${}^1\tilde{X}$	extended Dynkin diagram of X	
${}^2\tilde{A}_{2n}$		(n+1 vertices)
${}^2\tilde{A}_{2n-1}$		(n+1 vertices)
${}^2\tilde{D}_n$		(n vertices)
${}^2\tilde{E}_6$	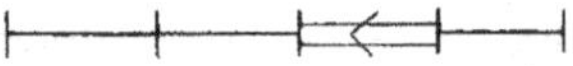	
${}^3\tilde{D}_4$		

We shall now choose a system of "épinglages" of the U_a's (cf. [BT4], 4.1). This is a system $(x_a)_{a \in \Phi}$ where, for all a, x_a is one of three things:

(i) an isomorphism $K \longrightarrow U_a$;

(ii) an isomorphism $K(Z^{1/e}) \longrightarrow U_a$ (here $Z^{1/e}$ denotes any e-th root of Z and, when e = 3, all cubic roots involved may be chosen equal);

(iii) an isomorphism $H \longrightarrow U_a$, where H is the product $K(Z^{1/2}) \times Z^{1/2}\cdot K$ endowed with the group structure

$$(*) \qquad (u,v)\cdot(u',v') = (u+u',\ v+v'+(u^{\sigma}u' - u'^{\sigma}u))$$

in which σ represents the nontrivial K-automorphism of $K(Z^{1/2})$ (observe that H is nothing else but the group H^{λ} of [BT4], 4.1.15, for $\lambda = 1/2$,) transformed by the automorphism $(x,y) \longmapsto (x,2y)$ of the underlying variety).

In all cases except ${}^2\tilde{A}_{2n}$ (i.e. when G is of type A_{2n} and e = 2), we take for (x_a) a coherent system of "épinglages" deduced from a Chevalley-Steinberg system (cf. [BT4], 4.1.16). In order to describe the system (x_a) in the case $^2\tilde{A}_{2n}$, let us choose an orthogonal basis $(a_i)_{1 \leq i \leq n}$ of the (relative) root lattice: thus, $\Phi = \{\pm a_i, \pm 2a_i, \pm a_i \pm a_j$ with $i \neq j\}$. For $a \in \underline{\underline{\Phi}}$, let θ_a denote the automorphism of the source of x_a defined as follows: if a contains a_i+a_j (resp. a_i-a_j; resp. $-a_i-a_j$), $\theta_a(k) = 2k$ (resp. $k/2$) and if a contains a_i (resp. $-a_i$), $\theta_a(u,v) = (2u,4v)$ (resp. (u,v)). Finally, we set $x_a = x'_a \circ \theta_a$, where $(x'_a)_{a \in \underline{\underline{\Phi}}}$ is a coherent system of "épinglagles", as in loc. cit.

Let us now describe a certain schematic root datum $(\mathfrak{T}, (\mathfrak{U}_a)_{a \in \underline{\underline{\Phi}}})$ in G over the ring $K = \underline{\underline{Z}}[Z,Z^{-1}]$ (cf. BT4, 3.1.1). The scheme is the "canonical group-scheme associated with the torus T", defined as in [BT4], 4.4.5 (as in [BT4], it can be shown that $\mathfrak{T}$ does not depend on the way T is expressed as a product of tori of the form $\prod_{L/K} \mathfrak{Mult}_L$) and the scheme $\mathfrak{U}_a$ is the "image by x_a" of:

the additive group-scheme canonically associated with the module K (resp. $K[Z^{1/e}]$) in case (i) (resp. (ii)) (cf. [BT4], 1.4.1);

the group-scheme whose underlying scheme is canonically associated with the module $\mathfrak{H} = K[Z^{1/2}] \times Z^{1/2} \cdot K$ and whose product operation is given by (*) in case (iii).

It is readily verified, using the appendix of [BT4], that the system $(\mathfrak{T}, (\mathfrak{U}_a))$ is indeed a schematic root datum. By Section 3.8.4 of [BT4], there exists a unique smooth connected group-scheme $\mathfrak{G}$ with generic fibre G containing the direct product

$$\prod_{a \in \underline{\underline{\Phi}}_-} \mathfrak{U}_a \times \mathfrak{T} \times \prod_{a \in \underline{\underline{\Phi}}_+} \mathfrak{U}_a$$

as an open subscheme ("big cell") (here, $\underline{\underline{\Phi}}_+ \subset \underline{\underline{\Phi}}$ denotes a system of positive root rays and $\underline{\underline{\Phi}}_- = -\underline{\underline{\Phi}}_+$). Finally, S being as above, the announced functor $\hat{G}_S$ is defined by

$$\hat{G}_S(R) = \mathfrak{G}(R((Z))) ,$$

this group being given the natural topology, induced by that of $R((Z))$.

Suppose now that R is a <u>perfect field</u> of characteristic e (which implies that $e = 2$ or 3). There is a "natural" isomorphism of each $\mathfrak{U}_a(R((Z)))$ onto $R((Z))$, namely

x_a^{-1} in case (i) ,

$x_a(r) \longmapsto r^e$ in case (ii) ,

$x_a(r,r') \longmapsto r'^2+r^4$ in case (iii) ,

and $\mathfrak{T}(R((Z)))$, which is a product of groups of the form $R((Z))^\times$ and $R((Z^{1/e}))^\times$, is clearly isomorphic to the group $\mathfrak{T}'(R((Z)))$ of rational points of a <u>split</u> torus $\mathfrak{T}'$. It is then readily verified (using [BT3], § 10, and the appendix of [BT4]), that, via those isomorphisms, the system $(\mathfrak{T}(R((Z))), (\mathfrak{U}_a(R((Z))))_{a\in\underline{\underline{\Phi}}})$ "is" the standard root datum of the group of rational points of an $R((Z))$-split simple group of type

C_n if $\underline{\underline{A}} = (\alpha_j(h_i))$ has type ${}^2\tilde{A}_{2n}$,

B_n if $\underline{\underline{A}}$ has type ${}^2\tilde{A}_{2n-1}$,

C_{n-1} if $\underline{\underline{A}}$ has type ${}^2\tilde{D}_n$,

F_4 if $\underline{\underline{A}}$ has type ${}^2\tilde{E}_6$,

G_2 if $\underline{\underline{A}}$ has type ${}^3\tilde{D}_4$.

This is the phenomenon already mentioned in § 7 for the special case of type ${}^2\tilde{E}_6$.

Let us return to the group-scheme $\mathfrak{G}$. In the classical cases ${}^2\tilde{A}_m$ and ${}^2\tilde{D}_n$, it can be given a more direct and more elementary description. Here, we shall only briefly treat the types ${}^2\tilde{A}_m$ (the case of ${}^2\tilde{D}_n$ is slightly more complicated because one must work with the spin group). According as $m = 2n-1$ or $2n$, set $I = \{\pm 1, \pm 2, \ldots, \pm n\}$ or $I = \{0, \pm 1, \ldots, \pm n\}$. Let V be the $K[Z^{1/2}]$-module $(K[Z^{1/2}])^I$ endowed with a coordinate system $\underline{\underline{z}} = (z_i)_{i\in I}$, let τ denote the K-automorphism of $K[Z^{1/2}]$ defined by $\tau(Z^{1/2}) = -Z^{1/2}$ and consider the hermitian form

$$h(\underline{\underline{z}};\underline{\underline{z}}') = \Sigma(z'^{\tau}_{-i}\, z_i + z'^{\tau}_i\, z_{-i}) \quad ,$$

where i runs from 1 to n or from 0 to n according as $m = 2n-1$ or $2n$. We represent by V_K the module V considered as a K-module; in it, we use the coordinate system $(\underline{x},\underline{y}) = (x_i, y_i)_{i \in I}$, where $x_i, y_i \in K$ and $z_i = x_i + y_i \cdot Z^{1/2}$. Separating the "real and imaginary parts" of h, we get $h = s + Z^{1/2} \cdot a$, where s and a are a symmetric and an alternating bilinear form in V_K respectively. Similarly, the determinant in End V can be written $\det_0 + Z^{1/2} \cdot \det_1$, where $\det_0$ and $\det_1$ are K-polynomials in End V considered as a K-module. Let q be the quadratic form $q(\underline{x},\underline{y}) = \frac{1}{2}s(\underline{x},\underline{y};\underline{x},\underline{y})$ in V_K. The multiplication by $Z^{1/2}$ is an automorphism J of the K-module V_K. Finally, the group-scheme $\mathbb{G}$ (corresponding to the type ${}^2\tilde{A}_m$) can be described as the subgroup-scheme of $\mathbb{GL}(V_K)$ defined by the equations $g \cdot a = a$, $g \cdot q = q$ (hence $g \cdot s = s$), ${}^gJ = J$, $\det_0 g = 1$, $\det_1 g = 0$. In other words, if $\mathbb{R}$ is a K-algebra, $\mathbb{G}(\mathbb{R})$ is the subgroup of all elements of $SL(V \otimes \mathbb{R}[Z^{1/2}])$ preserving the ($\mathbb{R}$-valued) "forms" a and q. (For the case $m = 2n$, see [Ti4], 7.4.)

Now, consider again the case $\mathbb{R} = R((Z))$, where R is a perfect field of characteristic 2 (in fact, any ring $\mathbb{R}$ such that the map $x \longmapsto x^2$ is a bijection of $\mathbb{R}[Z^{1/2}]$ onto $\mathbb{R}$ would do). Let V' (resp. V'') denote the $\mathbb{R}[Z^{1/2}]$-module, product of $2n+1$ (resp. $2n$) factors $\mathbb{R}[Z^{1/2}]$ indexed by $\{0,\pm 1,\ldots,\pm n\}$ (resp. $\{\pm 1,\ldots,\pm n\}$). In those modules, we use again coordinates z_i where i runs through the same index sets. In V', consider the quadratic form $q'(\underline{z}) = \sum_{i=1}^{n} z_{-i}z_i + Z^{1/2} \cdot z_0^2$, and in V'', the alternating bilinear form $a'(\underline{z};\underline{z}') = \sum_{i=1}^{n} (z'_{-i}z_i - z'_i z_{-i})$. If $m = 2n-1$, $V \otimes \mathbb{R}[Z^{1/2}]$ can be identified with V'', hence with a quotient of V', the "bilinearization" and the "real part" (K-part) of q' are the inverse images in V' of the "forms" $a_{\mathbb{R}}$ and $q_{\mathbb{R}}$ (with obvious notational conventions), and it is easy to verify that the projection $V' \longrightarrow V''$ induces an isomorphism $SO(q') \xrightarrow{\sim} \mathbb{G}(\mathbb{R})$. If $m = 2n$, $V \otimes \mathbb{R}[Z^{1/2}]$ can be identified with V', the bilinear form $h_{\mathbb{R}[Z^{1/2}]}$ is the inverse image of a' by the projection $V' \longrightarrow V''$ and, this time, the latter induces an isomorphism $\mathbb{G}(\mathbb{R}) \xrightarrow{\sim} Sp(a')$. Thus we have found again the two isomorphisms obtained earlier in a different way.

The description of the functor $\hat{G}_S$ associated to an <u>arbitrary</u> system S of affine type, i.e. a system satisfying (A1) but not necessarily (A2) and (A3) now amounts to a combination of extension problems. In particular, when $\Lambda = \coprod \underline{Z}.h_i$, one must define a central extension of the above functor $\hat{G}$ by the multiplicative group-scheme $\mathfrak{Mult}$ [1]; this is related to work of C. Moore [Mo2], H. Matsumoto [Ma3] and P. Deligne [De1]. Note that if, with the notation used throughout this appendix, we assume $e = 1$, we denote by S_{ad} the system obtained in the same way as S but replacing Λ by the dual of the lattice of roots and by $\mathfrak{G}_{ad}$ the split <u>adjoint</u> group-scheme of the same type as G , then the functor $\hat{G}_{S_{ad}}$ is <u>not</u> equal to $R \longmapsto \mathfrak{G}_{ad}(R((Z)))$ in general; for instance, $\hat{G}_{S_{ad}}(\underline{C})$ is the image of the canonical map

$$\mathfrak{G}(\underline{C}((Z))) \longrightarrow \mathfrak{G}_{ad}(\underline{C}((Z))) \ ,$$

whose cokernel is isomorphic to the center of G .

---

[1] As P. Deligne pointed out to me, the word "extension" must be understood here in a "schematic sense"; one should not expect the extension map to be surjective for rational points over an arbitrary ring R .

REFERENCES

[Bi] G.D. BIRKHOFF, *A theorem on matrices of analytic function*, Math, Ann. 74 (1913), 122-133.

[Bo] N. BOURBAKI, *Groupes et Algébras de Lie*, Chap. IV, V, VI, Hermann, Paris, 1968.

[BT1] A. BOREL et J. TITS, *Homomorphismes "abstracts" de groupes algébriques simples*, Ann. of Math., 97 (1973), 499-571.

[BT2] F. BRUHAT et J. TITS, *Groups algébriques simples sur un corps local*, Proc. Conf. on local fields (Driebergen, 1966), Springer, 1967, 23-36.

[BT3] F. BRUHAT et J. TITS, *Groupes réductifs sur un corps local, I: Données radicielles valuées*, Publ. Math. I.H.E.S. 41 (1972), 5-251.

[BT4] F. BRUHAT et J. TITS, *Groupes réductifs sur un corps local, II: Schémas en groupes, Existence d'une donnée radicielle valuée*, Publ. Math. I.H.E.S. 60 (1984), 5-184.

[Ch1] C. CHEVALLEY, *Sur certains groupes simples*, Tohoku Math. J. (2)7 (1955), 14-66.

[Ch2] C. CHEVALLEY, *Séminaise sur la classification des groupes de Lie algébriques*, 2 vol., Inst. H. Poincaré, Paris, 1958, mimeographed.

[Ch3] C. CHEVALLEY, *Certains schémas de groupes semi-simples*, Sém. N. Bourbaki, exposé 219 (1980-1981), Benjamin, New York, 1966.

[Cu] C.W. CURTIS, *Groups with a Bruhat decomposition*, Bull. Amer. Math. Soc. 70 (1964), 357-360.

[D21] P. DELIGNE, *Extention de* G(K) *par* $K_2(K)$, Séminaire I.H.E.S., 1979, non publié.

[De2] M. DEMAZURE, *Schémas en groupes réductifs*, Bull. Soc. Math. Fr. 93 (1965), 369-413; cf. aussi le *Séminaire de Géométrie algébrique*, de M. DEMAZURE er A. GROTHENDIECK, vol. 3, Springer Lecture Notes in Math. n° 153, 1970.

[DG] M. DEMAZURE et P. GABRIEL, *Groupes algébriques*, I, Masson, Paris, 1970.

[Ga] H. GARLAND, *The arithmetic theory of loop groups*, Publ. Math. I.H.E.S. 52 (1980), 5-136.

[GW] R. GOODMAN and N.R. WALLACH, *Structure and unitary cocyle representations of loop groups and the group of diffeomorphisms of the circle*, Jour. r. angew. Math. 347 (1984), 69-133.

[IM] N. IWAHORI and H. MATSUMOTO, *On some Bruhat decomposition and the structure of the Hecke ring of* p-*adic Chevalley groups*, Publ. Math. I.H.E.S. 25 (1965), 5-48.

[Ka1] V. KAC, *An algebraic definition of compact Lie groups*, Trudy M.I.E.M. 5 (1969), 36-47.

[Ka2] V. KAC, *Simple irreducible Lie algebras of finite growth*, Izvestija Akad. Nauk S.S.S.R. (ser. mat.) 32 (1968), 1923-1967; English transl., Math. USSR Izvestija 2 (1968), 1271-1311.

[Ka3] V. KAC, *Infinite dimensional Lie algebras*, Birkhäuser, Boston, 1983.

[KP1] V. KAC and D. PETERSON, *Regular functions on certain infinite dimensional groups, in* Arithmetic and Geometry (ed. M. ARTIN and J. TATE), Birkhäuser, Boston, 1983, 141-166.

[KP2] V. KAC and D. PETERSON, *Cohomology of infinite-dimensional groups and their flag varieties: algebraic part*, Colloque Elie Cartan, Lyon June 1984, to appear.

[Ma1] R. MARCUSON, *Tits's system in generalized non-adjoint Chevalley groups*, J. of Algebra 34 (1975), 84-96.

[Ma2] O. MATHIEU, *Sur la construction de groupes associés aux algèbres de Kac-Moody*, C.R. Acad. Sci. série I, 299 (1984), 161-164.

[Ma3] H. MATSUMOTO, *Sur les sous-groupes arithmétiques des groupes semi-simples déployés*, Ann. Sci. Ec. Normale Sup. 2 (4 - série) (1969), 1-62.

[Mo1] R.V. MOODY, *A simplicity theorem for Chevalley groups defined by generalized Cartan matrices*, preprint, 1982.

[Mo2] C. MOORE, *Group extensions of* p-*adic and adelic linear groups*, Publ. Math. I.H.E.S. 35 (1968), 5-70.

[Mo3] J. MORITA, *On adjoint Chevalley groups associated with completed Euclidean Lie algebras*, Comm. in Algebra 12 (1984), 673-690.

[MT] R. MOODY and K. TEO, *Tits systems with crystallographic Weyl groups*, J. Algebra 21 (1972), 178-190.

[RS] A. REIMAN and M. SEMENOV-TJAN-SHANSKII, *Reduction of Hamiltonian systems, affine Lie algebras and Lax equations*, I, Invent. Math.54 (1979), 81-100; II, Invent. Math.63 (1981), 423-432.

[SW] G. SEGAL and G. WILSON, *Loop groups and equations of KdV type*, Publ. Math. I.H.E.S. 61 (1985).

[Sh] I. SHAFAREVICH, *On some infinite dimensional groups*, II, Izvestija Akad. Nauk S.S.S.R. (ser. mat.) 45 (1981); English transl., Math. USSR, Izvestija 18 (1982), 214-226.

[Sl] P. SLODOWY, *Singularitäten, Kac-Moody Liealgebren, assoziierte Gruppen und Verallgemeinerungen*, Habilitationsschrift, Bonn, 1984.

[Ti1] J. TITS, *Groupes semi-simples isotropes*, Coll. théor. gr. alg., C.B.R.M. (Bruxelles, 1962), Librairie Universitaire, Louvain, et Gauthier-Villars, Paris, 1962, 137-147.

[Ti2] J. TITS, *Buildings of spherical type and finite BN-pairs*, Springer Lecture Notes in Math. n° 386 (1974).

[Ti3] J. TITS, Résumé de cours, Annaire du Collġe de France, 81 $^{\underline{e}}$ année (1980)-1981), 75-86.

[Ti4] J. TITS, Résumé de cours, Annuaire du Collège de France, 82 $^{\underline{e}}$ an 82 ^{e} année (1981-1982), 91-105.

[Ve1] J.-L. VERDIER, *Algèbres de Lie, systèmes hamiltoniens, courbes algébriques (d'après* M. ADLER *et* P. van MOERBEKE), Sém. N. Bourbaki, exposé n° 566 (1980-1981), Springer Lecture Notes in Math. n° 901 (1982), 85-94.

[Ve2] J.-L. VERDIER, *Les représentations des algèbres de Lie affines: Applications à quelques problèmes de physique (d'après* E. DATE, M. JIMPB, M. KASHIWARA, T. MIWA), Sém. N. Bourbaki, exposé n° 596 (1981-1982), Astérisque 92-93 (1983), 365-377.

[Wi] R. WILSON, *Euclidean Lie algebras are universal central extensions*, *in* Lie algebras and related topics, Springer Lecture Notes in Math. n° 933, 210-213.

MODULAR POINTS, MODULAR CURVES, MODULAR SURFACES AND MODULAR FORMS

D. Zagier

University of Maryland
College Park, MD 20742

Max-Planck-Institut für Mathematik,
D-5300 Bonn, FRG

This talk, instead of being a survey, will concentrate on a single example, using it to illustrate two themes, each of which has been a leitmotif of much recent work in number theory and of much of the work reported on at this Arbeitstagung (lectures of Faltings, Manin, Lang, Mazur-Soulé, Harder). These themes are:

i) special values of L-series as reflecting geometrical relationships, and

ii) the analogy and interplay between classical algebraic geometry over $\mathbb{C}$ and algebraic geometry (in one dimension lower) over $\mathbb{Z}$, and more especially between the theory of complex surfaces and the theory of arithmetic surfaces à la Arakelov-Faltings. In particular, we will see that there is an intimate relationship between the positions of modular curves in the homology groups of modular surfaces and the positions of modular points in the Mordell-Weil groups of the Jacobians of modular curves.

The particular example we will treat is the elliptic curve E defined by the diophantine equation

$$y(y-1) = (x+1)x(x-1); \qquad (1)$$

most of what we have to say applies in much greater generality, but by concentrating on one example we will be able to simplify or sharpen many statements and make the essential points emerge more clearly.

The exposition has been divided into two parts. In the first (§§1-5), which is entirely expository, we describe various theorems and conjectures on elliptic and modular curves, always centering our discussion on the example (1). In particular, we explain how one can construct infinitely many rational solutions of (1) by a construction due to Heegner and Birch, and how a result of Gross and the author and one of Waldspurger lead one to surmise a relationship between these solutions and the coefficients of a modular form of half-integral weight. The second part (§§6-9) is devoted to a proof of this relationship.

I would like to thank G. van der Geer and B. Gross for useful discussions on some of the material in this talk.

1. The elliptic curve E and its L-series.

Multiplying both sides of (1) by 4 and adding 1 we obtain the Weierstrass form

$$y_1^2 = 4x^3 - 4x + 1 \quad (y_1 = 2y - 1); \tag{2}$$

from this one calculates that the curve E has discriminant $\Delta = 37$ and j-invariant $j = 2^{12}3^3/37$. Of course, (1) and (2) are affine equations and we should really work with the projective equations $y^2z - yz^2 = x^3 - xz^2$ and $y_1^2z = 4x^3 - 4xz^2 + z^3$ whose points are the points of (1) or (2) together with a "point at infinity" (0:1:0). The points of E over any field k form a group with the point at infinity being the origin and the group law defined by $P + Q + R = 0$ if P,Q,R are collinear; the negative of a point (x,y) of (1) or (x,y_1) of (2) is $(x,1-y)$ or $(x,-y_1)$, respectively. In accordance with the philosophy of modern geometry, we try to understand E by looking at the groups $E(k)$ of k-rational points for various fields k.

$k = \mathbb{R}$: The set of real solutions of (1) is easily sketched; it consists of two components, $\alpha \le x \le \beta$ and $\gamma \le x$, where $\alpha = -1.107\ldots$, $\beta = 0.2695\ldots$, $\gamma = 0.8395\ldots$ are the roots of $4x^3 - 4x + 1 = 0$ (the group $E(\mathbb{R})$ is isomorphic to $S^1 \times \mathbb{Z}/2\mathbb{Z}$). We have the real period

$$\omega_1 = \int_{E(\mathbb{R})^0} \frac{dx}{y_1} = 2\int_\gamma^\infty \frac{dx}{\sqrt{4x^3-4x+1}} = 2.993458644\ldots; \tag{3}$$

the numerical value is obtained by using the formula $\omega_1 = \pi/M(\sqrt{\gamma-\alpha}, \sqrt{\gamma-\beta})$, where $M(a,b)$ denotes the arithmetic-geometric mean of Gauss ($M(a,b) = \lim_{n\to\infty} a_n = \lim_{n\to\infty} b_n$ for $a,b > 0$, where $\{a_0,b_0\} = \{a,b\}$, $\{a_{n+1}, b_{n+1}\} = \{\frac{a_n+b_n}{2}, \sqrt{a_nb_n}\}$).

$k = \mathbb{C}$: As well as the real period we have the imaginary period

$$\omega_2 = 2\int_\beta^\gamma \frac{dx}{\sqrt{4x^3-4x+1}} = 2.451389381\ldots i \tag{4}$$

(which can be calculated as $i\pi/M(\sqrt{\beta-\alpha}, \sqrt{\gamma-\alpha})$). The set of complex points of the (projective) curve E is isomorphic to the complex torus $\mathbb{C}/\mathbb{Z}\omega_1 + \mathbb{Z}\omega_2$ via the Weierstrass $\wp$-function:

$$\mathbb{C}/\mathbb{Z}\omega_1 + \mathbb{Z}\omega_2 \xrightarrow{\sim} E(\mathbb{C})$$

$$z \longmapsto \left(p(z), \frac{p'(z)+1}{2}\right),$$

$$p(z) = \frac{1}{z^2} + \sum_{m,n}{}' \left(\frac{1}{(z-m\omega_1-n\omega_2)^2} - \frac{1}{(m\omega_1+n\omega_2)^2}\right)$$

($\sum'_{m,n}$ means $\sum_{(m,n)\neq(0,0)}$), which satisfies

$$p'^2 = 4p^3 - g_2 p - p_3,$$

$$g_2 = 60 \sum_{m,n}{}' \frac{1}{(m\omega_1+n\omega_2)^4} = \frac{4\pi^4}{3\omega_2^4}\left(1 + 240 \sum_{n=1}^{\infty} \frac{n^3}{e^{2\pi i n\omega_1/\omega_2}-1}\right) = 4,$$

$$g_3 = 140 \sum_{m,n}{}' \frac{1}{(m\omega_1+n\omega_2)^6} = \frac{8\pi^6}{27\omega_2^6}\left(1 - 504 \sum_{n=1}^{\infty} \frac{n^5}{e^{2\pi i n\omega_1/\omega_2}-1}\right) = -1.$$

$K = \mathbb{Q}$: The Mordell-Weil group $E(\mathbb{Q})$ is infinite cyclic with generator $P_0 = (0,0)$, the first few multiples being

$$2P_0 = (1,0),\ 3P_0 = (-1,1),\ 4P_0 = (2,3),\ 5P_0 = (\tfrac{1}{4},\tfrac{5}{8}),\ 6P_0 = (6,-14)$$

and their negatives $-(x,y) = (x,1-y)$. If we write nP_0 as (x_n, y_n) and x_n as N_n/D_n with $(N_n, D_n) = 1$, then

$$\log\max(|N_n|, |D_n|) \sim cn^2 \qquad (|n| \to \infty)$$

with a certain positive constant c (in other words, the number of solutions of (1) for which x has numerator and denominator less than B is asymptotic to $2c^{-1/2}(\log B)^{1/2}$ as $B \to \infty$). This constant is called the <u>height</u> of P_0 and denoted $h(P_0)$; it can be calculated via an algorithm of Tate (cf. [14], [2]) as

$$h(P_0) = \sum_{i=1}^{\infty} 4^{-i-1} \log(1 + 2t_i^2 - 2t_i^3 + t_i^4),$$

where the t_i $(=1/x_{2^i})$ are defined inductively by

$$t_1 = 1,\ t_{i+1} = (1 + 2t_i^2 - 2t_i^3 + t_i^4)/(4t_i - 4t_i^3 + t_i^4),$$

and we find the numerical value

$$h(P_0) = 0.0511114082\ldots \,. \qquad (5)$$

Similarly one can define $h(P)$ for any $P \in E(\mathbb{Q})$; clearly $h(nP_0) =$

$n^2h(P_0)$.

<u>$k = \mathbb{Z}/p\mathbb{Z}$:</u> Finally, we can look at E over the finite field $k = \mathbb{Z}/p\mathbb{Z}$, $p \neq 37$ prime. Here $E(k)$ is a finite group of order $N(p)+1$, where

$$N(p) = \#\{x,y \pmod p \mid y^2 - y \equiv x^3 - x \pmod p\}.$$

We combine the information contained in all these numbers into the <u>L-series</u>

$$L_E(s) = \prod_{p\neq 37} \frac{1}{1+\frac{N(p)-p}{p^s}+\frac{p}{p^{2s}}} \cdot \frac{1}{1+\frac{1}{37^s}}; \qquad (6)$$

the special behavior of 37 is due to the fact that $\Delta \equiv 0 \pmod{37}$, so that the reduction of E over $\mathbb{Z}/37\mathbb{Z}$ is singular. Multiplying out, we obtain $L_E(s)$ as a Dirichlet series

$$L_E(s) = \sum_{n=1}^{\infty} \frac{a(n)}{n^s}, \qquad (7)$$

the first few $a(n)$ being given by

n	1	2	3	4	5	6	7	8	9	10	11	12	13	14	15
a(n)	1	-2	-3	2	-2	6	-1	0	6	4	-5	-6	-2	2	6

. (8)

Since clearly $N(p) \leq 2p$, the product (6) and the sum (7) converge absolutely for $\mathrm{Re}(s) > 2$; in fact, $|N(p)-p|$ is less than $2\sqrt{p}$ (Hasse's theorem), so we have absolute convergence for $\mathrm{Re}(s) > 3/2$. We will see in §3 that $L_E(s)$ extends to an entire function of s and satisfies the functional equation

$$\tilde{L}_E(s) := (2\pi)^{-s}37^{s/2}\Gamma(s)L_E(s) = -\tilde{L}_E(2-s); \qquad (9)$$

in particular, $L_E(s)$ vanishes at $s = 1$. The Birch-Swinnerton-Dyer conjecture relates the invariants of E over $\mathbb{R}$, $\mathbb{Q}$ and $\mathbb{Z}/p\mathbb{Z}$ by predicting that

$$\mathrm{ord}_{s=1} L_E(s) = \mathrm{rk}\, E(\mathbb{Q}) = 1$$

and that

$$\frac{d}{ds} L_E(s)\Big|_{s=1} = 2h(P_0)\cdot\omega_1\cdot S \qquad (10)$$

with a certain positive integer S which is supposed to be the order of the mysterious Shafarevich-Tate group Ш. Since the finiteness of

Ш is not known (for E or any other elliptic curve), this last statement cannot be checked. However, $L'_E(1)$ can be computed numerically (cf. §3), and its value 0.3059997738... strongly suggests (cf. (3), (5)) the equation

$$L'_E(1) = 2h(P_0)\omega_1, \tag{11}$$

i.e. (10) with S = 1; the truth of this equation follows from equation (18) below.

2. Twists of L_E; the numbers A(d).

Let p be a prime congruent to 3 (mod 4) which is a quadratic residue of 37 and consider the "twisted" L-series

$$L_{E,p}(s) = \sum_{n=1}^{\infty} \left(\frac{n}{p}\right) \frac{a(n)}{n^s} \tag{12}$$

($(\frac{\cdot}{p})$ = Legendre symbol). The proof of the analytic continuation of L_E will also show that each $L_{E,p}$ continues analytically and has a functional equation under $s \to 2-s$. Now, however, the sign of the functional equation is +, so we can consider the value (rather than the derivative) of $L_{E,p}$ at $s = 1$, and here one can show that

$$L_{E,p}(1) = \frac{2\omega_2}{i\sqrt{p}} A(p)$$

with ω_2 as in (4) and some integer A(p). The value $L_{E,p}(1)$ can be calculated numerically by the rapidly convergent series $L_{E,p}(1) = 2\sum_{n=1}^{\infty} (\frac{n}{p}) \frac{a(n)}{n} e^{-2\pi n/p\sqrt{37}}$ (cf. §4), so we can compute A(p) for small p. The first few values turn out to be

p	3	7	11	47	67	71	83	107	127	139	151	211	223
A(p)	1	1	1	1	36	1	1	0	1	0	4	9	9

. (13)

More generally, $L_{E,d}(s)$ can be defined for all d satisfying $(\frac{d}{37})=1$, $-d$ = discriminant of an imaginary quadratic field K (just replace $(\frac{\cdot}{p})$ in (12) by $(\frac{-d}{\cdot})$, the Dirichlet character associated to $K/\mathbb{Q}$), and we still have

$$L_{E,d}(1) = \frac{2\omega_2}{i\sqrt{d}} A(d) \tag{14}$$

for some $A(d) \in \mathbb{Z}$; the first few values not in (13) are

d	4	40	84	95	104	111	115	120	123	136	148
A(d)	1	4	1	0	0	1	36	4	9	16	9

. (15)

The most striking thing about the values in (13) and (15) is that they are all squares. This is easily understood from the Birch-Swinnerton-Dyer conjecture: the Dirichlet series $L_{E,d}$ is just the L-series of the "twisted" elliptic curve

$$E\langle d\rangle: -dy^2 = 4x^3 - 4x + 1, \tag{16}$$

so $A(d)$ should be either 0 (if $E\langle d\rangle$ has a rational point of infinite order) or (if $E\langle d\rangle(\mathbb{Q})$ is finite) the order of the Shafarevich-Tate group of $E\langle d\rangle$ and hence a perfect square (since this group, if finite, has a non-degenerate $(\mathbb{Q}/\mathbb{Z})$-valued alternating form). Surprisingly, even though we are far from knowing the Birch-Swinnerton-Dyer conjecture or the finiteness of $Ш(E\langle d\rangle)$, we can <u>prove</u> that $A(d)$ is a square for all d, and in fact prove it in <u>two different ways</u>: On the one hand, a theorem of Waldspurger leads to the formula

$$A(d) = c(d)^2, \tag{17}$$

where $c(d)$ $(\in \mathbb{Z})$ is the dth Fourier coefficient of a certain modular form of weight $\frac{3}{2}$. On the other hand, a theorem of Gross and myself gives the formula

$$L_E'(1)L_{E,d}(1) = \frac{4\omega_1\omega_2}{i\sqrt{d}} h(P_d) \tag{18}$$

for a certain explicitly constructed point ("Heegner point") P_d in $E(Q)$; writing P_d as $b(d)$ times the generator P_0 of $E(\mathbb{Q})$ and comparing equation (18) with (14) and (11), we obtain

$$A(d) = b(d)^2. \tag{19}$$

We thus have two canonically given square roots $b(d)$ and $c(d)$ of the integer $A(d)$, and the question arises whether they are equal. The object of this paper is to give a geometrical proof of the fact that this is so. First, however, we must define $b(d)$ and $c(d)$ more precisely, and for this we need the modular description of the elliptic curve E, to which we now turn.

3. <u>The modular curve</u> E.

The essential fact about the elliptic curve E is that it is a <u>modular curve</u>. More precisely, let Γ be the subgroup of $SL_2(\mathbb{R})$

generated by the group

$$\Gamma_0(37) = \{(\begin{smallmatrix} a & b \\ c & d \end{smallmatrix}) \in SL_2(\mathbb{Z}) \mid c \equiv 0 \pmod{37}\}$$

and the matrix $w_{37} = (\begin{smallmatrix} 0 & -1/\sqrt{37} \\ \sqrt{37} & 0 \end{smallmatrix})$. This group acts on the upper half-plane $\mathcal{H}$ in the usual way and the quotient $\mathcal{H}/\Gamma$ can be compactified by the addition of a single cusp ∞ to give a smooth complex curve of genus 1. We claim that this curve is isomorphic to $E(\mathbb{C})$; more precisely, there is a (unique) isomorphism

$$\mathcal{H}/\Gamma \cup \{\infty\} \xrightarrow{\sim} E(\mathbb{C}) \tag{20}$$

sending ∞ to $0 \in E(\mathbb{C})$ and such that the pull-back of the canonical differential $\frac{dx}{2y-1} = \frac{dx}{y_1}$ is $-2\pi i f(\tau)d\tau$, where

$$f(\tau) = q-2q^2-3q^3+2q^4-2q^5+6q^6-q^7+6q^9+\ldots \qquad (q = e^{2\pi i\tau}) \tag{21}$$

is the unique normalized cusp form of weight 2 on Γ, i.e. the unique holomorphic function f on $\mathcal{H}$ satisfying

$$f(\tfrac{a\tau+b}{c\tau+d}) = (c\tau+d)^2 f(\tau) \qquad (\tau \in \mathcal{H}, (\begin{smallmatrix} a & b \\ c & d \end{smallmatrix}) \in \Gamma) \tag{22}$$

and $f(\tau) = q + O(q^2)$ as $\text{Im}(\tau) \to \infty$. This claim is simply the assertion of the Weil-Taniyama conjecture for the elliptic curve under consideration, and it is well-known to specialists that the Weil-Taniyama conjecture can be checked by a finite computation for any given elliptic curve; moreover, the particular curve E was treated in detail by Mazur and Swinnerton-Dyer in [11]. Nevertheless, for the benefit of the reader who has never seen an example of a modular parametrization worked out, we will give the details of the proof of (20); our treatment is somewhat different from that in [11] and may make it clearer that the algorithm used would apply equally well to any elliptic curve. The reader who is acquainted with the construction or who is willing to take (20) on faith can skip the rest of this section.

We have two quite different descriptions of the isomorphism (20), depending whether we use the algebraic model (1) or the analytic model $\mathbb{C}/\mathbb{Z}\omega_1 + \mathbb{Z}\omega_2$ for $E(\mathbb{C})$. We start with the algebraic model. The problem is then to show the existence of two Γ-invariant and holomorphic functions $\xi(\tau)$ and $\eta(\tau)$ satisfying

$$\eta(\tau)^2 - \eta(\tau) = \xi(\tau)^3 - \xi(\tau), \quad -2\pi i f(\tau) = \frac{\xi'(\tau)}{2\eta(\tau)-1} \tag{23}$$

(this gives a map as in (20) with the right pull-back of $\frac{dx}{2y-1}$; that it is an isomorphism is then easily checked). Equations (23) imply that ξ and η have poles of order 2 and 3, respectively, at ∞, and recursively determine all coefficients of their Laurent expansions. Calculating out to 9 terms, we see that these expansions must begin

$$\begin{aligned} \xi(\tau) &= q^{-2}+2q^{-1}+5+9q+18q^2+29q^3+51q^4+82q^5+131q^6+\dots\,, \\ \eta(\tau) &= q^{-3}+3q^{-2}+9q^{-1}+21+46q+92q^2+180q^3+329q^4+593q^5+\dots\,. \end{aligned} \tag{24}$$

So far we have not used the fact that f is a modular form on Γ; we could have taken any power series $f(\tau) = q+\dots$ and uniquely solved (23) to get Laurent series $\xi(\tau) = q^{-2}+\dots$, $\eta(\tau) = q^{-3}+\dots$. However, since ξ and η are supposed to be Γ-invariant functions with no poles in $\mathcal{H}$, and since f is a modular form of weight 2, the two functions $f_4 = f^2\xi$ and $f_6 = f^3\xi$ must be holomorphic modular forms on Γ of weight 4 and 6, respectively. But the space $M_k(\Gamma)$ of modular forms of weight k on Γ is finite-dimensional for any k and one can obtain a basis for it by an algorithmic procedure (e.g., using the Eichler-Selberg trace formulas, but we will find a shortcut here), so we can identify f_4 and f_6 from the beginnings of their Fourier expansions. Once one has candidates f_4 and f_6, one <u>defines</u> $\xi = f_4/f^2$ and $\eta = f_6/f^3$; these are then automatically modular functions on Γ, and the verification of (23) reduces to the verification of the two formulae

$$f_6^{\,2} - f_6 f^3 = f_4^{\,3} - f_4 f^4, \quad f(2f_6 - f^3) = \frac{1}{2\pi i}(2f_4 f' - f f_4') \,, \tag{25}$$

which are identities between modular forms on Γ (of weights 12 and 8, respectively) and hence can be proved by checking finitely many terms of the Fourier expansions. In our case the dimension of $M_k(\Gamma)$ equals $[\frac{5k}{6}] + 3[\frac{k}{4}]$ for $k > 0$, k even, so $M_2(\Gamma)$ is generated by f while $M_4(\Gamma)$ and $M_6(\Gamma)$ have dimension 6 and 8, respectively. However, we will be able to identify f_4 and f_6 without calculating bases for these spaces. The space $M_2(\Gamma_0(37))$ is the direct sum of $M_2(\Gamma) = \mathbb{C}f$ and the 2-dimensional space of modular forms F of weight 2 on $\Gamma_0(37)$ satisfying $F(-\frac{1}{37\tau}) = -37\tau^2F(\tau)$. As a basis of this latter space we can choose the theta-series

$$\Theta(\tau) = \sum_{a,b,c,d\in\mathbb{Z}} q^{Q(a,b,c,d)} = 1+2q+2q^2+4q^3+2q^4+4q^5+8q^6+4q^7+10q^8+\dots,$$

$$\begin{aligned} Q(a,b,c,d) &= \tfrac{1}{8}(4b+c-2d)^2 + \tfrac{1}{4}(2a+c+d)^2 + \tfrac{37}{8}c^2 + \tfrac{37}{4}d^2 \\ &= a^2 + 2b^2 + 5c^2 + 10d^2 + ac + ad + bc - 2bd \end{aligned}$$

and the cusp form

$$h(\tau) = \tfrac{3}{4}\Theta(\tau) - \tfrac{1}{2}E_2^-(\tau) = q + q^3 - 2q^4 - q^7 + \ldots ,$$

where $E_2^-(\tau) = \frac{3}{2} + \sum_{\substack{d,n>0 \\ 37\nmid d}} dq^{nd}$ is an Eisenstein series. The four functions f^2, Θ^2, Θh and h^2 lie in the space

$$U = \{F \in M_4(\Gamma) \mid \mathrm{ord}_{\tau=A}(F) \geq 2,\ \mathrm{ord}_{\tau=B}(F) \geq 4\},$$

where A and B are the fixed points in $\mathcal{H}/\Gamma_0(37)$ of order 2 and 3, respectively, because any function in $M_2(\Gamma_0(37))$ must vanish at A and vanish doubly at B. For the same reason, $f_4 = f^2\xi$ lies in U (recall that ξ has no poles in $\mathcal{H}$); and since U has codimension 2 in $M_4(\Gamma)$ (a general function in $M_4(\Gamma)$ satisfies $\mathrm{ord}_A F = 2r$, $\mathrm{ord}_B F = 3s+1$ for some $r,s \geq 0$), these five functions must be linearly dependent. Looking at the first few Fourier coefficients, we find that f_4 must be given by

$$f_4(\tau) = (\Theta(\tau) - 3h(\tau))^2 - \tfrac{37}{4}h(\tau)^2 - \tfrac{1}{4}f(\tau)^2.$$

As to f_6, we observe that the function

$$\psi(\tau) = \eta(\tau)^2/\eta(37\tau)^2 + 37\eta(37\tau)^2/\eta(\tau)^2$$

is Γ-invariant and holomorphic in $\mathcal{H}$ and has a triple pole at ∞, so must be a linear combination of ξ, η and 1; looking at the first few Fourier coefficients we find that $\psi = \eta - 5\xi + 6$, so f_6 must be $\psi f^3 + 5f_4 f - 6f^3$. As explained above, once we have our candidates f_4 and f_6 it is a finite computation to check (25) and thus establish that $\tau \mapsto (f_4(\tau)f(\tau) : f_6(\tau) : f(\tau)^3)$ maps $\mathcal{H}/\Gamma \cup \{\infty\}$ to $E(\mathbb{C}) \subset \mathbb{P}^2(\mathbb{C})$ as claimed.

For the second description of the map (20), we define a function $\phi : \mathcal{H} \to \mathbb{C}$ by

$$\phi(\tau) = 2\pi i \int_\tau^{i\infty} f(\tau')d\tau' = -q + q^2 + q^3 - \tfrac{1}{2}q^4 + \tfrac{2}{5}q^5 - \ldots . \qquad (26)$$

From $\phi' = -2\pi i f$ and (22) it follows that the difference $\phi(\frac{a\tau+b}{c\tau+d}) - \phi(\tau)$ is a constant for all $\gamma = \binom{a\ b}{c\ d} \in \Gamma$. Call this constant $C(\gamma)$; clearly $C : \Gamma \to \mathbb{C}$ is a homomorphism. The theory of Eichler-Shimura implies that the image $\Lambda = C(\Gamma)$ is a lattice in $\mathbb{C}$ with $g_2(\Lambda)$ and $g_3(\Lambda)$ rational integers. Since we can calculate $\phi(\tau)$ and hence $C(\gamma)$

numerically (the series in (26) converges rapidly), we can calculate a basis of Λ numerically and get g_2 and g_3 exactly. The result $g_2 = 4$, $g_3 = -1$ shows that Λ is the lattice $\mathbb{Z}\omega_1 + \mathbb{Z}\omega_2$ of §1, and the identity $\phi(\gamma\tau) - \phi(\tau) = C(\gamma)$ shows that $\mathcal{H} \xrightarrow{\phi} \mathbb{C} \to \mathbb{C}/\Lambda$ factors through Γ. We thus obtain a map $\mathcal{H}/\Gamma \xrightarrow{\phi} E(\mathbb{C}) = \mathbb{C}/\Lambda$ such that the pull-back $\phi^*(dz)$ equals $-2\pi i f(\tau)d\tau$, as asserted. In practice, it is easier to calculate the image in E of a particular point $\tau \in \mathcal{H}$ by using (26) and reducing modulo Λ than by using the first description of the map (20).

4. Modular forms attached to E

The most important consequence of the modular description of the elliptic curve E is that the L-series of E equals the L-series of the modular form f, i.e. that the numbers $a(n)$ in (7) are precisely the Fourier coefficients in (21). This follows from the Eichler-Shimura theory (cf. [13]). As a consequence, the function $\tilde{L}_E$ defined in (9) has the integral representation

$$L_E(s) = \int_0^\infty f\left(\frac{it}{\sqrt{37}}\right) t^{s-1} dt = \int_1^\infty f\left(\frac{it}{\sqrt{37}}\right)(t^{s-1} - t^{1-s})dt,$$

from which the analytic continuation and functional equation are obvious. Differentiating and setting $s = 1$ we find

$$\frac{\sqrt{37}}{2\pi} L_E'(1) = \tilde{L}_E'(1) = 2\int_1^\infty f\left(\frac{it}{\sqrt{37}}\right)\log t\, dt = 2\sum_{n=1}^\infty a(n) G\left(\frac{2\pi n}{\sqrt{37}}\right),$$

with

$$G(x) = \int_1^\infty e^{-xt}\log t\, dt = \frac{1}{x}\int_x^\infty e^{iu}\frac{du}{u},$$

and since there are well-known expansions for $G(x)$, this can be used to calculate $L_E'(1) = 0.30599\ldots$ to any desired degree of accuracy, as mentioned in §1. Similarly, if $-d$ is the discriminant of an imaginary quadratic field in which 37 splits, then the "twisted" form $f^*(\tau) = \sum\left(\frac{-d}{n}\right)a(n)q^n$ is a cusp form of weight 2 and level $37d^2$ satisfying $f^*(-1/37d^2\tau) = -37d^2\tau^2 f^*(\tau)$, so

$$\tilde{L}_{E,d}(s) := (2\pi)^{-s}37^{s/2}d^s\Gamma(s)L_{E,d}(s) = \int_1^\infty f^*\left(\frac{it}{d\sqrt{37}}\right)(t^{s-1} + t^{1-s})dt,$$

from which we deduce the functional equation $\tilde{L}_{E,d}(s) = \tilde{L}_{E,d}(2-s)$ and the formula $L_{E,d}(1) = 2 \sum_{n=1}^{\infty} (\frac{-d}{n}) \frac{a(n)}{n} e^{-2\pi n/d\sqrt{37}}$ mentioned in §2. In particular, we can calculate the numbers $A(d)$ defined by (14) approximately and hence, since they are integers, exactly.

The other modular form which will be important to us is the form of weight $3/2$ associated to f under Shimura's correspondence. Around ten years ago, Shimura [12] discovered a relationship between modular forms of arbitrary even weight $2k$ and modular forms of half-integral weight $k+1/2$. This was studied subsequently by many other authors. In particular, Kohnen (in [8] for forms of level 1 and in [9] for forms of odd squarefree level) showed how Shimura's theory could be refined by imposing congruence conditions modulo 4 on the Fourier expansion so as to get a perfect correspondence between appropriate spaces of forms of weights $2k$ and $k+1/2$. The result in the case $k = 1$ and prime level is the following ([9], Theorem 2):

Theorem 1 (Shimura; Kohnen). *For N prime and $\varepsilon \in \{\pm 1\}$ let $S^{\varepsilon}_{3/2}(N)$ denote the space of all functions $g(\tau)$ satisfying*

i) $g(\tau)/\theta(\tau)^3$, where $\theta(\tau)$ is the standard theta-series $\sum_{n\in\mathbb{Z}} q^{n^2}$, is invariant under $\Gamma_0(4N)$;

ii) $g(\tau)$ has a Fourier development $\sum_{d>0} c(d)q^d$ with $c(d) = 0$ if $-d \equiv 2$ or $3 \pmod 4$ or $(\frac{-d}{N}) = -\varepsilon$.

Let $S^{\varepsilon}_2(\Gamma_0(N))$ denote the space of cusp forms of weight 2 on $\Gamma_0(N)$ satisfying $f(-1/N\tau) = \varepsilon N\tau^2 f(\tau)$. Then $\dim S^{\varepsilon}_{3/2}(N) = \dim S^{\varepsilon}_2(\Gamma_0(N))$, and for each Hecke eigenform $f = \sum a(n)q^n \in S^{\varepsilon}_2(\Gamma_0(N))$ there is a 1-dimensional space of $g \in S^{\varepsilon}_{3/2}(N)$ whose Fourier coefficients are related to those of f by

$$a(n)c(d) = \sum_{\substack{r|n \\ r>0}} (\frac{-d}{r}) c(\frac{n^2}{r^2}d) \qquad (n\in\mathbb{N},\ -d \text{ a fundamental discriminant}). \tag{27}$$

In our case $N = 37$, $\varepsilon = +1$ and the space $S^{+}_2(\Gamma_0(37))$ is one-dimensional, spanned by the function f of (21). Theorem 1 therefore asserts that there is a unique function

$$g(\tau) = \sum_{\substack{d>0 \\ -d \equiv 0 \text{ or } 1 \pmod 4 \\ (-d/37) = 0 \text{ or } 1}} c(d)e^{2\pi i d\tau}$$

such that $g(\tau)/\theta(\tau)^3$ is $\Gamma_0(148)$-invariant and the Fourier coefficients $c(d)$ (normalized, say, by $c(3) = 1$) satisfy (27). It is not an entirely trivial matter to calculate these coefficients; a method for doing so, and a table up to $d = 250$, were given in [3, pp. 118-120, 145] in connection with the theory of "Jacobi forms." We give a short table:

d	3	4	7	11	12	16	27	28	36	40	44	47	48	63	64	67	71	75	83	...	148
$c(d)$	1	1	-1	1	-1	-2	-3	3	-2	2	-1	-1	0	2	2	6	1	-1	-1	...	-3

(28)

We now come to the theorem of Waldspurger [15], mentioned in §2, which relates these coefficients to the values at $s = 1$ of the twisted L-series $L_{E,d}(s)$. Again we need a refinement due to Kohnen [10, Theorem 3, Cor. 1] which gives a precise and simple identity in the situation of Theorem 1:

Theorem 2 (Waldspurger; Kohnen). Let $f = \sum a(n)q^n \in S_2^{\varepsilon}(\Gamma_0(N))$, $g = \sum c(d)q^d \in S_{3/2}^{\varepsilon}(N)$ correspond as in Theorem 1. Let $-d$ be a fundamental discriminant with $(\frac{-d}{N}) = 0$ or ε and let $L_{f,d}(s)$ be the associated convolution L-series $\sum (\frac{-d}{N}) a(n) n^{-s}$. Then

$$L_{f,d}(1) = 3\pi \frac{\|f\|^2}{\|g\|^2} \frac{|c(d)|^2}{\sqrt{d}} \tag{29}$$

where

$$\|f\|^2 = \int_{\mathcal{H}/\Gamma_0(N)} |f(\tau)|^2 du\,dv, \qquad \|g\|^2 = \int_{\mathcal{H}/\Gamma_0(4N)} |g(\tau)|^2 v^{-1/2} du\,dv \qquad (\tau = u + iv) \tag{30}$$

are the norms of f and g in the Petersson metric. (Note that the identity is independent of the choice of g, since replacing g by λg ($\lambda \in \mathbb{C}^*$) multiplies both $\|g\|*$ and $|c(d)|^2$ by $|\lambda|^2$.)

Actually, the exact coefficient in (29) is not too relevant to us, for knowing that $L_{f,d}(1)$ is a fixed multiple of $c(d)^2/\sqrt{d}$ implies that the numbers $A(d)$ defined by (14) are proportional to $c(d)^2$, and calculating $A(3) = c(3)^2 = 1$ we deduce (17). Then going back and substituting (17) and (14) into (29) we deduce $3\pi\|f\|^2/\|g\|^2 = 2\omega_2/i$. We now show (since the result will be needed later) that

$$\|f\|^2 = \omega_1\omega_2/2\pi^2 i, \tag{31}$$

it then follows that $\|g\|^2 = 3\omega_1/4\pi$. To prove (31), we recall from §3 that there is an isomorphism ϕ from $\mathcal{H}/\Gamma \cup \{\infty\}$ to $E(\mathbb{C}) = \mathbb{C}/\Lambda$ with $\phi^*(dz) = 2\pi i f(\tau)d\tau$. Since $[\Gamma:\Gamma_0(37)] = 2$ we have

$$2\pi^2\|f\|^2 = 4\pi^2 \int_{\mathcal{H}/\Gamma} |f(\tau)|^2 du\,dv = \int_{\mathcal{H}/\Gamma} |-2\pi i f(\tau)|^2 du\,dv$$

$$= \int_{\mathbb{C}/\Lambda} dx\,dy = \omega_1\omega_2/i$$

as claimed.

5. Heegner points on E

In this section we describe a construction which associates to each integer $d > 0$ a point $P_d \in E(\mathbb{Q})$. These are the "modular points" of the title, since their construction depends on the modular description of E given in §3.

We assume first that $-d$ is a fundamental discriminant, i.e. the discriminant of an imaginary quadratic field K. We consider points $\tau \in \mathcal{H}$ of the form $\tau = \frac{b+i\sqrt{d}}{2a}$ with

$$a,b \in \mathbb{Z},\ a > 0,\ 37|a, \quad b^2 \equiv -d \pmod{4a}. \tag{32}$$

If $(\frac{-d}{37}) = -1$, there are no such τ and we set $P_d = 0$; otherwise the set of τ is invariant under Γ and there are h distinct points $\tau_1,\ldots,\tau_h$ modulo the action of Γ, where $h = h(-d)$ is the class number of K. The theory of complex multiplication shows that these points are individually defined over a finite extension H of $\mathbb{Q}$ (the Hilbert class field of K) and collectively over $\mathbb{Q}$ (i.e. their images in $\mathcal{H}/\Gamma$ are permuted by the action of the Galois group of H over $\mathbb{Q}$). Hence the sum $\phi(\tau_1) +\ldots+ \phi(\tau_h)$, where $\phi:\mathcal{H}/\Gamma \to E(\mathbb{C})$ is the map constructed in §3, is in $E(\mathbb{Q})$. Moreover this sum is divisible by u, where u is $\frac{1}{2}$ the number of units of K (= 1,2 or 3) if $37\nmid d$ and $u = 2$ if $37|d$; this is because each point $\tau_j \in \mathcal{H}$ is the fixed point of an element of Γ of order u. We define $P_d \in E(\mathbb{Q})$ by

$$uP_d = \sum_{j=1}^{h} \phi(\tau_j); \tag{33}$$

this is well-defined because $E(\mathbb{Q})$ is torsion-free. If d is not fundamental, we define P_d^0 the same way but with the extra condition $(a,b,\frac{b^2+d}{4a}) = 1$ in (32) (now $h(-d)$ is the class number of a certain non-maximal order of K, and the points $\tau_1,\ldots,\tau_h \in \mathcal{H}/\Gamma$ are defined over the corresponding ring class field), and then set $P_d = \sum_{e^2|d} P^0_{d/e^2}$.

The definition of P_d just given is a special case of a construction due to Heegner and Birch (cf. [1]) and in general would yield rational points in the Jacobian of $X_0(N)/w_N$ $(X_0(N) = \mathcal{H}\cup(\text{cusps})/\Gamma_0(N))$. From a modular point of view, a point $\tau \in \mathcal{H}/\Gamma_0(N)/w_N$ classifies isomorphism classes of unordered pairs of N-isogenous elliptic curves $\{E_1,E_2\}$ over $\mathbb{C}$ (namely $E_1 = \mathbb{C}/\mathbb{Z}+\mathbb{Z}\tau$, $E_2 = \mathbb{C}/\mathbb{Z}+N\mathbb{Z}\tau$, with the isogenies $E_1 \to E_2$, $E_2 \to E_1$ induced by $N\cdot \mathrm{id}_{\mathbb{C}}$ and $\mathrm{id}_{\mathbb{C}}$, respectively), and the points $\tau_1,\ldots,\tau_h$ correspond to those with complex multiplication by an order $\mathcal{O}$ of $\mathbb{Q}(\sqrt{-d})$ (namely $E_1 = \mathbb{C}/\mathfrak{a}$, $E_2 = \mathbb{C}/\mathfrak{n}\mathfrak{a}$, where $\mathfrak{a} = \mathbb{Z}+\mathbb{Z}\tau$ is a fractional $\mathcal{O}$-ideal and $\mathfrak{n}$ an intergral $\mathcal{O}$-ideal of norm N). A general formula for the heights of these "Heegner points" was proved recently by B. Gross and myself [4]; the result in our special case becomes

Theorem 3 (Gross-Zagier): <u>Suppose</u> $-d$ <u>is a fundamental discriminant with</u> $(\frac{-d}{37}) = 1$ <u>and let</u> $P_d \in E(\mathbb{Q})$ <u>be the point defined by (32). Then the height of</u> P_d <u>is given by</u>

$$h(P_d) = \frac{\sqrt{d}}{8\pi^2\|f\|^2} L_E'(1)L_{E,d}(1).$$

(To get this statement from [4], take $\chi = 1$ in Theorem 2 there, noting that $v_{f,1} = uP_d$ and $L(f,1,s) = L_E(s)L_{E,d}(s)$; the height in [4] is one-half that on E because it is calculated on $X_0(37)$ which is a double cover of E.)

In view of equation (31), Theorem 3 is equivalent to the formula (18) given in §2. As explained there, this formula gives both equation (10) for $L_E'(1)$ and the relationship (19) between $A(d)$ and the integers $b(d)$ defined by $P_d = b(d)P_0$. The equality $b(d)^2 = c(d)^2$ suggested comparing the values of $b(d)$ and $c(d)$. Note that the numbers $b(d)$ are numerically calculable: one finds the h Γ-inequivalent solutions of (32) by reduction theory, computes the corresponding values $\phi(\frac{b+i\sqrt{d}}{2a})$ by (26), adds the resulting complex numbers; modulo $\Lambda = \mathbb{Z}\omega_1+\mathbb{Z}\omega_2$, the result must be a multiple of the point $P_0 = -.92959\ldots + \frac{1}{2}\omega_2$. Thus for $d = 67$ we have $h = 1$ and

$$P_{67} = \phi(\tfrac{9+i\sqrt{67}}{74}) = .40936\ldots \equiv 6P_0 \pmod{\Lambda},$$

so $b(67) = 6$; for $d = 83$ we have $h = 3$,

$$P_{83} = \phi(\tfrac{55+i\sqrt{83}}{222}) + \phi(\tfrac{19+i\sqrt{83}}{74}) + \phi(\tfrac{55+i\sqrt{83}}{74}) = (.541\ldots + 1.225\ldots i)$$

$$+ (.194\ldots - .570\ldots i) + (.194\ldots + .570\ldots i) \equiv -P_0 \pmod{\Lambda},$$

so $b(83) = -1$; for $d = 148$ we have $h = 2$,

$$2P_{148} = \phi(\frac{i}{\sqrt{37}}) + \phi(\frac{1}{2} + \frac{i}{\sqrt{37}}) = .19189... - .60125... \equiv -6P_0 (\mathrm{mod}\ \Lambda),$$

so $b(148) = -3$. In this way one can make a table of the multiples $b(d)$. Such a table (up to $d = 150$) was computed by B. Gross and J. Buhler, while I was independently computing the Fourier coefficients $c(d)$ by the method mentioned in §4; the letter with their data arrived in Germany on the very morning that I had completed my computations and drafted a letter to them, and the perfect agreement of the two tables gave ample reason to conjecture the following:

Theorem 4. $b(d) = c(d)$ <u>for all</u> d.

The remainder of this paper is devoted to the proof of this result.

6. <u>Curves on Hilbert modular surfaces</u>

In view of the uniqueness clause in Theorem 1, what we need to do to prove Theorem 4 is simply to show that $\sum b(d)q^d$ belongs to $S^+_{3/2}(37)$, i.e. that the positions of the Heegner points in the Mordell-Weil group of E are the Fourier coefficients of a modular form of weight 3/2. This statement is reminiscent of a theorem of Hirzebruch and the author [7] according to which the positions of certain modular curves in the homology group of a modular surface are the Fourier coefficients of a modular form of weight 2. Since this result is not only very analogous to the one we want, but will actually be used to prove it, we recall the exact statement.

Let p be a prime congruent to $1 \pmod 4$ and let $\mathcal{O} = \mathbb{Z} + \mathbb{Z}\frac{1+\sqrt{p}}{2}$ be the ring of integers in $\mathbb{Q}(\sqrt{p})$. The group $PSL_2(\mathcal{O})$ (<u>Hilbert modular group</u>) acts on $\mathcal{H} \times \mathcal{H}$ by

$$M\circ(\tau_1,\tau_2) = (\frac{a\tau_1+b}{c\tau_1+d}, \frac{a'\tau_2+b'}{c'\tau_2+d'}) \qquad (M = \pm\begin{pmatrix} a & b \\ c & d\end{pmatrix} \in PSL_2(\mathcal{O}),\quad \tau_1,\tau_2 \in \mathcal{H}),$$

where $'$ denotes conjugation in $\mathbb{Q}(\sqrt{p})/\mathbb{Q}$. The quotient $\mathcal{H} \times \mathcal{H}/SL_2(\mathcal{O})$ can be naturally compactified by the addition of finitely many points ("cusps"), and when the singularities thus introduced are resolved by cyclic configurations of rational curves according to Hirzebruch's recipe [6] the resulting surface $Y = Y_p$ is a nearly smooth compact algebraic surface (it still has quotient singularities coming from the points in $\mathcal{H} \times \mathcal{H}$ with a non-trivial isotropy group in $PSL_2(\mathcal{O})$, so it is a rational homology manifold or "V-manifold"). The middle homology of Y splits as

$$H_2(Y) \;=\; H_2^c(Y) \oplus \langle S_1\rangle \oplus \ldots \oplus \langle S_r\rangle, \tag{34}$$

where $S_1,\ldots,S_r$ are (the homology classes of) the curves used in the resolutions of the cusp singularities and $H_2^c(Y)$ consists of homology classes orthogonal to the S_j; the homology groups in (34) are taken with coefficients in $\mathbb{Q}$.

For each integer $N > 0$ there is an algebraic curve $T_N \subset Y$ defined as follows. Consider all equations

$$A\tau_1\tau_2 + \frac{\lambda}{\sqrt{p}}\tau_1 - \frac{\lambda'}{\sqrt{p}}\tau_2 + B \;=\; 0 \tag{35}$$

with $A,B \in \mathbb{Z}$, $\lambda \in \mathcal{O}$, and $\lambda\lambda' + ABp = N$. Each one defines a curve in $\mathcal{H}\times\mathcal{H}$ isomorphic to $\mathcal{H}$ and the union of these curves is invariant under $SL_2(\mathcal{O})$; T_N is defined as closure in Y of the image of this union in $\mathcal{H}\times\mathcal{H}/SL_2(\mathcal{O})$. If $(\frac{N}{p}) = -1$, there are no solutions of $\lambda\lambda' + ABp = N$ and T_N is empty. If $(\frac{N}{p}) = +1$ then T_N is irreducible (all equations (35) are equivalent under $PSL_2(\mathcal{O})$) and isomorphic to the modular curve $X_0(N)$. The main result of [7] is

Theorem 5 (Hirzebruch-Zagier). _Let_ $[T_N^c]$ _denote the projection to_ $H_2^c(Y)$ _of the homology class of_ T_N _in the splitting (34). Then the power series_ $\sum_{N=1}^{\infty} [T_N^c]e^{2\pi i N\tau}$ _is a modular form of weight_ 2, _level_ p _and Nebentypus_ $(\frac{}{p})$.

Here "modular form of weight 2, level p and Nebentypus" means a modular form $F(\tau)$ satisfying $F(\frac{a\tau+b}{c\tau+d}) = (\frac{a}{p})(c\tau+d)^2F(\tau)$ for $\binom{a\ b}{c\ d} \in \Gamma_0(p)$; when we say that a power series with coefficicents in $H_2^c(Y)$ is such a form we mean that each component (with respect to a basis of $H_2^c(Y)$ over $\mathbb{Q}$) is. Alternatively, if $[X]$ is any homology class in $H_2(Y)$, then the power series $\sum (X\circ T_N^c)e^{2\pi i N\tau}$, where $(X\circ T_N^c)$ denotes the intersection pairing of $[X]$ and $[T_N^c]$, is a modular form of the specified type, now with ordinary numerical Fourier coefficients. In particular, this is true for $X = T_M$, one of our special curves on Y. In fact the proof of Theorem 5 in [7] consisted in calculating the intersection numbers $(T_M\circ T_N^c)$ explicitly and showing that they were the Fourier coefficients of a modular form. The formula obtained for $(T_M\circ T_N^c)$, in the case when N and M are coprime, was

$$(T_M\circ T_N^c) \;=\; \sum_{\substack{x^2<4NM \\ x^2\equiv 4NM \pmod p}} H\left(\frac{4NM-x^2}{p}\right) + I_p(MN) \tag{36}$$

where

$$H(d) = \sum_{e^2|d} h'(-d/e^2)$$

($h'(-d) = h(-d)$ for $d > 4$, $h'(-3) = 1/3$, $h'(-4) = 1/2$) and $I_p(n)$ is a certain arithmetical function whose definition we do not repeat. The proof of (36) was geometrical: the physical intersection points of T_M and T_N in $\mathcal{H} \times \mathcal{H}/PSL_2(\mathcal{O})$ are in 1:1 correspondence with certain equivalence classes of binary quadratic forms and are counted by the first term in (36), while the term $I_p(MN)$ counts the intersection points of T_M and T_N at infinity and the intersection of T_M with the combination of cusp-resolution curves S_j which was removed from T_N to get T_N^c.

7. Heegner points as intersection points of modular curves on modular surfaces

Now suppose that p is a prime satisfying $p \equiv 1 \pmod 4$, $(\frac{p}{37}) = 1$, and (for later purposes) $p > 2 \cdot 37$, say $p = 101$. As already mentioned, the curve T_{37} on Y_p is in this case isomorphic to $X_0(37) = \mathcal{H} \cup \{\text{cusps}\}/\Gamma_0(37)$. For instance, if $p = 101$ we can get an equation (35) for T_{37} by taking $A = B = 0$ and $\lambda = 21+2\sqrt{101}$, an element of $\mathcal{O}$ of norm 37; then the solution of (35) is given parametrically by $\{(\lambda\tau, \lambda'\tau),\ \tau \in \mathcal{H}\}$ and the matrices $M \in SL_2(\mathcal{O})$ which preserve this set are those of the form $\begin{pmatrix} a & b\lambda \\ c/\lambda & d \end{pmatrix}$ with $\begin{pmatrix} a & b \\ c & d \end{pmatrix} \in \Gamma_0(37)$, so we get a degree 1 map $\mathcal{H}/\Gamma_0(37) \to \mathcal{H} \times \mathcal{H}/SL_2(\mathcal{O})$ and hence a map $X_0(37) \to Y_{101}$. On Y_p we have an extra involution ι which is induced by the involution $(\tau_1, \tau_2) \mapsto (\tau_2, \tau_1)$ of $\mathcal{H} \times \mathcal{H}$, and this induces the involution w_{37} on $T_{37} = X_0(37)$, so our curve $E \simeq \mathcal{H} \cup \{\text{cusps}\}/\Gamma$ can be found on the quotient surface Y/ι. However, since all T_N are invariant under ι and there is no difference (except a factor of 2) between the intersection theory of ι-invariant curves on Y or of their images in Y/ι, we will continue to work on the surface Y rather than the quotient surface Y/ι, which has a one-dimensional singular locus.

In §5 we constructed for each $d > 0$ a set of $(1+(\frac{-d}{37}))H(d)$ points in $X_0(37)$, namely the set of roots of quadratic equations $a\tau^2+b\tau+c=0$ with $b^2-4ac=-d$ and $37|a$. (If d is of the form $3n^2$ or $4n^2$ then $H(d)$ is not an integer and we are using the convention that a fixed point of an element of order u in $\Gamma_0(37)$ is to be counted with multiplicity $\frac{1}{u}$ in $\mathcal{H}/\Gamma_0(37)$; from now on we will ignore this technicality.) Call this set $\mathcal{P}_d$. The point $P_d \in E(\mathbb{Q})$ was (one-half of) the sum of the images of the points of $\mathcal{P}_d$ in E. If we worked on $X_0(37)$, or on some other $X_0(M)$ of higher genus, we would have to take the sum in the Jacobian of the curve rather than on the curve itself, i.e. P_d would be the point of $Jac(X_0(M))$ represented by the divisor $\mathcal{P}_d - \deg(\mathcal{P}_d)\cdot(\infty)$ of degree 0.

The geometric content of (36) is that the intersection points of T_N and T_M in $\mathcal{H}\times\mathcal{H}/PSL(\mathcal{O})$ are the points of $\mathcal{P}_d$ for certain d, namely those of the form $\frac{148N-x^2}{p}$, i.e.

$$T_{37} \cap T_N = \bigcup_{\substack{|x|<\sqrt{148N} \\ x^2\equiv 148N \pmod p}} \mathcal{P}_{(148N-x^2)/p} \cup D_\infty \tag{37}$$

where D_∞ is contained in the part of Y_p at infinity (resolutions of the cusp singularities); here when we write union we of course mean for the points to be counted with appropriate multiplicities, i.e. we are working with divisors rather than just sets of points. If we simply <u>count</u> the points in (37), i.e. replace each $\mathcal{P}_d$ by its degree, we obtain the numbers (36), and Theorem 5 tells us that these are the Fourier coefficients of a modular form of weight 2, level p, and Nebentypus $(\frac{-}{p})$. If instead we <u>add</u> the points in (37) in the Jacobian of T_{37}, i.e. replace each $\mathcal{P}_d$ by P_d, then we will deduce from this that the corresponding statement holds:

Proposition: <u>For</u> $N > 0$ <u>define</u> $B(N)$ <u>by</u>

$$B(N) = \sum_{\substack{x^2<148N \\ x^2\equiv 148N \pmod p}} b\left(\frac{148N-x^2}{p}\right)$$

<u>with</u> $b(d)$ <u>as in</u> §5. <u>Then</u> $\sum B(N)q^N$ <u>is a modular form of weight</u> 2, <u>level</u> p <u>and Nebentypus</u> $(\frac{-}{p})$.

<u>Proof</u>. Let $\mathcal{M}$ denote the set of all modular forms of the specified type, so that Theorem 5 asserts

$$\sum (T_N^c \circ X)q^N \in \mathcal{M} \quad \text{for all} \quad [X] \in H_2(Y). \tag{38}$$

The space $\mathcal{M}$ is finite-dimensional and has a basis consiting of modular forms with rational Fourier coefficients. Hence there is an infinite set $\mathcal{R}$ of finite relations over $\mathbb{Z}$ defining $\mathcal{M}$, i.e. a set $\mathcal{R}$ whose elements are sequences

$$R = (r_0, r_1, r_2, \ldots), \quad r_N \in \mathbb{Z}, \quad r_N = 0 \text{ for all but finitely many } N$$

and such that

$$\sum_{N=0}^{\infty} C(N)q^N \in M \Leftrightarrow \sum_{N=0}^{\infty} r_N C(N) = 0 \qquad (\forall R \in R). \tag{39}$$

(For instance, one could find integers $N_1,\ldots,N_d$ with $d = \dim M$ and such that the $N_j{}^{th}$ Fourier coefficients of forms in M are linearly independent; then for each N we have a relation $C(N) = \sum_{j=1}^{d} \lambda_j C(N_j)$ with rational numbers $\lambda_1,\ldots,\lambda_d$, and we could take for R the set of these relations, each multiplied by a common denominator.) Equation (38) now implies that

$$\sum_{N=1}^{\infty} r_N (T_N^c \circ X) = 0$$

for all $R \in R$, and since this holds for <u>all</u> homology classes X, we must have $\sum r_N [T_N^c] = 0$ in $H_2(Y,\mathbb{Q})$. Since T_N^c is a linear combination of T_n and curves S_j coming from the cusp resolutions, this means

$$\sum_{N=1}^{\infty} r_N [T_N] + \sum_{j=1}^{r} s_j [S_j] = 0 \tag{40}$$

in $H_2(Y,\mathbb{Q})$ for some rational numbers $s_1,\ldots,s_r$. Multiplying by a further common denominator we can assume that the s_j are also integers and that the relation (40) holds in integral homology. But the Hilbert modular surface Y is known to be simply connected, so the exact sequence

$$0 = H^1(Y,\mathcal{O}) \to H^1(Y,\mathcal{O}^*) \to H^2(Y,\mathbb{Z}) \qquad (\mathcal{O} = \text{structure sheaf of } Y)$$

induced from $0 \to \mathbb{Z} \to \mathcal{O} \to \mathcal{O}^* \to 0$ shows that any divisor on Y which is homologous to 0 is linearly equivalent to 0. Hence the relation (39) implies that the divisor $\sum r_N T_N + \sum s_j S_j$ is the divisor of a meromorphic function on Y, i.e. there is a meromorphic function Φ on Y which has a zero or pole of order r_N on each T_N (resp. s_j on each S_j) and no other zeros or poles. If we restrict Φ to T_{37}, then it follows that the zeros and poles of Φ occur at the intersection points of T_{37} with other T_N and at the cusps, and in fact (by (38)) that

$$\text{divisor of } \Phi\Big|_{T_{37}} = \sum_{N\geq 1} r_N \sum_{\substack{x^2<148N \\ x^2\equiv 148N \pmod p}} P_{(148N-x^2/p)} + d_\infty$$

where d_∞ is a divisor with support concentrated at the cusps. Take the image in E, observing that the cusps map to 0, and add the points obtained; since the points of a principal divisor sum to zero and the points of P_d sum to $b(d)P_0$, we deduce $\sum r_N B(N) = 0$ with $B(N)$ as in the Proposition. The desired result now follows from equation (39).

8. Completion of the proof

We are now nearly done. For each $N > 0$ define

$$C(N) = \sum_{\substack{x^2<148N \\ x^2\equiv 148N \pmod p}} c\left(\frac{148N-x^2}{p}\right),$$

where $c(d)$ are the Fourier coefficients defined in §4. Then

$$G(z) := \sum_{N>0} C(N)q^N = \sum_{\substack{d>0 \\ x\in\mathbb{Z} \\ pd+x^2\equiv 0 \pmod{148}}} c(d)q^{(pd+x^2)/148}$$

$$= g(pz)\theta(z)\,|\,U_{148},$$

where $\theta = \sum q^{x^2}$ and U_m is the map which picks out every m^{th} coefficient of a Fourier expansion, i.e.

$$\phi(z)\,|\,U_m = \frac{1}{m}\sum_{j \pmod m} \phi\left(\frac{z+j}{m}\right).$$

Since g is a modular form of weight $3/2$ and θ one of weight $1/2$, and since U_m maps modular forms to modular forms of the same weight, it is clear that $G(z)$ is a modular form of weight 2; a routine calculation shows that it has level p and Nebentypus $(\frac{\cdot}{p})$. Hence both $G(z)$ and $F(z) = \sum B(N)q^N$ belong to the finite-dimensional space M. Moreover, since $b(d) = c(d)$ for small d by the calculations mentioned in §5, the first Fourier coefficients of F and G agree, and this suffices to show $F = G$. Specifically, with $p = 101$ the agreement of $c(d)$ and $b(d)$ for $d < 150$ implies the agreement of $B(N)$ and $C(N)$ for $1 \le N \le 100$, and this is more than enough to ensure that $F = G$ (it would suffice to have agreement up to $N = 9$). Hence $B(N) = C(N)$ for all N. We claim that this implies $b(d) = c(d)$ for all d. Indeed, suppose inductively that $b(d') = c(d')$ for all $d' < d$. If $(\frac{-d}{37}) = -1$ or $-d \equiv 2$ or $3 \pmod 4$ then $c(d)$ and $b(d)$ are both zero and there is nothing to prove. Otherwise we can find an integer n with

$$n^2 \equiv -pd \pmod{148}, \quad |n| \le 37.$$

Take $N = \frac{pd+n^2}{148}$. Then in the equations

$$B(N) = \sum_{\substack{x^2<148N \\ x^2\equiv 148N \pmod p}} b\left(\frac{148N-x^2}{p}\right), \quad C(N) = \sum_{\substack{x^2<148N \\ x^2\equiv 148N \pmod p}} c\left(\frac{148N-x^2}{p}\right)$$

the numbers $\pm n$ occur as values of x and all other values of x are larger in absolute value because $|n| \le 37 < \frac{1}{2}p$ by assumption. Thus $B(N)$ equals 1 or 2 times $b(d)$ plus a certain linear combination of $b(d')$ with $d' < d$, and $C(N)$ equals the same multiple of $c(d)$ plus the same linear combination of lower $c(d')$, so the equality $B(N) = C(N)$ and the inductive assumption $b(d') = c(d')$ imply that $b(d) = c(d)$ as desired.

9. Generalization to other modular curves

Our exposition so far was simplified by several special properties of the elliptic curve E: that it was actually isomorphic to a modular curve rather than just covered by one, that its Mordell-Weil group had rank one and no torsion, etc. We end the paper by discussing to what extent the results proved for E generalize to other curves.

First, we could replace E by an arbitrary elliptic curve whose L-series coincides with the L-series of a modular form f of weight 2 and some (say, prime) level N, with $f(-\frac{1}{N\tau}) = N\tau^2 f(\tau)$. Then we would again have a covering map $\phi: X_0(N)/w_N \to E$, Heegner points $P_d \in E(\mathbb{Q})$ for all $d>0$ (with $P_d = 0$ if $-d \not\equiv 0 \pmod{4N}$), and a relationship $c(d)^2 \sim h(P_d)$ for the Fourier coefficients $c(d)$ of a modular form in $S^+_{3/2}(N)$ corresponding to f as in Theorem 1. We could then ask whether all the P_d belong to a one-dimensional subspace $\langle P_0\rangle$ of $E(\mathbb{Q})/E(\mathbb{Q})_{tors}$ and, if so, whether the coefficients $b(d)$ defined by $P_d = b(d)P_0 \in E(\mathbb{Q}) \otimes \mathbb{Q}$ are proportional to the Fourier coefficients $c(d)$. More generally, we could forget elliptic curves entirely and simply start with a modular curve $X_0(N)$ or $X_0(N)/w_N$ (still, say, with N prime). The construction of §5 yields Heegner points P_d in the Jacobian of this curve over $\mathbb{Q}$. To avoid torsion we tensor with $\mathbb{Q}$ and write $V = \mathrm{Jac}(X_0(N)/w_N)(\mathbb{Q}) \otimes_{\mathbb{Z}} \mathbb{Q}$. The Hecke algebra acts on V the same way as it acts on cusp forms of weight 2, so $V \otimes \mathbb{R}$ splits as $\oplus_f V_f$, where the f are Hecke eigenforms $f = \sum_1^\infty a(n)q^n$ in $M_2^+(\Gamma_0(N))$ (normalized by $a(1) = 1$) and V_f is the subspace of $V \otimes \mathbb{R}$ on which the nth Hecke operator acts as multiplication by $a(n)$. For each f we define $P_{d,f}$ as the component of P_d in V_f. The Fourier coefficients $a(n)$ will be in $\mathbb{Z}$ if f corresponds to an elliptic curve E defined over $\mathbb{Q}$; in that case V_f is isomorphic to $E(\mathbb{Q}) \otimes_{\mathbb{Z}} \mathbb{Q}$ and we are back in the situation described before. In general the $a(n)$ will be integers in an algebraic number field $K_f \subset \mathbb{R}$, the Fourier coefficients $c(d)$ of the form in $S_{3/2}^+(N)$ corresponding to f can also be chosen to lie in K_f, and the main theorem of [4] combined with Theorem 2 tells us that $h(P_{d,f})$ is proportional to $c(d)^2$. This suggests that the right generalization of Theorem 4 is:

Theorem 6. _Let_ f, $c(d)$ _be as above. Then_ $P_{d,f} = c(d)P_0$ _for all_ d _and some_ $P_0 \in V_f$. _In particular, the projections_ $P_{d,f}$ _of the Heegner points all lie in a one-dimensional subspace of_ V_f.

Theorem 6 is equivalent (because of the uniqueness clause in Theorem 1 and the way the Hecke operators act on Heegner points) to the following apparently weaker theorem:

Theorem 6'. _The power series_ $\sum_{d>0} P_d q^d$ _is a modular form of weight_ $3/2$ _and level_ N.

(As with Theorem 5, this means that $\sum P_d q^d \in V[[q]]$ belongs to the subspace $V \otimes S_{3/2}^+(N)$ or, in more down-to-earth terms, that each component of this power series, with respect to a fixed basis of V over $\mathbb{Q}$, is a modular form in $S_{3/2}^+(N)$.)

How can we prove these theorems? The argument of §§6-7 permits us to embed our modular curve in the Hilbert modular surface Y_p for any prime $p \equiv 1 \pmod 4$ with $(\frac{N}{p}) = 1$ and to prove that the power series

$$\sum_M \left(\sum_{\substack{x^2 < 4NM \\ x^2 \equiv 4NM \ (\mathrm{mod}\ p)}} P_{(4NM-x^2)/p} \right) q^M$$

is a modular form (with coefficients in V) of weight 2, level p and Nebentypus. To deduce Theorem 6' we would need the following assertion:

Let $h(\tau)$ _be a power series of the form_

$$\sum_{\substack{d>0 \\ -d\equiv\text{square} \pmod{4N}}} b(d)q^d$$

with N *prime, and suppose that the power series*

$$h(p\tau)\theta(\tau)|U_N = \sum_{M>0}\left(\sum_{\substack{x^2<4NM \\ x^2\equiv 4NM \pmod p}} b\left(\frac{4NM-x^2}{p}\right)\right)q^M$$

is a modular form of weight 2, *level* p *and Nebentypus* $(\frac{\cdot}{p})$ *for every prime* $p \equiv 1 \pmod 4$ *with* $(\frac{N}{p}) = 1$. *Then* h *belongs to* $S^+_{3/2}(N)$.

This assertion is extremely likely to be true. The argument of §8 proves it — even if the hypothesis on $h(p\tau)\theta(\tau)|U_N$ is made for only one prime $p > 2N$ — under the additional assumption that one possesses a candidate $g = \sum c(d)q^d \in S^+_{3/2}(N)$ for h with $c(d) = b(d)$ for sufficiently many values of d. Thus the method of proof we used for $N = 37$ can be used for any other fixed value of N if we do a finite amount of computation. To get a general proof of Theorems 6 and 6' along these lines one would need either to prove the assertion above or else to generalize the geometric proof in some way (perhaps by using Hilbert modular surfaces of arbitrary discriminant, for which the intersection theory has been worked out by Hausmann [5]).

In any case, however, we would like to have a proof of Theorem 6 using only intrinsic properties of the modular curve, rather than its geometry as an embedded submanifold of an auxiliary modular surface. Such a proof has been given by B. Gross, W. Kohnen and myself. It is a direct generalization of the main result of [4]: instead of a formula for the height $h(P_d)$ of a Heegner point, we give a formula for the height <u>pairing</u> $(P_d, P_{d'})$ of two Heegner points, where $(\ ,\ ): V \times V \to \mathbb{R}$ is the bilinear form associated to the quadratic form h. The formula implies that $\sum_{d>0} (P_d, P_{d'})q^d$ belongs to $S^+_{3/2}(N)$ for each discriminant d', and Theorem 6' follows.

Finally, we mention that the correct generalization of Theorem 4 to composite levels N should be formulated using the theory of "Jacobi forms" developed in [3] rather than the theory of modular forms of half-integral weight. This, too, will be carried out in the joint work with Kohnen and Gross mentioned above.

Bibliography

[1] B. Birch, Heegner points of elliptic curves, Symp. Mat., Ist. di Alta Mat. 15(1975), 441-445.

[2] J. Buhler, B. Gross and D. Zagier, On the conjecture of Birch and Swinnerton-Dyer for an elliptic curve of rank 3, to appear in Math. Comp. (1985).

[3] M. Eichler and D. Zagier, The Theory of Jacobi Forms, to appear in Progress in Mathematics, Birkhäuser, Boston-Basel-Stuttgart (1985).

[4] B. Gross and D. Zagier, Points de Heegner et dérivées de fonctions L, C.R. Acad. Sci. Paris 297(1983), 85-87.

[5] W. Hausmann, Kurven auf Hilbertschen Modulflächen, Bonner Math. Schriften 123(1980).

[6] F. Hirzebruch, Hilbert modular surfaces, L'Ens. Math. 19(1973), 183-281.

[7] F. Hirzebruch and D. Zagier, Intersection numbers of curves on Hilbert modular surfaces and modular forms of Nebentypus, Inv. Math. 36(1976), 57-113.

[8] W. Kohnen, Modular forms of half-integral weight on $\Gamma_0(4)$, Math Ann. 248(1980), 249-266.

[9] W. Kohnen, New forms of half-integral weight, J. reine Angew. Math. 333(1982), 32-72.

[10] W. Kohnen, Fourier coefficients of modular forms of half-integral weight, to appear in Math. Ann. (1985).

[11] B. Mazur and H.P.F. Swinnerton-Dyer, Arithmetic of Weil curves, Inv. Math. 25(1974), 1-61.

[12] G. Shimura, On modular forms of half-integral weight, Ann. of Math. 97(1973), 440-481.

[13] H.P.F. Swinnerton-Dyer and B. Birch, Elliptic curves and modular functions, in Modular Functions of One Variable IV, Lecture Notes in Math. 476, Springer, Berlin-Heidelberg-New York (1975), 2-32.

[14] J. Tate, Letter to J-P. Serre, Oct. 1, 1979.

[15] J.-L. Waldspurger, Sur les coefficients de Fourier des formes modulaires de poids demi-entier, J. Math. pures et appl. 60 (1981), 375-484.

25. Mathematische Arbeitstagung 1984

Ad-hoc - Vorträge

EIGENVALUES OF THE DIRAC OPERATOR

Michael Atiyah
Mathematical Institute
Oxford OX1 3LB

§1. The Theorems

In recent years mathematicians have learnt a great deal from physicists and in particular from the work of Edward Witten. In a recent preprint [3], Vafa and Witten have proved some striking results about the eigenvalues of the Dirac operator, and this talk will present their results. I shall concentrate entirely on the mathematical parts of their preprint leaving aside the physical interpretation which is their main motivation.

The mathematical context is the following. We fix a compact Riemannian spin manifold M of dimension d, and denote by D the Dirac operator of M acting on the spin bundle S. In addition if we are given a hermitian vector bundle V with a connection A we can define the extended Dirac operator:

$$D_A : S \otimes V \to S \otimes V.$$

In terms of an orthonormal basis e_j of tangent vectors D_A is given locally by $\sum_{j=1}^{d} e_j \nabla_j$, where ∇_j is the covariant derivative in the e_j-direction and e_j acts on spinors by Clifford multiplication. In particular D_A depends on A only in the 0-order term, i.e. if B is a second connection on V, then $D_A - D_B$ is a multiplication operator not involving derivatives.

The operator D_A is self-adjoint and has discrete eigenvalues λ_j, both positive and negative, which we will suppose indexed by increasing absolute value so that

$$|\lambda_1| \leq |\lambda_2| \leq \dots .$$

The questions which Vafa and Witten address themselves to concern the way in which the λ_j depend on A (and V): the metric on M is assumed fixed throughout. More precisely they are interested in getting uniform upper bounds. The simplest and most basic of their results is

THEOREM 1. *There is constant* C *(depending on* M *but not on* V *or* A)*. such that* $|\lambda_1| \le C$.

More generally there is a uniform estimate for the n-th eigenvalue:

THEOREM 2. *There is a constant* C' *(depending on* M *but not on* V, A *or* n) *such that* $|\lambda_n| \le C'n^{1/d}$.

Remarks. 1) The asymptotic formula $\lambda_n \sim n^{1/d}$ is a very general result for eigenvalues of elliptic operators, but Theorem 2 is much more precise.

2) Theorem 1 does not hold for the Laplace operator Δ_A of V. To see this just consider $d = 2$ and V to be a line-bundle of constant curvature F : then $\lambda_1 = |F| \to \infty$ with the Chern class of V. This emphasizes that the uniformity in Theorems 1 and 2 is with respect to the continuous parameter A and also with respect to the discrete parameters describing the topological type of V.

3) The inequalities in Theorems 1 and 2 go in the opposite direction to the Kato inequalities for eigenvalues of Laplace type operators. This had, in principle, been conjectured by physicists on the grounds of Fermion-Boson duality.

For odd-dimensional manifolds there are even stronger results, namely:

THEOREM 1*. *If* d *is odd, there exists a constant* C_* *so that every interval of length* C_* *contains an eigenvalue of* D_A.

THEOREM 2*. _If_ d _is odd, there exists a constant_ $C_*^!$ _so that every interval of length_ $C_*^! n^{1/d}$ _contains_ n _eigenvalues._

Note that Theorems 1* and 2* are definitely false in _even_ dimensions. To see this recall that, when d is even, S decomposes as $S^+ \oplus S^-$ and D_A is of the form

$$(1.1) \qquad D_A = \begin{pmatrix} 0 & D_A^- \\ D_A^+ & 0 \end{pmatrix}$$

so that $D_A^2 = D_A^- D_A^+ \oplus D_A^+ D_A^-$, and the non-zero eigenvalues of the two factors $D_A^- D_A^+$ and $D_A^+ D_A^-$ coincide. If V has large positive curvature then typically $D_A^- D_A^+$ will have a zero-eigenvalue of large multiplicity while $D_A^+ D_+^-$ will be a 'large' positive operator. Hence D_A^2 will have a large gap between its 0-eigenvalue and its first non-zero eigenvalue. Moreover this gap tends to infinity with the size of the curvature of V. When $d = 2$ and V is a line-bundle of constant curvature it is just the first Chern class of V which determines the size of the first gap.

§2. _The even proof_

Although the theorems we have just stated appear purely _analytical_ results, involving upper bounds on eigenvalues, it is a remarkable feature of the work of Vafa and Witten that the proofs are essentially _topological_. To understand how this comes about I will consider first the case when the dimension d is _even_. Then, as observed at the end of §1, the spinors decompose and D_A takes the form given in (1.1). In particular a 0-eigenvalue of D_A arises whenever either D_A^+ or D_A^- has a non-trivial nullspace N_A^+ or N_A^- respectively. Next recall that the _index_ of D_A^+ is defined as

$$\text{index } D_A^+ = \dim N_A^+ - \dim N_A^-$$

so that a non-zero value for index D_A^+ forces D_A to have a 0-eigenvalue. On the other hand index D_A^+ is a purely topological invariant, given by an explicit formula [1] involving characteristic cohomology classes of V and M. Hence, whenever the index, computed topologically, is non-zero we have a 0-eigenvalue for D_A (for all connections A on the given bundle V) and so trivially Theorem 1 holds.

For d even Theorem 1 therefore has significant content only for those bundles V for which the index formula gives zero. To treat these the key idea is now the following. Suppose we can find a connection A_o on V so that

(i) D_{A_o} has a 0-eigenvalue.

(ii) $\|D_A - D_{A_o}\| \leq C$

then it will follow that the smallest eigenvalue of D_A does not exceed C. Now we cannot actually find such a connection on V itself but we can find one on some multiple $NV = V \otimes C^N$ of V, and this will do equally well since the only effect of taking multiple copies of D_A is to increase the multiplicity of each eigenvalue.

We now proceed as follows. First choose a bundle W' so that the index of D^+ on $S^+ \otimes V \otimes W'$ is non-zero. From the index formula (and the assumption that the index of D^+ on $S \otimes V$ is zero) it is enough to take W' to be the pull-back to M of a generating bundle on S^{2d} (i.e. with $c_{2d} = (d-1)!$) by a map $M \to S^{2d}$ of degree 1: this makes the index equal to dim V. Thus for any connection B' on W' (which combines with A to give a connection say A' on $V \otimes W'$) the operator $D_{A'}^+$ has a non-zero index. Hence $D_{A'}$ has a zero-eigenvalue.

Next choose an orthogonal complement W" to W', i.e. a bundle so that

(2.1) $W' \oplus W'' \cong M \times C^N$

and fix a connection B" on W" (defining, together with A, a connection A" on $V \otimes W''$). The operator

$$D_{A'\oplus A''} = D_{A'} \oplus D_{A''}$$

still of course has a zero-eigenvalue (since $D_{A'}$ has). On the other hand $A_o = A' \oplus A''$ is a connection on

$$V \otimes (W' \oplus W'') \cong V \otimes C^N = NV$$

and so can be compared with the connection NA (once we have fixed the isomorphism (2.1)). Comparing the corresponding Dirac operators we find

(2.2) $D_{A_o} - D_{NA} = B$

where B is the matrix valued 1-form which describes the connection $B' \oplus B''$ in the trivialization given by (2.1). Since B is quite independent of V and A we get a uniform constant $C = ||B||$ and this completes the proof of Theorem 1 in the even case.

Note that the simple formula (2.2), which is essential for the proof, depends on the fact that the highest order part of D_A is independent of A.

To prove Theorem 2 (for even d) we proceed in a similar manner but this time we pull back the bundle W' (and its complement W") from S^d by using maps of degree n. The index formula then shows that $D^+_{A'}$ has index n dim V. Theorem 2 then follows easily if one can show that the constants $C = ||B||$ grow like $n^{1/d}$. When M is a torus T and $n = r^d$ (with r an integer) this follows by using the covering map $T^d \to T^d$ given by $x \to rx$: since B is a 1-form (with matrix values) it picks up a factor r. For general M one applies this construction to a small box in M and the case of general n follows by interpolation.

§3. The odd proof

If we replace M by $M \times S^1$, where S^1 is the circle, the eigenvalues λ_j of D_A get replaced by $\pm\sqrt{\lambda_j^2 + m^2}$ where m runs over the integers. The smallest eigenvalues are therefore the same on M and on $M \times S^1$. This means that theorem 1 for d even, when applied to $M \times S^1$, immediately yields Theorem 1 for d odd. A similar but more careful count of eigenvalues shows that Theorem 2 for d even also implies Theorem 2 for d odd.

Notice also that conversely, if we first establish Theorems 1 and 2 for d odd, they then follow for d even. In fact for d odd we want to establish directly the much stronger results given by Theorems 1* and 2*. The reason why the odd case yields stronger results is roughly the following. In §2, for d even, we used the index theorem, together with a deformation argument relating a connection A to another connection A_o. In the odd case the analogue of the index theorem is itself concerned with 1-parameter families, as we shall now recall.

Suppose that D_t is a periodic one-parameter family of self-adjoint elliptic operators, with the parameter $t \in S^1$. The eigenvalues λ_j are now functions of t and when t goes once round the circle the λ_j have, as a set, to return to their original position. However λ_j need not return to λ_j: we may get a shift, e.g. λ_j might return to λ_{j+n} for some integer n. This integer n is called the <u>spectral flow</u> of the family and it is a topological invariant of the family. It represents the number of negative eigenvalues which have become positive (less the number of positive eigenvalues which have become negative).

The spectral flow, like the index, is given by an explicit topological formula [2]. Moreover, for the first order differential operators (e.g. Dirac operators) this formula is actually related to an index formula as follows. If D_t is the family, defined on a manifold M, consider the single operator

$$\mathscr{D} = \frac{\partial}{\partial t} + D_t$$

defined on $M \times S^1$. Note that

$$\mathscr{D}^* = -\frac{\partial}{\partial t} + D_t$$

so that $\mathscr{D}$ is not self-adjoint. Then one has [2]

(3.1) spectral flow of $\{D_t\}$ = index of $\mathscr{D}$.

As an illustrative example consider the case when M is also a circle with angular variable x and take

$$D_t = -i\frac{\partial}{\partial x} + t.$$

The eigenvalues are $n + t$ with n integral and so, as t increases from 0 to 1, we get a spectral flow of precisely one. The periodicity of D_t is expressed by the conjugation property:

$$D_{t+1} = e^{-ix} D_t e^{ix} .$$

The operator $\mathscr{D}$ acts naturally on the functions $f(x,t)$ such that

$$\begin{aligned} (3.2) \qquad & f(x+1,t) = f(x,t) \\ & f(x,t+1) = e^{-ix} f(x,t). \end{aligned}$$

In fact these equations describe sections of a certain line-bundle on the torus $S^1 \times S^1$.

Functions satisfying (3.2) have a Fourier series expansion

$$(3.3) \quad f(x,t) = \Sigma f_n(t)e^{inx}$$

where $f_n(t+1) = f_{n+1}(t)$.

Solving the equation $\mathcal{D}f = 0$ leads to the relations

$$f_n'(t) + (n+t)f_n(t) = 0$$

and so

$$f_n(t) = C_n \exp\{-\frac{(n+t)^2}{2}\} .$$

In view of the conditions (3.3) C_n is independent of n. Thus $\mathcal{D}$ has a one-dimensional null-space spanned by the theta function

$$f(x,t) = \exp\left(\frac{-t^2}{2}\right)\sum_n \exp(inz - n^2/2)$$

where $z = x - it$. A similar calculation shows that $\mathcal{D}^*f = 0$ has no L^2-solution, so that index $\mathcal{D} = 1$ which checks with the spectral flow.

After this digression about spectral flow we return to consider the Dirac operators D_A on a manifold M of odd dimension d. Let $S^d \to U(N)$ be a generator of $\pi_d(U(N))$, where we take N in the stable range, i.e. $N \geq \frac{d+1}{2}$, and now compose with a map $M \to S^d$ of degree one to give a map $F : M \to U(N)$. Consider F as a multiplication operator on the bundle $S \otimes NV = S \otimes V \otimes C^N$, on which the Dirac operator D_{NA} is defined. Since the matrix parts of F and D_{NA} act on different factors in the tensor product they commute, and so

$$[D_{NA}, F] = X$$

is independent of A. This multiplication operator X acts

essentially on $S \otimes C^N$ (trivially extended to $S \otimes V \otimes C^N$), and is locally given by

$$X = \Sigma e_i \; F^{-1} \; \partial_i F.$$

In particular $\|X\| = C$ is a uniform constant independent of V and A.

Consider now the linear family of connections

$$A_t = (1 - t)A + t\,F(A)$$

joining A to its gauge transform $F(A)$. The corresponding family of Dirac operators is

(3.4) $D_t = D_A + tX.$

By construction $D_o = D_A$ and $D_1 = F^{-1}D_A F$ is unitarily equivalent to D_o. Thus we have a periodic family of self-adjoint operators with a spectral flow. Moreover the general formula for the spectral flow (e.g. via the index formula on $M \times S^1$) shows that in our case, because of the construction of F, we have spectral flow equal to one. It follows that, for some value of t, the operator D_t has a zero-eigenvalue. Hence as before (3.4) shows that the smallest eigenvalue of D_A does not exceed C.

The use of spectral flow to prove Theorem 1 for odd d is so far quite similar to the use of the index to prove Theorem 1 for even d. However, spectral flow has the advantage that 0 is not a distinguished point of the spectrum, i.e. the spectral flow of a family is unchanged by adding a constant. Replacing 0 by some other value μ and repeating our argument then shows that there is an eigenvalue of D_A within C of μ, and this is the content of Theorem 1*.

Theorem 2* follows by extending the argument using maps $F : M \to U(N)$ of higher degree, on the same lines as Theorem 2 was proved in the even case.

Finally it is worth pointing out that the upper bounds on the eigenvalues of Dirac operators given by those methods are fairly sharp. In fact Vafa and Witten actually determine the best bound when M is a flat torus. For this they use the index theorem for multi-parameter families of elliptic operators - not just the spectral flow of a one-parameter family.

References

1. M.F. Atiyah and I.M. Singer, The index of elliptic operators III, Ann. of Math. 87 (1968), 546-604.

2. M.F. Atiyah, V.K. Patodi and I.M. Singer, Spectral asymmetry and Riemannian geometry III, Math. Proc. Camb. Phil. Soc. 79 (1976), 71-99.

3. C. Vafa and E. Witten, Eigenvalue inequalities for fermions in gauge theories. Princeton University preprint, April 1984. [Commun. in Math. Physics, 95, No.3 (1984), 257-276.]

MANIFOLDS OF NON POSITIVE CURVATURE

W. Ballmann
Mathematisches Institut
Wegelerstraße 10
5300 Bonn 1

This is mainly a report on recent and rather recent work of the author and others on Riemannian manifolds of nonpositive sectional curvature. The names of the other people involved are M. Brin, K. Burns, P. Eberlein and R. Spatzier.

Denote by M^n a complete connected smooth Riemannian manifold, by K_M the sectional curvature of M and by d the distance on M induced by the Riemannian metric. We always assume $K_M \leq 0$, that is, $K_M(\sigma) \leq 0$ for every tangent plane σ of M.

One of the significant consequences of the assumption $K_M \leq 0$ is as follows. Let γ_1 and γ_2 be unit speed geodesics in the universal covering space $\tilde{M}$ of M such that $\gamma_1(0) = \gamma_2(0)$. Then for $t, s \geq 0$

$$d^2(\gamma_1(t), \gamma_2(s)) \geq t^2 + s^2 - 2ts \cdot \cos(\dot{\gamma}_1(0), \dot{\gamma}_2(0))$$

with equality if and only if $\gamma_1|[0,t]$ and $\gamma_2|[0,s]$ belong to the boundary of a totally geodesic and flat triangle. It follows that the exponential map $\exp: T_p\tilde{M} \longrightarrow \tilde{M}$ is a diffeomorphism for each $p \in \tilde{M}$. In particular, M is a $K(\pi,1)$; the homotopy type of M is determined by $\Gamma = \pi_1(M)$. As we will see below, there are also strong relations between the structure of Γ and the geometry of M.

One of the principal aims in the study of nonpositively curved manifolds is to specify the circumstances under which assertions about negatively curved manifolds become false - if they become false - under the weaker assumption of nonpositive sectional curvature. For example, a theorem of Milnor [Mi] asserts that Γ has exponential growth if M is compact and negatively curved. As for the weaker assumption $K_M \leq 0$, Avez [Av] showed that Γ has exponential growth if and only if M is not flat.

In general, one expects some kind of flatness in M if some property of negatively curved manifolds is not shared by M. Hence it is only natural to try to measure the flatness of M. In the case of locally symmetric spaces, the rank is such a measure. The question arises, whether such a notion can be introduced in a meaningful way for general manifolds of nonpositive sectional curvature. This is indeed the content of Problem 65 in Yau's list [Y]. We state this problem in a slightly modified form and in two parts.

a) DEFINE THE RANK OF M AND SHOW THAT Γ CONTAINS A FREE ABELIAN SUBGROUP OF RANK k IF M IS COMPACT OF RANK k.

Note that in the case M is compact and locally symmetric, the (usual) rank of M is given by the maximal number k such that Γ contains a free abelian subgroup of rank k. See also Theorem 1 below.

b) SHOW THAT Γ CONTAINS A FREE ABELIAN SUBGROUP OF RANK 2 IF M HAS A 2-FLAT.

Here a k-flat is defined to be a totally geodesic and isometrically immersed Euclidean space of dimension k.

As in the case of locally symmetric spaces, the rank of M should be an integer between 1 and $n = \dim(M)$. Further properties of this notion, which one expects, are as follows.

P1) IF M IS LOCALLY SYMMETRIC, THEN THE RANK OF M SHOULD COINCIDE WITH ITS USUAL RANK.

P2) FLAT MANIFOLDS OF DIMENSION n SHOULD HAVE RANK n. NEGATIVELY CURVED MANIFOLDS SHOULD HAVE RANK ONE.

Vice versa, manifolds of rank one should resemble negatively curved manifolds.

P3) THE RANK OF $\tilde{M}$ SHOULD BE EQUAL TO THE RANK OF M. THE RANK OF A RIEMANNIAN PRODUCT $M_1 \times M_2$ SHOULD BE THE SUM OF THE RANKS OF M_1 AND M_2.

Note that $M_1 \times M_2$ still has nonpositive sectional curvature. If M_1 and M_2 are compact, then $M_1 \times M_2$ does not carry a metric of negative sectional curvature, see Theorem 1 below.

Of course, there may be different satisfactory solutions to problem a). One candidate for the rank of M, and maybe the most obvious one, is the following:

$$\text{Rank}(M) = \max\{k \mid M \text{ contains a } k\text{-flat}\}.$$

At this point it is only conjectural that this notion of rank solves problem a). Also note that with this definition of rank, problem b) is part of problem a). With respect to Rank (M) , the following results are known.

<u>Theorem 1</u> (Gromoll-Wolf [GW], Lawson-Yau [LY]). If M^n is compact, then every abelian subgroup of Γ is free abelian of rank at most n . If Γ contains a free abelian subgroup of rank k , then M contains a totally geodesic and isometrically immersed flat k-torus.

This result is the extension of the theorem of Preissmann [Pr] which states that every abelian subgroup of Γ is infinite cyclic if M is compact and negatively curved. Theorem 1 implies that

$$\text{Rank}(M) \geq \max\{k \mid \Gamma \text{ contains a free abelian subgroup of rank } k\}$$

if M is compact. Problem a) now consists in showing that equality holds.

We say that M satisfies the visibility axiom if any two distinct points in the ideal boundary of $\tilde{M}$ can be joined by a geodesic [EO]. For example, compact negatively curved manifolds satisfy the visibility axiom.

<u>Theorem 2</u> (Eberlein [E1]). If M is compact, then M satisfies the visibility axiom if and only if M does not contain a 2-flat, that is, Rank (M) = 1.

Thus problem b) can be reformulated as saying that M satisfies the visibility axiom if and only if every abelian subgroup of Γ is infinite cyclic.

We now discuss a different notion of rank which was introduced in [BBE]. We need some definitions. Denote by SM the unit tangent bundle of M. For $v \in SM$, let γ_v be the geodesic which has v as initial velocity vector. Along γ_v consider the space $J^P(v)$ of all parallel Jacobi fields. Note that by the assumption $K_M \leq 0$, a parallel field X along γ_v, which is linearily independent of $\dot{\gamma}_v$, is such a parallel Jacobi field if and only if $K_M(\dot{\gamma}_v(t) \wedge X(t)) = 0$ for all t. Now set

$$\text{rank}(v) = \dim(J^P(v)) \quad \text{and}$$

$$\text{rank}(M) = \min\{\text{rank}(v) \mid v \in SM\}.$$

Note that $\text{rank}(M) = 1$ if M has a point p such that the sectional curvatures of all tangent planes at p are negative. In particular, $\text{rank}(M) = 1$ if M is a compact surface of negative Euler characteristic

The above definition of rank was motivated by the results in the papers [B1], [B2], and [BB] which deal primarily with geodesic flows on manifolds of rank one. (Formally, the general assumption in [B1] and [B2] is that M has a geodesic which does not bound a flat half plane, but in view of Theorem 4 below this is equivalent to $\text{rank}(M) = 1$.) The geodesic flow g^t operates on SM, and by definition $g^t(v) = \dot{\gamma}_v(t)$. The geodesic flow leaves invariant the Liouville measure of SM.

We now state some of the properties of manifolds of rank one.

<u>Theorem 3.</u> Suppose $\text{rank}(M) = 1$.

i) [BB] If M is compact, then g^t is ergodic.

ii) [B1] If M has finite volume, then g^t has a dense orbit.

iii) (Eberlein [B2]) If M has finite volume, then tangent vectors to closed geodesics are dense in SM.

Part i) of this theorem generalizes, at the same time, the celebrated theorem of Anosov that the geodesic flow on a compact negatively curved manifold is ergodic [An] and the result of Pesin that the geodesic flow on a compact surface of negative Euler characteristic is ergodic [Pe]. The proof of part i) makes essential use of the results of Pesin [Pe] and of the results in [B1].

As for manifolds of higher rank, the following result is one of the basic ingredients in all the further developments.

Theorem 4 [BBE]. If the volume of M is finite or if M is analytic, then

$$\text{rank}(M) = \max\{ k \mid \text{each geodesic of } M \text{ is contained in a } k\text{-flat} \}.$$

In particular, $\text{rank}(M) \leq \text{Rank}(M)$. There are examples where this inequality is strict, see the introduction of [BBE]. In an earlier version of Theorem 4, Burns proved that each geodesic in M bounds a flat half plane if $\text{rank}(M) \geq 2$, see [Bu].

The counterpart to Theorem 3 in the higher rank case is as follows.

Theorem 5. Suppose that $\text{rank}(M) = k \geq 2$ and that K_M has a lower bound $-a^2$.

i) [BBE] If M has finite volume, then g^t is not ergodic.

ii) [BBS] If M has finite volume, then g^t has k-1 independent differentiable first integrals on an open, dense, and g^t-invariant subset of SM.

iii) [BBS] If M is compact, then tangent vectors to totally geodesic and isometrically immersed flat k-tori are dense in SM.

It follows from iii) that Γ contains free abelian subgroups of rank k if $\text{rank}(M) = k$. In particular, problem a) is solved with this notion of rank.

There are some immediate questions related to the assumptions in Theorems 3 and 5. Namely, is it possible to delete the assumption that K has a lower bound in Theorem 5 and the assumption that M is compact in part i) of Theorem 3? I believe that the answer is yes in both cases. That the compactness assumption can be deleted in part iii) of Theorem 5 is a consequence of the following result.

<u>Theorem 6 [B3, BS].</u> Suppose that rank $(M) \geq 2$, K_M has a lower bound $-a^2$ and M has finite volume. If $\tilde{M}$ is irreducible, then M is a locally symmetric space of noncompact type.

Actually, Burns-Spatzier [BS] need the stronger assumption that M is compact. Under the further assumptions M compact and dim $(M) \leq 4$, Theorem 6 was proved earlier by the author in joint work with Heintze [BH]. All these proofs are along completely different lines, up to the fact that they are based on the results in [BBE] and [BBS].

The use of Theorem 6 lies in the fact that, for many purposes, it will be sufficient to prove a given assertion in the rank one case and the symmetric space case separately in order to get a conclusion in the general case. Using this device and results of Prasad-Raghunathan [PR], the author in collaboration with Eberlein defined algebraically a number rank (Γ) , the rank of the fundamental group Γ of M , and showed that rank (Γ) = rank (M) . Using various other previous results of Eberlein and a recent result of Schroeder one obtaines the following conclusion.

<u>Theorem 7 [BE].</u> Suppose that $K_M \geq -a^2$ and M has finite volume. Then M is an irreducible locally symmetric space of noncompact type of rank $k \geq 2$ if and only if the following three conditions are satisfied:

- i) Γ does not contain a normal abelian subgroup (except $\{e\}$)
- ii) no finite index subgroup of Γ is a product
- iii) rank $(\Gamma) = k$.

Here a Riemannian manifold N is called irreducible if no finite covering of N is a Riemannian product. Theorem 7 can be used to extend the rigidity results of Mostow [Mo] and Margulis [Ma]. Namely, using their results and Theorem 7 we obtain:

<u>Theorem 8.</u> Suppose that $K_M \geq -a^2$ and M has finite volume. Suppose M* is an irreducible locally symmetric space of noncompact type and higher rank with finite volume. If the fundamental groups of M and M* are isomorphic, then M and M* are isometric up to normalizing constants.

Under the stronger assumption that M is compact, Theorem 8 was proved earlier by Gromov [GS] and, in a special case, by Eberlein [E2].

References

[An] D.V. Anosov, Geodesic Flows on Closed Riemannian Manifolds with Negative Curvature, Proc. Steklov Inst. Math. 90, Amer. Math. Soc., Providence, Rhode Island, 1969.

[Av] A. Avez, "Variétés riemanniennes sans points focaux", C.R. Acad. Sc. Paris 270 (1970), 188 - 191.

[B1] W. Ballmann, "Einige neue Resultate über Mannigfaltigkeiten nicht positiver Krümmung", Bonner math. Schriften 113 (1978), 1 - 57.

[B2] W. Ballmann, "Axial isometries of manifolds of non-positive curvature", Math. Ann. 259 (1982), 131 - 144.

[B3] W. Ballmann, in preparation.

[BB] W. Ballmann and M. Brin, "On the ergodicity of geodesic flows", Erg. Th. Dyn. Syst. 2 (1982),311 - 315.

[BBE] W. Ballmann, M. Brin and P. Eberlein, "Structure of manifolds of nonpositive curvature. I", Preprint, Bonn - College Park - Chapel Hill 1984.

[BBS] W. Ballmann, M. Brin and R. Spatzier, "Structure of manifolds of nonpositive curvature. II", Preprint, Bonn - College Park - Berkeley 1984.

[BE] W. Ballmann and P. Eberlein, in preparation.

[BH] W. Ballmann and E. Heintze, unpublished.

[Bu] K. Burns, "Hyberbolic behavior of geodesic flows on manifolds with no focal points", Erg. Th. Dyn. Syst. 3 (1983), 1 - 12.

[BS] K. Burns and R. Spatzier, in preparation.

[E1] P. Eberlein, "Geodesic flow in certain manifolds without conjugate points", Trans. AMS 167 (1972), 151 - 170.

[E2] P. Eberlein, "Rigidity of lattices of nonpositive curvature", Erg. Th. Dyn. Syst. 3 (1983), 47 - 85.

[EO] P. Eberlein and B. O'Neill, "Visibility manifolds", Pac. J. Math. 46 (1973), 45 - 109.

[GW] D. Gromoll and J. Wolf, "Some relations between the metric structure and the algebraic structure of the fundamental groups in manifolds of nonpositive curvature", Bull. AMS 77 (1971), 545-552.

[GS] M. Gromov and V. Schroeder, Lectures on Manifolds of Nonpositive Curvature, in preparation.

[LY] H.B. Lawson and S.-T. Yau, "Compact manifolds of nonpositive curvature", J. Differential Geometry 7 (1972), 211 - 228.

[Ma] G.A. Margulis, "Discrete groups of motions of manifolds of non-positive curvature", AMS Translations 109 (1977), 33 - 45.

[Mi] J. Milnor, "A note on curvature and fundamental group", J. Differential Geometry 2 (1968), 1 - 7.

[Mo] G.D. Mostow, Strong Rigidity of Locally Symmetric Spaces, Annals of Math. Studies No. 78, Princeton University Press, Princeton, New Jersey, 1973.

[Pe] Ja.B. Pesin, "Geodesic flows on closed Riemannian manifolds without focal points", Math. USSR Izv. 11 (1977), 1195 - 1228.

[PR] G. Prasad and M.S. Raghunathan, "Cartan subgroups and lattices in semi-simple groups", Ann. of Math. 96 (1972), 296 - 317.

[Pr] A. Preissmann, "Quelques propriêtês globales des espaces de Riemann", Comment. Math. Helvetici 15 (1943), 175 - 216.

[Y] S.-T. Yau, Seminar on Differential Geometry, Annals of Math. Studies No. 102, Princeton University Press and University of Tokyo Press, Princeton, New Jersey, 1982.

Metrics with Holonomy G_2 or Spin (7)

by

Robert L. Bryant

§1. The Holonomy of Riemannian Manifolds

In this section, all objects are assumed smooth unless stated otherwise, M will denote a connected, simply connected n-manifold and g will denote a Riemannian metric on M. If $\gamma: [0,1] \to M$ is a path in M, then the Levi-Civita connection of g induces a well-defined parallel translation along γ, $P_\gamma: T_{\gamma(0)}M \to T_{\gamma(1)}M$ which is an isometry of vector spaces. For every $x \in M$, we let H_x denote the set of all P_γ where γ ranges over all paths with $\gamma(0) = \gamma(1) = x$. It is well-known, see [1], that the simple connectivity of M implies that H_x is a connected, closed Lie subgroup of $SO(T_xM)$, the group of oriented isometries of T_xM with itself. Moreover $P_\gamma(H_{\gamma(0)}) = H_{\gamma(1)}$ for any path γ. It follows that by choosing an isometry $i: T_xM \simeq \mathbb{R}^n$, we can identify H_x with a subgroup $H \subseteq SO(n)$. The conjugacy class of H in $O(n)$ is independent of the choice of x or i. By abuse of language we speak of H as the holonomy of g.

The holonomy group is a measure of the curvature of g. For example, if H preserves an orthogonal decomposition $\mathbb{R}^n = \mathbb{R}^{n_1} \oplus \mathbb{R}^{n_2}$, then $g = g_1 + g_2$ locally where g_i is a local metric on $\mathbb{R}^{n_i}$. It follows that, in order to determine which subgroups of $SO(n)$ can be holonomy groups of Riemannian metrics, it suffices to determine the subgroups $H \subseteq SO(n)$ which act irreducibly on $\mathbb{R}^n$ and are holonomy groups of Riemannian metrics. By examining the Bianchi identities and making extensive use of representation theory, Berger [2] proved the following classification theorem.

<u>Theorem (Berger):</u> Let (M^n,g) be a connected, simply connected Riemannian n-manifold and suppose that its holonomy group $H \subseteq SO(n)$ acts irreducibly on $\mathbb{R}^n$. Then either (M,g) is locally symmetric or else H is one of the following subgroups of $SO(n)$

(i) $SO(n)$

(ii) $U(m)$ if $n = 2m > 2$

(iii) $SU(m)$ if $n = 2m > 2$

(iv) $Sp(1)Sp(m)$ if $n = 4m > 4$

(v) $Sp(m)$ if $n = 4m > 4$

(vi) G_2 if $n = 7$

(vii) $Spin(7)$ if $n = 8$

(viii) $Spin(9)$ if $n = 16$

After noting that the above list is exactly the list of subgroups of $SO(n)$ which act transitively on $S^{n-1} \subseteq \mathbb{R}^n$, Simons [3] gave a direct proof that the holonomy of an irreducible non-symmetric metric on M^n acts transitively on S^{n-1}.

It is natural to ask which of the possibilities on Berger's list actually do occur. It is easy to show that the "generic" metric on M^n has holonomy $SO(n)$. If $n = 2m$, a matric with holonomy a subgroup of $U(m)$ is, of course, a Kähler metric. Such a metric is given in local coordinates on $\mathbb{C}^m$ in the form

$$g_f = (\partial^2 f/\partial z^i \, \partial \bar{z}^j) dz^i \circ d\bar{z}^j$$

where f is a smooth function on $\mathbb{C}^m$ satisfying the condition that its complex hessian $H_f = (\partial^2 f/\partial z^i \, \partial \bar{z}^j)$ be positive definite. For a "generic" f with $H_f > 0$, the metric g_f will have holonomy $U(m)$. Every metric on M^{2m} with holonomy $H \subseteq SU(m)$ can be put in the above form locally where f satisfies the complex Monge-Ampere equation $\det(H_f) = 1$. Again, the "generic" solution of this equation yields a metric whose holonomy is exactly SU(m). Since Sp(m) ⊆ SU(2m), we ca even construct metrics whose holonomy is Sp(m) on M^{4m} (m > 1) locally by selecting a linear map $J: \mathbb{C}^{2m} \to \mathbb{C}^{2m}$ satisfying $J^2 = -I$ and $J = -{}^tJ$ and considering the g_f where f satisfies the system of equations $\bar{H}_f\, J\, H_f = J$. Even though this is an overdetermined system of equations for f, enough solutions can be found to exhibit local metrics with holonomy exactly Sp(m). A similar construction with complex contact structures on $\mathbb{C}^{2m+1}$ allows one to exhibit metrics locally on $\mathbb{R}^{4m}$ with holonomy Sp(1) · Sp(m). It must be emphasized that it is the encoding of holonomy properties into the Cauchy-Riemann equations (which are completely understood locally) that allows the construction of metrics in cases (ii)-(v) to be reduced to a managable partial differential equations problem.

There remain the "exceptional" cases (vi)-(viii). In a surprising paper, Alekseevski [4] showed that any metric on M^{16} with holonomy Spin(9) was necessarily locally symmetric. Thus, case (viii) can be removed from Berger's list. It is worth remarking that cases (vi) and (vii) do not occur as symmetric spaces [6]. This raises the possibility that these two cases do not occur at all. As of this writing, no examples of cases (vi) or (vii) are known. Nevertheless, there is extensive literature on the properties of these elusive metrics. See [7], [8], and [9] and the bibliographies contained therein.

In this lecture, we shall outline a proof of the existence of local metrics in cases (vi) and (vii). The details, which involve an analysis of a differential system to be constructed below will be published elsewhere. For the appropriate concepts from differential systems and Cartan-Kähler theory, the reader may consult [10].

§2. Linear Algebra, H-structure, and Differential Systems

Our strategy will be to describe a set of differential equations whose solutions will represent metrics on M^n with the desired holonomy. We begin by giving a somewhat non-standard description of G_2.

Let $\omega^1, \omega^2, \ldots, \omega^7$ be an oriented orthonormal coframing of $\mathbb{R}^7$. We define the 3-form

$$\varphi = \omega^{123} + \omega^{145} + \omega^{167} + \omega^{246} - \omega^{257} - \omega^{356} - \omega^{347}$$

where ω^{ijk} is an abbreviation for $\omega^i \wedge \omega^j \wedge \omega^k$.

<u>Proposition 1</u>: $G_2 = \{A \in GL(7) | A^*(\varphi) = \varphi\}$ where G_2 is the 14-dimensional simple Lie group of compact type.

We will not prove Proposition 1 here. It is interesting to note that a dimension count shows that the orbit of φ in $\Lambda^3(\mathbb{R}^7)$ under $GL(7)$ is open. (In fact, there are exactly two open $GL(7)$ orbits in $\Lambda^3(\mathbb{R}^7)$. The stabilizer of a form $\tilde{\varphi}$ in the other open orbit is the simple Lie group of non-compact type of dimension 14.) The form φ was discovered by Chevalley [5]. Bonan [7] showed that

$$\Lambda^{G_2} = \text{span}\{1, \varphi, *\varphi, *1 = (1/7)\varphi \wedge *\varphi\}$$

where $\Lambda^{G_2} \subseteq \Lambda(\mathbb{R}^7)$ is the subring of G_2-invariant exterior forms.

If V is a seven dimensional vector space, we will say that $\alpha \in \Lambda^3(V^*)$ is <u>positive</u> if there exists a linear isomorphism $L: V \to \mathbb{R}^7$ so that $\alpha = L^*(\varphi)$. The set $\Lambda^3_+(V^*) \subseteq \Lambda^3(V^*)$ of positive forms is clearly an open subset of $\Lambda^3(V^*)$. If $\alpha \in \Omega^3(M^7)$ we say that α is positive iff $\alpha|_x$ is positive for all $x \in M^7$. We let $E \subseteq \Lambda^3(T^*M)$ denote the open submanifold of positive 3-forms. $\pi: E \to M$ is a smooth fiber bundle with fibers isomorphic to $GL(7)/G_2$. The sections of E are the positive forms on M and are also obviously in 1-1 correspondence with the set of G_2 reductions of the tangent bundle of M, i.e., G_2-structures on M. Since

$G_2 \subseteq SO(7)$, it follows that each G_2-structure on M induces a canonical underlying orientation and Riemannian metric.

On the other hand, if (M^7,g) is an oriented Riemannian manifold with holonomy G_2, it is easy to see that there is a unique parallel positive 3-form α_g on M whose underlying orientation and metric are the given ones.

Proposition 2: Let α be a positive 3-form on M, and let $*\alpha$ be the dual 4-form with respect to the underlying metric and orientation. Then α is parallel with respect to the underlying metric's Levi-Civita connection iff $d\alpha = d{*\alpha} = 0$.

Proposition 2 is due to Gray [8] in the context of vector cross products. It follows from this that every positive 3-form α which satisfies the system of partial differential equations $d\alpha = d{*\alpha} = 0$ has an underlying metric whose holonomy is a subgroup of G_2 and conversely every metric whose holonomy is a subgroup of G_2 arises from such an α.

The conditions $d\alpha = d{*\alpha} = 0$ form a quasi-linear first order system for the 35 (= $\dim \Lambda^3_+(V^*)$) unknown coefficients of α. The system is quasi-linear because coefficients of $*\alpha$ are algebraic functions of the coefficients of α. A priori, this appears to be 56 (= $\dim(\Lambda^4(\mathbb{R}^7) \oplus \Lambda^5(\mathbb{R}^7))$ equations for the 35 unknowns. However, there is a (miraculous) identity

$$({*d\beta}) \wedge \beta + ({*d{*\beta}}) \wedge {*\beta} = 0$$

valid for any positive β where the $*$ is the Hodge star of the underlying $SO(7)$ structure. It can be shown that the remaining 49 = 56 - 7 equations are independent.

This overdetermined system is invariant under the diffeomorphism group of M and hence cannot be elliptic. However, it can be shown that it is transversely elliptic, i.e., elliptic when restricted to a local slice of the action of Diff(M) on $\Omega^3_+(M)$.

Our first main result is

Theorem 1: The system $d\alpha = d{*}\alpha = 0$ for $\alpha \in \Omega^3_+(M)$ is involutive with Cartan characters $(s_1, s_2, \ldots, s_7) = (0,0,1,4,10,13,7)$. In particular, the "generic" solution has the property that its underlying metric has holonomy exactly G_2.

We remark that Theorem 1 is essentially a calculation. One describes the appropriate differential system with independence condition on $E \subseteq \Lambda^3(T^*M)$ and calculates both the integral elements and the Cartan characters to arrive at the result. Note that this system is real analytic in local coordinates. The transversality property actually implies that any solution is real analytic in some coordinate system anyway, so the application of Cartan-Kähler theory is vindicated. Details will appear elsewhere.

We now turn to the analogous case $H = \mathrm{Spin}(7)$. Write $\mathbb{R}^8 = \mathbb{R}^1 \oplus \mathbb{R}^7$ and augment the given coframing of $\mathbb{R}^7$ by an ω^0. We then define the 4-form on $\mathbb{R}^8$

$$\phi = \omega^0 \wedge \varphi + \psi = {*}\phi$$

where $\psi = {*}\varphi \in \Lambda^4(\mathbb{R}^7)$.

Proposition 3: $\mathrm{Spin}(7) = \{A \in GL(8) \mid A^*(\phi) = \phi\}$ where $\mathrm{Spin}(7) \subseteq SO(8)$ is isomorphic to the universal cover of $SO(7)$.

Proposition 3 is not difficult to prove assuming Proposition 1. The form ϕ was discovered by Bonan [7] who showed that

$$\Lambda^{\mathrm{Spin}(7)} = \{1, \phi = {*}\phi, {*}1 = (1/14)\phi^2\}$$

where $\Lambda^{\mathrm{Spin}(7)} \subseteq \Lambda(\mathbb{R}^8)$ is the subring of Spin(7)-invariant exterior forms on $\mathbb{R}^8$. The GL(8)-orbit of $\phi \in \Lambda^4(\mathbb{R}^8)$ is not open but is, of course, a smooth submanifold of $\Lambda^4(\mathbb{R}^8)$. We shall say that an $\alpha \in \Lambda^4(V^*)$ is <u>admissible</u> if there exists a linear isomorphism $L: V \to \mathbb{R}^8$ so that $\alpha = L^*(\phi)$. If $\alpha \in \Omega^4(M^8)$, we shall say that α is admissible if $\alpha|_x$ is admissible for all $x \in M^8$. We let $F \subseteq \Lambda^4(T^*M)$ denote the submanifold of admissible 4-forms. $\pi: F \to M^8$ is a smooth fiber bundle with fibers isomorphic to GL(8)/Spin(7). Clearly the space of sections of F, i.e. the space of admissible 4-forms on M, is in 1-1 correspondence with the space of Spin(7)-structures on M. Since $\mathrm{Spin}(7) \subseteq \mathrm{SO}(8)$, we see that each admissible α on M canonically induces an orientation and metric on M.

On the other hand, if (M^8,g) is an oriented Riemannian manifold with holonomy Spin(7), it is easy to see that there is a unique <u>parallel</u> admissible 4-form α_g on M whose underlying orientation and metric are the given ones.

<u>Proposition 4</u>: Let α be an admissible 4-form on M. Then α is parallel with respect to the Levi-Civita connection of the underlying metric iff $d\alpha = 0$.

Proposition 4 is actually more elementary than the corresponding Proposition 2, but seems to have been overlooked. It follows from this that every admissible 4-form α which satisfies $d\alpha = 0$ has an underlying metric whose holonomy is a subgroup of Spin(7) and conversely every metric whose holonomy is a subgroup of Spin(7) arises from such an α.

Since F is not an open subset of a vector bundle over M, the condition $d\alpha = 0$ is only a <u>quasi-linear</u> first order system of 56 ($= \dim \Lambda^5(\mathbb{R}^7)$) equations for the 43 ($= \dim(\mathrm{GL}(8)/\mathrm{Spin}(7))$) unknown coefficents of the section $\alpha: M \to F$. It can be shown that these 56 equations are algebraically independent. Again, this over-determined system is invariant under the diffeomorphism group of M and can be shown to be transversely elliptic.

The analogue of Theorem 1 for Spin(7) is

Theorem 2: The system $d\alpha = 0$ for sections $\alpha: M \to F$ is involutive with Cartan characters $(s_1, s_2, \ldots, s_8) = (0,0,0,1,4,10,20,8)$. In particular, the "generic" solution has the property that its underlying metric has holonomy exactly Spin(7).

Theorem 2 is also a calculation with the appropriate differential system with independence condition on $F \subseteq \Lambda^4(T^*M)$. Details will appear elsewhere.

§3. Closing Remarks

The methods of §2 only yield the weakest positive result. Namely, that there exist local metrics on $\mathbb{R}^7$ and $\mathbb{R}^8$ which are not locally symmetric and have holonomy equal to G_2 and Spin(7) respectively. This at least shows that Berger's list cannot be shortened any further. Of course, in many respects this is quite unsatisfactory.

In the first place, we do not know a single example of such a metric in either case. The search for such metrics is led by Gray [8] but has so far proved fruitless.

In the second place, we do not know if there exists a complete metric even on $\mathbb{R}^7$ or $\mathbb{R}^8$ with holonomy G_2 or Spin(7). This problem reminds us, in some respects, of the conjecture that a complete Kähler metric on $\mathbb{C}^m$ which has holonomy a subgroup of $SU(m)$ is actually flat [11].

Finally, we do not know if there exists a compact example of either kind. Nevertheless, the descriptions of such metrics afforded by Theorems 1 and 2 allow one to prove a good number of theorems about possible examples. In a forthcoming joint work by the author and Reese Harvey it is shown that a compact (M^7,g) with holonomy G_2 must be orientable, spin, and have finite fundamental group. The first Pontriagin class of M^7 must be non-zero and the deformation theory of the solutions of $d\alpha = d^*\alpha = 0$ is unobstructed, the dimension of the local moduli space being $b_3 > 0$ where b_3 is the third Betti number of M. Similar results are obtained for 8-manifolds with holonomy Spin(7). The difficulty of explicitly writing down such a metric can be appreciated by contemplating the fact that no explicit example of a Calabi-Yau metric on a K-3 surface is known as of this writing.

BIBLIOGRAPHY

1. Kobayaski, S. and Nomizu, K., Foundations of Differential Geometry, Wiley and Sons, New York, 1963 and 1969.

2. Berger, M., Sur les Groupes d'Holonomie Homogène des Variétés à Connexion Affine et des Variétés Riemanniennes, Bull. Soc. Math. France, 83 (1955), 279-300.

3. Simons, J., On Transitivity Holonomy Systems, Ann. of Math., 76 (1962), 213-234.

4. Alekseevski, D. V., Riemannian Spaces with Unusual Holonomy Groups, Funkcional Anal. i Priloven 2 (1968), 1-10. Translated in Functional Anal. Appl.

5. Chevalley, C., Algebraic Theory of Spinors, 1954.

6. Helgason, S., Differential Geometry, Lie Groups, and Symmetric Spaces, Academic Press, 1978.

7. Bonan, E., Sur les Variétés Riemanniennes à Groupe d'Holonomie G_2 ou Spin(7), C. R. Acad. Sci. Paris 262 (1966), 127-129.

8. Gray, A., Weak Holonomy Groups, Math. Z. 123 (1971), 290-300.

9. Fernandez, M. and Gray, A., Riemannian Manifolds with Structure Group G_2, Annali di Mat. pura ed appl. (IV), 32 1982, 19-45.

10. Chern, S. S. et al, Exterior Differential Systems, to appear.

11. Yau, S. T., ed., Problem Section in Seminar on Differential Geometry, Annals of Math. Studies, no. 102, Princeton, University Press, 1982.

ON RIEMANNIAN METRICS ADAPTED TO THREE-DIMENSIONAL CONTACT MANIFOLDS

by

S.S. Chern
Mathematical Sciences Research Institute
1000 Centennial Drive
Berkeley, California 94720

and

R.S. Hamilton
Department of Mathematics
University of California, San Diego
La Jolla, California 92093

0. Introduction It was proved by R. Lutz and J. Martinet [8] that every compact orientable three-dimensional manifold M has a contact structure. The latter can be given by a one-form ω, the contact form, such that $\omega \wedge d\omega$ never vanishes; ω is defined up to a non-zero factor. A Riemannian metric on M is said to be adapted to the contact form ω if: 1) ω has the length 1; and 2) $d\omega = 2{*}\omega$, $*$ being the Hodge operator. The Webster curvature W, defined below in [9], is a linear combination of the sectional curvature of the plane ω and the Ricci curvature in the direction perpendicular to ω.

Adapted Riemannian metrics have interesting properties. The main result of the paper is the theorem:

Every contact structure on a compact orientable three-dimensional manifold has a contact form and an adapted Riemannian metric whose Webster curvature is either a constant ≤ 0 or is everywhere strictly positive.

The problem is analogous to Yamabe's problem on the conformal transformation of Riemannian manifolds Most recently, R. Schoen has proved Yamabe's conjecture in all cases, including that of positive scalar curvature [9]. It is thus an interesting question whether in the second case of our theorem the Webster curvature can be made a positive constant.

---

1),2) Research supported in part by NSF grants DMS84-03201 and DMS84-01959.

After our theorem was proved, we learned that a similar theorem on CR-manifolds of any odd dimension has been proved by Jerison and Lee. [7] As a result, our curvature was identified with the Webster curvature. We feel that our viewpoint is sufficiently different from Jerison-Lee and that the three-dimensional case has so many special features to merit a separate treatment.

In an appendix, Alan Weinstein gives a topological implication of the vanishing of the second fundamental form in (54). For an interesting account of three-dimensional contact manifolds, cf. [2].

1. **Contact Structures.** Let M be a manifold and B a subbundle of the tangent bundle TM. There is a naturally defined anti-symmetric bilinear form Λ on B with values in the quotient bundle TM/B

(1) $\Lambda: B \times B \longrightarrow TM/B$

defined by the Lie bracket;

(2) $\Lambda(V,W) \equiv [V,W] \bmod B.$

It is easy to verify that the value of $\Lambda(V,W)$ at a point $p \in M$ depends only on the values of V and W at p. The bundle B defines a foliation if and only if it satisfies the Frobenius integrability condition $\Lambda = 0$. Conversely, a <u>contact structure</u> on M is a subbundle B of the tangent bundle of codimension 1 such that Λ is non-singular at each point $p \in M$. This can only occur when the dimension of M is odd.

It is an interesting problem to find some geometric structure which can be put on every three-manifold, since this would be helpful in studying its topology. Along these lines we have the following remarkable theorem of Lutz and Martinent (see [8], [10]).

1.1 **Theorem.** *Every compact orientable three-manifold possesses a contact structure.*

There are many different contact structures possible, since the set of B with $\Lambda \neq 0$ is open. Even on S^3 there are contact structures for which the bundles B_1 and B_2 are topologically distinct. Nevertheless the notion of a contact structure is rather flabby, in the following sense. We say B is conjugate to B_* if there is a diffeomorphism $\varphi: M \longrightarrow M$ which has $\varphi(B) = B_*$. Then we have the following result due to Gray (see [4]).

1.2 Theorem. *Given a contact structure* B, *any other contact structure* B_* *close enough to* B *is conjugate to it.*

2. Metrics adapted to contact structures. A contact form ω is a 1-form on M which is nowhere zero and has the contact bundle B for its null space. In a three-manifold a non-zero 1-form ω is a contact form for the contact structure B = Null ω if and only if $\omega \wedge d\omega \neq 0$ at every point. The contact structure B determines the contact form up to a scalar multiple. The choice of a contact form ω also determines a vector field V in the following way.

2.1 Lemma. *There exists a unique vector field* V *such that* $\omega(V) = 1$ *and* $d\omega(V,W) = 0$ *for all* $W \in TM$.

Proof. Choose V_0 with $\omega(V_0) = 1$. Since $d\omega \wedge \omega \neq 0$, the form $d\omega$ is non-singular on B. Therefore there exists a unique $V_1 \in B$ with

$$d\omega(V_1, W) = d\omega(V_0, W)$$

for all $W \in B$. Let $V = V_0 - V_1$. Then $\omega(V) = \omega(V_0) - \omega(V_1) = 1$, and $d\omega(V,W) = 0$ for all $W \in B$. Since V is transverse to B and $d\omega(V,V) = 0$, we have $d\omega(V,W) = 0$ for all $W \in TM$.

Locally any two non-zero vector fields are conjugate by a diffeomorphism. However, this fails globally, since a vector field may have closed orbits while a nearby vector field does not. It is a classical result that locally any two contact forms are conjugate by a diffeomorphism. But globally two nearby contact forms may not be conjugate, since the vector fields they determine may not be.

A choice of a Riemannian metric on a contact manifold determines a choice of the contact form ω up to sign by the condition that ω have length 1. Let $*$ denote the Hodge star operator. We make the following definition.

2.2 Definition. A Riemannian metric on a contact three-manifold is said to be <u>adapted</u> to the contact form ω if ω is of length one and satisfies the structural equation

$$d\omega = 2 *\omega. \tag{3}$$

Such metrics have nice properties with respect to the contact structure. For example, we have the following results.

2.3 Lemma. *If the metric is adapted to the form* ω, *then the vector field* V *determined by* ω *is the unit vector field perpendicular to* B.

Proof. Let V be the unit vector field perpendicular to B. Then $\omega(V) = 1$, and for all vectors W in B we have $d\omega(V,W)=2*\omega(V,W)=0$. Hence V is the vector field determined by the contact form ω.

2.4 Lemma. *If the metric is adapted to the contact form* ω, *then the area form on* B *is given by* $\frac{1}{2}\, d\omega$.

Proof. The area form on B is $*\omega$.

A CR structure on a mainfold is a contact structure together with a complex structure on the contact bundle B; that is, an involution $J: B \longrightarrow B$ with $J^2 = -I$ where I is the identity. If M has dimension 3 then B has dimension 2, and a complex structure on B is equivalent to a conformal structure; that is knowing how to rotate by $90°$. Hence, a Riemannian metric on a contact three-manifold also produces a CR structure. CR structures have been extensively studied since they arise naturally on the boundaries of complex manifolds. The following observation will be basic to our study.

2.5 Theorem. *Let* M *be an oriented three-manifold with contact structure* B. *For every choice of contact form* ω *and a* CR *structure* J *there exists a unique Riemannian metric* g *adapted to the contact form* ω *and inducing the* CR *structure* J.

Proof. The form ω determines the unit vector field V perpendicular to B. The metric on B is determined by the conformal structure J and the volume form $*\omega\big|B = \frac{1}{2}\, d\omega\big|B$.

3. **Structural equations.** We begin with a review of the structural equations of Riemannian geometry. Let $\omega_\alpha, 1 \leq \alpha, \beta \leq \dim M$, be an orthonormal basis of 1-forms on a Riemannian manifold M. Then there exists a unique anti-symmetric matrix of 1-forms $\varphi_{\alpha\beta}$ such that the structural equations

$$(4) \qquad d\omega_\alpha + \varphi_{\alpha\beta} \wedge \omega_\beta = 0.$$

hold on M. The forms $\varphi_{\alpha\beta}$ describe the Levi-Civita connection of the metric in the moving frame ω_α. We can also view the ω_α as intrinsically defined 1-forms on the principal bundle of orthonormal bases. Then the $\varphi_{\alpha\beta}$ are also intrinsically defined as 1-forms on this principal bundle, and the collection $\{\omega_\alpha, \varphi_{\alpha\beta}\}$ forms an orthonormal basis of 1-forms in the induced metric on the principal bundle. The curvature tensor $R_{\alpha\beta\gamma\delta}$ is defined by the structural equation

$$(5) \quad d\varphi_{\alpha\beta} = -\varphi_{\alpha\gamma} \wedge \varphi_{\gamma\beta} + R_{\alpha\beta\gamma\delta}\, \omega_\gamma \wedge \omega_\delta, \quad 1 \leq \alpha, \beta, \gamma, \delta \leq \dim M,$$

where the summation convention applies.

In three-dimensions it is natural to replace a pair of indices in an anti-symmetric tensor by the third index. Thus we will write φ_{12}

$= \varphi_3$ and $R_{1212} = K_{33}$, etc. Here $K_{\alpha\beta}$ are the components of the Einstein tensor

(6) $$K_{\alpha\beta} = \frac{1}{2} Rg_{\alpha\beta} - R_{\alpha\beta},$$

which has the property that, for any unit vector V, K(V,V) is the Riemannian sectional curvature of the plane $V^\perp$. The structural equations then take the following form.

3.1 Structural equations in three dimensions.

$$d\omega_1 = \varphi_2\wedge\omega_3 - \varphi_3\wedge\omega_2,$$

(7) $$d\omega_2 = \varphi_3\wedge\omega_1 - \varphi_1\wedge\omega_3,$$

$$d\omega_3 = \varphi_1\wedge\omega_2 - \varphi_2\wedge\omega_1,$$

and

$$d\varphi_1 = \varphi_2\wedge\varphi_3+K_{11}\omega_2\wedge\omega_3+K_{12}\omega_3\wedge\omega_1+K_{13}\omega_1\wedge\omega_2,$$

(8) $$d\varphi_2 = \varphi_3\wedge\varphi_1+K_{21}\omega_2\wedge\omega_3+K_{22}\omega_3\wedge\omega_1+K_{23}\omega_1\wedge\omega_2,$$

$$d\varphi_3 = \varphi_1\wedge\varphi_2+K_{31}\omega_2\wedge\omega_3+K_{32}\omega_3\wedge\omega_1+K_{33}\omega_1\wedge\omega_2, K_{\alpha\beta}=K_{\beta\alpha}.$$

If the metric is adapted to the contact from ω, we choose the frames such that $\omega_3 = \omega$. As a consequence K_{33} is the sectional curvature of the plane $V^\perp$ and $\frac{1}{2}(K_{11}+K_{22})$ is the Ricci curvature in the direction V. The Webster curvature is defined by

(9) $$W = \frac{1}{8}(K_{11}+K_{22}+2K_{33}+4)$$

and has remarkable properties.

We proceed to illustrate these equations with three examples which are very relevant to our discussion, the sphere S^3, the unit tangent bundle of a compact orientable surface of genus > 1, and the

Heisenberg group H^3.

3.2 **Example.** The sphere S^3 is defined by the equation

$$x^2+y^2+z^2+w^2 = 1 \tag{10}$$

in R^4. Differentiating we get

$$\omega_0 = xdx + y\,dy + z\,dz + w\,dw = 0. \tag{11}$$

A specific choice of an orthonormal basis in the induced metric is

$$\begin{aligned} \omega_1 &= x\,dy - y\,dx + z\,dw - w\,dz, \\ \omega_2 &= x\,dz - z\,dx + y\,dw - w\,dy, \\ \omega_3 &= x\,dw - w\,dx + y\,dz - z\,dy. \end{aligned} \tag{12}$$

The reader can verify that if $\langle dx,dx\rangle = 1$, $\langle dx,dy\rangle = 0$, etc., then $\langle\omega_1,\omega_1\rangle = 1$, $\langle\omega_1,\omega_2\rangle = 0$, etc., and that $\langle\omega_0,\omega_0\rangle = 1$, $\langle\omega_0,\omega_1\rangle = 0$, etc. Taking exterior derivative we have

$$d\omega_1 = 2\omega_2\wedge\omega_3, \quad d\omega_2 = 2\omega_3\wedge\omega_1, \quad d\omega_3 = 2\omega_1\wedge\omega_2. \tag{13}$$

and hence in this basis

$$\varphi_1 = \omega_1, \quad \varphi_2 = \omega_2, \quad \varphi_3 = \omega_3. \tag{14}$$

which makes

$$K_{11} = 1, \quad K_{22} = 1, \quad K_{33} = 1, \tag{15}$$

and the other entries are zero. The Webster curvature $W = 1$.

3.3 **Example.** The unit tangent bundle of a compact orientable surface of genus $\neq 1$.

Let N be a compact orientable surface of genus g. If N is equipped with a Riemannian metric, its orthonormal coframe θ_1, θ_2, and the connection form θ_{12} satisfy the structural equations

$$d\theta_1 = \theta_{12}\wedge\theta_2, \quad d\theta_2 = \theta_1\wedge\theta_{12}, \quad d\theta_{12} = -K\theta_1\wedge\theta_2, \tag{16}$$

where K is the Gaussian curvature. Suppose $g\neq 1$. We can choose the metric such that

$$K = \epsilon = \begin{cases} +1, & \text{when } g=0, \\ -1, & \text{when } g>1. \end{cases} \tag{17}$$

The unit tangent bundle T_1N of N, as a three-dimensional manifold, has the metric

$$\frac{1}{4}(\theta_1^2 + \theta_2^2 + \theta_{12}^2). \tag{18}$$

Putting

$$\omega_1 = \frac{1}{2}\theta_1, \quad \omega_2 = \frac{1}{2}\theta_2, \quad \omega_3 = -\frac{1}{2}\epsilon\theta_{12}, \tag{19}$$

we find

$$d\omega_1 = 2\epsilon\omega_2\wedge\omega_3, \quad d\omega_2 = 2\epsilon\omega_3\wedge\omega_1, \quad d\omega_3 = 2\omega_1\wedge\omega_2, \tag{20}$$

and

$$\varphi_1 = \omega_1, \quad \varphi_2 = \omega_2, \quad \varphi_3 = (2\epsilon-1)\omega_3. \tag{21}$$

It follows that

$$K_{11} = K_{22} = 1, \quad K_{33} = 4\epsilon-3, \tag{22}$$

all other $K_{\alpha\beta}$'s being zero. By (9) we get

$$W = \epsilon.$$

This includes the example in §3.2 when $g = 0$, for the unit tangent bundle of S^2 is the real projective space RP^3, which is covered by S^3, and our calculation is local. On the other hand, T_1N, for $g > 1$, has a contact structure and an adapted Riemannian metric with $W = -1$.

3.4. Example. The Heisenberg group.

We can make C^2 into a Lie group by identifying (z,w) with the matrix

(23)
$$\begin{bmatrix} 1 & 0 & 0 \\ z & 1 & 0 \\ w & -\bar{z} & 1 \end{bmatrix}.$$

The subgroup given by the variety

(24)
$$z\bar{z} + w + \bar{w} = 0$$

is the Heisenberg group H^3. The group acts on itself by the translations

(25)
$$\begin{aligned} z &\longrightarrow z + a, \\ w &\longrightarrow w - \bar{a}z + b, \end{aligned}$$

which leave invariant the complex forms

(26)
$$dz \text{ and } dw + \bar{z}\, dz.$$

Hence an invariant metric is given by

(27)
$$ds^2 = |dz|^2 + |dw + \bar{z}\, dz|^2.$$

Introduce the real coordinates

(28)
$$z = x + iy \qquad w = u + iv.$$

Then the variety (24) is

(29) $$x^2 + y^2 + 2u = 0$$

and differentiation gives

(30) $$du + x\,dx + y\,dy = 0.$$

Then an orthonormal basis of 1-forms in the metric above is given by

(31) $$\omega_1 = dx, \quad \omega_2 = dy, \quad \omega_3 = dv + x\,dy - y\,dx,$$

and we compute

(32) $$\begin{cases} d\omega_1 = 0, & d\omega_2 = 0, & d\omega_3 = 2\omega_1 \wedge \omega_2, \\ \varphi_1 = \omega_1, & \varphi_2 = \omega_2, & \varphi_3 = -\omega_3, \\ K_{11} = 1, & K_{22} = 1, & K_{33} = -3, \end{cases}$$

and the other entries are zero. By (9) we have W=0. All these examples give metrics adapted to a contact form $\omega = \omega_3$, since in an orthonormal basis $*\omega_3 = \omega_1 \wedge \omega_2$.

In general, given a metric adapted to a contact form ω, we shall restrict our attention to orthonormal bases of 1-forms ω_1, ω_2, ω_3 with $\omega_3 = \omega$. Considering the dual basis of vectors, we only need to choose a unit vector in B. These form a principal circle bundle, and all of our structural equations will live naturally on this circle bundle. It turns out to be advantageous to compare the general situation to that on the Heisenberg group. Therefore, we introduce the forms ψ_1, ψ_2, ψ_3 and the matrix L_{11}, $L_{12}, \ldots, L_{33}$ defined by

(33) $$\begin{cases} \varphi_1 = \psi_1 + \omega_1, & \varphi_2 = \psi_2 + \omega_2, & \varphi_3 = \psi_3 - \omega_3, \\ K_{11} = L_{11} + 1, & K_{22} = L_{22} + 1, & K_{33} = L_{33} - 3, \\ K_{12} = L_{12}, & K_{13} = L_{13}, & K_{23} = L_{23}. \end{cases}$$

Thus the ψ and L all vanish on the Heisenberg group. We then compute the following.

3.5. Structure equations for an adapted metric. They are:

$$(34)\quad \begin{cases} d\omega_1 = \psi_2\wedge\omega_3-\psi_3\wedge\omega_2, \\ d\omega_2 = \psi_3\wedge\omega_1-\psi_1\wedge\omega_3, \\ d\omega_3 = 2\omega_1\wedge\omega_2, \end{cases}$$

and

$$(35)\quad \begin{cases} \psi_1\wedge\omega_2 - \psi_2\wedge\omega_1 = 0, \\ \psi_1\wedge\omega_1 + \psi_2\wedge\omega_2 = 0, \end{cases}$$

and

$$(36)\quad \begin{cases} d\psi_1 = \psi_2\wedge\psi_3+L_{11}\omega_2\wedge\omega_3+L_{12}\omega_3\wedge\omega_1+L_{13}\omega_1\wedge\omega_2, \\ d\psi_2 = \psi_3\wedge\psi_1+L_{21}\omega_2\wedge\omega_3+L_{22}\omega_3\wedge\omega_1+L_{23}\omega_1\wedge\omega_2, \\ d\psi_3 = \psi_1\wedge\psi_2+L_{31}\omega_2\wedge\omega_3+L_{32}\omega_3\wedge\omega_1+L_{33}\omega_1\wedge\omega_2. \end{cases}$$

Proof. The equation $d\omega_3 = 2\omega_1\wedge\omega_2$ comes from the condition $d\omega = 2{*}\omega$ that the metric is adapted to the contact form ω. Then the corresponding structural equation yields $\psi_1\wedge\omega_2 - \psi_2\wedge\omega_1 = 0$. Using $dd\omega_3 = 0$ we compute $\psi_1\wedge\omega_1 + \psi_2\wedge\omega_2 = 0$ also.

3.6. Corollary. We can find functions a and b on the principal circle bundle so that

$$(37)\quad \begin{cases} \psi_1 = a\omega_1+b\omega_2, \\ \psi_2 = b\omega_1-a\omega_2. \end{cases}$$

Proof. This follows algebraically from the equations (35).

It is even more convenient to write these equations in complex form. We make the following substitutions.

3.7. Complex substitutions.

On account of the complex structure in B it is convenient to use the complex notation. We shall set:

$$(38)\quad \begin{cases} \Omega = \omega_1+i\omega_2, & \omega = \omega_3, \\ \Psi = \psi_1+i\psi_2, & \psi = \psi_3, \\ \iota = a+ib, & \\ p = \frac{1}{2}(L_{11}+L_{22}), \quad q = \frac{1}{2}(L_{11}-L_{22}), \quad r = L_{12}, & \\ s = q+ir, & \\ z = \frac{1}{2}(L_{13}+iL_{23}), & \\ t = L_{33}, & \\ W = \frac{1}{4}(t-a^2-b^2), & \end{cases}$$

where W is the <u>Webster</u> <u>curvature</u>, to be verified below. Note that $\Psi = \iota\bar{\Omega}$. Thus Ω and ω give a basis for the 1-forms on M, while ι and ψ define the connection.

3.8 Complex structural equations.

$$(39)\quad \begin{aligned} d\Omega &= i(\psi\wedge\Omega-\iota\bar{\Omega}\wedge\omega), \\ d\omega &= i\Omega\wedge\bar{\Omega}, \end{aligned}$$

and

$$(40)\quad \begin{cases} d\psi = i[2W\Omega\wedge\bar{\Omega}+(z\bar{\Omega}-\bar{z}\Omega)\wedge\omega], \\ d\iota \equiv i(2\iota\psi+z\Omega-s\omega) \bmod \bar{\Omega}, \\ p+|\iota|^2 = 0. \end{cases}$$

Proof. This is a direct computation. Note that the real functions p,W and the complex functions z,s give the curvature of the metric.

The equation $p+|\iota|^2 = 0$ has the important consequence that we can compute the Webster curvature W from the $K_{\alpha\beta}$. The result is the expression for W in (9).

The following notation will be useful. If f is a function on a Riemannian manifold with frame ω_α, then

$$\text{(41)} \qquad df = D_\alpha f \cdot \omega_\alpha,$$

where $D_\alpha f$ is the derivative of f in the direction of the dual vector field V_α. If f is a function on the principal bundle then we can still define $D_\alpha f$ as the derivative in the direction of the horizontal lifting of V_α. In this case we will have

$$\text{(42)} \qquad df \equiv D_\alpha f \cdot \omega_\alpha \mod \varphi_{\alpha\beta}.$$

If the function f represents a tensor then $D_\alpha f$ are its covariant derivatives, and the extra terms in $\varphi_{\alpha\beta}$ depend on what kind of tensor is represented. In the example if T is a covariant 1-tensor and

$$\text{(43)} \qquad f = T(V_\gamma),$$

then,

$$\text{(44)} \qquad df = D_\alpha f \cdot \omega_\alpha + T(V_\beta)\,\varphi_{\beta\gamma},$$

while if T is a covariant 2-tensor and

$$\text{(45)} \qquad f = T(V_\gamma, V_\delta),$$

then

$$\text{(46)} \qquad df = D_\alpha f \cdot \omega_\alpha + T(V_\beta, V_\delta)\,\varphi_{\beta\gamma} + T(V_\gamma, V_\beta)\,\varphi_{\beta\delta},$$

and so on. In the complex notation we write

$$\text{(47)} \qquad df = \partial f \cdot \Omega + \bar{\partial} f \cdot \bar{\Omega} + D_v f \cdot \omega$$

as the definition of the differential operators ∂f, $\bar{\partial} f$, and $D_V f$. As usual

(48)
$$\begin{cases} \partial f = \frac{1}{2}(D_1 f - iD_2 f), \\ \bar{\partial} f = \frac{1}{2}(D_1 f + iD_2 f), \\ D_V f = D_3 f, \end{cases}$$

reflecting the transition from real to complex notation. If f is a function on the principal circle bundle coming from a symmetric k-tensor on B then

(49)
$$df = \partial f \cdot \Omega + \bar{\partial} f \cdot \bar{\Omega} + D_V f \cdot \omega + ikf\psi.$$

For example, the function ι represents a trace-free symmetric 2-form on B, and the structural equation for ι tells us

3.9. Lemma.

(50)
$$\bar{\partial}\iota = iz \text{ and } D_V\iota = -is.$$

4. Change of basis. We start with the simplest change of basis, namely rotation through an angle θ. We take θ to be a function on M and study what happens on the principal circle bundle. The new basis ω_1^*, ω_2^*, ω_3^* is given by $\omega_3^*=\omega_3=\omega$ and

(51)
$$\omega_1^* = \cos.\theta\ \omega_1 - \sin.\theta\omega_2,$$
$$\omega_2^* = \sin.\theta\ \omega_1 + \cos.\theta\ \omega_2$$

or in complex terms $\omega^* = \omega$ and

(52)
$$\Omega^* = e^{i\theta}\ \Omega.$$

Then from the structural equations we immediately find that

4.1. Lemma.

$$
\psi^* = \psi + d\theta, \tag{53}
$$
$$
\iota^* = \iota\, e^{2i\theta}.
$$

Now a function or tensor on the principal circle bundle comes from one on M by the pull-back if and only if it is invariant under rotation by θ. Thus we see that the curvature form $d\psi^* = d\psi$ is invariant and hence lives on M. The form $\Omega\wedge\bar{\Omega}$ is also invariant, so $W = W^*$ is invariant and W is a function on M. This W is the scalar curvature introduced by Webster (see [11]). Likewise $|\iota|^2$ is invariant and hence a function on M. The function ι defines a tensor $\iota\bar{\Omega}^2$ which is invariant. Hence its real and imaginary parts

$$
a(\omega_1^2-\omega_2^2) + 2b\,\omega_1\omega_2, \tag{54}
$$
$$
b(\omega_1^2-\omega_2^2) - 2a\,\omega_1\omega_2
$$

define trace-free symmetric bilinear forms on B (they differ by rotation). This form is called the torsion tensor by Webster (see [11]); it is analogous to the second fundamental form for a surface.

We now consider more interesting changes of basis. First we change the CR structure while leaving the contact form ω fixed. In order to keep the metric adapted to the contact form we must leave $\omega_1\wedge\omega_2$ invariant. This gives a new basis

$$
\begin{aligned}
\omega_1^* &= A\omega_1+B\omega_2,\\
\omega_2^* &= C\omega_1+D\omega_2,\\
\omega_3^* &= \omega_3
\end{aligned} \tag{55}
$$

with $AD-BC = 1$. An infinitesimal change of basis is given by the tangent to a path at $t = 0$. Thus an infinitesimal change of the basis which changes CR structure but leaves the contact form invariant and keeps the metric adapted is given by

$$\omega_1' = g\omega_1 + h\omega_2,$$
$$\omega_2' = k\omega_1 + l\omega_2,$$
$$\omega_3' = 0$$

with $g+l = 0$. Since the rotations are trivial we may as well take $h=k$. This gives

$$\text{(56)} \qquad \begin{aligned} \omega_1' &= g\omega_1 + h\omega_2, \\ \omega_2' &= h\omega_1 - g\omega_2, \\ \omega_3' &= 0. \end{aligned}$$

In complex notation if $f = g+ih$ then

$$\text{(57)} \qquad \Omega' = f\bar{\Omega} \text{ and } \omega' = 0.$$

For future use we compute the infinitesimal change ψ' in ψ and ι' in ι from the structural equations (39), (40). We find that f transforms as a 2-tensor

$$\text{(58)} \qquad df = \partial f \cdot \Omega + \bar{\partial} f \cdot \bar{\Omega} + D_V f \cdot \omega + 2if\psi$$

and that

4.2. Lemma.

$$\text{(59)} \qquad \begin{aligned} \iota' &= -i\, D_V f, \\ \psi' &= i(\partial f \cdot \bar{\Omega} - \bar{\partial}\bar{f} \cdot \Omega) - (\iota\bar{f} + \bar{\iota} f)\omega \end{aligned}$$

using the fact that we know $\psi \wedge \Omega$ and ψ is real.

On the other hand we may wish to fix the CR structure and change the contact form while keeping the metric adapted. In this case let $\omega_3^* = f^2\omega_3$ where f is a positive real function. Excluding rotation we find that to keep the metric adapted we need

$$\begin{aligned} \omega_1^* &= f\cdot\omega_1 - D_2 f\cdot\omega_3, \\ \omega_2^* &= f\cdot\omega_2 + D_1 f\cdot\omega_3, \qquad (60) \\ \omega_3^* &= f^2\omega_3. \end{aligned}$$

In complex notation

$$\begin{aligned} \Omega^* &= f\Omega + 2i\,\bar{\partial} f\cdot\omega, \qquad (61) \\ \omega^* &= f^2\omega. \end{aligned}$$

For an infinitesimal variation we differentiate to obtain

$$\begin{aligned} \Omega' &= f'\,\Omega + 2i\,\bar{\partial} f'\cdot\omega, \qquad (62) \\ \omega' &= 2f'\,\omega. \end{aligned}$$

Hence changes of metric fixing the CR structure are given by a potential function f, much the same way as changes of metric fixing a conformal structure. The main difference is that the derivatives of f enter the formula for the new basis.

As a consequence of ddf = 0 we have

$$\partial\bar{\partial} f - \bar{\partial}\partial f + iD_V f = 0. \qquad (63)$$

We also define the sub-Laplace operator

$$\Box f = 2(\partial\bar{\partial} f + \bar{\partial}\partial f) = (D_1 D_1 f + D_2 D_2 f). \qquad (64)$$

Then a straightforward computation substituting in the structural equations yields

(4.3. **Lemma.**)

4.3. Lemma.

$$(65)\qquad \begin{aligned} \psi^* &= \psi+3i\left[\frac{\partial f}{f}\,\Omega - \frac{\bar\partial f}{f}\,\bar\Omega\right] - \left[\frac{\Box f}{2f} + 6\,\frac{\partial f\bar\partial f}{f^2}\right], \\ \iota^* &= \iota - 2\,\frac{\bar\partial\bar\partial f}{f} - 6\left[\frac{\bar\partial f}{f}\right]^2. \end{aligned}$$

Differentiating the first we get

$$d\psi^* \equiv d\psi - 2i\,\frac{\Box f}{f}\,\Omega\wedge\bar\Omega \mod \omega,$$

which shows the remarkable relation given by

4.4 Lemma.

$$(66)\qquad f^3W^* = fW - \Box f.$$

4.5. Corollary. In an infinitesimal variation

$$(67)\qquad W' = -\Box f' - 2f'W.$$

5. Energies. Let μ be the measure on M

$$(68)\qquad \mu = \omega_1\wedge\omega_2\wedge\omega_3 = \frac{i}{2}\,\Omega\wedge\bar\Omega\wedge\omega$$

induced by the metric. Here are two interesting energies which we may form. The first is

$$(69)\qquad E_W = \int_M W\,\mu,$$

which is analogous to the energy

$$(70)\qquad E = \int_M R\,\mu$$

in the Yamabe problem. The second is

(71) $$E_\iota = \int_M |\iota|^2 \mu,$$

which is a kind of Dirichlet energy.

In this section we shall study the critical points of these energies.

First we observe that for computational reasons it is easier to integrate over the principal circle bundle P. The measure there is

(72) $$\nu = \omega_1 \wedge \omega_2 \wedge \omega_3 \wedge \psi_3 = \tfrac{i}{2}\, \Omega \wedge \bar{\Omega} \wedge \omega \wedge \psi.$$

If f is a function on the base M then

(73) $$\int_P f\, \nu = 2\pi \int_M f\, \mu,$$

so nothing is lost.

Next we observe that we can integrate by parts.

5.1. Lemma. For any f on P

(74) $$\int_P \partial f \cdot \nu = 0 \text{ and } \int_P D_V f \cdot \nu = 0.$$

Proof. The first follows from

$$\int_P d(f \bar{\Omega} \wedge \omega \wedge \psi) = 0$$

and the second follows from

$$\int_P d(f \Omega \wedge \bar{\Omega} \wedge \psi) = 0,$$

since $d\Omega \equiv 0 \mod \bar{\Omega}, \omega$ and $d\omega \equiv 0 \mod \Omega, \bar{\Omega}$ and $d\psi \equiv 0 \mod \Omega, \bar{\Omega}, \omega$.

5.2. Theorem. *The energy E_W is critical over all contact forms with a fixed CR structure and fixed volume if and only if W is constant. It is critical* *over all CR structures with a fixed contact form if and only if* $\iota = 0$.

Proof. We compute the infinitesimal variation E_W'. Fixing the CR structure and varying the potential f of the contact form with $\omega^* = f^2\omega$ gives $\nu' = 4f'\nu$ and

$$E_W' = \int_P (-\Box f' + 2f' W)\,\nu = 2\int f' W\,\nu,$$

since $\Box$ integrates away. The volume is fixed when $\int f'\,\nu = 0$. Thus, $E_W' = 0$ precisely when W is constant.

Fixing the contact form and varying the CR structure we use the following.

5.3 Lemma.

(75) $$E_W = \frac{1}{2}\int_P d\psi\wedge\omega\wedge\psi.$$

Proof. We use the structural equation to see

$$d\psi\wedge\omega = 2iW\Omega\wedge\bar{\Omega}\wedge\omega$$

and integrate by parts to get the result. Then we have

$$E_W' = \frac{1}{2}\int_P d\psi'\wedge\omega\wedge\psi + d\psi\wedge\omega\wedge\psi'$$

(using $\omega' = 0$), and this gives

$$E_W' = \frac{1}{2}\int_P \psi'\wedge\Omega\wedge\bar{\Omega}\wedge\psi.$$

Then using Lemma 4.2 we get

$$E_W' = -\frac{1}{2}\int_P (\iota\bar{f}+\bar{\iota}f)\,\nu.$$

so that the CR structure is critical for fixed ω precisely when $\iota=0$.

Next we consider the energy E_ι.

5.4. Theorem. *The energy* E_ι *is critical over all* CR *structures with fixed contact form if and only if* $D_V\iota = 0$, *which is equivalent to* $s = 0$, *or* $K_{11} = K_{22}$ *and* $K_{12} = 0$. *The energy* E_ι *is critical over all contact forms with fixed* CR *structure and fixed volume if and only if*

(76) $$2i(\partial z - \bar\partial \bar z) + 3p = \text{constant}.$$

Proof. The energy E_ι is given by

$$E_\iota = \int_P |\iota|^2 \nu,$$

so its first variation is

$$E_\iota' = \int_P (\iota' \bar\iota + \iota \bar\iota')\nu + |\iota|^2 \nu'.$$

When ω is fixed, $\omega' = 0$ and $\nu' = 0$. By Lemma 4.2 we have the result that if $\Omega' = f\,\bar\Omega$ then $\iota' = -iD_V f$, and this gives

5.5. Lemma.

$$E_\iota' = 2\,\mathrm{Im} \int_P \bar f\, D_V\iota\ \nu.$$

Since f is any real function on M, we see $E_\iota' = 0$ when $D_V\iota = 0$. Then $s = 0$ by Lemma 3.9 and $K_{11} = K_{22}$ and $K_{12} = 0$ by substitution (38).

This condition says that, at each point of M, the sectional curvature of all planes perpendicular to the contact plane B are equal.

If, on the other hand, we fix the CR structure and vary the contact form by a potential f, we have from Lemma 4.3

$$\iota^* = \iota - 2\,\frac{\bar\partial\bar\partial f}{f} - 6\left[\frac{\bar\partial f}{f}\right]^2.$$

Taking an infinitesimal variation

$$\iota' = -2\bar\partial\bar\partial f', \quad \nu' = 3f'\,\nu$$

Then the variation in E_ι is

$$E_\iota{}' = \int_P \{-2(\iota\,\partial\partial f' + \bar\iota\,\bar\partial\bar\partial f') + 3|\iota|^2 f'\}\,\nu,$$

from which we see that $E_\iota{}' = 0$ precisely when

$$2(\partial\partial\iota + \bar\partial\bar\partial\bar\iota) - 3|\iota|^2$$

is constant. Since $\partial\iota = iz$ by Lemma 3.9, and $|\iota|^2 + p = 0$, this gives the equation (76).

6. Changing Webster Scalar Curvature. The problem of fixing the CR structure and changing the Webster scalar curvature is precisely analogous to the Yamabe problem of fixing the conformal structure and changing the scalar curvature, except the problem is subelliptic, and the estimates and constants for the 3-dimensional CR case look like the 4-dimensional conformal case. The first result is the following.

6.1. Theorem. *Let M be a compact orientable three-manifold with fixed CR structure. Then we can change the contact form so that the Webster scalar curvature W of the adapted Riemannian metric is either positive or zero or negative everywhere.*

Proof. We have $f^3W^* = fW - \Box f$ from Lemma 4.4. We take f to be the eigenfunction of $W - \Box$ with lowest eigenvalue λ_1. By the strict maximum principle for subelliptic equations (see Bony [1]) we conclude that f is strictly positive. Since $Wf - \Box f = \lambda_1 f$ we have

$f^2W^* = \lambda_1$. Hence W^* always has the same sign as λ_1.

Next we show that in the negative curvature case we can make W whatever we want, in particular, a negative constant.

6.2. Theorem. *Let* M *be a compact orientable three-manifold with a fixed* CR *structure. If some contact form has negative Webster scalar curvature, then every negative function* $W<0$ *is the Webster scalar curvature of one and only one contact form* ω.

Proof. Let C be the space of all contact forms and let $\mathcal{F}$ be the space of functions. We define the operator P by

$$P:C \longrightarrow \mathcal{F}, \quad P(\omega) = W.$$

Let $\mathcal{F}^-$ be the space of negative functions and let C^- be the space of contact forms with negative Webster curvature. Then

$$P:C^- \longrightarrow \mathcal{F}^- \tag{77}$$

is also defined. We claim the P in (77) is a global diffeomorphism. This follows from the following observations.

a) C^- is not empty.

b) P is locally invertible.

c) P is proper (the inverse image of a compact set is compact).

d) $\mathcal{F}^-$ is simply connected.

We then argue that (a) allows us to start inverting somewhere, (b) allows us to continue the inverse along paths, (c) says that the inverse doesn't stop until we run out of $\mathcal{F}^-$, and (d) tells us that the inverse is independent of the path and hence unique.

Before we start the proof we remark on a few technical details. There are two possible approaches to the proof. One is to work with C^{∞} functions and quote the Nash-Moser theorem (see [5] for an exposition) using the ideas in [6] to handle the subelliptic estimates. The other is to work with the Folland-Stein spaces S_k^p (see [3]) which measure k derivatives in the direction of the contact structure in L^p norm. We can take $\omega \epsilon S_{k+2}^p$ and $W \epsilon S_k^p$ provided pk>8 so that $W \epsilon C^{\circ}$ by the appropriate Sobolev inclusion. The easiest case analytically is to take p = 2, which necessitates $k \geqslant 5$.

We proceed with the proof. Observation (a) follows from the hypothesis. To see (b) we compute the derivative of P, and apply the inverse function theorem.

In fact, from Corollary 4.5 we write

$$\bar{\Box} f' + 2\bar{W} f' = -W',$$

by putting dashes on the original metric. The operator $\bar{\Box} + 2\bar{W}$ has zero null space by the maximum principle, since $\bar{W} < 0$. Since it is self-adjoint, it must also be onto and hence invertible. This proves that DP is invertible when $\bar{W} < 0$, and so P is locally invertible on all of C^-.

To see assertion (c) that P is proper, we apply the maximum principle to the equation

$$f^3 W = f\bar{W} - \bar{\Box} f.$$

Where f is a maximum $\bar{\Box} f \leqslant 0$, and where f is a minimum $\bar{\Box} f \geqslant 0$. Since W and $\bar{W}$ are both negative we get the estimate

$$(78) \qquad \left[(\bar{W}/W)_{min}\right]^{\frac{1}{2}} \leqslant f_{min} \leqslant f_{max} \leqslant \left[(\bar{W}/W)_{max}\right]^{\frac{1}{2}}.$$

Notice that the estimate fails if W and $\bar{W}$ are positive. Having control of the maximum and minimum of f, it is easy to control the higher derivatives using the equation and the subelliptic Garding's inequality

(79) $$\|f\|_{S^P_{k+2}} \leq C\,(\|\bar{\Box} f\|_{S^P_k} + \|f\|_{L^P}).$$

In the C^∞ case this shows P is proper. For given any compact set of M, we have uniform bounds on W_{max} and W_{min} and all $\|W\|_{S^P_k}$. This gives bounds on f_{max} and f_{min} and all $\|f\|_{S^P_k}$ for all f in the preimage, so the preimage is compact since C^∞ is a Montel space. To work in the Banach space S^P_k we also need the following observation. Suppose we have a sequence of contact forms ω_n with $W_n \longrightarrow \bar{W} < 0$ in S^P_k. The previous estimates give bounds on ω_n in S^P_{k+2}, which implies convergence of a subsequence in S^P_k. Let $\omega_n \longrightarrow \bar{\omega}$, and write $\omega_n = f_n^2\bar{\omega}$. The maximum principle estimate shows $f_n \longrightarrow 1$ in C°. Then using the equation we get the estimate

(80) $$\|f_n - 1\|_{S^P_{k+2}} \leq C\,\|W_n - \bar{W}\|_{S^P_k};$$

this shows $\omega_n \longrightarrow \bar{\omega}$ in S^P_{k+2}, and proves P is proper.

The assertion (d) that $\mathcal{T}^-$ is simply connected follows by shrinking along straight line paths to $W = -1$. This completes the proof of the theorem.

7. Minimizing Torsion. We consider finally the problem of minimizing the energy

(81) $$E_\iota = \int_P |\iota|^2\, \nu$$

representing the L^2 norm of the torsion by the heat equation with the contact form ω fixed. From Lemma 5.5 we have the result that if we take a path of Ω's depending on t with $\Omega' = f\bar{\Omega}$ then

$$E_\iota' = 2\,\mathrm{Im} \int_P \bar{f}\, D_V \iota\ \nu.$$

Following the gradient flow of E_ι we let $f = i\, D_V \iota$. This gives heat equation for E_ι. Since $D_V\iota = -$ is by Lemma 3.9 we get the following results.

7.1. Heat Equation Formulas.

$$\Omega' = i\, D_V\iota \cdot \bar{\Omega},$$

$$E_\iota' = -2\int_P |s|^2 \,\nu, \tag{82}$$

$$\iota' = D_V^2\iota.$$

These equations show that if the solution exists for all time then the energy E_ι decreases and the curvature $s \longrightarrow 0$. The equation $\iota' = D_V^2\iota$ is a highly degenerate parabolic equation, since the right hand side involves only the second derivative in the one direction V. Nevertheless, it is not a bad equation, since the maximum principle applies. This shows that the maximum absolute value of ι decreases. The equation is in fact just the ordinary heat equation restricted to each orbit in the flow of V. Physically we can imagine the manifold P to be made of a bundle of wires insulated from each other, with the heat flowing only along the wires. When the orbits of V are closed, the analysis should be fairly straightforward. When the orbits of V are dense, things are much more complicated, and probably lead to small divisor problems.

A regular foliation is one where each leaf is compact and the space of leaves is Hausdorff. In this case we always have a Seifert foliation, one where each leaf has a neighborhood which is a finite quotient of a bundle. In three dimensions the Seifert foliated manifolds are well-understood by the topologists, and provide many of the nice examples. We conjecture the following result.

7.2. Conjecture. Let M be a compact three-manifold with a fixed contact form ω whose vector field V induces a Seifert foliation. There there exists a CR structure on M such that the associated metric has $s = 0$, i.e., the sectional curvature of all planes at a given point perpendicular to the contact bundle $B = \text{Null}\ \omega$ are equal. The metric is obtained as the limit of the heat equation flow as $t \longrightarrow \infty$.

REFERENCES

[1] J.M. Bony, Principe du maximum, inégalité de Harnack, et unicité du problème de Cauchy pour les opérateurs elliptiques dégénerés, Ann. Inst. Fourier 19(1969), 277-304.

[2] A. Douady, Noeuds et structures de contact en dimension 3, d'après Daniel Bennequin, Séminaire Bourbaki, 1982/83, no.° 604.

[3] G.B. Folland and E.M. Stein, Estimates for the $\bar{\partial}_b$-complex and analysis on the Heisenberg group, Comm. Pure and App. Math 27(1974), 429-522.

[4] J.W. Gray, Some global properties of contact structures, Annals of Math 69(1959), 421-450.

[5] R. Hamilton, The inverse function theorem of Nash and Moser, Bull. Amer. Math. Soc. 7(1982), 65-222.

[6] R. Hamilton, Three-manifolds with positive Ricci curvature, J. Diff. Geom. 17(1982), 255-306.

[7] D. Jerison and J. Lee, A subelliptic, non-linear eigenvalue problem and scalar curvature on CR manifolds, Microlocal Analysis, Amer. Math Soc. Contemporary Math Series, 27(1984), 57-63.

[8] J. Martinet, Formes de contact sur les variétés de dimension 3, Proc. Liverpool Singularities Symp II, Springer Lecture Notes in Math 209(1971), 142-163.

[9] R. Schoen, Conformal deformation of a Riemannian metric to constant scalar curvature, preprint 1984.

[10] W. Thurston and H.E. Winkelnkemper, On the existence of contact forms, Proc. Amer. Math. Soc. 52(1975), 345-347.

[11] S.M. Webster, Pseudohermitian structures on a real hypersurface, J. Diff. Geom. 13(1978), 25-41.

APPENDIX

by
Alan Weinstein

THREE-DIMENSIONAL CONTACT MANIFOLDS WITH VANISHING TORSION TENSOR

In a lecture on some of the material in the preceding paper, Professor Chern raised the question of determining those 3-manifolds admitting a contact structure and adpated Riemannian metric for which the torsion invariant $c^2=a^2+b^2$ is identically zero. (See §3. A variational characterization of such structures is given in Theorem 5.2.) The purpose of this note is to show that the class of manifolds in question consists of certain Seifert fiber manifolds over orientable surfaces, and that the first real Betti number $b_1(M)$ of each such manifold M is even. These results are not new; see our closing remarks.

By a simple computation, it may be seen that the matrix $\begin{bmatrix} a & b \\ b & -a \end{bmatrix}$ (see Corollary 3.5) represents the Lie derivative of the induced metric on the contact bundle B with respect to the contact vector field V. We thus have:

A.1. Lemma. *The invariant* c^2 *is identically zero if and only if* V *is a Killing vector field.* (*In other words,* M *is a "K-contact manifold"; see* [1].)

We would like the flow generated by V to be periodic. If this is not the case, we can make it so by changing the structures in the following way. Let G be the closure, in the automorphism group of M with its contact and metric structures, of the 1-parameter group generated by V. G must be a torus, so in its Lie algebra we can find Killing vector fields V' arbitrarily close to V and having periodic flow. Let ω' be the 1-form which annihilates the subbundle B' perpendicular to V' and which satisfies $\omega'(V') \equiv 1$. For V' sufficiently close to V, ω' will be so close to the original contact form ω that it is itself a contact form. Since the flow of V' leaves the metric invariant, it leaves the invariant the form ω', from which it follows that V' is the contact vector field associated with ω'.

Having made the changes described in the previous paragraph, we may revert to our original notation, dropping primes, and assume that the flow of V is periodic. A rescaling of ω will even permit us to assume that the least period of V is 1. (Note that, by Gray's theorem [2], we could actually assume that the new contact structure equals the one which was originally given.)

Suppose for the moment that the action of $S^1 = \mathbb{R}/\mathbb{Z}$ generated by V is free. Then M is a principal S^1 bundle over the surface M/S^1. The form ω is a connection on this bundle; since ω is a contact form, the corresponding curvature form on M/S^1 is nowhere vanishing. Thus M/S^1 is an orientable surface, and the Chern class of the fibration $M \longrightarrow M/S^1$ is non-zero. By the classification of surfaces, $b_1(M/S^1)$ is even; by the Gysin sequence, $b_1(M) = b_1(M/S^1)$ and is therefore even as well.

We are left to consider the case where the action of S^1, although locally free, is not free. The procedure which we will follow is that of [8]. Let $\Gamma \subsetneq S^1$ be the (finite) subgroup generated by the isotropy groups of all the elements of M. Then M is a branched cover of M/Γ, and M/Γ is a principal bundle over M/S^1 with fiber the circle S^1/Γ. The branched covering map $M \longrightarrow M/\Gamma$ induces isomorphisms on real cohomology, so it suffices to show that $b_1(M/\Gamma)$ is even. To see this, we consider the fibration $S^1/\Gamma \longrightarrow M/\Gamma \longrightarrow M/S^1$. The quotient spaces M/Γ and M/S^1 are V-manifolds in the sense of [4], and we have a fibre bundle in that category. The base M/S^1 is actually a topological surface which is orientable since it carries a nowhere-zero 2-form on the complement of its singular points. Now the contact form may once again be considered as a connection on our V-fibration, and so, just as in the preceding paragraph, we may conclude that $b_1(M/\Gamma)$ is even.

Remarks. A K-contact manifold is locally a 1-dimensional bundle over an almost-Kähler manifold. When the base is Kähler, the contact manifold is called <u>Sasakian.</u> Using harmonic forms, Tachibana

[5] has shown that the first Betti number of a compact Sasakian manifold is even. On the other hand, since every almost complex structure on a surface is integrable, every 3-dimensional K-contact mainfold is Sasakian, and hence our result follows from Tachibana's theorem. In higher dimensions, compact symplectic manifolds with odd Betti numbers in even dimension are known to exist [3] [7], and circle bundles over them will carry K-contact structures, while having odd Betti numbers in even dimension.

The paper [6] contains a study of which Seifert fiber manifolds over surfaces actually admit S^1-invariant contact structures.

Acknowledgments. This research was partially supported by NSF Grant DMS84-03201. I would like to thank Geoff Mess for his helpful advice.

REFERENCES

1. D. Blair, Contact Manifolds in Riemannian Geometry, Lecture Notes in Math., vol. 59 (1976).
2. J. Gray, Some global properties of contact structures, Ann. of Math. 69 (1959), 421-450.
3. D.McDuff, Examples of simply connected symplectic manifolds which are not Kähler, preprint, Stony Brook, 1984.
4. I. Satake, The Gauss-Bonnet theorem for V-manifolds, J. Math. Soc. Japan 9 (1957), 464-492.
5. S. Tachibana, On harmonic tensors in compact Sasakian spaces, Tohoku Math. J. 17 (1965), 271-284.
6. C.B. Thomas, Almost regular contact manifolds, J. Diff. Geom. 11 (1976), 521-533.
7. W. Thurston, Some simple examples of symplectic manifolds. Proc. A.M.S. 55 (1976), 467-468.
8. A. Weinstein, Symplectic V-manifolds, periodic orbits of hamiltonian systems, and the volume of certain riemannian manifolds, Comm. Pure Appl. Math. 30 (1977), 265-271.

4-MANIFOLDS WITH INDEFINITE INTERSECTION FORM

S.K. Donaldson
All Souls College
Oxford, England

In writing up this lecture I shall not concentrate so much on describing problems of 4-manifold topology; instead I shall explain how a simple topological construction has applications in two different directions. First I will recall that, just as bundles over a single space have homotopy invariants, so do families of bundles, and that these define corresponding invariants in families of connections. Next I will sketch the way in which such a topological invariant, when endowed with a geometric realisation, becomes important for studying holomorphic bundles over algebraic varieties. Last I will indicate how this same homotopy invariant of families of connections, combined with arguments involving moduli spaces of self-dual connections over a Riemannian 4-manifold, gives restrictions on the possible homotopy types of smooth 4-manifolds and I will speculate on possible future progress in this area.

Topology of bundles.

This is standard material that may be found in [2] for example. Consider a fixed manifold X and a family of bundles over X parametrised by some auxiliary space T, so we have a bundle P over the product $X \times T$ with structure group G (compact and connected, say). Take first the case when T is a point so we have a single bundle over X, determined up to equivalence by a homotopy class of maps from X to BG. This may be non-trivial, detected for example by characteristic classes in the cohomology of X. If we choose a connection A on the bundle the real characteristic classes can be represented by explicit differential forms built from the curvature of the connection. Equally if D is an elliptic differential operator over X then using a connection it may be extended to act on objects (functions, forms, spinors etc.) twisted by a vector bundle associated to P. This has an integer valued index:

$$\text{index}\,(D_A) = \dim \ker D_A - \dim \operatorname{coker} D_A$$

which is a rigid invariant of the bundle, independent of the connection. So these are two ways in which the underlying homotopy may be represented geometrically, by curvature and by differential operators. The Chern-Weil and Atiyah-Singer theorems then give formulae relating the three.

In the same way for a general family parametrised by T the bundle P is classified by a homotopy class of maps from T to the mapping space $\mathrm{Maps}(X,BG)$, and at the other extreme from the case T = point we have a universal family parametrised by this mapping space. Again we may always choose a connection over $X \times T$, which we may think of as a family of connections parametrised by T, and conversely any family of equivalence classes of connections on some bundle essentially arises in this way. (This is precisely true if we work with based maps and bundles, removing base points gives small technical differences which can safely be ignored here). Equivalently we have the infinite dimensional space $\mathcal{B}$ of all equivalence classes of connections obtained by dividing the affine space of connections $\mathcal{A}$ by the bundle automorphism group $\mathcal{G}$. $\mathcal{B}$ has the homotopy type of $\mathrm{Maps}(X,BG)$.

Again we may construct topological invariants of such families of bundles. In cohomology we can use the characteristic classes again. There is a slant product:

$$H^{p+q}(X \times T) \otimes H_q(X) \longrightarrow H^p(T)$$

so that characteristic classes of bundles over $X \times T$ contracted with, or integrated over, homology classes in the base manifold X yield cohomology classes in families of connections. In particular if G is, say, a unitary group we obtain in this way a map:

$$\mu : H_2(X) \longrightarrow H^2(T)$$
$$\mu(\alpha) = c_2(P)/\alpha$$

(A simpler example is to take the Jacobian parametrising complex line bundles over a Riemann surface. Operating in the same way with the first Chern class gives the usual correspondence between the 1-dimensional homology of the surface and the cohomology of the Jacobian). We can do the corresponding thing in K-theory and realise the resulting elements in the K-theory of T by using differential operators again. For example if the base manifold X is the 2-sphere then a unitary bundle over $S^2 \times T$ defines an element of $K(S^2 \times T)$ which maps to $K(T)$ by the inverse of the Bott periodicity map. If we take the Dirac operator D over S^2 then a family of connections gives a family of Dirac operators $\{D_t\}$ parametrised by T and, after suitable stabilisation the index of this family [2] defines the required class:

$$\text{index } D_t = [\text{Ker } D_t] - [\text{coker } D_t] \in K(T)$$

Of course we obtain other classes in this way and the Atiyah-Singer index theorem for families gives formulae relating these to the underlying homotopy. In particular we may understand our class above from either point of view via the formula:

$$c_1(\text{index } D_t) = \mu(\text{fundamental class of } S^2) .$$

Stable bundles on algebraic curves and surfaces.

Here I only want to say enough to fit into our overall theme; more details and references may be found in [4], but I learnt the point of view we are adopting now from lectures of Quillen.

There is a general algebraic theory dealing with the action of a complex reductive group $G^{\mathbb{C}}$ on a vector space $\mathbb{C}^{n+1}$ via a linear representation. Equivalently we may take the induced action on $\mathbb{CP}^n$ and the hyperplane bundle H over it. In that theory there is a definition of a "stable" point. Now suppose that $\mathbb{C}^{n+1}$ has a fixed Hermitian metric, inducing metrics on H and on $\mathbb{CP}^n$, and picking out a maximal compact subgroup $G \subset G^{\mathbb{C}}$ whose action preserves these metrics. There is a general theory dealing with the metrical properties of these actions and relating them to the purely complex algebraic properties. Roughly speaking if we restrict to the stable points then a transversal to the $G^{\mathbb{C}}$-action on $\mathbb{CP}^n$ is induced by taking the points in $\mathbb{C}^{n+1}$, or equivalently H^{-1}, which minimise the norm in their $G^{\mathbb{C}}$ orbits. The corresponding variational equations cutting out the transversal take a simple form and are the zeros of a map:

$$m : \mathbb{CP}^n \longrightarrow J^* \qquad [7]\ ,\ [8]\ .$$

Large parts of this theory can be developed abstractly from general properties of Lie groups and the fact that the curvature form of the Hermilian line bundle H gives the Kähler symplectic form on $\mathbb{CP}^n$.

Atiyah and Bott [1] observed that the theory of holomorphic structures on a vector bundle E over an algebraic curve C could be cast in the same form, except with an infinite dimensional affine space in place of a projective space. For a holomorphic structure on E is given by a $\bar{\partial}$-operator and these are parametrised by a complex affine space A . The infinite dimensional group $G^{\mathbb{C}}$ of complex linear automorphisms of E acts by conjugation and the quotient set is by definition the set of equivalence classes of holomorphic (or algebraic) bundles, topologically equivalent to E . Independently, and from another point of view, stability of algebraic bundles had been defined in algebraic geometry; the definition uses the notion of the degree of a bundle - the integer obtained by evaluating the first Chern class on the fundamental cycle.

If now E has a fixed Hermitian metric then a $\bar{\partial}$-operator induces a unique unitary connection. Regarded as connections the symmetry group of the affine space A is reduced to the subgroup $G \subset G^{\mathbb{C}}$ of unitary automorphisms, and this subgroup preserves the natural metric form on the space of connections A derived from integration over C . We would have all the ingredients for the abstract theory described above if we had a Hermitian line bundle L over A with curvature generating this metric form, and acted on by $G^{\mathbb{C}}$.

It was explained above that over a space of connections we obtain virtual bundles from the associated elliptic operators. In particular we can take the Dirac operator over the algebraic curve C , which is the same as the $\bar{\partial}$-operator after tensoring with a square root $K_C^{1/2}$ of the canonical bundle, so the kernel and cokernel form the usual sheaf cohomology. Moreover we get a genuine line bundle if we take the highest exterior power or determinant of the relevant vector spaces. Thus we get a complex line bundle L_C over A :

$$L_C = \chi(E \otimes K_C^{1/2}) = \det H^0(E \otimes K_C^{1/2}) \otimes \det H^1(E \otimes K_C^{1/2})^{-1}$$

acted upon by $G^{\mathbb{C}}$, and realising via the first Chern class the cohomology class obtained under our map μ from the fundamental cycle of the curve C, as in Section I.

Now Quillen has defined Hermitian metrics [9] on such determinant line bundles and computed the associated curvature to be precisely the metric form above. Thus all the ingredients for applying the general theory are present - the map m cutting out a transversal to the stable orbits is given by the curvature of a connection and the preferred points, minimising Quillens analytic torsion norm, are given by the projectively flat unitary connections.

We can study algebraic bundles over any projective variety; in particular over an algebraic surface X. Now the definition of stability requires the choice of a polarisation - the first chern class of an ample line bundle L over X. This means that the degree of a bundle is defined, in the normal way. We can represent this polarising class by a Kähler form ω, the curvature of some metric on L. Then the same theory holds; we do not find flat connections on stable bundles but connections whose curvature is orthogonal to the Kähler metric at each point. The relation with metrics on cohomology is less well established but the relevant line bundle should probably be of a form such as:

$$L_X = \chi(E \otimes K_X^{1/2} \otimes L^{1/2}) \otimes \chi(E \otimes K_X^{1/2} \otimes L^{-1/2})^{-1} .$$

Suppose that L has a section s cutting out a curve $C \subset X$, we can think in the sense of currents of C as a degenerate form of a metric. There is an exact sequence:

$$0 \longrightarrow E \otimes K_X^{1/2} \otimes L^{1/2} \xrightarrow{\ s\ } E \otimes K_X^{1/2} \otimes L^{1/2} \longrightarrow E|_C \otimes K_C \longrightarrow 0$$

whose long exact sequence in cohomology gives an isomorphism $L_X \cong L_C$; moreover one can compute formulae for the difference in norms, one defined relative to C and one to the metric ω on X , compared under this isomorphism, with an explicit difference term given by integrals involving Chern-Weil polynomials in the curvature. These are useful for throwing problems back to the curve from the surface.

All this should probably be understood in the following way. Topologically we have a map μ from $H_2(X)$ to the cohomology of any family of connections over X . If we wish to define stable bundles then we need a polarisation $[\omega]$ of X which via this map μ and Poincaré Duality induces a corresponding "polarising class" in the infinite dimensional space of connections. We may represent the original class in various explicit ways; by a metric or by a line bundle or by a curve, and to each such representation on X we get a corresponding representation in the space of connections. The usual formulae for homologies between the representations on X go over to corresponding formulae on the connections which we can use in our arguments involving stable bundles. But the existence of these formulae underlines that the basic correspondence between the geometry of the base manifold and its stable bundles is generated by our simple construction of Section I.

Connections over smooth 4-manifolds.

Self dual connections are solutions to a differential equation which is special to 4-dimensions. On an oriented Riemannian 4-manifold the 2-forms decompose into the ± 1 eigenspaces of the star operator; the same is true for bundle valued forms, and a connection is self-dual if its curvature lies in the $+1$ eigenspace. If the manifold is an algebraic surface with the standard orientation reversed these are the connections whose existence characterised stable bundles in the previous section. (For on a Kähler surface the self dual 2-forms are made up of the (0,2) and (2,0) forms and the span of the Kähler form). Correspondingly these solutions of differential equations in Riemannian geometry behave rather like objects in algebraic geometry; in particular the solutions, up to equivalence by bundle automorphisms, are parametrised by finite dimensional moduli spaces rather as the Jacobian parametrises line bundles over a Riemann surface. Moreover these moduli spaces have applications in differential topology.

At present there is no general theory of smooth 4-manifolds. A central problem is to understand the relationship between homotopy and differentiable structures and to quantify the gap between them. For simply connected 4-manifolds the homotopy type is easily understood - the sole invariant is the intersection form on the integral 2-dimensional homology. Likewise the classification of topological 4-manifolds up to homeomorphism has been established by Freedman [6], and is virtually the same as that up to homotopy. Now while there are many integral definite forms, and so corresponding topological 4-manifolds, it was proved by methods similar to those described below [5] that none of these arise from smooth manifolds beyond the obvious examples given by diagonalisable forms. The interesting remaining class of forms are the even (which corresponds to spin manifolds) indefinite forms which are all of the shape:

$$n\, E_8 + m \begin{pmatrix} 0 & 1 \\ 1 & 0 \end{pmatrix}$$

For smooth manifolds n must be even by Rohlin's Theorem and the simplest known example, beyond $S^2 \times S^2$ which has form $\begin{pmatrix} 0 & 1 \\ 1 & 0 \end{pmatrix}$, is the smooth 4-manifold underlying a complex K3-surface, having 2 E_8's and 3 $\begin{pmatrix} 0 & 1 \\ 1 & 0 \end{pmatrix}$'s in the intersection form. By taking connected sums with $S^2 \times S^2$ one can always increase m so that the problem of realisation of these forms is to discover the minimal value of m for each given n. It is hoped that a proof that for positive n the value of m must be at least 3 (implying in particular that the K3 surface is smoothly indecomposable, hence genuinely the simplest "non-obvious" smooth 4-manifold) using the methods described below, will appear very shortly.

First a word on the formal structure of these proofs. We need some way of distinguishing the forms which are obviously realised when n is zero from the case when n is positive. The relevant property that emerges is that a direct sum $H_1 \oplus H_2 \oplus \ldots \oplus H_k$ of copies of the "hyperbolic" form $\begin{pmatrix} 0 & 1 \\ 1 & 0 \end{pmatrix}$ is distinguished by the fact that the symmetric power:

$$(H_1 \oplus \ldots \oplus H_k)^{k+1}$$

is identically zero mod 2. For example when $k = 1$ this says that for any four integral elements $\alpha_1, \ldots \alpha_4$:

$$(\alpha_1 \cdot \alpha_2)(\alpha_3 \cdot \alpha_4) + (\alpha_1 \cdot \alpha_3)(\alpha_2 \cdot \alpha_4) + (\alpha_1 \cdot \alpha_4)(\alpha_2 \cdot \alpha_3) = 0 \mod 2$$

So our proofs are really to establish such identities when α_i are integral homology classes and $(\cdot)$ is the intersection pairing.

These identities are obtained by pairing two kinds of information and, since we are interested in the differences between homotopy and differentiable structures, it is probably important to stress the contrast in the ways that these arise. By definition our moduli space M of self dual connections on some bundle parametrises a family of connections and we have seen in the first section above that we can produce cohomology classes in such parameter spaces. Alternatively we can think of the moduli space as a subset of the infinite dimensional space $\mathcal{B}$ of all equivalence classes of connections, cut out by the non-linear differential equations giving the self duality condition. Since we regard the homotopy type of the base manifold X^4 as known we may regard the homotopy type of this infinite dimensional parameter space of connections as known. For example we have our map; defined in an elementary way:

$$\mu : H_2(X^4) \longrightarrow H^2(\mathcal{B}_X)$$

and in fact this generates a copy of the polynomial algebra on $H_2(X^4)$ within $H^*(\mathcal{B}_X)$.

Our moduli space M sits within this infinite dimensional space. At present we may regard this as largely unknown and mysterious, except for properties that can be understood by linearisation, for example the dimension of the space. What we do know is that the moduli space carries a fundamental class in homology; or rather, as we shall see, that it may be truncated, typically, to a manifold with boundary ∂M so we may assert:

$$< \phi, [\partial M] > = 0 \quad \text{for any } \phi \text{ in } H^*(\mathcal{B}_X)$$

To produce a suitable cohomology class ϕ we may use our map μ - this builds in the two dimensional homology that we wish to study; likewise we may produce more subtle cohomology classes coming, from our present point of view, from the index of the 4-dimensional Dirac operator on a spin 4-manifold. But all this is homotopy, the smooth structure and the difference between differentiable and topological manifolds enters by the existence of the relative homology class carried by the moduli space of solutions to the differential equation.

Here is an explicit example, directly relevant to the case when we study 4-manifolds with one negative eigenvalue in their intersection form. Take the complex projective plane with its standard orientation reversed; then we may study the self-dual connections via the stable holomorphic bundles as above, and in particular if we consider rank 2 bundles with $c_1 = 0$; $c_2 = -2$ then the appropriate moduli space has been described by Barth [3] as follows. To the original projective plane P we may associate the dual plane P^* , so points of one plane are lines in the other. The conic curves in P^* are parametrised by a copy of $\mathbb{CP}^5$; the non-singular conics form an open subset, the complement of a divisor which is naturally identified with the symmetric product $\mathrm{sym}^2(P)$ (since a singular conic is made up of two lines). According to Barth the moduli space of algebraic bundles may be identified with these non-singular conics, which we may obviously truncate by removing an open neighbourhood of $\mathrm{sym}^2(P)$ to get a manifold M with boundary ∂M made up; loosely speaking, of a circle bundle over $\mathrm{sym}^2(P)$ with fibre L say.

We can understand our map μ very easily in this example, and doing so explicitly will illustrate the general case. Let ℓ be a line in P (so representing a generator of $H_2(P)$). Then it follows essentially immediately from our discussion of the previous sections and the description by Barth of the "jumping lines" of a bundle that a representative for the cohomology class $\mu[\ell]$ is given by the hyperplane V_ℓ in $\mathbb{CP}^5$ consisting of conics through the point ℓ in P^*. Consider four general lines $\ell_1, \ell_2, \ell_3, \ell_4$ in P. The eight dimensional cohomology class $\mu(\ell_1)\cdot\mu(\ell_2)\cdot\mu(\ell_3)\cdot\mu(\ell_4)$ is represented by the projective line $V_{\ell_1} \cap V_{\ell_2} \cap V_{\ell_3} \cap V_{\ell_4}$. The intersection of this with our truncated moduli space M is a surface with boundary three copies of the loop L , corresponding to the three point-pairs:

$$((\ell_1 \cap \ell_2),(\ell_3 \cap \ell_4)),((\ell_1 \cap \ell_3),(\ell_2 \cap \ell_4)),((\ell_1 \cap \ell_4),(\ell_2 \cap \ell_3))$$

in P . If we proceed analogously on any (simply connected) 4-manifold with one negative part of the intersection form then we have a broadly similar moduli space - a non-compact manifold of real dimension 10 . If we consider a cup product $\mu(\alpha_1)\mu(\alpha_2)\mu(\alpha_3)\mu(\alpha_4)$ for any 4 surfaces α_i then we are led in the same way to consider a set of point pairs of the form:

$$((\alpha_i \cap \alpha_j),(\alpha_k \cap \alpha_l))$$

and the number of such pairs, modulo 2, is just the expression in terms of the intersection pairing given above. The key additional fact is that for a manifold with a spin structure (unlike $\mathbb{CP}^2$) the corresponding loop L is essential in the space of connections, detected by a mod 2 cohomology class w_1 , thus we argue in the manner above with the cohomology class $\phi_q = w_1 \cdot \mu(\alpha_1) \cdot \mu(\alpha_2) \cdot \mu(\alpha_3) \cdot \mu(\alpha_4)$.

Finally I will make two general remarks. Following Taubes [10] the structure of these boundaries to moduli spaces can be understood reasonably explicitly in terms of a number of "instantons" - connections concentrated near a finite set of points on the manifold. In the complex algebraic version we should probably think of these as being bundles obtained from deformations of ideal sheaves, rather in the way that the symmetric products of an algebraic curve map into the Jacobian. The ways that these instantons can be oriented relative to each other give the structure of the "link" L in the moduli space itself and this depends upon the values of the anti self-dual harmonic 2-forms at the points. The possibilities become rapidly more complicated as the number m of negative parts of the intersection form grows larger, and roughly speaking what distinguishes the cases $m = 0,1,2$ is that the codimension of the "special divisors", on which the forms are aligned in exceptional ways, is sufficiently high. It would seem to be possible that the behaviour of these harmonic forms (which of course globally reflect the cohomology, via Hodge Theory) contains differential topological information about the 4-manifold. In the complex case these anti self-dual forms are made up of the Kähler symplectic form and the holomorphic 2-forms and these are well known to carry a lot of information about the complex structure. Rather similarly the "periods" of the harmonic forms, the relation with the integral structure, also enter into the Riemannian theory via line bundles and Hodge Theory.

I have emphasised here that no global properties of these moduli spaces beyond existence are really used in these arguments, and indeed the number of explicit examples that are known is rather small. On the other hand we have seen that we may easily construct cohomology classes over these moduli spaces and that we have obtained information by pairing these with the relative homology class carried by the manifold. It seems that the moduli spaces should carry an absolute homology class with respect to cohomology with sufficiently small support, which can then be paired to give integer valued invariants. Moreover these should be independent of the Riemannian metric on the 4-manifold in the usual way that the homology class carried by the fibre of a map is a deformation invariant.

Of course there are many ways in which rigid integer valued invariants can be produced by analytic methods - integration of forms or indices of operators; but as I recalled in the first section these can all be understood entirely from homotopy, via the usual formulae. This is not obviously the case for our moduli spaces. For example if we take the case described above on the projective plane then we see that $(\mu[\ell])^5\,[M] = 1$, given by the intersection of five hyperplanes. It is not clear that this could be predicted from the homotopy type of $\mathbb{CP}^2$ alone. Again, the fact that these cohomology classes appear so naturally in the complex algebraic theory gives extra motivation in this direction.

References:

[1] Atiyah, M.F. and Bott, R. "The Yang-Mills equations over Riemann surfaces". Trans. Roy. Soc. London A 308 (1982) 523-615.

[2] Atiyah, M.F. and Singer, I.M. "The index of elliptic operators IV". Annals of Math. 93 (1971) 119-138.

[3] Barth, W. "Moduli of vector bundles on the projective plane". Inventiones math. 42 (1977) 63-91.

[4] Donaldson, S.K. "Anti self dual Yang-Mills connections over complex algebraic surfaces and stable vector bundles". To appear in Proc. Lond. Math. Soc.

[5] Donaldson, S.K. "An application of gauge theory to four dimensional topology". Journal Diff. Geom. 18 (1983) 279-315.

[6] Freedman, M.H. "The topology of four dimensional manifolds". Journal Diff. Geom. 17 (1982) 357-453.

[7] Kirwan, F.C. "The cohomology of quotients in symplectic and algebraic geometry". Princeton U.P. to appear.

[8] Ness, L. "A stratification of the null cone by the moment map". To appear in the Amer. Journal Math.

[9] Quillen, D. Lecture at 1982 Arbeitstagung, Bonn.

[10] Taubes, C.H. "Self-dual connections on manifolds with indefinite intersection form". To appear in Journal of Diff. Geom.

ARITHMETISCHE KOMPAKTIFIZIERUNG DES MODULRAUMS DER ABELSCHEN VARIETÄTEN

G. Faltings
Fachbereich Mathematik
Universität-Gesamthochschule Wuppertal
Gaußstr. 20
5600 Wuppertal 1

INHALTSVERZEICHNIS

§ 1 EINLEITUNG

Die Konstruktion des Modulraumes A_g der prinzipal polarisierten abelschen Varietäten der Dimension g über $\mathbb{Z}$ ist seit langem bekannt. Man erhält je nach Geschmack einen groben Modulraum oder ein algebraisches Feld, nach Einführung von Level-Strukturen sogar einen feinen Modulraum. Es sind auch Methoden der Kompaktifizierung bekannt über den komplexen Zahlen (siehe [AMRT], [N]), doch fehlte bis jetzt die Beschreibung einer solchen über $\mathbb{Z}$. Dies geschieht in dieser Arbeit. Genauer gesagt, konstruieren wir ein algebraisches Feld, welches eigentlich über $\mathbb{Z}$ ist, das A_g als offene Teilmenge enthält, und über dem eine universelle semiabelsche Varietät existiert. Der Rand wird ziemlich genau beschrieben, und man erhält für Level-n-Strukturen sogar einen algebraischen Raum.

Dabei wird für unsere Zwecke ein algebraisches Feld gegeben durch ein Schema $\underline{S}$, von endlichem Typ über $\mathbb{Z}$, sowie eine endliche Abbildung $\underline{R} \to \underline{S} \times_{\mathbb{Z}} \underline{S}$, welche $\underline{R}$ zu einem Gruppoid über $\underline{S}$ macht, und für die die Projektionen von $\underline{R}$ auf $\underline{S}$ étale sind. Man erkennt leicht die Äquivalenz zur Definition in [DM], und wer will, kann sich nach Einführung von Level-n-Strukturen darauf beschränken, daß $\underline{R}$ abgeschlossenes Unterschema von $\underline{S} \times_{\mathbb{Z}} \underline{S}$ ist, wobei man dann bei algebraischen Räumen landet ([A]). Bei der Konstruktion von $\underline{S}$ benutzt man M. Artin's Deformations-Theorie ([A]) sowie eine leichte Verallgemeinerung von D. Mumford's Konstruktion degenerierender abelscher Varietäten. Als $\underline{R}$ nimmt man einfach die Normalisierung des von dem Modulproblem A_g gelieferten Gruppoids. Daß dies die gewünschten Eigenschaften hat, folgt aus einer Betrachtung degenerierender abelscher Varietäten, indem man zeigt, daß man die in Mumford's Konstruktion auftretenden Perioden aus den Koeffizienten der θ-Reihe ablesen kann.

Schließlich sei noch erwähnt, daß anders als im Fall der Kurven die Kompaktifizierung nicht kanonisch ist, sondern von der Wahl einer Kegelzerlegung der positiv semidefiniten quadratischen Formen in g Variablen abhängt. Dies ist auch bei der komplexen toroidalen Kompaktifizierung der Fall, und in der Tat liefern unsere Methoden über $\mathbb{C}$ gerade diese Modelle.

Der Aufbau der Arbeit ist wie folgt:

Zunächst betrachten wir degenerierende abelsche Varietäten und ordnen ihnen quadratische Formen zu. Dies wird zum einen benutzt, um später die Kompaktheit zu zeigen, und motiviert zum anderen die Wahl der Daten, welche bei der verallgemeinerten Mumford-Konstruktion eingehen.

Diese folgt dann im nächsten Kapitel. Dabei sind alle auftretenden Schwierigkeiten im wesentlichen schon von Mumford in [M4] gelöst worden. Wir brauchen dies nur noch von Tori auf semiabelsche Varietäten zu verallgemeinern.

Danach bereitet die Konstruktion von $\underline{S}$ und $\underline{R}$ keine großen Probleme mehr. Ihr ist das vierte Kapitel gewidmet, worauf dann die Anwendungen folgen:

Wir betrachten Level-Strukturen, Modulformen (unter anderem eine arithmetische Behandlung der minimalen Kompaktifizierung), étale Garben und Kohomologie, die Torelli-Abbildung sowie die Beziehungen zur komplexen Theorie. Eine weitere Anwendung wäre es, den ersten Teil des Beweises der Mordell-Vermutung zu vereinfachen (siehe [F]), und es bleibt zu hoffen, daß eine arithmetische Theorie der Siegel'schen Modulformen in Zukunft noch einiges Schöne hervorbringt.

Der Leser wird bemerken, daß alle wesentlichen Grundideen von D. Mumford übernommen worden sind, und dieser hätte sicher auch noch die Resultate dieser Arbeit erhalten, wenn er sich nicht anderen Interessen zugewandt hätte. Einer seiner Schüler, Ching-Li Chai, hat kürzlich ebenfalls eine arithmetische Kompaktifizierung des Modulraums A_g beschrieben. (Siehe [C]). Nach den mir voliegenden Informationen hat er auch Mumford's Konstruktion verallgemeinert (entsprechend unserem § 3), benutzt aber für die Konstruktion der Kompaktifizierung Theta-Funktionen und Aufblasungen. Dies hat den Vorteil großer Explizitheit und den Nachteil, daß man keine universelle semiabelsche Varietät erhält, und daß man nur über $\mathbb{Z}[^1/2]$ kompaktifiziert. Auf jeden Fall hat er seine Resultate unabhängig von mir und früher erhalten, so daß ihm bei Überschneidungen der Vorrang gebührt. Da er sehr viel mehr Sorgfalt auf die Ausarbeitung der Details verwendet als der Verfasser dieser Arbeit, konnten seine Ergebnisse bisher noch nicht erscheinen.

§ 2 DEGENERIERENDE ABELSCHE VARIETÄTEN

a) Sei R ein normaler kompletter lokaler Ring mit maximalem Ideal $\mathfrak{m}$, Restklassenkörper $k = R/\mathfrak{m}$ und Quotientenkörper K. Wir nehmen an, daß K eine Charakteristik verschieden von zwei hat, doch ist es durchaus zugelassen, daß die Charakteristik von k zwei ist. s und η seien der spezielle und der generische Punkt von $\mathrm{Spek}(R)$.

G sei eine semiabelsche Varietät über $\mathrm{Spek}(R)$, d.h. G ist ein glattes algebraisches Gruppen-Schema über R, von endlichem Typ, dessen Fasern zusammenhängend sind und Erweiterungen von abelschen Varietäten durch Tori. Die Darstellung vereinfacht sich sehr, wenn die spezielle Faser G_s selbst ein Torus ist. Wir empfehlen, sich die Argumente zuerst an diesem Spezialfall klar zu machen. Der allgemeine

Fall erfordert keine neuen Ideen, sondern nur eine Reihe von Notationen und Definitionen. Wir setzen voraus, daß G_η eine abelsche Varietät ist, und daß der maximale Torus von G_s zerfällt. Dann ist die formale Komplettierung $\hat{G}$ eine Erweiterung einer formalen abelschen Varietät $\hat{A}$ (entsprechend einem A über R) durch einen formalen Torus $\hat{T} \cong \hat{G}_m^r$. Es gibt eine Gruppe $\tilde{G}$, mit $\hat{\tilde{G}} \cong \hat{G}$, so daß $\tilde{G}$ eine Erweiterung von A durch $T = G_m^r$ ist.

$$0 \to T \to \tilde{G} \to A \to 0$$

Sei $X = X(T) \cong \mathbb{Z}^r$ die Charaktergruppe von T. Dann wird $\tilde{G}$ gegeben durch einen Morphismus $X \to \mathrm{Pic}^0(A)(R)$

$$\mu \longmapsto \mathcal{O}_\mu \quad ,$$

welcher jedem $\mu \in X$ das zugehörige Geradenbündel auf A zuordnet. Es gibt kanonische Isomorphismen

$$\mathcal{O}_\mu \otimes \mathcal{O}_\nu \cong \mathcal{O}_{\mu+\nu} \quad .$$

b) Wir nehmen weiter an, daß auf G ein Geradenbündel $\underline{L}$ gegeben ist, welches auf der generischen Faser G_η eine prinzipale Polarisation definiert. ([M1], Ch. 6, § 2). Dann besitzt $\underline{L}$ eine kanonische kubische Struktur, oder äquivalent dazu, definiert $m^*(\underline{L}) \otimes pr_1{}^*(\underline{L})^{-1} \otimes pr_2^*(\underline{L})^{-1}$ eine Biextension von $G \times G$ durch G_m (siehe [MB], I, § 2).

Das formale Geradenbündel $\hat{\underline{L}}$ ist dann samt seiner kubischen Struktur Pullback eines $\hat{\underline{M}}$ auf $\hat{A}$, welches eine prinzipale Polarisation für $\hat{A}$ definiert. $\hat{\underline{M}}$ kommt von einem $\underline{M}$ auf A, und $\tilde{\underline{L}}$ auf $\tilde{G}$ sei das Pullback von $\underline{M}$. Dann ist $\hat{\tilde{\underline{L}}}$ isomorph zu $\hat{\underline{L}}$, wobei der Isomorphismus die kubische Struktur respektiert. Allerdings ist dieser Isomorphismus nicht eindeutig, sondern kann mit einem Charakter $\chi : \tilde{G} \to G_m$ modifiziert werden. Es sei noch bemerkt, daß ein solcher Charakter eindeutig bestimmt ist durch seine Einschränkung $\mu \in X$ auf T, und daß man auf diese Weise genau alle μ's erhält, welche im Kern der Abbildung $X \to \mathrm{Pic}^0(A)$ liegen.

Bisweilen werden wir voraussetzen, daß $\underline{L}$ symmetrisch ist, d.h., daß $[-1]^*(\underline{L}) \cong \underline{L}$ ($[-1] = -id : G \to G$). Dann ist auch $[-1]^*(\underline{M}) \cong \underline{M}$, doch sind diese Isomorphismen im allgemeinen nicht miteinander verträglich. Wenn man sie so normalisiert, daß sie auf der Faser in Null die Identität sind, so unterscheiden sich die Symmetrien auf $\hat{\underline{L}}$ und dem Pullback von $\hat{\underline{M}}$ um einen Charakter χ wie oben.

c) $\underline{M}$ definiert einen Isomorphismus $A \xrightarrow{\sim} \mathrm{Pic}^0(A)$, und somit erhält man eine Abbildung

$$\underline{c} : X \to A(R) \quad ,$$

mit $\underline{c}(\mu)^*(\underline{M}) \cong \underline{M}_\mu \cong \underline{M} \otimes \mathcal{O}_\mu$ (Schnitte von $\underline{M}_\mu$ über einer offenen Teilmenge von A entsprechen Schnitte von $\hat{\underline{L}}$ über dem Urbild, welche sich unter T gemäß μ transformieren). Für das folgende müssen wir diese Isomorphismen geeignet normalisieren:

<u>Definition</u>:

Ein zulässiges System von Isomorphismen besteht aus Isomorphismen

i) $\underline{c}(\mu)^*(\mathcal{O}_\nu) \cong \mathcal{O}_\nu$

ii) $\underline{M}_\mu \cong \underline{c}(\mu)^*(\underline{M})$, so daß

a) Die Isomorphismen in i) sind linear in μ und ν

b) Für $\mu, \nu \in X$ kommutiert das Diagramm

$$\begin{array}{ccccc}
\underline{c}(\mu+\nu)^*(\underline{M}) & \xrightarrow{\sim} & \underline{c}(\mu)^*(\underline{c}(\nu)^*(\underline{M})) & \xrightarrow{\sim} & \underline{c}(\mu)^*(\underline{M}_\nu) \\
\downarrow \wr & & & & \downarrow \wr \\
\underline{M}_{\mu+\nu} & \xleftarrow{\sim} & \underline{M}_\mu \otimes \mathcal{O}_\nu & \xleftarrow{\sim} & \underline{c}(\mu)^*(\underline{M}) \otimes \underline{c}(\mu)^*(\mathcal{O}_\nu)
\end{array}$$

Man sieht leicht, daß zulässige Systeme von Isomorphismen existieren. Je zwei unterscheiden sich dadurch, daß man die Isomorphismen in ii) mit einem $q(\mu) \in R^*$ multipliziert. $q : X \longrightarrow R^*$ muß die Eigenschaft haben, daß $b(\mu,\nu) = q(\mu+\nu)/(q(\mu)\cdot q(\nu))$ bilinear ist in μ und ν . Die Isomorphismen in i) werden dann mit $b(\mu,\nu)$ multipliziert. Wir wählen von nun an ein festes zulässiges System von Isomorphismen.

c) Da $\underline{L}$ auf G eine prinzipale Polarisation definiert, ist $\Gamma(G,\underline{L})$ ein R-Modul vom Rang 1 (in der Tat sogar ein divisorielles Ideal).

Sei $\theta_{\underline{L}} \in \Gamma(G,\underline{L})$ ein nicht verschwindendes Element. Genauso ist $\Gamma(A,\underline{M})$ frei, mit einem Erzeugenden $\theta_{\underline{M}}$. Für $\mu \in X$ erzeugt dann $\theta_{\underline{M}}{}^{\mu} = \underline{c}(\mu)^*\theta_{\underline{M}}$ $\Gamma(A,\underline{M}_{\mu}) = \Gamma(A,\underline{c}(\bar{\mu})^*(\underline{M}))$.

Den formalen Schnitt $\hat{\theta}_L \in \Gamma(\hat{G},\hat{\underline{L}})$ kann man nun nach T-Eigenfunktionen entwickeln:

$$\hat{\theta}_{\underline{L}} = \sum_{\mu \in X} a(\mu) \cdot \hat{\theta}_{\underline{M}}{}^{\mu} .$$

Dabei sind die Koeffizienten $a(\mu) \in R$, und sie konvergieren gegen Null in der $\underline{m}$-adischen Topologie. Bei Wechsel des zulässigen Systems von Isomorphismen erhalten sie einen Faktor $q(\mu)$. Damit ist klar, daß der Inhalt des folgenden Satzes nicht von dieser Wahl abhängt:

Satz 1:

i) $a(\mu) \neq 0$ für alle $\mu \in X$.

ii) $b(\mu,\nu) = a(\mu+\nu)a(0)/(a(\mu)a(\nu)) \in K^*$ ist bilinear in μ und ν

iii) Falls $\mu \neq 0$, so ist $b(\mu,\mu) \in \underline{m}$

iv) Wenn ein Geradenbündel $\underline{L}_1$ auf G dieselbe prinzipale Polarisation definiert, so daß $\hat{\underline{L}}_1 \cong \hat{\underline{L}}$, so liefert $\underline{L}_1$ dieselbe Bilinearform $b(\mu,\nu)$.

Bemerkung:

Die Aussage iii) folgt aus i) und ii):
Diese liefern, daß $a(n\mu)a(-n\mu) = a(0)^2 \cdot b(n\mu)^{n^2}$. Da die linke Seite in R liegt und für $n \to \infty$ $\mathfrak{m}$-adisch gegen Null konvergiert, liegt auch $b(\mu,\mu)$ in R und ist keine Einheit.

Beweis von Satz 1:

Man bettet R geeignet in einen kompletten diskreten Bewertungsring ein und darf dann annehmen, daß $\dim(R) = 1$. Es steht uns frei, den Grundkörper zu erweitern, d. h., R durch die Normalisierung in einer endlichen Erweiterung zu ersetzen. Wir dürfen dann zum Beispiel annehmen, daß alle 2-Teilungspunkte von G_η K-rational sind. Wir behandeln zunächst den folgenden Spezialfall:

$\underline{L} \cong [-1]^*\underline{L}$ ist symmetrisch, und die Symmetrie ist die Identität auf allen 2-Teilungspunkten von T. (Wir identifizieren $T[2] = \hat{T}[2] \subseteq \tilde{G}[2] = \hat{G}[2] \subseteq G[2]$ die 2-Teilungs-Untergruppen.)

Dann ist auch $\underline{M}$ symmetrisch, und der Charakter χ von $\tilde{G}$, welcher den Unterschied der Symmetrien zwischen $\hat{\underline{L}}$ und $\hat{\underline{M}}$ beschreibt, ist gleich Eins auf $T[2]$. Somit ist $\chi|T = 2\mu_0$, mit einem $\mu_0 \in X$.

Dann ist $[-1]^*\theta_{\underline{L}} = \pm\theta_{\underline{L}}$, und so ergibt sich, daß $a(2\mu_0-\mu) = a(\mu)\cdot$ (Einheit aus R^*). Wir zeigen zunächst, daß Funktion $b,c : X \to K$ existieren mit

(*) $a(\mu)a(\nu) = b(\mu+\nu)c(\mu-\nu)$. Dies ist äquivalent zu der folgenden Behauptung:

(*)' Sei $\rho \in X$. Dann gibt es Funktionen b_ρ, c_ρ auf X mit $a(\mu+\nu+\rho)a(\mu-\nu) = b_\rho(\mu)c_\rho(\nu)$.

Dazu zunächst etwas Terminologie:

Ein Schnitt $f \in \Gamma(\hat{G} \times \hat{G}, \hat{\underline{L}}^2 \otimes \hat{\underline{L}}^2)$ heißt ein Produkt, falls $f = g \otimes h$ mit $g,h \in \Gamma(\hat{G},\hat{\underline{L}}^2)$. Analog für Schnitte von $\hat{\underline{M}}^2$. Das fundamentale Beispiel eines zerlegten Schnittes ergibt sich wie folgt:

Betrachte die Isogenie

$$\phi_G : G \times G \to G \times G$$
$$(x,y) \longmapsto (x+y, x-y)$$

Bekanntlich ist $\phi_G^*(\underline{L} \otimes \underline{L}) = \underline{L}^2 \otimes \underline{L}^2$. Sei $H \subseteq G[2]$ eine endliche flache Untergruppe der Ordnung 2^d ($d = \dim(G)$), so daß $H(K) \subseteq G[2](K)$ ein maximal isotroper Unterraum für die durch die Polarisation gegebene symplektische Form ist. Dann kann man H äquivariant auf $\underline{L}^2$ operieren lassen, und descente liefert ein Geradenbündel $\underline{L}_1$ auf $G_1 = G/M$, welches auf $G_{1,\eta}$ eine prinzipale Polarisation definiert. Jeder $H \times H$-invariante Schnitt von $\underline{L}^2 \otimes \underline{L}^2$ liefert dann einen globalen Schnitt von $\underline{L}_1 \otimes \underline{L}_1$ und ist damit ein Produkt. Beispiele für $H \times H$-invariante Schnitte erhalten wir wie folgt:

$$\phi_G^*(\theta_{\underline{L}} \otimes \theta_{\underline{L}})(x,y) = \theta_{\underline{L}}(x+y) \otimes \underline{\theta}_L(x-y)$$

ist schon invariant unter der Diagonal-Aktion von H . Dann ist

$$\sum_{z \in H(R)} \underline{\theta}_L(x+y+z) \otimes \underline{\theta}_L(x-y+z)$$

$H \times H$-invariant, und damit ein Produkt. Man kann die H-Aktion auf $\underline{L}$ noch mit einem Charakter $\varepsilon : H \to \{\pm 1\}$ twisten, und erhält, daß auch

$$\sum_{z \in H(R)} \varepsilon(z) \underline{\theta}_L(x+y+z) \otimes \underline{\theta}_L(x-y+z)$$

ein Produkt ist.

Wir wählen nun ein H , welches $T[2]$ umfaßt. Dann ist $H/T[2] = H_1 \subseteq A[2]$ maximal isotrop, und jedes solche H_1 kann man auf diese Weise erhalten. Man läßt nun H so auf $\underline{L}^{\otimes 2}$ operieren, daß man den Überblick über die zulässigen Isomorphismen nicht verliert. Dazu gehe man folgendermaßen vor:

Lasse H_1 auf $\underline{M}^2$ operieren. Dies liefert ein $\underline{M}_1$ auf $A_1 = A/H_1$. Die Abbildung

$$\underline{c}_1 : X \xrightarrow{c} A \to A_1$$

definiert dann eine Erweiterung

$$0 \to T = T_1 \to \tilde{G}_1 \to A_1 \to 0$$

via Geradenbündeln $\mathcal{O}_{1,\mu}$ und $\underline{M}_{\frac{1}{2}1,\mu} = \underline{M}_1 \otimes \mathcal{O}_{1,\mu}$ auf A_1 . Beim Pullback nach A geht $\underline{M}_{1,\mu}$ über in $\underline{M}^2 \otimes \mathcal{O}_{2\mu}$, und $\mathcal{O}_{1,\mu}$ in $\mathcal{O}_{2\mu}$. Man erhält ein kommutatives Diagramm

$$\begin{array}{ccccccccc} 0 & \longrightarrow & T & \longrightarrow & \tilde{G} & \longrightarrow & A & \longrightarrow & 0 \\ & & \downarrow 2\cdot & & \downarrow & & \downarrow & & \\ 0 & \longrightarrow & T=T_1 & \longrightarrow & \tilde{G}_1 & \longrightarrow & A_1 & \longrightarrow & 0 \end{array}$$

und es ist $\tilde{G}_1 \cong G_1/H$. Weiter sieht man sofort, daß man ein zulässiges System von Isomorphismen für $\tilde{G}_1, \underline{M}_1$ wählen kann, welches bei Pullback verträglich ist mit dem für $\tilde{G}, \underline{M}$ (z. B. ist das Pullback von $\underline{c}_1(\mu)^*(\mathcal{O}_{1,\nu})$ gleich $\underline{c}(\mu)^*\mathcal{O}_{2\nu}$ und isomorph zu $\mathcal{O}_{2\nu}$, u.s.w.)

Die Operation von H_1 auf $\underline{M}^2$ liefert eine Operation von H auf $\underline{\tilde{L}}^2$ und $\underline{\hat{L}}^2$. Diese Operation ist algebraisch, d. h., kommt von einer Operation von H auf $\underline{L}^2$: Je zwei Operationen von H auf $\underline{L}^2$ oder $\underline{\hat{L}}^2$ differieren um einen Charakter $\varepsilon : H(R) \to \{\pm 1\}$, und mindestens eine formale Operation ist algebraisch.

Wir können nun (*)' zeigen:

Sei $\rho \in X$. Wähle $\varepsilon : H(R) \to \{\pm 1\}$ mit $\varepsilon | T[2] = \rho | T[2]$. Dann ist der folgende Schnitt von $\underline{\hat{L}}^2 \otimes \underline{\hat{L}}^2$ ein Produkt:

$$\sum_{z \in H(R)} \varepsilon(z)\ \hat{\theta}_{\underline{L}}(x+y+z) \otimes \hat{\theta}_{\underline{L}}(x-y+z)$$

$$= \sum_{\substack{z \in H(R)\\ \mu,\nu \in X}} \varepsilon(z)\ a(\mu)a(\nu)\ \hat{\theta}_{\underline{M}}^{\mu}(x+y+z) \otimes \hat{\theta}_{\underline{M}}^{\nu}(x-y+z)$$

Bei festem μ,ν verschwindet die Summe über $H(R)$ (sogar schon über $T[2]$) , außer wenn es $\alpha,\beta \in X$ gibt mit $\mu = \rho+\alpha+\beta$, $\nu = \alpha-\beta$. Also ergibt sich

$$\sum_{\alpha,\beta \in X} a(\rho+\alpha+\beta)\ a(\alpha-\beta) \sum_{z \in H(R)} \varepsilon(z)\,\theta_{\underline{M}}^{\rho+\alpha+\beta}(x+y+z) \otimes \theta_{\underline{M}}^{\alpha-\beta}(x-y-z)$$

Die innere Summe läßt sich umschreiben als

$$(\underline{c}(\alpha)^* \otimes \underline{c}(\beta)^*) \sum_{z \in H_1(R)} \varepsilon(z)\ z^*(\phi_A^*(\theta_{\underline{M}}^{\rho} \otimes \theta_{\underline{M}}))$$

mit
$$\phi_A : A \times A \to A \times A$$
$$(x,y) \longmapsto (x+y, x-y)$$

Dabei ist $\theta_A{}^{\rho} \otimes \theta_A$ der einzige Schnitt von $\underline{c}(\rho)^*(\underline{M}) \otimes \underline{M}$, $\phi_A{}^*(\theta_A{}^{\rho} \otimes \theta_A)$ ein Schnitt von $(\underline{M}^2 \otimes \mathcal{O}_{\rho}) \otimes (\underline{M}^2 \otimes \mathcal{O}_{\rho})$, und die innere Summe wieder ein Produkt, etwa von der Form $g \otimes h$. Wir erhalten schließlich, daß

$$\sum_{\alpha,\beta\in X} a(\rho+\alpha+\beta)\ a(\alpha-\beta)\ \underline{c}(\alpha)^*(g) \otimes \underline{c}(\beta)^*(h)$$

ein Produkt ist.

Da $\underline{c}(\alpha)^*(g)$ ein Schnitt ist von $M^2 \otimes \mathcal{O}_{2\alpha+\rho}$, und $\underline{c}(\beta)^*(h)$ ein Schnitt von $\underline{M}^2 \otimes \mathcal{O}_{2\beta+\rho}$, folgt daraus die Behauptung (*)', indem man obige Summe schreibt als

$$(\sum_{\alpha\in X} b_\rho(\alpha)\ \underline{c}\ (\alpha)^*(g)) \otimes (\sum_{\beta\in X} c_\rho(\beta)\ \underline{c}(\beta)^*(h)) \ .$$

Damit sind (*)' und (*) gezeigt, d. h.

$$a(\mu)\ a(\nu) = b(\mu+\nu)\ c(\mu-\nu) \ .$$

d) Wir zeigen zunächst Teil i) von Satz 1:

Wir wissen schon, daß $a(\mu) = 0 \iff a(2\mu_0-\mu) = 0$. Wir behaupten zunächst, daß eine Untergruppe $Y \subset X$ existiert mit

$$a(\mu) \neq 0 \iff \mu \in \mu_0 + Y \ .$$

Ersetzt man $a(\mu)$ durch $a(\mu+\mu_0)$, so darf man annehmen, daß $\mu_0 = 0$. Sei $Y = \{\mu | a(\mu) \neq 0\}$. Da $\hat{\theta}_G \neq 0$, ist Y nicht leer, und es ist $Y = -Y$.

i) $0 \in Y$: Sei $\mu \in Y \Rightarrow a(\mu)\ a(\mu) = b(2\mu)\ c(0) \neq 0$,
$a(\mu)\ a(-\mu) = b(0)\cdot c(2\mu) \neq 0$, somit ist
$b(0) \neq 0$, $c(0) \neq 0$ und $a(0)^2 = b(0)\cdot c(0) \neq 0$.

ii) $\mu,\nu \in Y \Rightarrow b(2\mu) \neq 0$, $c(2\nu) \neq 0$ (siehe i)
$\Rightarrow a(\mu+\nu)\ a(\mu-\nu) = b(2\mu)\ c(2\nu) \neq 0 \Rightarrow \mu\pm\nu \in Y$

iii) $Y = X$:
Andernfalls gäbe es ein endlich flaches Untergruppenschema $N \subseteq T$, $N \neq (0)$, so daß alle $\mu \in Y$ auf N identisch den Wert 1 annehmen, und $\hat{\theta}_G$ ist ein Eigenvektor für die Aktion von N auf $\hat{\underline{L}}$. Für

Elemente $x_1,\dots,x_n \in G(R)$ mit $\sum_{j=1}^n x_j = 0$ ist $\bigotimes_{j=1}^n x_j^*(\theta_G)$ ein globaler Schnitt von $\otimes x_j^*(\underline{L}) \cong \underline{L}^{\otimes n}$, welcher ein Eigenvektor für N ist. G(R) ist Zariski-dicht in G_η, und es ist wohlbekannt, daß für $n \geq 3$ die oben definierten Schnitte von $\underline{L}^{\otimes n}$ eine projektive Einbettung von G_η liefern. Andererseits muß diese Einbettung über $(G/N)_\eta$ faktorisieren, was ein Widerspruch ist.

Damit ist zunächst i) gezeigt. ii) ist nun ganz einfach: Aus der Identität

$$a(\mu)\ a(\nu) = b(\mu+\nu)\ c\ (\mu-\nu)$$

folgt für $\lambda,\mu,\nu \in X$:

$$a(\lambda+\mu+\nu)\cdot a(\lambda+\mu)^{-1}\cdot a(\lambda+\nu)^{-1}\cdot a(\mu+\nu)^{-1}\cdot$$
$$\cdot a(\lambda)\cdot a(\mu)\cdot a(\nu)\ a(0)^{-1} = 1$$

(Berechne $a(\lambda+\mu+\nu)\cdot a(\lambda)$, $a(\lambda+\mu)\ a(\lambda+\nu)$, $a(\mu)\ a(\nu)$ und $a(\mu+\nu)\cdot a(0)$ nach obiger Identität), und dies ist Behauptung ii).

e) Wir kommen nun zu beliebigen $\underline{L}$'s . Diese erhält man durch Translation mit einem Element aus G(K) aus einem $\underline{L}$ der bisher betrachteten Art (symmetrisch, Symmetrie = 1 auf T[2]) . Wenn dieses Element in G(R) liegt, so induziert es einen Automorphismus von $\hat{G}$, und man rechnet alles direkt nach. Im allgemeinen kann man es jedenfalls ausdehnen zu einem Element aus $G^*(R)$, wobei G^* das Néron-Modell von G bezeichne. G ist die Zusammenhangskomponente der Eins von G^*, und $\underline{L}$ und θ_G dehnen sich aus auf G^*. Allerdings hat die Ausdehnung $\underline{L}^*$ von $\underline{L}$ im allgemeinen keine kubische Struktur mehr.

Wie bisher kann man θ_{G^*} auf jeder Komponente von $\hat{G}^*$ nach $\hat{T}$-Eigenfunktionen entwickeln. Man erhält dann Koeffizienten $\{\tilde{a}(\mu), \mu \in X\}$, welche von der Komponente abhängen. Wir müssen zeigen, daß sie alle verschieden von Null sind, und daß $\tilde{a}(\mu+\nu)\ \tilde{a}(0)/(a(\mu)a(\nu)) = \tilde{b}(\mu,\nu)$ bilinear und unabhängig von der Komponente ist. Wir wissen schon, daß nicht alle $\tilde{a}(\mu)$ verschwinden. Wir schließen mit unserem alten Trick: Wähle $H \subseteq G \subseteq G^*$ wie vorher. Dann ist für jeden Charakter $\varepsilon : H(R) \to \{\pm 1\}$ und jedes $x_0 \in G^*(R)$

$$\sum_{z\in H} \varepsilon(z)\,\theta_{G^*}(x+x_0+y+z) \otimes \theta_{G^*}(x-y+z)$$

wieder ein Produkt, und es ergibt sich, daß $\tilde{a}(\mu)\ \tilde{\tilde{a}}(\nu) = \tilde{b}(\mu+\nu)\ \tilde{c}\ (\mu-\nu)$, mit geeigneten Funktionen $\tilde{b},\tilde{c}$ auf X. Dabei sind $\tilde{a},\tilde{\tilde{a}}$ die Koeffizienten zu verschiedenen Komponenten. Da für die Funktion a zur Einskomponente schon alles nähere bekannt ist, folgt leicht die Behauptung, und Satz 1 ist vollständig bewiesen.

f) Schließlich benötigen wir noch ein Resultat, nach dem $b(\mu,\nu)$ und die Polarisation auf A die Polarisation von G_η bestimmen. Es seien dazu gegeben zwei G's, G_1 und G_2, so daß $\hat{G}_1 \cong \hat{G}_2$ und damit $A_1 \cong A_2 \cong A$. Weiter nehmen wir an, daß Geradenbündel $\underline{L}_1$ und $\underline{L}_2$ auf G_1 bzw. G_2 existieren, welche prinzipale Polarisationen auf den generischen Fasern liefern und auch dieselbe Polarisation auf A ergeben. (d. h. $\underline{M}_1$ und $\underline{M}_2$ unterscheiden sich um eine Translation). Schließlich sollen $\underline{L}_1$ und $\underline{L}_2$ dieselbe Bilinearform b liefern.

<u>Satz 2</u>:

Unter diesen Umständen ist der formale Isomorphismus $\hat{G}_1 \xrightarrow{\sim} \hat{G}_2$ algebraisch, d. h., er wird induziert von einem Isomorphismus polarisierter abelscher Varietäten $G_{1,\eta} \xrightarrow{\sim} G_{2,\eta}$.

<u>Beweis</u>: Es ist stets erlaubt, zu einer endlichen Erweiterung von K überzugehen. Weiter dürfen wir annehmen, daß $\underline{M}_1 \cong \underline{M}_2 \cong \underline{M}$. Dann sind $\hat{\underline{L}}_1$ und $\hat{\underline{L}}_2$ isomorph zum Pullback von $\underline{M}$ auf $\hat{G}_1 \cong \hat{G}_2 = \hat{G}$. Diese Isomorphismen respektieren die kubische Struktur, sind aber nicht unbedingt eindeutig. Sie liefern aber kanonische Isomorphismen

$$\hat{\underline{L}}_1 \otimes [-1]^*\hat{\underline{L}}_1 \cong \hat{\underline{L}}_2 \otimes [-1]^*\hat{\underline{L}}_2 = \text{Pullback}$$

von $\underline{M} \otimes [-1]^*\underline{M}$.

Wir zeigen, daß sich bei diesem Isomorphismus die algebraischen Schnitte $\Gamma(G_1, \underline{L}_1 \otimes [-1]^*\underline{L}_1)$ und $\Gamma(G_2, \underline{L}_2 \otimes [-1]^*\underline{L}_2)$ entsprechen. Genauer gesagt zeigen wir, daß man ein Erzeugendensystem der algebraischen Schnitte von $\underline{L}_1 \otimes [-1]^*\underline{L}$ oder von $\underline{L}_2 \otimes [-1]^*\underline{L}_2$ erhält durch die Reihen

$$\sum_{\beta \in X} b(\rho+\beta,\beta)\ \underline{c}(\beta)^*(f) \quad .$$

Dabei durchläuft $\rho \in X$ ein Vertretersystem für $X/2X$, und f eine Basis der globalen Schnitte von $\underline{M} \otimes [-1]^*\underline{M}$. Wähle wie bisher $H \subseteq \hat{G}[2]$ endlich und flach, maximal isotrop, mit $H \supseteq T[2]$. Dann operiert H auf $\underline{L}_1 \otimes [-1]^*\underline{L}_1$ und $\underline{L}_2 \otimes [-1]^*\underline{L}_2$, ähnlich wie bisher. Es ist bekannt, daß für $j = 1,2$ $\Gamma(G_j, \underline{L}_j \otimes [-1]^*\underline{L}_j)$ eine Basis aus H-Eigenvektoren besitzt, wobei jeder Charakter $\varepsilon: H(R) \to \{\pm 1\}$ genau einmal vorkommt. Wenn $\theta_{\underline{L}_j} \in \Gamma(G_j, \underline{L})$ ein nicht verschwindendes Element ist, so liegt für $y \in G_j(R)$

$$\sum_{z \in H(R)} \varepsilon(z)\theta_{\underline{L}_j}(x+y+z)\theta_{\underline{L}_j}(-x+y+z)$$

im ε-Eigenraum, und man kann durch Wahl von y in einer Zariski-dichten Menge von $G_j(R)$ erreichen, daß dies $\neq 0$ wird. Rechnen wir nun formal:

$$\hat{\theta}_{\underline{L}_j}(x) = \sum_{\mu \in X} a_j(\mu)\theta_{\underline{M}}^{\mu}(x) \qquad (\text{dabei } \underline{L}_j \cong \text{Pullback von } \underline{M})$$

Wähle $\rho \in X$ mit $\rho|\hat{T}[2] = \varepsilon|T[2]$

$\Rightarrow$

$$\sum_{z \in H(R)} \varepsilon(z)\hat{\theta}_{\underline{L}_j}(x+y+z)\theta_{\underline{L}_j}(-x+y+z)$$

$$= \sum_{\substack{\alpha,\beta \in X \\ z \in H(R)}} \varepsilon(z)a_j(\rho+\alpha+\beta)a_j(\alpha-\beta)\theta_{\underline{M}}^{\rho+\alpha+\beta}(x+y+z)\theta_{\underline{M}}^{\alpha-\beta}(-x+y+z)$$

$$= \sum_{\substack{\alpha,\beta \in X \\ z \in H(R)}} \varepsilon(z)a_j(\rho+\alpha)a_j(\alpha)b(\rho,\beta)b(\beta,\beta)\cdot$$

$$\cdot\underline{c}(\beta)^*(\theta_{\underline{M}}^{\rho+\alpha}(x+y+z) \otimes \theta_{\underline{M}}^{\alpha}(-x+y+z)) \quad .$$

$$= \sum_{\beta \in X} b(\rho+\beta,\beta)\ \underline{c}(\beta)^*(\sum_{\substack{\alpha \in X \\ z \in H(R)}} \varepsilon(z)a_j(\rho+\alpha)a_j(\alpha)\underline{\theta}_M^{\rho+\alpha}(x+y+z)\theta_{\underline{M}}^{\alpha}(-x+y+z))$$

Die innere Summe ist ein Schnitt von $\underline{M}_\rho \otimes [-1]^*\underline{M}$, (welcher noch von y abhängt), der sich unter H (bei einer geeignet zu definierenden Operation von H auf diesem Bündel) als Eigenvektor transformiert. Der zugehörige Charakter ist unabhängig von $j = 1,2$, und somit sind die inneren Summen Vielfache voneinander, für $j = 1,2$. Es folgt, daß der formale Isomorphismus $\hat{\underline{L}}_1 \otimes [-1]^*\hat{\underline{L}}_1 \cong \hat{\underline{L}}_2 \otimes [-1]^*\hat{\underline{L}}_2$ einen Isomorphismus

$$\Gamma(G_{1,\eta}, \underline{L}_1 \otimes [-1]^*\underline{L}_1) \cong \Gamma(G_{2,\eta}, \underline{L}_2 \otimes [-1]^*\underline{L}_2)$$

induziert, und damit auch einen Isomorphismus

$$\bigoplus_{n\geq 0} \Gamma(G_{1,\eta}, \underline{L}_1{}^n \otimes [-1]^*\underline{L}_1{}^n) \cong \bigoplus_{n\geq 0} \Gamma(G_{2,\eta}, \underline{L}_2{}^n \otimes [-1]^*\underline{L}_2{}^n)$$

Dies liefert aber unmittelbar die Behauptung.

§ 3 Mumford's Konstruktion

a) In diesem Kapitel liefern wir eine Art Umkehrung der vorhergehenden Betrachtungen. Dazu sei R ein exzellenter normaler Ring, $I \subseteq R$ ein Ideal, so daß R komplett ist in der I-adischen Topologie.

Ferner geben wir vor:

a) eine abelsche Varietät A über R , zusammen mit einem amplen Geradenbündel $\underline{M}$ auf A .

b) Eine Erweiterung $\tilde{G}$ von A durch einen Torus $T \cong G_m^r$:
$0 \to T \to \tilde{G} \to A \to 0$

c) Eine Bilinearform b auf $X = X(T)$, mit Werten in K^* (K = Quotientenkörper von R):
$b : X \times X \to K^*$.
Dabei sei $b(\mu,\mu) \in I$, falls $\mu \neq 0$.

Unser Ziel ist es, eine semiabelsche Varietät G über R zu konstruieren, so daß b die Koeffizienten der zugehörigen θ-Reihe liefert (Falls $\underline{M}$ eine prinzipale Polarisation definiert). Etwas allgemeiner ist das Datum c) folgendes:

i) eine Untergruppe $Y \subseteq X$ von endlichem Index

ii) eine lineare Abbildung $i : Y \to \widetilde{G}(K)$

iii) eine lineare Abbildung $\underline{c} : X \to A(R)$,
so daß das folgende Diagramm kommutiert:

$$\begin{array}{ccc} Y & \longrightarrow & \widetilde{G}(K) \\ \downarrow & & \downarrow \\ A(R) & \longrightarrow & A(K) \end{array}$$

iv) Ein System von Isomorphismen
$\underline{c}(\mu)^*(\mathcal{O}_\nu) \cong \mathcal{O}_\nu$, $\mu \in Y, \nu \in X$,
linear in μ und ν

v) Ein System von Isomorphismen
$\underline{M}_\mu = \underline{M} \otimes \mathcal{O}_\mu \cong \underline{c}(\mu)^*(\underline{M})$, $\mu \in Y$,
so daß für $\mu, \nu \in Y$ das folgende Diagramm kommutiert:

$$\begin{array}{ccccc} \underline{c}(\mu+\nu)^*(\underline{M}) & \xrightarrow{\sim} & \underline{c}(\mu)^*(\underline{c}(\nu)^*(\underline{M})) & \xrightarrow{\sim} & \underline{c}(\mu)^*(\underline{M}_\nu) \\ \downarrow\wr & & & & \downarrow\wr \\ \underline{M}_{\mu+\nu} & \xleftarrow{\sim} & \underline{M}_\mu \otimes \mathcal{O}_\nu & \xleftarrow{\sim} & \underline{c}(\mu)^*(\underline{M}) \otimes \underline{c}(\mu)^*(\mathcal{O}_\nu) \end{array}$$

vi) Eine Bilinearform
$b : Y \times X \to K^*$
symmetrisch auf Y , mit $b(\mu,\mu) \in I$, falls $\mu \in Y$, $\mu \neq 0$.

Die Abbildungen in ii), iv) und vi) sollen kompatibel sein in dem folgenden Sinne: b entspricht einer Abbildung $Y \xrightarrow{b} T(K)$. Dann sei für $\mu \in Y$ $i(y)b(y)^{-1} \in \widetilde{G}(R)$, und die Isomorphismen in iv) werden definiert durch Translation mit diesem Element.

Wir definieren auf den Daten noch eine Äquivalenzrelation, wie folgt: Sei $c : Y \times X \to R^*$ eine Bilinearform, so daß eine Funktion $q : Y \to R^*$ existiert mit

$$c(\mu,\nu) = q(\mu+\nu)\ q(0)/(q(\mu)\cdot q(\nu)) \qquad \mu,\nu \in Y$$

Dann erlauben wir, daß man die Isomorphismen in v) mit $q(\mu)$, die in iv) mit $c(\mu,\nu)$ multipliziert, und schließlich b durch $b \cdot c$ ersetzt. Aus der Definitheit der Form b folgt sofort, daß $i : Y \to G(K)$ eine Injektion ist.

b) Wir werden einen Quotienten $G = \tilde{G}/i(Y)$ definieren, so daß G eine abelsche Varietät ist, mit einer Polarisation vom Grad(M)$\cdot$[X:Y] . Dies wurde von Mumford in [M4] durchgeführt für den Fall, daß $\tilde{G} = T$ ein Torus ist.

Von nun an folgen wir den Ausführungen in [M4] , § 2,3,4:

<u>Definition</u>: ([M4], Definition 2.1). Ein relativ komplettes Modell besteht aus

a) Ein Schema P über A , integer, lokal von endlichem Typ
b) Eine offene Einbettung $i : \tilde{G} \subseteq \tilde{P}$
c) Ein invertierbares Geradenbündel $\underline{L}$ auf $\tilde{P}$
d) Eine Aktion von $\tilde{G}$ auf $(\tilde{P},\underline{L})$, welche die Translationsoperation von $\tilde{G}$ auf $\tilde{G}$ fortsetzt. Bezeichnung: (T_g, T_g^*)
e) Eine Operation von Y auf $(P, \underline{L} \otimes$ Pullback von $\underline{M}$) Bezeichnung: (S_μ, S_μ^*)

Diese mögen erfüllen:

i) Es gibt $U \subset \tilde{P}$ offen, von endlichem Typ, so daß $\tilde{P} = \bigcup_{\mu \in Y} S_\mu(U)$.

ii) Sei $v \in \mathbb{R}(\tilde{G})$ eine Bewertung auf dem Funktionenkörper $K(\tilde{G})$ von $\tilde{G}$, welche ≥ 0 ist auf R . Sei $x \in A$ das Zentrum von v auf A (A ist komplett), und für $\nu \in X$ sei χ_ν ein lokales (in x) Erzeugendes von $\mathcal{O}_\nu$. χ_ν ist eindeutig bestimmt bis auf eine Einheit in $\mathcal{O}_{A,x}$,und ein Element aus $K(\tilde{G})$. Dann gilt:

v hat Zentrum auf $\tilde{P}$ <=> $\forall\, \nu \in X,\ \exists \mu \in Y$ mit $v(\chi_\nu \cdot b(\mu,\nu)) \geq 0$ (es reicht dabei, $\nu \in Y$ zu betrachten).

iii) Auf $\tilde{G}_\eta$ operieren Y und $\tilde{G}$ durch Translationen. (Es folgt, daß S_μ und T_g auf $\tilde{P}$ kommutieren, und daß $S_\mu^*(\underline{M}) \cong \underline{c}(\mu)^*(\underline{M})$) .

iv) Sei $\mu \in Y$. Dann ist $\underline{c}(\mu)^*(\underline{M}) \cong \underline{M} \otimes \mathcal{O}_\mu$, und somit induziert S_μ^* einen Isomorphismus $\tilde{S}_\mu^* : S_\mu^*(\underline{L}) \xrightarrow{\sim} \underline{L} \otimes \mathcal{O}_{-\mu}$. Weiter ist für $g \in \tilde{G}$, mit Bild $x \in A$, kanonisch $x^*(\mathcal{O}_{-\mu}) \cong \mathcal{O}_{-\mu}$, via Translation mit g . Dann kommutiert das folgende Diagramm

$$\begin{array}{ccccc}
T_g{}^*S_\mu{}^*(\underline{L}) & \xrightarrow[\sim]{\widetilde{S}_\mu{}^*} & T_g{}^*(\underline{L}\otimes \mathcal{O}_{-\mu}) & = & T_g{}^*(\underline{L})\otimes T_g{}^*(\mathcal{O}_{-\mu}) \\
\| & & & & \downarrow \wr\, T_g{}^* \\
S_\mu{}^*T_g{}^*(\underline{L}) & & & & \underline{L}\otimes T_g{}^*(\mathcal{O}_{-\mu}) \\
\downarrow \wr\, T_g{}^* & & & & \| \\
S_\mu{}^*(\underline{L}) & \xrightarrow[\sim]{\widetilde{S}_\mu{}^*} & \underline{L}\otimes \mathcal{O}_{-\mu} & \xleftarrow{\sim} & \underline{L}\otimes x^*(\mathcal{O}_{-\mu})
\end{array}$$

v) $\underline{L}\otimes\underline{M}$ ist ample auf $\widetilde{P}$.

Bemerkung:

Die Bedingung iv) wird etwas einfacher, falls $g \in T$, also $x = 0$. Der Isomorphismus $x^*(\mathcal{O}_{-\mu}) \xrightarrow{\sim} \mathcal{O}_{-\mu}$ ist Multiplikation mit $\mu(g)^{-1}$, und man kann die Kommutativität schreiben als

$$T_g{}^*\widetilde{S}_\mu{}^* = \mu(g)\, \widetilde{S}_\mu{}^*\, T_g{}^* \; .$$

Bemerkung: Man kann die Kompatibilität iv) umformulieren: Es reicht, sie auf $\widetilde{G}_\eta \subseteq \widetilde{P}_\eta \subseteq \widetilde{P}$ zu verifizieren. Zunächst liefern die T_g^* eine äquivariante Operation von $\widetilde{G}$ auf $\underline{L}|\widetilde{G}$, und damit wird $\underline{L}|\widetilde{G}$ $\widetilde{G}$-linear trivial: $\underline{L}|\widetilde{G} = \mathcal{O}_{\widetilde{G}}$. (kanonisch bis auf Einheit aus R^*). Ebenso operiert $\widetilde{G}$ auf dem Pullback von $\mathcal{O}_\nu$ auf $\widetilde{G}$, und es ist kanonisch Pullback $(\mathcal{O}_\nu) = \mathcal{O}_{\widetilde{G}}$. (Achtung: $T \subseteq \widetilde{G}$ operiert kanonisch auf dem Pullback von $\mathcal{O}_\nu$, doch unterscheidet sich diese Operation um den Charakter ν von der Einschränkung der $\widetilde{G}$-Operation). Sei nun $\mu \in Y$, $i(\mu) \in \widetilde{G}(K)$. Dann ist auf $\widetilde{G}_\eta$ $S_\mu = T_{i(\mu)}$, somit $S_\mu{}^*(\underline{L}|\widetilde{G}_\eta) \cong \underline{L}|G_\eta$. Der Isomorphismus $S_\mu{}^* : S_\mu{}^*(\underline{L}) \xrightarrow{\sim} \underline{L}\otimes\mathcal{O}_{-\mu} \xrightarrow{\sim} \underline{L}$ wird auf $\widetilde{G}_\eta$ gegeben durch eine globale Einheit auf $\widetilde{G}_\eta$. Dann bedeutet Bedingung iv), daß diese Funktion konstant ist, also gegeben durch ein $a(\mu) \in K^*$. Analog ist $S_\mu{}^* = a(\mu)\cdot \underline{c}(\mu)^*$:

$$S_\mu{}^*(\underline{M}\otimes\underline{L}|\widetilde{G}_\eta) \cong S_\mu{}^*(\underline{M}|\widetilde{G}_\eta) \xrightarrow{\sim} \underline{c}(\mu)^*(\underline{M}|\widetilde{G}_\eta)$$
$$\xrightarrow{\sim} (\underline{M}\otimes\mathcal{O}_\mu|\widetilde{G}_\eta) \xrightarrow{\sim} (\underline{M}\otimes\underline{L}\otimes\mathcal{O}_\mu)|\widetilde{G}_\eta \xrightarrow{\sim} (\underline{M}\otimes\underline{L})|\widetilde{G}_\eta \; .$$

Es ist $a(0) = 1$, $a(\mu+\nu) = a(\mu)\, a(\nu)\, b(\mu,\nu)$. (Man beachte, daß auf

$\tilde{G}_n$ die beiden Isomorphismen Pullback$(\underline{c}(\mu)^*(\mathcal{O}_\nu) \xrightarrow{\sim} T^*_{i(\mu)}$ (Pullback $(\mathcal{O}_\nu)) \xrightarrow{\sim}$ Pullback $(\mathcal{O}_\nu)$ sich um $b(\mu,\nu)$ unterscheiden: Der eine Isomorphismus ist Pullback des entsprechenden Isomorphismus auf A, der andere kommt von der Operation von $i(\mu)$ auf Pullback $(\mathcal{O}_\nu))$.

Beispiel ([M4], 2.3-2.5.)

Wir konstruieren unter bestimmten Voraussetzungen ein relativ komplettes Modell. Diese sind:

Sei $\Sigma \subset X$ ein Erzeugendensystem, $0 \in \Sigma = -\Sigma$. Wähle eine Funktion $a : Y \to K^*$ mit $a(0) = 1$, $a(\mu+\nu) = b(\mu,\nu)\; a(\mu)a(\nu)$. (Man zeigt leicht, daß ein solches a existiert). Es gelte

$$(*) \quad a(\mu)\; b(\mu,\alpha) \in R \quad \text{für} \quad \mu \in Y, \alpha \in \Sigma.$$

Im allgemeinen kann man kein solches $a(\;)$ finden. Dies hängt damit zusammen, daß wir auch ein Geradenbündel auf der abelschen Varietät konstruieren wollen, welches die Polarisation induziert. In den uns interessierenden Fällen wird dies aber kein Problem sein. Zum Beispiel kann man $(*)$ stets erfüllen, wenn R faktoriell ist, und man Y durch $n \cdot Y$ ersetzt, n genügend groß.

Betrachte die beiden folgenden quasikohärenten graduierten Algebren $\mathcal{S}_1$ und $\mathcal{S}_2$ über A:

$$\mathcal{S}_1 = \mathcal{O}_A \oplus \bigoplus_{n\geq 1} \mathcal{O}_{\tilde{G}}\, \theta^n = \mathcal{O}_A \oplus \bigoplus_{\substack{n\geq 1\\ \nu\in X}} \mathcal{O}_\nu \cdot \theta^n$$

$$\mathcal{S}_2 = \mathcal{O}_A \oplus \bigoplus_{n\geq 1} \mathcal{O}_{\tilde{G}}\, \underline{M}^n \cdot \theta^n = \mathcal{O}_A \oplus \bigoplus_{\substack{n\geq 1\\ \nu\in X}} (\underline{M}^n \otimes \mathcal{O}_\nu) \cdot \theta^n$$

$\tilde{G}$ operiert offensichtlich auf $\mathcal{S}_1$ (θ bleibt fest), und Y operiert auf $\mathcal{S}_2 \otimes_R K$, nach der Regel

$$S_\mu^*(\phi \cdot f_\nu \theta^n) = a(\mu)^n b(\mu,\nu)\, \underline{c}(\mu)^*(\phi)\, \underline{c}(\mu)^*(f_\nu) \cdot \theta^n$$

$$= a(\mu)^n \underline{c}(\mu)^*(\phi)\, i(\mu)^*(f_\nu \theta^n) \in \underline{M}^n \otimes \mathcal{O}_{\nu+n\mu} \otimes_R K \cdot \theta^n$$

($\phi \in \underline{M}^n$, $f_\nu \in \mathcal{O}_\nu$ lokale Schnitte)

$\mathcal{R}_1 \subseteq \mathcal{S}_1 \otimes_R K$ sei der Unterring, welcher erzeugt wird von

$$\{a(\mu)\ b(\mu,\alpha)\ \mathcal{O}_{\mu-\alpha}\cdot\theta \mid \mu \in Y, \alpha \in \Sigma\}$$

und $\mathcal{R}_2 \subseteq \mathcal{S}_2 \otimes_R K$ werde erzeugt von

$$\{S_\mu{}^*(\underline{M} \otimes \mathcal{O}_\alpha \cdot \theta\ ) \mid \mu \in Y, \alpha \in \Sigma\}$$

Dann ist $\tilde{P} = \mathrm{Proj}_A(\mathcal{R}_1) = \mathrm{Proj}_A(\mathcal{R}_2)$ ein relativ komplettes Modell: $\tilde{G}$ operiert auf $(\mathrm{Proj}_A(\mathcal{R}_1), \mathcal{O}(1)) = (\tilde{P},\underline{L})$, und Y auf $(\tilde{P},\underline{L} \otimes \underline{M})$. Für $\alpha \in \Sigma, \theta \in \Gamma(A,\underline{M} \otimes \mathcal{O}_\alpha)$ ist $\theta\cdot\Theta$ ein globaler Schnitt in $\Gamma(\tilde{P};\underline{L} \otimes M)$, und $U_{0,\alpha,\theta\cdot\Theta} = P\text{-}V(\theta\cdot\Theta)$ ist affin und von endlichem Typ $/R$. Da die θ's $\underline{M} \otimes \mathcal{O}_2$ erzeugen, überdecken die offenen Mengen $U_{\mu,\alpha,\theta\cdot\Theta} = S_\mu(U_{0,\alpha,\theta\cdot\Theta})$ ganz P , und $U_{0,0,\theta\cdot\Theta} \cong \tilde{G}\text{-}V(\theta)$.

Wenn man Θ eine Basis von $\Gamma(A,\underline{M} \otimes \mathcal{O}_\alpha)$ durchlaufen läßt, erhält man auf diese Weise eine Überdeckung von $\tilde{P}$ wie in i). Die Bedingung ii) zeigt man wie in [M4], iii) ist leicht, v) schon gezeigt, und iv) rechnet man einfach nach.

Von nun an bezeichne $(\tilde{P},\underline{L},\ldots)$ ein relativ komplettes Modell. Es folgen nun eine Reihe von Tatsachen, welche den Sätzen aus [M4], § 3, und 4 entsprechen:

<u>[M4], 3.1:</u>

Sei $\mu \in Y$, und $f = b(\mu,\mu) \in R$. Das Pullback von $\mathcal{O}_\mu$ auf $\tilde{G}$ besitzt ein kanonisches Erzeugendes $h_\mu \in \Gamma(\tilde{G},\mathcal{O}_\mu)$. (Das direkte Bild von $\mathcal{O}_{\tilde{G}}$ bei $\tilde{G} \to A$ ist die direkte Summe aller $\mathcal{O}_\nu, \nu \in X)$. Dann dehnt sich h_μ aus zu einem regulären Schnitt des Pullbacks von $\mathcal{O}_\mu$ auf $\tilde{P}_f = \tilde{P} \otimes_R R_f$, welcher dort $\mathcal{O}_\mu$ erzeugt.

<u>Beweis:</u> Aus den Verträglichkeitsbedingungen zu Anfang dieses Kapitels folgt $i(\mu) \in \tilde{G}(R_f)$. Auf $\tilde{P}_f$ stimmen dann S_μ und $T_{i(\mu)}$ überein, und die Isomorphismen

$$\tilde{S}_\mu{}^* : S_\mu{}^*(\underline{L}) \xrightarrow{\sim} \underline{L} \otimes \mathcal{O}_{-\mu} \quad \text{und} \quad T_{i(\mu)}{}^* : T_{i(\mu)}{}^*(\underline{L}) \xrightarrow{\sim} \underline{L}$$

liefern einen globalen Schnitt von $\mathcal{O}_\mu$ über $\tilde{P}_f$, welcher das Bündel dort erzeugt.

Die Einschränkung dieses Schnittes auf $\tilde{G}$ transformiert sich unter T gemäß $-\mu$, genauso wie h_μ. Also stimmen die beiden bis auf eine Einheit überein.

[M4], 3.2:

$\tilde{P}_\eta = \tilde{G}_\eta$

[M4], 3.3:

Jede irreduzible Komponente von $\tilde{P}_0 = \tilde{P} \otimes_R (R/I)$ ist eigentlich über R/I.

Beweis: Sei Z eine irreduzible Komponente von $\tilde{P}_0$, v eine Bewertung des Funktionenkörpers $K(Z)$, $v \geq 0$ auf R. Wähle eine Bewertung v_1 von $K(\tilde{G})$ ($v_1 \geq 0$ auf R) mit Zentrum Z, und sei v_2 das Kompositum von v und v_1. Für $\mu \in Y$ sei $h_\mu \in \Gamma(\tilde{G}, \mathcal{O}_\mu)$ das kanonische erzeugende Element. Nach unserem Analogon zu [M4]3.1 ist für $n >> 0$ $b(\mu,\mu)^n \cdot h_\mu$ regulär im generischen Punkt von Z, und verschwindet dort. Sei $x \in A$ das Zentrum von v_2 auf A, und ℓ_μ ein lokales Erzeugendes von $\mathcal{O}_\mu$ nahe x. Dann ist $b(\mu,\mu)^n \cdot h_\mu \cdot$ (Pullback von ℓ_μ) regulär und gleich Null im generischen Punkt von Z, hat also bei v_2 Bewertung >0. Somit ist auch $v_2\,(\chi_\mu\, b(n\mu,\mu)) > 0$, im Sinne der Bedingung ii) bei der Definition eines relativ kompletten Modells. $(h_\mu \in R^* \cdot \chi_\mu \cdot \ell_\mu)$. Dies gilt für alle $\mu \in Y$. Somit hat v_2 ein Zentrum auf $\tilde{P}$ und v eins auf Z. Mache weiter wie in [M4].

[M4], 3,5:

Sei $U_0 = U \otimes_R {}^R/I$, (U wie in Bedingung i) an $\tilde{P}$) $\bar{U}_0$ ist eigentlich über R/I.

[M4], 3.6:

Es gibt eine endliche Teilmenge $S \subseteq Y$, so daß für $\mu,\nu \in Y$, $\mu-\nu \notin S$

$$S_\mu(\overline{U}_0) \cap S_\nu(\overline{U}_0) = \emptyset \quad .$$

Beweis:

Seien $F \subset \tilde{P}$ die Fixpunkte unter $T \subset \tilde{G}$. Für jede zusammenhängende Teilmenge $F' \subseteq F$ operiert T auf $\underline{L}|F'$ via einen Charakter $\nu \in X$, und auf $\underline{L}|S_\mu(F')$ via $\nu+\mu$. Also ist für $\mu \neq 0$ $F' \cap S_\mu(F') = \emptyset$. Weiter wie in [M4].

[M4], 3.7:

Y operiert frei auf $\tilde{P}_0$.

[M4], 3.8:

$\tilde{P}_0$ ist zusammenhängend

Beweis:

Genauso wie in [M4]: $\tilde{G}_0 \subseteq \tilde{P}_0$ definiert eine Zusammenhangskomponente von $\tilde{P}_0$. Wenn es eine zweite gibt, wähle eine diskrete Bewertung v von $K(\tilde{G})$ mit Zentrum in dieser Zusammenhangskomponente. Sei $R' = \{f \in K(A) \,|\, v(f) \geq 0\}$. Ersetze $\tilde{P}$ durch den Abschluß der generischen Faser von $\tilde{P} \times_A \mathrm{Spek}(R')$, und wende [M4] , 3.9. an.

[M4], 3.10:

Für $n \geq 1$ existiert ein Schema P_n , projektiv über A/I^n , mit amplem Geradenbündel $\mathcal{O}(1)$, und ein étaler surjektiver Morphismus

$$\pi : \tilde{P} \otimes_R (R/I^n) \to P_n \quad ,$$

welcher $(P_n, \mathcal{O}(1))$ zum Quotienten unter Y macht von $(\tilde{P} \otimes_R (R/I^n), \underline{L} \otimes \underline{M})$.

Die P_n definieren ein formales Schema $\mathcal{P}$ über R , welches algebraisch ist. Wir erhalten also ein projektives P über R , mit $\hat{P} = \mathcal{P}$. P trägt ein amples Geradenbündel $\mathcal{O}(1)$. Außerdem hat man ein abgeschlossenes Unterschema $B \subseteq P$, so daß $\hat{B}$ = Quotient von $(\hat{P} - \bigcup_{\mu \in Y} S_\mu(\tilde{G}))^\wedge/Y$. Sei $G = P-B$. Es ist $\hat{G} = \hat{\tilde{G}}$.

[M4], 4.2:

G ist glatt über R .

[M4], 4.3:

P ist irreduzibel

Definition ([M4], 4.4)

Eine semiabelsche Untergruppe $\tilde{G}_1 \subseteq \tilde{G}$, $0 \to T_1 \to \tilde{G} \to A_1 \to 0$ heißt integrabel, falls gilt:

i) $Y_1 = i^{-1}(G_1(R))$ hat denselben Rang wie der Torus T_1 von $\tilde{G}_1$.

ii) $\underline{c}(Y_1) \subseteq A_1(R)$

iii) Für $\nu \in X, \nu | T_1 = 1$ und $\mu \in Y_1$ ist der Isomorphismus $\underline{c}(\mu)^* \mathcal{O}_\nu \cong \mathcal{O}_\nu$ auf A_1 die Identität $(\mathcal{O}_\nu | A_1 \cong \mathcal{O})$

Beispiele integrabler Untergruppen erhält man etwa durch Graphen der Multiplikation $m : \tilde{G} \times \tilde{G} \to \tilde{G}$ oder der Inversenabbildung $[-1] : \tilde{G} \to \tilde{G}$.

Es gilt:

Jede integrable Untergruppe $\tilde{G}_1 \subseteq \tilde{G}$ definiert ein abgeschlossenes Unterschema $G_1 \subseteq G$, wie folgt:

a) Sei W_1 der Abschluß von $\tilde{G}_1$ in $\tilde{P}$. Dann ist W_1 Y_1-invariant.

b) Sei $\mathcal{W}_1$ die I-adische Komplettierung von W_1 . $\mathcal{W}_1$ ist ebenfalls Y_1-invariant, und die Vereinigung $\mathcal{W}_2 = \bigcup_{\mu \in Y/Y_1} S_\mu(W_1)$ ist lokal endlich (dies definiert $\mathcal{W}_2$ als reduziertes Unterschema von $\mathcal{P}$).

c) Sei $\mathcal{W}_3 = \mathcal{W}_2/Y \subset \mathcal{P}$

d) Sei $W_3 \subseteq P$ definiert durch $\hat{W}_3 = \mathcal{W}_3$.

e) $G_1 = W_3 \cap G$.

Nur Schritt b) ist nicht trivial. Er folgt aus einer Variante von [M4];

Prop. 4.5, wobei man im Beweis benutzt, daß für $\mu \in Y$ und $n >> 0$ $b(\mu,\mu)^n \cdot h_\mu$ (h_μ = kanonisches Erzeugendes auf $\tilde{G}$ des Pullbacks von $\mathcal{O}_\mu$) auf U ein regulärer Schnitt des Pullbacks von $\mathcal{O}_\mu$ ist, welcher auf U_0 verschwindet. Man beachte auch, daß für $\nu \in X$ mit $\nu|T_1 = 1$ das Pullback des Geradenbündels $\mathcal{O}_\nu$ auf W_1 kanonisch trivial ist, wobei die Trialisierung auf $\tilde{G}_1$ durch h_ν gegeben wird.

Es folgt:

<u>[M4], 4.8, 4.9</u>:

G ist ein Gruppenschema über R , und $G_\eta = P_\eta$ ist abelsche Varietät.

Weiter können wir die Struktur der Torsions-Untergruppen von G bestimmen:

Die Multiplikation auf $\tilde{G}$ setzt sich fort zu einer Multiplikation auf $\tilde{G}^* = \bigcup_{\mu \in Y} S_\mu(\tilde{G}) \subseteq \tilde{P}$, und jedes $\mu \in Y$ definiert ein $\sigma_\mu \in \tilde{G}^*(R)$.
Für $n \geq 1$ sei $Z_\mu^{(n)} \subseteq G^*$ das Faserprodukt

$$\begin{array}{ccc} Z_\mu^{(n)} & \longrightarrow & G^* \\ \downarrow & & \downarrow n \\ \{\sigma_\mu\} & \longrightarrow & G^* \end{array}$$

Für $\nu \in Y$ liefert Translation mit σ_ν einen Isomorphismus $Z_\mu^{(n)} \xrightarrow{\sim} Z_{\mu+n\nu}^{(n)}$, und die disjunkte Vereinigung $\coprod_{\mu \in Y/nY} Z_\mu^{(n)}$ wird zu einem Gruppenschema über R .

<u>[M4], 4.10</u>:

Der Kern $G^{(n)}$ der Multiplikation mit n auf G ist isomorph zu $\coprod_{\mu \in Y/nY} Z_\mu^{(n)}$.

<u>Beweis</u>:

Wie in [M4] .

[M4], 4.11:

Sei $\underline{p} \subseteq R$ ein Primideal, $Y_1 = \{\mu \in Y | b(\mu,\mu) \notin \underline{p}\}$, $s_1 \in \mathrm{Spek}(R)$ der zugehörige Punkt. Dann ist Y_1 eine Untergruppe von Y, Y/Y_1 ist torsionsfrei, und es gibt exakte Sequenzen von Gruppenschemata über dem Körper $k(s_1)$:

$$0 \to \tilde{G}_{s_1}^{(n)} \to G_{s_1}^{(n)} \to Y_1/nY_1 \to 0$$

$$0 \to \tilde{G}_{s_1}^{tor} \to G_{s_1}^{tor} \to Y_1 \otimes (\mathbb{Q}/\mathbb{Z}) \to 0$$

[M4], 4,12:

G ist semiabelsch

c) Das ample Geradenbündel $\mathcal{O}(1)$ auf P erfüllt $\mathcal{O}(\hat{1})|\hat{G} \cong \hat{\underline{L}} \otimes \hat{\underline{M}}$ $(\hat{G} \cong \hat{\tilde{G}})$. Da $\tilde{G}$ auf $\underline{L}|\tilde{G}$ operiert, ist $\underline{L}|\tilde{G}$ kanonisch trivial, und somit erhält $\hat{\underline{L}} \otimes \hat{\underline{M}} \cong \hat{\underline{M}}$ eine kubische Struktur. Wir zeigen, daß diese mit der kubischen Struktur auf $\mathcal{O}(\hat{1})$ übereinstimmt. Dies ist der Fall, wenn die kubische Struktur auf $\mathcal{O}(\hat{1})$ verträglich ist mit der T-Operation (T operiert auf $\hat{\underline{L}}$ und (trivial) auf $\hat{\underline{M}}$). Dies ergibt sich aus den nun folgenden Überlegungen:

Für $J \subseteq \{1,2,3\}$ erhält man durch Addition der Koordinaten in J einen Morphismus $m_J : \tilde{G}^3 \to \tilde{G}$, und zusammen ein

$$\underline{m} = \Pi m_J : \tilde{G}_1 = \tilde{G}^3 \to \tilde{G}^8 .$$

Der Graph von $\underline{m}$ ist eine integrable Untergruppe von $\tilde{G}^{11}$. Man findet dann ein relativ komplettes Modell $\tilde{P}_1$ für $\tilde{G}_1$, so daß $\underline{m}$ sich fortsetzt zu $\underline{m} : \tilde{P}_1 \to \tilde{P}^8$.

Wegen der kubischen Struktur auf $\underline{M}$ ist $\underline{m}^*(\bigotimes_J \underline{M}^{\pm 1})$ trivial (die Exponenten ergeben sich als $(-1)^{|J|}$ auf dem Faktor J). Dann ist

$$\underline{m}^*(\bigotimes_J \mathcal{O}(1)^{\pm 1}) \cong \underline{m}^*(\bigotimes_J (\hat{\underline{L}} \otimes \hat{\underline{M}})^{\pm 1}) \cong \underline{m}^*(\bigotimes_J \underline{L}^{\pm 1})$$

Darauf operieren G_1 und $Y_1 = Y^3$. Es gilt nun, daß die Operationen von $T_1 = T^3$ und Y_1 kommutieren:

Für $(g_1,g_2,g_3) \in T_1$ und $(\mu_1,\mu_2,\mu_3) \in Y_1$ ist

$$\prod_J \left(\prod_{j,k\in J} \mu_j(g_k)\right)^{\pm 1} = 1$$

Da die kubische Struktur auf $\hat{\mathcal{O}}(1)$ durch ihre Y-Invarianz eindeutig bestimmt ist, folgt die Behauptung.

Weiter können wir den Grad der durch $\mathcal{O}(1)$ auf G_η definierten Polarisation berechnen:

Er ist gleich dem Rang des torsionsfreien R-Moduls $\Gamma(P,\mathcal{O}(1)) = \Gamma(\hat{P},\hat{\underline{L}}\otimes\hat{\underline{M}})^Y$. (Y-Invarianten). Wir ersetzen zunächst $\hat{P}$ durch seine Normalisierung, was nichts an diesen Invarianten ändert. Sei $\hat{\theta}\in\Gamma(\hat{P},\hat{\underline{L}}\otimes\hat{\underline{M}})$. Da $T\subseteq G$ auf $\underline{L}$ und (trivial) auf $\underline{M}$ operiert, kann man $\hat{\theta}$ nach T-Eigenfunktionen entwickeln:

$$\hat{\theta} = \sum_{\nu\in X}\theta(\nu)$$

Aus den Kommutationsregeln zwischen Y und T folgt, daß für $\mu\in Y$ $S_\mu^*(\theta(\nu)) \in \underline{L}\otimes\underline{M}_{\mu+\nu}$.

Wenn also θ Y-invariant ist, muß gelten:

$$\theta(\mu+\nu) = S_\mu{}^*(\theta(\nu))$$

Andererseits kann man für jedes $\nu\in X$ den Rang des ν-Eigenraums in $\Gamma(\hat{P},\underline{L}\otimes\underline{M})$ abschätzen:

$$\Gamma(\hat{\tilde{P}},\hat{\underline{L}}\otimes\hat{\underline{M}})^\nu \subseteq \Gamma(\hat{\tilde{G}},\hat{\underline{L}}\otimes\hat{\underline{M}})^\nu \cong \Gamma(A,\underline{M}\otimes\mathcal{O}_\nu)$$

(Da $\hat{L}|\hat{G}$ trivial).

Insgesamt folgt:

Rang $(\Gamma(P,\mathcal{O}(1))) \leq [X:Y]\cdot\text{Grad }(\underline{M})$ $(\text{Grad }(M)^2 = $Grad der von $\underline{M}$ gelieferten

Abbildung $A \to A^{dual}$).

Wir zeigen, daß hier Gleichheit gilt. Dazu bezeichne $H = \mathrm{Ker}(Y) \subseteq T$ das durch X/Y definierte multiplikative Unterschema von T. H operiert auf $(\tilde{P}, \hat{\underline{L}} \otimes \underline{M})$, und diese Operation kommutiert mit Y. Wir können dann zum H-Quotienten übergehen. $(\tilde{P}/H$, G/H u.s.w.) und annehmen, daß $Y = X$. Auf $\hat{G} \cong \hat{\tilde{G}}$ sind $\mathcal{O}(\hat{1})$ und der Pullback von $\hat{\underline{M}}$ isomorph als kubische Bündel. Dann ist bekannt, daß der Grad der Polarisation auf G_η mindestens so groß ist wie der von $\underline{M}$ auf A. Wir hatten jedoch schon eine Abschätzung in die andere Richtung.

Es folgt insgesamt (für beliebiges Y):

$\mathcal{O}(1)$ definiert eine Polarisation vom Grad $[X:Y] \cdot \mathrm{Grad}(\underline{M})$

Da in allen unseren Abschätzungen nun die Gleichheit gilt, können wir auch eine Basis von $\Gamma(G_\eta, \mathcal{O}(1))$ angeben: Sei $\nu \in X, \theta(\nu) \in \Gamma(A, \underline{M} \otimes \mathcal{O}_\nu)$. Dann können wir $\theta(\nu)$ auffassen als Schnitt von $\underline{L} \otimes \underline{M}$ über $\tilde{G}$, welcher sich unter T gemäß ν transformiert. Es folgt: Es gibt ein $r \in R$, $r \neq 0$, so daß sich $r \cdot \theta(\nu)$ ausdehnt zu einem regulären globalen Schnitt aus $\Gamma(\hat{P}, \hat{\underline{L}} \otimes \hat{\underline{M}})^\nu$.

Dann existiert ein $\theta \in \Gamma(P, \mathcal{O}(1))$ mit

$$\hat{\theta} = \sum_{\mu \in Y} S_\mu^*(r \cdot \theta(\nu))$$

Wenn ν ein Vertretersystem für X/Y durchläuft, und $\theta(\nu)$ eine Basis von $\Gamma(A, \underline{M} \otimes \mathcal{O}_\nu)$, so erhält man auf diese Weise eine Basis von $\Gamma(G_\eta, \mathcal{O}(1))$.

Eine andere Schreibweise ist übrigens

$$\hat{\theta} = \sum_{\mu \in Y} a(\mu)\, b(\mu,\nu)\, \underline{c}(\mu)^* \, (r \cdot \theta(\nu)) \quad ,$$

mit einer Funktion $a : Y \to K^*$,

$$a(0) = 1, \; a(\mu+\nu) = a(\mu)\, a(\nu)\, b(\mu,\nu) \quad .$$

Dies erinnert schon an das vorherige Kapitel. Es bleibt uns noch eine

Kleinigkeit:

Wir haben bis jetzt ein relativ komplettes Modell nur unter der Annahme konstruiert, daß

(*) $\underline{M}$ ist sehr ample, und $a(\mu)\ b(\mu,\alpha) \in R$ für

$\mu \in Y$, $\alpha \in \Sigma$. Wir wollen noch aufzeigen, wie man dies fallen lassen kann:

Wir setzen voraus, daß eine Funktion $a(\ ) : X \to K^*$ existiert mit $a(0) = 1$, $a(\mu+\nu) = a(\mu)a(\nu)b(\mu,\nu)$. Außerdem existiere ein $r \in R$, $r \neq 0$, so daß $r \cdot a(\mu) \in R$ für alle $\mu \in X$. Ein solches $a(\ )$ läßt sich in den für uns wichtigen Fällen finden, zum Beispiel wenn R regulär ist.

(*) ist dann immer erfüllt, wenn man für ein genügend großes n $\underline{M}$ ersetzt durch $\underline{M}^n$, Y durch $n \cdot Y$, $i : Y \to \tilde{G}(K)$ durch $i \cdot \frac{1}{n} : nY \to \tilde{G}(K)$, und $a(\ )$ und $b(\ )$ durch ihre Einschränkungen auf nY bzw. $(nY) \times X$. Falls (*) schon erfüllt ist, läuft dies darauf hinaus, $\mathcal{O}(1)$ durch $\mathcal{O}(n)$ zu ersetzen.

Wähle nun n_1, n_2 genügend groß, so daß man zwei semiabelsche Varietäten G_1 und G_2 erhält, mit Geradenbündeln $\underline{N}_1$ und $\underline{N}_2$. Dann ist $(G_1, \underline{N}_1^{n_2}) \cong (G_2, \underline{N}_1^{n_1})$, und man erhält

$$G \cong G_1 \cong G_2$$

mit Geradenbündel $\underline{N}$, $\underline{N}_1 = \underline{N}^{n_1}, \underline{N}_2 = \underline{N}^{n_2}$. Wir benötigen noch Information über die globalen Schnitte $\Gamma(G,\underline{N})$. Wir kennen schon die globalen Schnitte von $\underline{N}_1 = \underline{N}^{n_1}$ und $\underline{N}_2 = \underline{N}^{n_2}$: Man erhält zum Beispiel eine Basis von $\Gamma(G_\eta, \underline{N}^{n_1})$ wie folgt: Durchlaufe ν ein Vertretersystem von $X/n_1 Y$, und $\theta(\nu)$ eine Basis von $\Gamma(A, \underline{M}^{n_1} \otimes \mathcal{O}_\nu)$. Dann gibt es für ein passendes $r \in R$, $r \neq 0$, eine Basis aus Elementen $\theta \in \Gamma(G, \underline{N}^{n_1})$ mit

$$\hat{\theta} = r \cdot \sum_{\mu \in n_1 Y} a(\mu) b(\mu,\nu)\ \underline{c}(\mu)^*(\theta(\nu))$$

Es liegt dann nahe, daß man eine Basis von $\Gamma(G,\underline{N})$ erhält aus θ's mit

$$\hat{\theta} = r\cdot\sum_{\mu\in Y} a(\mu)b(\mu,\nu)\ \underline{c}(\mu)^*(\theta(\nu)) \quad ,$$

$$\nu\in X/Y, \quad \theta(\nu)\in\Gamma(A,\underline{M}\otimes\mathcal{O}_\nu) \quad .$$

Und in der Tat rechnet man nach, daß diese $\hat{\theta}$ folgende Bedingung erfüllen: Für m_1, m_2 ganz und positiv mit $m_1n_1-m_2n_2>0$ ist

$$(\hat{\theta})^{m_1n_1-m_2n_2}\cdot\Gamma(G;\underline{N}^{m_2n_2})\subseteq\Gamma(G,\underline{N}^{m_1n_1}) \quad .$$

Daraus folgt, daß dies algebraische Schnitte sind.

Bemerkung:

Die Funktion $a(\mu)$ hängt von der Wahl des relativ kompletten Modells ab (siehe die Bemerkung nach der Definition eines solchen). Wirklich wichtig ist nur die Bilinearform b mit $a(\mu+\nu) = a(\mu)a(\nu)b(\mu,\nu)$. Der Leser wird sich leicht überzeugen, daß wir in der Tat gezeigt haben, daß man bei passender Wahl des kompletten Modells alle Funktionen $a(\ )$ erhält, welche dieser Gleichung genügen, und für die ein $r\in R$, $r\neq 0$ existiert mit $r\cdot a(\mu)\in R$. Zwei verschiedene unterscheiden sich um einen Homomorphismus $Y\rightarrow K^*$.
Wir formulieren nun das Hauptergebnis dieses Kapitels. Der Einfachheit halber betrachten wir nun prinzipale Polarisationen.

Satz 3:

Sei R exzellent normal, I-adisch komplett, Quotientenkörper K, $\tilde{G}$ über R eine semiabelsche Varietät,

$$0\rightarrow T\rightarrow\tilde{G}\rightarrow A\rightarrow 0 \quad ,$$

$T\cong G_m^r$ zerfallender Torus, Charaktergruppe $X\cong\mathbb{Z}^r$. A = abelsche Varietät.

Das Geradenbündel $\underline{M}$ auf A definiere eine prinzipale Polarisation, mit charakteristischer Abbildung $\underline{c}: X\rightarrow A(R)$. Wähle ein zulässiges System von Isomorphismen dafür, sowie eine symmetrische Bilinearform

$b : X \times X \to K^*$, so daß $b(\mu,\mu) \in I$ für $\mu \neq 0$. Dann existiert eine semiabelsche Varietät G über R , so daß die generische Faser G abelsch ist, und ein Geradenbündel $\underline{N}$ auf G , welches auf G_η eine prinzipale Polarisation definiert.

Es gilt:

i) $(\hat{G},\hat{\underline{N}}) \cong (\hat{G}$, Pullback $(\hat{\underline{M}}))$ mit kubischer Struktur.

ii) Sei $\theta_{\underline{N}} \in \Gamma(G,\underline{N}), \theta_{\underline{N}} \neq 0$, $\theta_{\underline{M}} \in \Gamma(A;\underline{M})$ ein erzeugendes Element. Dann ist

$\hat{\theta}_{\underline{N}} = \sum_{\mu \in X} a(\mu)\, \underline{c}(\mu)^*(\theta_{\underline{M}})$, mit $a(\mu) \in R$, $a(\mu) \neq 0$, und

$a(\mu+\nu)\; a(0) = a(\mu)\; a(\nu) b(\mu,\nu)$

iii) Im geeignet zu erklärenden Sinne ist

$G = \tilde{G}/i(X)$, wobei

$i : X \to \tilde{G}(K)$ wie folgt zu erklären ist:

Das zulässige System von Isomorphismen liefert eine Liftung von $\underline{c}$ zu einer linearen Abbildung $X \to \tilde{G}(R)$, und b liefert $X \to T(K)$. i ist das Produkt dieser beiden Abbildungen.

d) Abschließend benötigen wir noch einige Anmerkungen zur Kodaira-Spencer Klasse:

Die exakte Sequenz auf G

$$0 \to \Omega^1_R \otimes_R \mathcal{O}_G \to \Omega^1_{\tilde{G}} \to \Omega^1_{G/R} \to 0 \quad (\Omega^1_R = \Omega^1_{R/\mathbb{Z}})$$

liefert eine Abbildung

$$\kappa : \underline{t}^*_G = \Gamma(G,\Omega^1_{G/R}) \to H^1(G,\Omega^1_R \otimes_R \mathcal{O}_{\tilde{G}}) \;.$$

Weiter gibt die erste Chern-Klasse $c(\underline{N}) \in H^1(G,\Omega^1_{G/R})$ einen Morphismus $\underline{t}_G \otimes \Omega^1_R \to H^1(G,\Omega^1_R \otimes_R \mathcal{O}_{\tilde{G}})$, welcher im generischen Punkt ein Isomorphismus wird. Man kann dann κ auffassen als Bilinearform

$$\kappa : \underline{t}^*_G \times \underline{t}^*_G \to \Omega^1_K \;.$$

Es ist bekannt, daß κ symmetrisch ist. Außerdem enthält $\underline{t}^*_{\tilde{G}} = \underline{t}^*_{G}\ \underline{t}^*_{A}$, den dualen Tangentialraum zu A. $\kappa/\underline{t}^*_A \times \underline{t}^*_A$ ist die Kodaira-Spencer Klasse zu A, und $\kappa/\underline{t}^*_A \times \underline{t}^*_{\tilde{G}}$ beschreibt zusätzlich die Deformation der Erweiterung $0 \to T \to \tilde{G} \to A \to 0$. Sie entspricht dem Problem, eine translationsinvariante Differentialform aus $\underline{t}^*_{\tilde{G}}$ zu einem T-invarianten Schnitt aus $\Gamma(G,\Omega^1_{\tilde{G}})$ zu liften.

Wir nehmen nun an, daß $(\tilde{G},\underline{M})$ und das verträgliche System von Isomorphismen schon über einem Unterring $R_0 \subseteq R$ definiert sind, und betrachten, statt der absoluten Differentiale, Differentiale relativ R_0. $\kappa : \underline{t}^*_{\tilde{G}} \times \underline{t}^*_G \to \Omega_{K/R_0}$ verschwindet dann auf $\underline{t}_A \times \underline{t}^*_G$, und definiert eine Bilinearform $\underline{t}^*_T \times \underline{t}^*_T \to \Omega^1_{K/R_0}$. Es ist $\underline{t}^*_T \cong X \otimes R$, wobei $\mu \in X$ dem Differential $d\log(\mu) = d\mu/\mu$ auf T entspricht. κ wird also zu einer symmetrischen Bilinearform auf $X \times X$.

<u>Lemma</u>:

$\kappa(\mu,\nu) = d\log(b(\mu,\nu))$

<u>Beweis</u>:

Zunächst kann man die durch $c_1(\underline{N})$ vermittelte Abbildung

$$\underline{t}_T \subseteq \underline{t}_G \to H^1(G,\mathcal{O}_G)$$

noch etwas anders beschreiben: Es ist die Tangentialabbildung zur durch $\underline{N}$ definierten Abbildung $G \to \mathrm{Pic}^0(G)$. Die entsprechende Abbildung $T \subseteq \tilde{G} \to \mathrm{Pic}(\tilde{P})$ ist trivial, da T auf $\underline{L} \otimes \underline{M}$ operiert, also $g^*(\underline{L} \otimes \underline{M}) \cong \underline{L} \otimes M$ für $g \in T$. Allerdings ist dieser Isomorphismus nicht invariant unter $Y = X$. Vielmehr führt ein $\mu \in X$ einen Faktor $\mu(g)$ ein. Man erhält dann leicht die folgende Beschreibung von

$$\underline{c}_1(\underline{N})\,|\underline{t}_T : \qquad \underline{t}_T = \mathrm{Hom}(X,R) \to H^1(G,\mathcal{O}_G) :$$

Sei $l : X \to R$ eine Linearform. Da $\hat{P} = \hat{\tilde{P}}/X$, definiert l eine Klasse in $H^1(\hat{\tilde{P}},\mathcal{O}_{\hat{\tilde{P}}}) = H^1(P,\mathcal{O}_P)$. Deren Einschränkung auf G ist das Bild von l. Andererseits sei $\mu \in X$.

Wähle $\lambda \in \underline{t}_G{}^* = \underline{t}_{\tilde{G}}{}^*$ mit $\lambda|T = d(\log(\mu))$. Da $\tilde{G}$ über R_0 definiert ist, liftet man λ kanonisch zu einer $\tilde{G}(R_0)$-invarianten Form in $\Gamma(\tilde{G},\Omega^1_{\tilde{G}/R_0})$, und wegen $\tilde{G}_\eta = \tilde{P}_\eta$ auch zu einer solchen Form in $\Gamma(\tilde{P}\cdot\Omega^1_{\tilde{P}/R_0}) \otimes_R K$.

Diese Form ist nicht notwendig invariant unter Translation mit $i(\nu)$, für $\nu \in X$. Vielmehr ändert sie sich um $d\log(b(\mu,\nu))$, da $i : X \to \tilde{G}(R)$ bis auf Faktoren aus $\tilde{G}(R_0)$ übereinstimmt mit der durch b definierten Abbildung $X \to T(K)$. Das Lemma folgt nun leicht.

§ 4 KONSTRUKTION VON A_G

a) Wir kommen nun zum Hauptziel unserer Bemühungen, nämlich der Konstruktion eines über $\mathbb{Z}$ eigentlichen algebraischen Feldes, welches A_g als offene Teilmenge enthält. Wie schon in der Einleitung erwähnt, ist für uns ein algebraisches Feld eine Art Quotient $\underline{S}/\underline{R}$, wobei $\underline{S}$ ein Schema von endlichem Typ über $\mathbb{Z}$ ist, und $\underline{R} \to \underline{S} \times_{\mathbb{Z}} \underline{S}$ eine endliche Abbildung, welche $\underline{R}$ zu einem Gruppoid macht über $\underline{S}$. Außerdem wird vorausgesetzt, daß die Projektionen von $\underline{R}$ auf $\underline{S}$ étale sind. Man überzeugt sich leicht von der Äquivalenz dieser Definition mit der in [DM] :

Wenn man jedem Schema T die "descente-Daten zu $\mathrm{Hom}(T,S/R)$ " zuordnet, bestehend aus étalen Überdeckungen $T' \to T$ und Abbildungen $T' \to S$, $T' \times_T T' \to R$, mit geeigneten Kompatibilitätsbedingungen, so erhält man ein Gruppoid über T und ein algebraisches Feld im Sinne von [DM] .

Zur Konstruktion von $\underline{S}$ benötigen wir einige zusätzliche Daten:

b) Sei $X = \mathbb{Z}^g$, $B(X)$ bezeichne die symmetrischen Bilinearformen auf X ($B(X) = \mathrm{Hom}(S^2(X),\mathbb{Z})$) , und $B^+(X)_{\mathbb{R}} \subseteq B(X) \otimes_{\mathbb{Z}} \mathbb{R}$ den Kegel der positiv semi-definiten Formen.

Wir fixieren eine Zerlegung

$$B^+(X)_{\mathbb{R}} = \cup\sigma$$

mit rationalen Kegeln $\sigma \subseteq B^+(X)_{\mathbb{R}}$. (Die σ sind die konvexe Hülle endlich vieler rationaler Halbgeraden in $B^+(X)_{\mathbb{R}}$) .

Diese erfülle

i) Jede Seite eines σ in der Zerlegung kommt ebenfalls vor.

ii) Die Inneren $\mathring{\sigma}$ (im Sinne konvexer Mengen, nicht der Topologie!) sind disjunkt.

iii) Unter $GL(g,\mathbb{Z})$ gibt es nur endlich viele Konjugationsklassen von σ's .

Die Zerlegung heißt glatt, wenn zusätzlich gilt:

iv) Jedes σ wird aufgespannt von einer Teilmenge einer Basis von $B(X)$.

Es ist bekannt, daß man durch weiteres Unterteilen aus jeder Zerlegung eine glatte machen kann. In unserem Fall werden glatte Zerlegungen zu glatten Kompaktifizierungen führen.

Weiter setzen wir voraus:

v) Für jedes σ existiert eine lineare Abbildung $l_\sigma : X \to S^2(X)$ so daß für alle $\mu \in X$ $r_\sigma(\mu) = \mu \otimes \mu + l_\sigma(\mu) \in 2 \cdot S^2(X)$, und so daß für fast alle μ $r_\sigma(\mu) \in \sigma^\vee$.

Bedingung v) ist automatisch, wenn die Zerlegung glatt ist (also iv) $\Rightarrow$ v) , und kann sonst durch Unterteilen realisiert werden. Sie wird später die Existenz einer quadratischen Funktion a_σ sicherstellen, welche die Bilinearform b_σ (weiter unten) liefert:

$$a_\sigma(\mu) = \frac{1}{2}(\mu \otimes \mu + l_\sigma(\mu)) \in \mathbb{Z}\,[S^2(X)] \ .$$

Weiter gilt für jedes Quotientengitter $X \twoheadrightarrow X_1$, daß man durch Schneiden mit $B(X_1) \subseteq B(X)$ eine Kegelzerlegung von $B^+(X_1)_{\mathbb{R}}$ erhält.

Sei S der Torus mit Charaktergruppe $S^2(X) = B(X)^*$. Dann definiert jedes σ eine Torus-Einbettung $S \subseteq S_\sigma$, wobei S_σ affin ist mit Algebra $\mathbb{Z}[B(X)^* \cap \sigma^\vee]$. S operiert auf S_σ , und besitzt einen einzigen abgeschlossenen Orbit. Dessen Stabilisator ist der Untertorus von S , welcher zu $<\sigma> \subseteq B(X)$ gehört. Dabei sei $<\sigma>$ das von σ aufgespannte Untergitter. Wenn τ eine Seite von σ ist, so ist S_τ eine offene Teilmenge von S_σ.

Unsere Kompaktifizierung wird die Eigenschaft haben, daß sie lokal in der étalen Topologie isomorph ist zu einem S_σ.

Für jedes σ bezeichne $X \twoheadrightarrow X_\sigma$ den maximalen Quotienten mit $\sigma \subseteq B^+(X_\sigma)_{\mathbb{R}}$. Dann ist $<\sigma> \subseteq B(X_\sigma) \subseteq B(X)$. $S(\sigma)$ bezeichne den Torus mit Charaktergruppe $B(X_\sigma)^*$, und $S(\sigma) \subseteq S(\sigma)_\sigma$ die Torus-Einbettung zu $\sigma \subseteq B(X_\sigma)_{\mathbb{R}}$. Die universelle symmetrische Bilinearform

$$X_\sigma \times X_\sigma \to B(X_\sigma)^* = S^2(X_\sigma)$$

definiert eine symmetrische Bilinearform

$$b_\sigma : X_\sigma \times X_\sigma \to K_\sigma^*$$

(K_σ = Quotientenkörper von R_σ, R_σ = affiner Ring zu

$$S(\sigma)_\sigma = \mathbb{Z}[B(X_\sigma)^* \cap \sigma^\vee]) \quad ,$$

so daß $b_\sigma(\mu,\mu) \in R_\sigma$ für $\mu \in X_\sigma$, und für $\mu \neq 0$ $b_\sigma(\mu,\mu)$ auf dem abgeschlossenen $S(\sigma)$-Orbit von $S(\sigma)_\sigma$ verschwindet.
Sei r_σ der Rang von X_σ.

c) Wähle einen abgeschlossenen Punkt $s \in S(\sigma)_\sigma$, welcher im abgeschlossenen Orbit von $S(\sigma)$ liegt, und eine prinzipal polarisierte abelsche Varietät der Dimension $g-r_\sigma$ über dem algebraischen Abschluß von $k(s)$. Diese besitzt eine verselle Deformation, definiert über der strikten Henselisierung eines Polynomrings über $\mathbb{Z}$. Die Erweiterungen dieser versellen Deformation durch $T_\sigma = \mathbb{G}_m^{r_\sigma}$ werden parametrisiert durch das r_σ-fache Produkt des Duals der universellen abelschen Varietät. Sei R_0 die strikte Henselisierung in einem abgeschlossenen Punkt dieses Produkts, und R die strikte Henselisierung von $R_0 \otimes_{\mathbb{Z}} R_\sigma$ im Punkte s.

Dann ist $R_0 \subseteq R$, über R_0 existiert eine semiabelsche Varietät $\widetilde{G}$, $0 \to T_\sigma \to \widetilde{G} \to A \to 0$, so daß die Kodaira-Spencer Abbildung einen

Isomorphismus definiert $\underline{t}_{\tilde{G}} \times \underline{t}_A \xrightarrow{\sim} \Omega^1_{R_0/\mathbb{Z}}$. Wenn K_σ den Quotientenkörper von R bezeichnet, so existiert weiter die Bilinearform

$$b_\sigma : X_\sigma \times X_\sigma \to K^*_\sigma \quad ,$$

$b_\mu(\mu,\mu) \in \underline{m}$ = maximales Ideal, falls $\mu \neq 0$. Sei $\underline{M}$ ein amples Geradenbündel auf A , welches dort die prinzipale Polarisation definiert, und wähle ein zulässiges System von Isomorphismen für $\tilde{G}$, definiert über R_0 .

Satz 3 liefert dann über $\hat{R}$ (=$\underline{m}$-adische Komplettierung von R) eine abelsche Varietät G mit einem Geradenbündel $\underline{N}$, so daß $(\hat{G},\hat{\underline{N}}) \cong (\hat{\tilde{G}},\hat{\underline{M}})$, und daß die zugehörige Bilinearform (nach Satz 1) gleich b_σ ist. Nach dem Approximationssatz von M. Artin (siehe z. B. [A]) kann man annehmen, daß G und $\underline{N}$ schon über R definiert sind.

d) Aus der Toruseinbettung $S(\sigma) \subseteq S(\sigma)_\sigma$ erhält man eine Stratifikation $\text{Spek}(R) = U = \bigcup_{\tau \subseteq \sigma} U_\tau$, wobei τ über die Seiten von σ läuft. Für jedes $\tau \subseteq \sigma$ ist X_τ ein Quotient von X_σ , und $S(\tau)$ ein Untertorus von $S(\sigma)$. Wähle ein Komplement S'_τ , so daß $S(\sigma) = S(\tau) \times S'_\tau$. Dann ist $S(\tau)_\tau \times S'_\tau$ eine offene Teilmenge von $S(\sigma)$, somit R_σ ein Unterring von $R_\tau \otimes \mathbb{Z}[S']$. Weiter erhält man eine Zerlegung $b_\sigma = b_\tau \otimes b'$, wobei $b' : X_\sigma \times X_\sigma \to \mathbb{Z}[S']$ als Werte nur Einheiten annimmt.

Sei $s_1 \in U_\tau$ ein Punkt. Die Faser von $\tilde{G}$ in $s_1, \tilde{G}_{s_1}$, ist Erweiterung einer abelschen Varietät A_1 durch einen Torus T_1 . Der Torus T_1 ist in natürlicher Weise ein Untertorus von T_σ , zerfällt also, und die Charaktergruppe X_1 von T_1 ist ein Quotient von X_σ . Es liegt nahe zu vermuten, daß $X_1 = X_\tau$ (also $T_1 = T_\tau$) . Wir werden gleich sehen, daß dies in der Tat der Fall ist. Auf jeden Fall schließt man schon aus unserer Variante von [M4], 4.11 (siehe § 3), daß X_1 den gleichen Rang hat wie X_τ .

Es reicht dann, die vermutete Gleichheit $X_1 = X_\tau$ für den Fall zu zeigen, daß s_1 einer der generischen Punkte von U_τ ist. Wegen U_τ normal kann man dann ähnlich wie in [F], § 2, Lemma 1, T_1 als Untertorus in $G|\overline{U}_\tau$ einbetten. Aufgrund der bekannten Starrheitseigenschaften von Tori kann man diese Einbettung auf die formale Komplettierung von U längst $\overline{U}_\tau$ fortsetzen.

Sei $I \subset R$ das Ideal, welches den Abschluß von $\overline{U}_\tau$ definiert. Dann kann man formal längs I und längs $\underline{m}$ komplettieren. Wir unterscheiden dies durch Indizes: ^{1} und ^{2}. Dann gibt es exakte Sequenzen

$$0 \to \hat{T}_1^1 \to \hat{G}^1 \to \hat{G}_1^1 \to 0 ,$$

$$0 \to \hat{T}^2 \to \hat{G}^2 \to \hat{A}^2 \to 0 ,$$

und

$$0 \to (\hat{T}^2/\hat{T}_1{}^2) \to \hat{G}_1{}^2 \to \hat{A} \to 0 .$$

Die formale Komplettierung von $\underline{N}$ ist jeweils Pullback eines Geradenbündels auf den letzten Termen dieser Sequenzen.

Sei $\theta \in \Gamma(G,\underline{N})$, $\theta \neq 0$. Dann können wir θ formal entwickeln:

i) Auf $\hat{G}^1$:

$$\hat{\theta}^1 = \sum_{\mu \in X_1} \theta_1(\mu)$$

Dabei sind die $\theta_1(\mu)$ μ-Eigenfunktionen unter $\hat{T}_1$, und konvergieren gegen Null in der I-adischen Topologie

ii) Auf $\hat{G}_2$:

$$\hat{\theta}^2 = \sum_{\mu \in X_\sigma} a(\mu)\, \underline{c}(\mu)^*(\theta_{\underline{M}}) ,$$

wobei $a(\mu+\nu)\, a(0) = a(\mu)\, a(\nu)\, b(\mu,\nu)$.

Durch Vergleich folgt:

$$\hat{\theta}_1{}^2(\nu) = \sum_{\mu \mapsto \nu} a(\mu)\, \underline{c}(\mu)^*(\theta_{\underline{M}})$$

für $\nu \in X_1$ (man summiert über das Urbild von ν) .

Angenommen nun, es sei $X_1 \neq X_\tau$. Dann gibt es $\mu \in X_\sigma$, welches Bild $\neq 0$ in X_1 , aber Bild 0 in X_τ hat. Damit konvergiert einerseits $a(n\mu)a(-n\mu) = a(0)^2 \cdot b(\mu,\mu)^{n^2}$ I-adisch gegen Null, so daß $b(\mu,\mu) \in I$.

Andererseits ist $b(\mu,\mu) = b'(\mu,\mu) =$ Einheit mod I. Also ist tatsächlich $X_1 = X_\tau$.

Da $b_\sigma = b_\tau \cdot b'$, kann man $a(\ )$ analog zerlegen:

$$a(\mu) = a_1(\mu)\, a'(\mu) \quad ,$$

wobei:

i) $a_1(\mu)$ hängt nur vom Bild von μ in $X_1 = X_\tau$ ab, und
$a_1(\mu+\nu)\, a_1(0) = a_1(\mu) a_1(\nu) b_\tau(\mu,\nu)$.

ii) $a'(0) = 1$, $a'(\mu+\nu) = a'(\mu)\, a'(\nu)\, b'(\mu,\nu)$.

Damit folgt:

$$\hat{\theta}_1^2(\nu) = a_1(\nu) \sum_{\mu \to \nu} a'(\mu)\, \underline{c}(\mu)^*(\theta_{\underline{M}}) \quad , \quad \nu \in X_\tau$$

Man überlegt sich übrigens leicht, daß man erreichen kann, daß $a'(\mu) \in \hat{K}_\sigma$ regulär ist auf dem Pullback von $U_\tau \subseteq \mathrm{Spek}(R)$ in $\mathrm{Spek}(\hat{R})$. (Dies gilt schon, falls $\mu \in \mathrm{Kern}(X_\sigma \to X_\tau)$, und sonst modifiziere man a_1 und a' mit einer linearen Abbildung $X_\tau \to \hat{K}^*$.)

Wir wenden dies wie folgt an:

Sei wieder $s_1 \in U_\tau$. Durch das Degenerieren von $(G,\underline{N})$ in s_1 erhält man nach Satz 1 eine symmetrische Bilinearform b_1 auf $X_\tau \times X_\tau$, mit Werten im Quotientenkörper der Komplettierung des lokalen Ringes von U in s_1. Wir behaupten, daß b_1/b_τ als Werte Einheiten (in s_1) hat:

Dazu darf man zunächst R durch $\hat{R}$ ersetzen. Sei R_1 dann der lokale Ring zu s_1, mit Komplettierung $\hat{R}_1$. Die Reihe

$$\hat{\theta}^1 = \sum_{\mu \in X_\tau} \theta_1(\mu)$$

induziert dann die entsprechende Zerlegung über $\hat{R}_1$. Da

$$\hat{\theta}_1^2(\nu) = a_1(\nu) \sum_{\mu \to \nu} a'(\mu)\, \underline{c}(\mu)^*(\theta_{\underline{M}}) \quad ,$$

und $a'(\mu)$ = Einheit in $\hat{R}_1$, folgt daß $\theta_1(\nu) = a_1(\nu)$ (regulärer Schnitt, $\neq 0$ in s_1) . Es folgt, daß b_1 und die Bilinearform zu a_1 (dies ist b_τ) sich nur um Einheiten unterscheiden.

Schließlich liefert G_1 auf der formalen Komplettierung von U längs $\overline{U}_\tau$ eine Kodaira-Spencer Abbildung

$$\underline{t}^*{}_G \times \underline{t}^*{}_{G_1} \to \Omega^1_{R/R_\tau} \otimes_R \hat{R}_1$$

Diese induziert einen Isomorphismus.
Bild von $(\underline{t}^*{}_G \otimes \hat{\underline{t}}^*{}_{G_1})$ in $S^2(\underline{t}^*{}_G) \otimes_R \hat{R}_1$
$\xrightarrow{\sim} \Omega^1_{R/R_\tau} \otimes_R \hat{R}_1$.

Beweis:

Sie induziert einen Isomorphismus

$$\mathrm{Bild}(\underline{t}^*{}_G \otimes \hat{\underline{t}}^*{}_A) \xrightarrow{\sim} \Omega^1_{R/R_\sigma} \otimes_R \hat{R}_1 .$$

Die induzierte Abbildung

$$S^2(\underline{t}^*{}_{(T/T_1)}) \to \Omega^1_{R/R_\sigma} \otimes_R \hat{R}_1$$

ist gegeben durch

$$\mu \otimes \nu \longmapsto d(\log(b_\sigma(\mu,\nu)) = d\log(b'(\mu,\nu)),$$

für $\mu, \nu \in X(T/T_1) = \mathrm{Kern}(X_\sigma \to X_\tau)$.

Die $d\log(b'(\mu,\nu))$ bilden aber eine Basis von $\Omega^1_{\mathbb{Z}[s']/\mathbb{Z}}$ oder auch $\Omega^1_{R_\sigma/R_\tau} \otimes_R \hat{R}_1$, und es folgt alles.

e) Bis jetzt war R einfach die strikte Henselisierung von $R_0 \otimes_{\mathbb{Z}} R_\sigma$. Es ist dann induktiver Limes von endlich erzeugten $\mathbb{Z}$-Algebren, und es ist "alles" schon über einer solchen definiert. Wir erhalten dann ein Paar von endlich erzeugten $\mathbb{Z}$-Algebren, welches wir wieder $R_0 \subseteq R$ nennen, so daß R étale ist über $R_0 \otimes R_\sigma$, so daß $(\tilde{G},\underline{M})$ definiert sind über

R_0 , und so daß $(G,\underline{N})$ über R existieren. Außerdem erhält man durch strikte Lokalisierung von R in einem abgeschlossenen Punkt $s \in U_\sigma \subseteq U = \mathrm{Spek}(R)$ die bisherige Situation.

e) Lemma:

Indem man U gegebenenfalls durch eine kleinere étale Umgebung von s ersetzt, kann man folgendes erreichen:

i) Sei $\mathrm{Spek}(R) = U = \bigcup_{\tau \subseteq \sigma} U_\tau$ die Stratifikation, und $\hat{U}^\tau$ die formale Komplettierung von U längs $\bar{U}_\tau$. Dann existiert auf $\bar{U}^\tau$ eine exakte Sequenz

$$0 \to \hat{T}^\tau_\tau \to \hat{G}^\tau \to \hat{G}^\tau_\tau \to 0 \ , \quad \hat{G}^\tau_\tau \ \text{abelsch,}$$

und $\underline{\hat{N}}^\tau$ ist Pullback eines Geradenbündels $\underline{\hat{N}}^\tau_\tau$ auf $\hat{G}^\tau_\tau$

ii) Sei $\theta \in \Gamma(G,\underline{N}), \theta \neq 0$. Auf $\hat{U}^\tau$ entwickelt man θ nach $\hat{T}^\tau_\tau$-Eigenfunktionen:

$$\hat{\theta}^\tau = \sum_{\nu \in X_\tau} \hat{\theta}^\tau_\tau(\nu) \ .$$

Weiter zerlege man $b_\sigma = b_\tau \cdot b'$. Dann existiert ein

$$a_o \in \hat{R}^\tau = \Gamma(\hat{U}^\tau, \mathcal{O}_{\hat{U}^\tau}) \ , \ a_o \neq 0 \ ,$$

so daß $\hat{\theta}^\tau_\tau(\nu)\hat{\theta}^\tau_\tau(-\nu) = b_\tau(\nu,\nu)\cdot a_o \cdot f_\nu$, wobei der Schnitt $f_\nu \in \Gamma(\hat{G}^\tau_\tau, \underline{\hat{N}}^{\tau 2}_\tau)$ auf U_τ in keinem Punkt identisch auf der Faser verschwindet.

iii) Die Kodaira-Spencer Klasse von $\hat{G}^\tau_\tau$ induziert auf $\hat{U}_\tau$ einen Isomorphismus

$$\text{Bild von } \underline{t}^*_G \otimes \underline{t}^*_{G_\tau} \ \text{ in } \ S^2(\underline{t}^*_G) \otimes \mathcal{O}_{\hat{U}_\tau}$$
$$\xrightarrow{\sim} \Omega^1_{R/R_\tau} \otimes \mathcal{O}_{\hat{U}_\tau}$$

Beweis:

Punkt i) ist klar. Für ii) wählt man a_0 so, daß die Bedingung für $\nu = 0$ erfüllt ist. Dann gilt sie auch in allen Punkten $s_1 \in U_\tau$, welche s in ihrem Abschluß enthalten (nach den vorherigen Überlegungen über Bilinearformen). Außerdem reicht es, sie für eine gewisse endliche Anzahl von ν's zu verifizieren. (Wegen des Zusammenhangs mit Bilinearformen).

Damit läßt sich auch ii) erledigen. iii) geht genauso.

Korollar:

Sei $s_1 \in U$ ein abgeschlossener Punkt, $s_1 \in U_\tau$. Dann ist die Komplettierung der strikten Henselisierung von U in s isomorph zur Komplettierung eines der vorher konstruierten Ringe R (für τ statt σ), wobei sich die G's und $\underline{N}$'s entsprechen.

Beweis:

Sei R_1 der lokale Ring in s_1, $\hat{R}_1$ seine Komplettierung. Betrachte die Zerlegung in $\hat{G}$ in s_1 :

$$0 \to \hat{T}_1 \to \hat{G} \to \hat{A} \to 0 \ , \ \hat{T}_1 = \hat{T}_\tau \ .$$

Wenn R_2 eine verselle Deformation ist von $\hat{A}$ (mit Polarisation) und der Erweiterung durch T_1 , so ist die Abbildung $R_2 \hat{\otimes} R_\tau \to \hat{R}_1$ étale, nach Teil iii) des Lemmas. Weiter gilt für die zur Degeneration gehörige Bilinearform $b_1 : X_\tau \times X_\tau \to \hat{K}_1^*$, daß b_1/b_τ Werte in den Einheiten $\hat{R}_1^*$ annimmt. Wenn man die Abbildung $R_\tau \to \hat{R}_1$ mit einem geeigneten Element aus $S(\tau)(\hat{R}_1)$ twistet, darf man annehmen, daß $b_1 = b_\tau$, und alles hat seine Ordnung.

f) Jetzt können wir den Hauptsatz zeigen:

Satz 4:

Es existiert ein Schema $\underline{S}$, von endlichem Typ über $\mathbb{Z}$, glatt und mit geometrisch normalen Fasern, und ein Gruppoid $\underline{R} \to \underline{S} \times_{\mathbb{Z}} \underline{S}$, endlich, mit étalen Projektionen auf $\underline{S}$, so daß gilt:

i) Über $\underline{S}$ existiert eine semiabelsche Varietät G ,abelsch in den generischen Punkten, und ein Geradenbündel $\underline{N}$, welches in den generischen Punkten eine prinzipale Polarisation definiert.

ii) Seien G_1 und G_2 die beiden semiabelschen Varietäten der Dimension g , die man durch Pullback mit den Projektionen auf $\underline{R}$

erhält. Dann gibt es einen Isomorphismus $G_1 \cong G_2$, welcher in den generischen Punkten die Polarisation respektiert und mit der Gruppoid-Struktur verträglich ist.

iii) Sei $\underline{S}^0 \subseteq \underline{S}$ die dichte offene Teilmenge, über der G abelsch ist, $\underline{R}^0 = pr_1^{-1}(\underline{S}^0) = pr_2^{-1}(\underline{S}^0)$. Dann ist $R^0 \xrightarrow{\sim} \underline{\mathrm{Isom}}(G_1, G_2\ ;\ \text{Polarisation})$. Das durch $(\underline{S}^0, \underline{R}^0)$ definierte algebraische Feld ist A_g .

iv) Das durch $(\underline{S}, \underline{R})$ definierte algebraische Feld ist eigentlich über $\mathbb{Z}$.

v) $\underline{S}$ und $\underline{R}$ besitzen Stratifikationen $S = \cup \underline{S}_\sigma$, $R = \cup \underline{R}_\sigma$, parametrisiert durch die Konjugationsklassen der $\sigma \subseteq B^+(X)_{\mathbb{R}}$ unter $GL(g,\mathbb{Z})$. Für einen abgeschlossenen Punkt $s \in \underline{S}_\sigma$ ist die strikte Henselisierung von $\underline{S}$ in s isomorph zur strikten Henselisierung von s_σ in einem Punkt des abgeschlossenen S-Orbits. Dieser Isomorphismus erhält die Stratifikationen. $\underline{S}^0$ ist das Stratum zu $\sigma = (0)$.

Beweis:

Wir wählen $\underline{S}$ als endliche Vereinigung von den vorher konstruierten U's . Jedes solche U gehörte zu einem σ , und es gab eine étale Abbildung $U \to S(\sigma)_\sigma \times$ (Modulfeld der semiabelschen $0 \to G^r{}_m \to \tilde{G} \to A \to 0$). Wir fordern, daß für je ein σ in einer $GL(g,\mathbb{Z})$-Konjugationsklasse die zugehörigen U's die Menge (abgeschlossener $S(\sigma)$-Orbit) $\times$ (Modulfeld) überdecken. Da das Modulfeld quasikompakt ist, und da es nur endlich viele Konjugationsklassen von σ's gibt, reichen dazu endlich viele U's . Wir haben nun $\underline{S}$, und erhalten zwei semiabelsche Varietäten G_1 und G_2 durch Pullback auf $\underline{S} \times_{\mathbb{Z}} \underline{S}$. Über $\underline{S}^0 \times \underline{S}^0$ ist $\underline{R}^0 = \underline{\mathrm{Isom}}_{\underline{S}_0}(G_1, G_2;\ \text{Polarisation})$ endlich. Sei $\underline{R}$ die Normalisierung von $\underline{R}^o$ über $\underline{S} \times_{\mathbb{Z}} \underline{S}$. Dann ist $\underline{R}$ normal und endlich über $\underline{S} \times_{\mathbb{Z}} \underline{S}$. Außerdem existiert über $\underline{R}$ ein Isomorphismus $G_1 \cong G_2$ (siehe [F], § 2, Lemma 1), und die Aussagen i), ii), iii) und v) folgen, wenn wir zeigen, daß die Projektionen von $\underline{R}$ auf $\underline{S}$ étale sind und die Stratifikationen respektieren. Dazu sei $s \in \underline{R}$ ein abgeschlossener Punkt, mit Projektionen s_1 und s_2 auf S . Sei R die Komplettierung der strikten Henselisierung von $\underline{R}$ in s , und entsprechend für $R_1 \hat{\otimes} R_2$. Dann ist

R endlich über $R_1 \hat{\otimes} R_2$ (das komplette Tensorprodukt ist zu nehmen über der Komplettierung der strikten Henselierung von $\mathbb{Z}$ in einem Primideal $p\mathbb{Z}$, $p \neq 0$), normal, enthält R_1 und R_2, und es ist $G_1 \otimes_{R_1} R = G_2 \otimes_{R_2} R = G_{12}$. Zu den degenerierenden polarisierten abelschen Varietäten gehören Gitter $X_1 = X_2 = X_{12}$ und symmetrische Bilinearformen.

$b_1 : X_1 \times X_1 \to K_1{}^*$

$b_2 : X_2 \times X_2 \to K_2{}^*$

$b_{12}: X_{12} \times X_{12} \to K^*$

(K_1,K_2,K sind die Quotientenkörper). Es ist $b_{12} = b_1 \cdot b_1' = b_2 \cdot b_2'$ wobei b_1', b_2' : $X_{12} \times X_{12} \to R^*$ als Werte Einheiten annehmen.

Weiter hat man Polyeder $\sigma_1 \subseteq B^+(X_1)_{\mathbb{R}}$ und $\sigma_2 \subseteq B^+(X_2)_{\mathbb{R}}$. Es ist $\sigma_1 = \sigma_2$:

Angenommen $\sigma_{12} = \sigma_1 \cap \sigma_2$ ist eine echte Seite von σ_1 (oder analog von σ_2). Dann gibt es endlich viele Elemente $\mu_j, \nu_j \in X_1$, so daß für jede Bilinearform $b \in \sigma_{12}$ $\Sigma\, b(\mu_j, \nu_j) = 0$, aber daß diese Summe positiv wird für jedes b in $\sigma_1 - \sigma_{12}$, und negativ für $b \in \sigma_2 - \sigma_{12}$. Es sind dann $\Pi b_1(\mu_j, \nu_j)$ und $\Pi b_2^{-1}(\mu_j, \nu_j)$ Elemente aus R_1 bzw. R_2, wobei das erste Produkt keine Einheit ist. Ihr Produkt ist aber eine Einheit in R, und das geht nicht.

Sei also $\sigma_1 = \sigma_2 = \sigma$, und damit $X_1 = X_2 = X_{12} = X_\sigma$.

Es sei wieder R_0 die Basis einer versellen Deformation von G_s über $\underline{k(s)}$ (d.h., Deformation des abelschen Teils und der Erweiterung durch den Torus). Dann sind R_1 und R_2 Komplettierungen von strikten Henselisierungen von $R_0 \otimes R_\sigma$, und $\tilde{G}_1$ und $\tilde{G}_2$ sind Pullback eines $\tilde{G}$ über R_0. Man wähle ein zulässiges System von Isomorphismen für $\tilde{G}$ über R_0, und erhält dann einen Isomorphismus $R_1 \cong R_2$, bei dem sich b_1 und b_2 entsprechen. Nach Satz 2 ist dann auch $G_1 \cong G_2$, und der Isomorphismus über R entsteht einfach durch Basiserweiterung. Da R über die Normalisierung von $\underline{\mathrm{Isom}}_{S_O}$ (G_1,G_2; Polarisation) definiert wurde, ist $R = R_1 = R_2$. Die Projektionen sind also étale in $s \in \underline{R}$. Es bleibt die Aussage iv) . Dazu benutzt man ein Bewertungs-Kriterium: Sie V ein kompletter diskreter Bewertungsring, mit algebraisch abgeschlossenem Restklassenkörper k und Quotientenkörper K, und

$$\phi_1 : \mathrm{Spek}(K) \to \underline{S}$$

ein Morphismus der den einzigen Punkt von Spek(K) in einen der generischen Punkte von $\underline{S}$ abbildet. Dann gibt es eine endliche Erweiterung $K' \supseteq K$, mit Normalisierung V' von V, und

$$\phi_2 : \mathrm{Spek}(V') \to \underline{S} \quad ,$$

so daß man $\phi_1 \times \phi_2 : \mathrm{Spek}(K') \to \underline{S} \times_{\mathbb{Z}} \underline{S}$ zu einem K'-wertigen Punkt von $\underline{R}$ liften kann.

Es ist also folgendes zu zeigen:

Das Pullback $\phi_1{}^*(G)$ von G unter ϕ_1 ist eine prinzipal polarisierte abelsche Varietät. Wir brauchen ϕ_2, so daß $\phi_2{}^*(G)|K'$ isomorph zu $\phi_1{}^*(G)$ ist.

Dazu wähle man K' so groß, daß $\phi_1{}^*(G)$ semistabile Reduktion hat über K', und ersetzt V durch V', K durch K'.

Sei $v : K^* \to \mathbb{Z}$ die Bewertung. Das Néron-Modell von $\phi_1{}^*(G)$ definiert nach Satz 1 eine symmetrische Bilinearform $b : X_V \times X_V \to K^x$ auf einem Gitter X_V, und $v \circ b$ ist positiv definit. Es gibt also ein σ in der Kegel-Zerlegung, so daß $v \circ b$ im Inneren von σ liegt, und $X_V \cong X_\sigma$. (σ, X_σ) ist eindeutig bestimmt bis auf Konjugation mit $GL(g,\mathbb{Z})$.

Sei wieder R_0 eine verselle Deformation der speziellen Faser des Néron-Modells. Wir erhalten eine Abbildung $R_0 \to V$, so daß die universelle Überlagerung des Néron-Modells über R_0 definiert ist. Wenn man dann über R_0 ein zulässiges System von Isomorphismen wählt, definiert b einen Morphismus $R_0 \otimes R_\sigma \to V$ bzw. $\mathrm{Spek}(V)$ $\mathrm{Spek}(R_0) \times_{\mathbb{Z}} S(\sigma)_\sigma$, wobei der abgeschlossene Punkt von Spek(V) in den abgeschlossenen Orbit von $S(\sigma)_\sigma$ abgebildet wird. Nach Konstruktion von S kann man diese Abbildung liften in eines der U's.

Dies liefert ϕ_2, so daß das Néron-Modell von $\phi_1{}^*(G)$ und $\phi_2{}^*(G)$ dieselbe formale Komplettierung in V haben und dieselbe symmetrische Bilinearform definieren. Nach Satz 2 sind sie isomorph.

g) Damit ist der Beweis von Satz 4 beendet. Wir notieren hier nur noch das Korollar, daß die geometrischen Fasern von A_g über $\mathbb{Z}$ irreduzibel sind: Dies folgt aus der analytischen Theorie in Charakteristik 0, und der Rest ist genauso wie in [DM].

§ 5 LEVEL-N-STRUKTUREN

a) Alle unsere Überlegungen lassen sich auch mit Level-Strukturen durchführen. Wir wählen eine natürliche Zahl n. Da sich bekanntlich Level-n-Strukturen schlecht mit Charakteristiken vertragen, welche n teilen, arbeiten wir über $\mathbb{Z}\left[{}^1/n, e^{2\pi i/n}\right]$. Über diesem Grundring hat man eine kanonische symplektische Form $<,>$ auf $(\mathbb{Z}/n\mathbb{Z})^{2g}$, mit Werten in μ_n = n.te Einheitswurzeln. Eine Level-n-Struktur auf einer prinzipal polarisierten abelschen Varietät der Dimension g ist ein Isomorphismus $A^{(n)} \xrightarrow{\sim} (\mathbb{Z}/n\mathbb{Z})^{2g}$ ($A^{(n)}$ = n-Teilungspunkte), welcher die symplektische Struktur erhält. Die verschiedenen Level-n-Strukturen auf A sind konjugiert unter $Sp(2g, \mathbb{Z}/n\mathbb{Z})$. Es gibt ein algebraisches Feld $A_{g,n}$ über $Z\left[{}^1/n, e^{2\pi i/n}\right]$, welches die abelschen Varietäten mit Level-n-Struktur klassifiziert. Für $n \geq 3$ ist $A_{g,n}$ sogar ein algebraischer Raum.

Sei $\overline{A}_{g,n}$ die Normalisierung von $\overline{A}_g$ in $A_{g,n}$. $A_{g,n}$ wird gegeben durch ein Paar $(\underline{S}_n, \underline{R}_n)$. Dabei ist $\underline{S}_n$ die Normalisierung von $\underline{S}_n$ in der durch Hinzufügen von Level-n-Strukturen über $\underline{S}^0$ definierten Überlagerung, und $\underline{R}_n$ wieder die Normalisierung von $\underline{R}_n{}^0$.

b) Wir wollen ein lokales Modell finden für die Überlagerung $\underline{S}_n \to \underline{S}$. Dazu betrachtet man wieder den Torus S mit Charaktergruppe $B(X)^*$, und die Torus-Einbettungen $S \subseteq S_\sigma$. Die Multiplikation mit n auf S induziert eine verzweigte Überlagerung $n : S_\sigma \longrightarrow S_\sigma$.
Es gilt nun:

Satz 5:

i) Die Überlagerung $\underline{S}_n \to \underline{S}$ ist lokal in der étalen Topologie isomorph zu $n : S_\sigma \longrightarrow S_\sigma$. Insbesondere erhält auch $\underline{S}_n$ eine Stratifizierung, hat geometrisch normale Fasern, und ähnliches.

ii) Für $n \geq 3$ ist $\underline{R}_n \to \underline{S}_n \times_{\mathbb{Z}\left[{}^1/n, e^{2\pi i/n}\right]} \underline{S}_n$ eine abgeschlossene Einbettung, und $\overline{A}_{g,n}$ ein algebraischer Raum.

Beweis:

i) Sei R die Komplettierung der strikten Henselisierung von $\underline{S}$ in einem abgeschlossenen Punkt, G/R die universelle semiabelsche Varietät, $\tilde{G}$, $0 \to T_\sigma \to \tilde{G} \to A \to 0$, X_σ und $b_\sigma : X_\sigma \times X_\sigma \to K^*$ wie üblich. Die Behauptung i) läuft darauf hinaus zu zeigen, daß die n-Teilungspunkte von G_η den Körper $K(\sqrt[n]{b})$ erzeugen. Aus unserem Analogon zu [M4], 4.11(siehe § 3) ergibt sich eine exakte Sequenz von Gruppenschemata über K

$$0 \to \tilde{G}^{(n)} \to G^{(n)} \to X_\sigma / n\cdot X_\sigma \to 0 \quad .$$

Dabei ist $\tilde{G}^{(n)}$ étale vom Rang n^{2g-r_σ}, und die Faser über der Klasse modulo $n\cdot X_\sigma$ von $\mu \in X_\sigma$ ist isomorph zu $Z_\mu^{(n)}(\overline{K}) = \{g \in \tilde{G}(\overline{K}),\ n\cdot g = i(\mu)\}$. Wenn $b_\sigma : X \to T_\sigma(K)$ die zu $b_\sigma(,)$ gehörige Injektion ist, so ist $i(\mu) \in \tilde{G}(R)\cdot b_\sigma(\mu)$. Da $\tilde{G}(R)$ n-divisibel ist, ist der von den Koordinaten der n-Teilungspunkte erzeugte Körper gleich $K(\sqrt[n]{b_\sigma(X)})$, und es folgt die Behauptung.

ii) Wie bisher sind die Projektionen von $\underline{R}_n$ auf $\underline{S}_n$ étale, und somit ist $\underline{R}_n \to S_n \times_{\mathbb{Z}\left[{}^1/n, e^{2\pi i/n}\right]} \underline{S}_n$ unverzweigt.

Es reicht dann zu zeigen, daß über zwei geometrischen Punkten s_1 und s_2 von $\underline{S}_n$ höchstens ein geometrischer Punkt von $\underline{R}_n$ liegen kann. Es mögen wieder R_1, R_2 und R die Komplettierungen der strikten Henselisierungen von $\underline{S}_n$ bzw. $\underline{R}_n$ in s_1, s_2 und s bezeichnen. Es ist $R_1 \cong R$ und $R_2 \cong R$, und der induzierte Isomorphismus $R_1 \cong R_2$ ist unabhängig von der Wahl von s (und R), da schon bekannt ist, daß $\underline{R}_n^0$ sich abgeschlossen einbettet in $\underline{S}_n^0 \times_{\mathrm{Spek}(\mathbb{Z}\left[{}^1/n, e^{2\pi i/n}\right])} \underline{S}_n^0$. Es folgt, daß es nur ein s in der Faser über (s_1, s_2) geben kann.

Korollar:

Die geometrischen Fasern von $A_{g,n}$ über $\mathbb{Z}\left[{}^1/n, e^{2\pi i/n}\right]$ sind irreduzibel.

§ 6 MODULFORMEN UND MINIMALE KOMPAKTIFIZIERUNG

a) Auf $\overline{A}_g$ erhält man in natürlicher Weise eine Reihe von Vektorbündeln. Dabei ist ein Vektorbündel auf $\overline{A}_g$ gegeben durch ein Vektorbündel auf $\underline{S}$, dessen beide Pullbacks zu $\underline{R}$ isomorph sind (unter Erfüllung geeigneter Bedingungen).

1.) $\underline{t}_G$, $\underline{t}^*_G$, d.h. Tangential und Kotangentialbündel des universellen G's, und ihre Tensorpotenzen u.s.w. Sei $\omega_G = \Lambda^g \underline{t}^*_G$.

2.) Im allgemeinen ist $\Omega^1_{\underline{S}/\mathbb{Z}}$ nicht lokal frei. Als Ersatz dient besser das Bündel Θ, definiert wie folgt: Für jede Torus-Einbettung $S \subseteq S_\sigma$ sei Θ_{S_σ} das Unterbündel des direkten Bildes von $\Omega^1_{S/\mathbb{Z}}$, welches von den $d\mu/\mu$ erzeugt wird (μ = Charakter von S). Da $\underline{S}$ lokal in der étalen Topologie isomorph ist zu einem S_σ, erhält man Θ auf $\underline{S}$. Wenn die Kegelzerlegung glatt ist, so ist Θ die Garbe der Differentialformen mit logarithmischen Polen in ∞. Die Kodaira-Spencer Klasse liefert einen Isomorphismus $\kappa : S^2(\underline{t}^*_G) \xrightarrow{\sim} \Theta$. (Dies folgt aus der ziemlich expliziten Bestimmung von κ).

3.) Lokal existiert ein Geradenbündel $\underline{N}$ auf G, welches über $\underline{S}^0$ die prinzipale Polarisation definiert. Wenn $e : S \longrightarrow G$ den Nullschnitt bezeichnet, kann man annehmen, daß $e^*(\underline{N}) \cong \mathcal{O}_S$ trivial ist.

$\underline{N} \otimes [-1]^*\underline{N} = \underline{H}$ ist dann ein wohldefiniertes Geradenbündel auf G, mit $e^*(\underline{H}) \cong \mathcal{O}_{\underline{S}}$. Das direkte Bild $p_*(\underline{H})$ auf $\underline{S}$ ($p : G \to \underline{S}$ die Projektion) ist eine kohärente reflexive Garbe vom Rang 2^g. Ihre lokale Struktur ist recht interessant, siehe z. B [MB]. Man kann übrigens ihre lokale Struktur mit Hilfe unserer Überlegungen beim Beweis von Satz 2 aufhellen.

Auf jeden Fall ist $p_*(\underline{H})$ lokal frei auf $\underline{S}^0$. Wenn wir Level-2-Strukturen einführen, also zu $\underline{S}^0_2$ übergehen, so ergibt sich eine irreduzible Darstellung der Θ-Gruppe auf $p_*(\underline{H})$, und $p_*(\underline{H}) \cong \underline{K}^{2^g}$, für ein Geradenbündel $\underline{K}$ auf $\underline{S}^0_2$ oder auf $A_{g,2}$. Aus dem Satz von Riemann-Roch folgt, daß in $\mathrm{Pic}(A_{g,2}) \otimes_{\mathbb{Q}} \underline{K}$ und ω^{-1} dasselbe Bild haben. Eine Potenz von $\underline{K}$ stimmt somit mit einer Potenz von ω^{-1} überein.

b) Eine Modulform vom Gewicht k zur Gruppe $S_p(2g,\mathbb{Z})$ ist ein globaler Schnitt von ω^k über $\overline{A}_g$. Man definiert entsprechend Modulformen über $\mathbb{C}$, $\mathbb{Z}/_{p\mathbb{Z}}$ oder allgemeiner über einem beliebigen Ring. Man kann auch Schnitte von ω^k über A_g betrachten. Diese lassen sich wie folgt interpretieren:

Jeder abelschen Varietät A von der Dimension g wird ein $f_A \in \Gamma(A,(\Omega_A^g)^k)$ zugeordnet. Die verschiedenen f_A's entsprechen sich bei Basiswechsel oder Automorphismen von A. Jeder Schnitt von ω^k über A_g besitzt dann Fourierentwicklungen:

Sei wieder $X = \mathbb{Z}^g$, $\sigma \subseteq B^+(X)_{\mathbb{R}}$ ein konvexer rationaler Polyeder stabil unter Homothetien. Dabei wird nicht vorausgesetzt, daß σ in der vorher gewählten Kegelzerlegung auftaucht. Wir können jedoch wieder $X_\sigma, r_\sigma = \mathrm{Rang}(X_\sigma)$, $S, S(\sigma)$ und $S(\sigma)_\sigma$ definieren.

Sei A eine abelsche Varietät der Dimension $g-r_\sigma$. A sei prinzipal polarisiert, mit Dual A^v. Über $(A^v)^{r_\sigma}$ existiert dann die universelle Erweiterung von A durch $T_\sigma = \mathbb{G}_m^{r_\sigma}$:

$$0 \to T_\sigma \to \widetilde{G} \to A \to 0 \quad .$$

Sei $\hat{R}_\sigma$ die Komplettierung des Ringes R_σ zu $S(\sigma)_\sigma$ in der I-adischen Topologie, wobei das Ideal I den abgeschlossenen $S(\sigma)$-Orbit in $S(\sigma)_\sigma$ beschreibt. Dann liefert die Mumford-Konstruktion eine semiabelsche Varietät G über $\mathrm{Spek}(R_\sigma) \times (A^v)^{r_\sigma}$, (zunächst über dem formalen Schema, aber man macht alles algebraisch, da die Konstruktion gleich eine Kompaktifizierung dieser Varietät liefert), welche gute Reduktion hat auf dem Urbild des offenen $S(\sigma)$-Orbits $S(\sigma) \subseteq S(\sigma)_\sigma$.

Wenn $\mu_1,\ldots,\mu_{r_\sigma}$ eine Basis von X_σ ist, so ist

$$\Lambda^g \underline{t}^*_G \cong \hat{R}_\sigma(d(\log(\mu_1) \wedge \ldots \wedge d\log(\mu_{r_\sigma})) \otimes \Lambda^{g-r_\sigma}(\underline{t}^*_A)$$

und f_G hat eine Entwicklung

$$f_G = \sum_{\chi \in B(X_\sigma)^*} \chi \cdot f_\chi \cdot (d\log(\mu_1) \wedge \ldots \wedge d\log(\mu_{r_\sigma}))^k \quad ,$$

mit

$$f_\chi \in (\Lambda^{g-r_\sigma}(\underline{t}^*_A))^k$$

Die f_χ liefern globale Schnitte von ω^k über A_{g-r_σ} und es gibt ein χ_o, so daß f_χ verschwindet für $\chi \notin \chi_o + \sigma^V$.

Wenn $\sigma_1 \subseteq \sigma$ ein in σ enthaltener Polyeder ist, so stimmen die f_χ für σ_1 mit denen für σ überein. Schließlich sind sie noch invariant unter der Gruppe $GL(X_\sigma)$, genauer gesagt, für $a \in GL(X_\sigma)$ ist $f_{a(\chi)} = \det(a)^k \cdot f_\chi$ (beachte die Operation von a auf

$$d \log(\mu_1) \wedge \ldots \wedge d \log(\mu_{r_\sigma}) \; !) .$$

Wenn $r_\sigma > 1$, so folgt dann schon automatisch, daß $f_\chi \neq 0$ nur gelten kann, wenn $\chi \in B^+(X_\sigma)^V_{\mathbb{R}}$. (Koecher-Prinzip): χ definiert eine Linearform auf $B(X_\sigma)_{\mathbb{R}}$. Für jeden rationalen konvexen Polyeder $\sigma_1 \subseteq B^+(X_\sigma)_{\mathbb{R}}$ gibt es χ_1, so daß $a(\chi) \in \chi_1 + \sigma_1^V$ für alle $a \in GL(X_\sigma)$. Wenn $C \subseteq \sigma_1(\mathbb{R})$ eine kompakte Teilmenge ist, so ist dann die durch χ definierte Linearform nach unten beschränkt auf $\bigcup_{a \in GL(X_\sigma)} a(C)$.

Wenn $r_\sigma > 1$ ist, so kann man aber durch geeignete Wahl von C erreichen, daß die konvexe Hülle der obigen Menge gleich $B^+(X_\sigma)_{\mathbb{R}}$ ist. Also liegt χ in $B^+(X_\sigma)^V$.

Aus unserer Konstruktion von $\bar{A}$ folgt, daß f sich genau dann zu einem regulären Schnitt von ω^{k^g} auf $\bar{A}_g$ ausdehnt, wenn für alle σ in der gewählten Kegelzerlegung die Koeffizienten f_χ verschwinden, falls $\chi \notin \sigma^V$. Für jedes einzelne σ ist dies äquivalent zur Regularität von f auf dem Stratum $\underline{S}_\sigma$, und f ist schon auf ganz $\underline{S}_\sigma$ regulär, wenn dies in einem Punkt von $\underline{S}_\sigma$ gilt.

Daraus folgt, daß man nur die σ's mit $\dim(\sigma) = \frac{g(g+1)}{2}$ maximal betrachten muß. Dort erhält man eine Entwicklung

$$f = \sum_{\chi \in B(X)^*} \chi \cdot f_\chi \cdot (d \log(\mu_1) \wedge \ldots \wedge d \log(\mu_g))^k ,$$

mit $f_\chi \in \mathbb{Z}$.

f definiert eine Modulform $\Longleftrightarrow$ $f_\chi = 0$ für $\chi \notin B^+(X)^V$. (Dies ist automatisch, falls $g \geq 2$).

Man kann statt $\mathbb{Z}$ auch andere Grundringe wählen, wie $\mathbb{C}$, $\mathbb{Z}/_{n\mathbb{Z}}$ u.s.w. Beim Grundring $\mathbb{C}$ erhält man bis auf einen Faktor $(2\pi i)^{gk}$ die klassische Fourierentwicklung einer Modulform, indem man etwa $B^+(X)^\vee$ identifiziert mit den halbganzen symmetrischen positiv definierten Matrizen. Es folgt zum Beispiel, daß der Raum der Modulformen über $\mathbb{C}$ eine Basis besitzt, deren Elemente ganze Fourierkoeffizienten haben, und daß eine $\mathbb{C}$-Modulform genau dann über $\mathbb{Z}$ definiert ist, wenn alle Fourierkoeffizienten in $(2\pi i)^{gk}\cdot\mathbb{Z}$ liegen.

Wir notieren noch eine weitere Eigenschaft der Modulformen: Sei wieder

$$f = \sum_{\chi\in B^*(X_\sigma)^\vee} \chi\cdot f_\chi (d\log(\mu_1)\wedge\ldots\wedge d\log(\mu r))^k .$$

Sei $\langle\sigma\rangle \subseteq B(X_\sigma)$ das von σ aufgespannte Untergitter. Dann ist $B^+(X_\sigma)^\vee \cap \langle\sigma\rangle^\perp = \{0\}$, da σ eine positiv definite Form enthält. Es folgt, daß f konstant ist auf dem abgeschlossenen $S(\sigma)$-Orbit in $S(\sigma)_\sigma$.

Der konstante Wert wird gegeben durch die Modulform f_0, vom Gewicht k, auf $\overline{A}_{g-r_\sigma}$. Man überlegt sich leicht, daß die Fourierkoeffizienten der Entwicklung von f_0 parametrisiert werden durch $\chi\in B^+(\mathrm{Kern}(X\to X_\sigma))^\vee \subseteq B^+(X)^\vee$, und daß man für solche χ dieselben Koeffizienten für f und f_0 erhält. (Betrachte f auf Produkten $A_1\times A_2$ der Dimensionen $(r_\sigma, g-r_\sigma)$).

Entsprechendes gilt auch für Modulformen mit Level-Struktur (über $Z\left[{}^1/n,\ e^{2\pi i/n}\right]$ zum Beispiel): Dort wird die Fourier-Entwicklung parametrisiert durch $\chi\in\frac{1}{n}B(X_\sigma)^\vee$, und zu jedem σ gehören mehrere Fourier-Reihen.

c) Beispiele für Modulformen erhält man durch θ-Reihen : Wähle $\underline{a},\underline{b}\in(\frac{1}{n}\mathbb{Z}/\mathbb{Z})^g$. Dann ist bis auf 4.-te Einheitswurzeln

$$\theta(Z;\underline{a},\underline{b}) = e^{i\pi\underline{a}^t\underline{b}} \sum_{\underline{m}\in\mathbb{Z}^g} e^{i\pi((\underline{m}+\underline{a})Z^t(\underline{m}+\underline{a}))}\, e^{2i\pi\underline{m}^t\underline{b}}$$

eine Modulform vom Gewicht $^1/2$ zum Level $2n^2$, und ein Produkt $4k$ solcher θ's liefert eine Modulform vom Gewicht $2k$. Die Fourierkoeffizienten liegen in $\mathbb{Z}\left[e^{i\pi/n^2}\right]$, so daß die entsprechenden Modulformen über diesem Ring definiert sind. (Bis auf $2\pi i$'s). Die Fourierkoeffizienten von $\theta(Z,\underline{a},\underline{b})$ werden parametrisiert durch $\chi \in \frac{1}{2n^2} B(X)^* = \frac{1}{2n^2} S^2(X)$. Sie sind verschieden von Null nur für χ von der Form

$$\{\tfrac{1}{2}(\underline{m}+a) \otimes (\underline{m}+\underline{a}), \quad \underline{m} \in X = \mathbb{Z}^g\}.$$

Es folgt, daß für einen Kegel $\sigma \subseteq B^+(X)$ der konstante Term eines Produktes von θ's verschwindet, außer wenn alle vorkommenden $\underline{a}$'S im Kern von $^1/n \cdot X/X \to {}^1/n\ X_\sigma/X_\sigma$ liegen. In diesem Fall ist der konstante Term wieder ein Produkt solcher θ's, mit $g-r_\sigma$ statt g.

Die $\theta(Z;\underline{a},\underline{b})$ hängen mit den θ-Nullwerten zusammen: Wir arbeiten von nun an in Charakteristik $\neq 2$. Beim a) unter 3.) konstruierten Isomorphismus $p_*(\underline{H}) \cong \underline{K}^{2^g}$, mit $\underline{K} \equiv \omega^{-1}$ modulo Torsion in $\mathrm{Pic}(\underline{A}_{g,2})$, entsprechen die $\theta(Z;\underline{a},\underline{b})$ mit $\underline{a},\underline{b} \in \frac{1}{2}X/X$ im geeignet zu definierenden Sinne einer Basis von $p_*(H)$. Entsprechendes gilt für die direkten Bilder $p_*(\underline{H}^{2^l})$.

Es ist nun bekannt, daß diese Basen $\underline{H}$ erzeugen, und daß die θ-Nullwerte sogar eine projektive Einbettung von A_g definieren (genauer gesagt des groben Modulraums). Da wir das Verhalten der θ-Reihen am Rande auch kennen, so folgt leicht:

<u>Satz 6</u>:

Wähle n, und sei $A = \bigoplus_{m\geq 0} \Gamma(\overline{A}_{g,n},\omega^m) \otimes \mathbb{Z}\left[^1/2\right]$. Dann wird für genügend großes m ω^m über $\overline{A}_{g,n} \otimes \mathbb{Z}\left[^1/2\right]$ von seinen globalen Schnitten erzeugt. Die dadurch definierte Abbildung

$$\overline{A}_{g,n} \otimes \mathbb{Z}\left[^1/2\right] \to \mathrm{Proj}(A)$$

hat als Bild ein projektives normales Schema $A^*_{g,n}$ über $\mathbb{Z}\left[e^{2\pi i/n},{}^1/2n\right]$. Sie definiert eine offene Einbettung des groben Modulraums zu $A_{g,n}$ in $A^*_{g,n}$, und das Komplement hat auf jeder Faser Dimension $g(g-1)/2$. Genauer hat das Bild jedes Stratums $\bar{A}_{n,\sigma}$ Dimension $(g-r_\sigma)(g-r_\sigma+1)/2$.

Das weitere Studium der arithmetischen Theorie der Siegel'schen Modulformen verdient sicher noch einige Aufmerksamkeit: Vermutlich gelten die obigen Resultate auch in Charakteristik zwei, und es sollte auch Anwendungen auf Kongruenzen geben. Dies würde aber wohl den Rahmen der hiesigen Ausführungen sprengen.

Eine weitere Verfolgung der Ansätze von L. Moret-Bailly ([MB]) scheint hier geboten.

§ 7 ETALE GARBEN

a) Durch die modulare Interpretation erhält man sofort étale Garben auf $\bar{A}_g$, nämlich die direkten Bilder $R^j p_*(\mathbb{Z}/m\mathbb{Z})$ $(p: G \to \bar{A}_g$ die universelle semiabelsche Varietät). Der Einfachheit halber formulieren wir die Aussagen für $\mathbb{Q}_l$-Garben, doch gelten entsprechende Varianten für $\mathbb{Z}_l$ oder $\mathbb{Z}/l^m\mathbb{Z}$. Außerdem nehmen wir an, daß unsere Kegelzerlegung glatt ist, so daß alle Strata $\bar{A}_{g,\sigma}$ glatt über $\mathbb{Z}$ sind.

Wähle eine Primzahl l und einen Level n. Wir arbeiten grundsätzlich über $Z\left[e^{2\pi i/n},{}^1/nl\right]$.

Auf $A_{g,n}$ ist die Garbe $R^1 p_*(\mathbb{Q}_l)$ lokal konstant und besitzt eine nicht ausgeartete symplektische Form mit Werten in $\mathbb{Q}_l(-1)$ ((-1) = Tate-Twist). Am Rand ist sie zahm verzweigt: $\bar{A}_{g,n} - A_{g,n}$ ist ein Divisor mit normalen Überkreuzungen. Seine irreduziblen Komponenten sind die Strata $\bar{A}_{g,n,\sigma}$ für $\sigma \subseteq B^+(X)_{\mathbb{R}}$ ein eindimensionaler Kegel der Zerlegung. (Für $n = 1$ entsprechen sie sogar eindeutig den Konjugationsklassen dieser σ unter $GL(X)$). Die Operation der zugehörigen Monodromie erhält man aus der Beschreibung der l-Torsionspunkte einer degenerierenden abelschen Varietät, die wir in § 3 (entsprechend [M4], 4.11) gegeben haben: Sei s_σ ein Erzeugendes der Halbgruppe $\langle\sigma\rangle \cap B^+(X)$. s_σ ist eine positiv definite symmetrische Bilinearform $s_\sigma : X_\sigma \times X_\sigma \to \mathbb{Z}$ und definiert ein unipotentes Element aus $Sp(2g,\mathbb{Z})$. (Wenn s_σ durch eine symmetrische Matrix S gegeben wird, ist die $\begin{pmatrix}1 & S\\ 0 & 1\end{pmatrix}$. Dieses Element liefert die gewünschte Monodromie-Transformation.

b) Da $R^1p^*(\mathbb{Q}_l)$ zahm verzweigt ist im Unendlichen, sind die direkten Bildgarben (über $\mathbb{Z}\left[e^{2\pi i/n}, {}^1/_{ln}\right]$) lokal konstant. (Siehe zum Beispiel [L]). Dasselbe gilt für aus $R^1p^*(\mathbb{Q}_l)$ abgeleitete Garben:

Satz 7:

Sei $\rho : Csp(2g,\mathbb{Q}) \to GL(V)$ eine algebraische Darstellung der Gruppe Csp der symplektischen Ähnlichkeiten auf einem endlich dimensionalen $\mathbb{Q}$-Vektorraum V . Für $n \geq 3$ bezeichne $\underline{F}_\rho$ die zugehörige étale $\mathbb{Q}_l$-Garbe auf $A_{g,n}$, $\psi : A_{g,n} \to Spek(\mathbb{Z}\left[e^{2\pi/n}, \frac{1}{n}\right])$ die Projektion. Dann sind alle direkten Bilder $R^q\psi_*(\underline{F}_\rho)$, $R^q\psi_!(\underline{F}_\rho)$ lokal konstant.

Bemerkung:

Es wäre wünschenswert,auch die Eichler-Shimura Relation zu verallgemeinern. Dies scheint jedoch sehr kompliziert zu sein. Auch hier ist noch ein weiteres Feld für zukünftige Untersuchungen.

§ 8 DIE TORELLI-ABBILDUNG

a) Zu jeder glatten Kurve vom Geschlecht g gehört kanonisch ihre Jacobische, eine prinzipal polarisierte abelsche Varietät der Dimension g . Wenn man die glatte Kurve in eine singuläre stabile Kurve degenerieren läßt, ergibt sich eine semiabelsche Varietät, und es liegt nahe, die Gegebenheiten der allgemeinen Theorie in § 2,3 hier näher zu beschreiben, Wir untersuchen dabei zugleich das Verhalten der Torelli-Abbildung $M_g \to A_g$ zwischen den Modulräumen am Rande.

b) Sei wieder R ein kompletter normaler lokaler Ring, K der Quotientenkörper, $C \to Spek(R)$ eine stabile Kurve, so daß die generische Faser C_η glatt ist. Zur speziellen Faser C_s konstruiert man einen

Graphen $\mathbb{G}$, dessen Ecken V den irreduziblen Komponenten von C_s entsprechen, und dessen Kanten E die singulären Punkte auf C_s parametrisieren. Eine Kante hat als Endpunkte die beiden Ecken, die den irreduziblen Komponenten entsprechen, auf denen der singuläre Punkt liegt.

Wir nehmen weiter an, daß alle irreduziblen Komponenten von C_s geometrisch irreduzibel sind und alle Doppelpunkte rational über dem Restklassenkörper k von R . Dies gilt immer, wenn k algebraisch abgeschlossen ist.

Jeder Kante $e \in E$ ordnet man ein Hauptideal $I_e \subsetneq R$ zu: Die Komplettierung $\hat{\mathcal{O}}_{C,e}$ des lokalen Rings im Punkt zu e ist isomorph zu $R[[S,T]]/(ST-f_e)$, und I_e werde erzeugt von f_e .

e) Sei $\Gamma = \pi_1(\mathbb{G})$ die Fundamentalgruppe des Graphen $\mathbb{G}$, und $\tilde{\mathbb{G}} \longrightarrow \mathbb{G}$ die universelle Überlagerung, mit Gruppe Γ . Dann gibt es eine Überlagerung formaler Schemata $\hat{\tilde{C}} \to \hat{C}$, ebenfalls mit Gruppe Γ .

Sei $X = \Gamma^{ab} = H_1(\mathbb{G},\mathbb{Z})$, $Y = X^* = H^1(\mathbb{G},\mathbb{Z})$, und T der zerfallende Torus mit Charaktergruppe X . Die Elemente aus $T(R)$ entsprechen den Homomorphismen $\Gamma \to R^*$. Jeden solchen Homomorphismus kann man benutzen, um eine äquivariante Operation von Γ auf dem trivialen Geradenbündel $\mathcal{O}_{\hat{\tilde{C}}}$ zu definieren, und damit ein Geradenbündel auf $\hat{C}$ oder auch C .

Da $\Gamma(\hat{\tilde{C}},\mathcal{O}_{\hat{\tilde{C}}}) = R$, erhält man so einen Isomorphismus

$$T(R) \xrightarrow{\sim} \mathrm{Kern}(\mathrm{Pic}(C) \longrightarrow \mathrm{Pic}(\hat{\tilde{C}}))$$

c) Sei $G = \mathrm{Pic}^0(C/R)$. Dann ist G eine semiabelsche Varietät über R , und der eben angegebene Isomorphismus stammt aus einer exakten Sequenz

$$0 \to \hat{T} \to \hat{G} \to \hat{A} \to 0 \quad ,$$

mit A abelsch über R . Sei wieder $\tilde{G}$ die entsprechende Erweiterung

$$0 \to T \to \tilde{G} \to A \to 0 \quad .$$

Es ist $\tilde{G}(R) = G(R)$. Man erhält auch eine Abbildung $T(K)\tilde{G}(R) \to \mathrm{Pic}^0(C_\eta)$: Es reicht, die Restriktion $T(K) \to \mathrm{Pic}^0(C_\eta)$ anzugeben. Ein Element aus $\Gamma(K)$ wird gegeben durch einen Homomorphismus $\Gamma \to K^*$, welcher eine äquivariante Operation von Γ auf $\mathcal{O}_{\hat{C}} \otimes_R K$ definiert. Wähle ein Γ-invariantes gebrochenes kohärentes Ideal $J \subseteq \mathcal{O}_{\hat{C}} \otimes_R K$. Dies ergibt eine kohärente torsionsfreie Garbe auf $\hat{C}$ oder auch C , und auf der generischen Faser ein Geradenbündel vom Grad 0 .

Wir definieren nun eine Abbildung von X in den Kern des obigen Morphismus: Für $e \in E$ sei $\mathcal{O}_{C,e}$ der zugehörige lokale Ring, und f_e ein erzeugendes Element von I_e . Ein $x \in X$ wird gegeben durch ganze Zahlen $x_e \in \mathbb{Z}$, für alle $e \in E$, so daß für alle $p \in V$ $\sum_{e \to p} \pm x_e = 0$. Dabei wählt man eine Orientierung aller $e \in E$, und die Summe geht über alle e's mit Anfangs-oder Endpunkt p , wobei das Vorzeichen je nach Orientierung zu wählen ist. Wir definieren dann eine symmetrische Bilinearform

$$b : X \times X \subset K^*$$

durch

$$b(x,y) = \prod_{e \in E} f_e^{x_e y_e} \quad .$$

Dann ist $b(x,x) \in \underline{m}$ = maximales Ideal R , falls $x \neq 0$. $b(\ )$ entspricht einem Homomorphismus

$$b : X \to T(K) \quad .$$

Es gilt nun

<u>Satz 8</u>:

Es sei $\mathrm{Char}(K) \neq 2$

i) Es gibt einen Homomorphismus $\underline{c} : X \to \tilde{G}(R) = \mathrm{Pic}^0(C)$, so daß für alle $x \in X$ $b(x)\,\underline{c}(x) \in \tilde{G}(K)$ das triviale Geradenbündel auf C_η definiert.

ii) Sei $\underline{M}$ ein amples Geradenbündel auf A , welches die prinzipale Polarisation definiert. Dann ist die Zusammensetzung $X \xrightarrow{\underline{c}} \tilde{G}(R) \to A(R)$ die zur Erweiterung $\tilde{G}$ von A durch T

gehörige Abbildung. Ihre Liftung via $\underline{c}$ definiert ein zulässiges System von Isomorphismen für $(\tilde{G},A,\underline{M})$.

iii) G ist isomorph zu der semiabelschen Varietät, welche die Mumford-Konstruktion mit den obigen Daten liefert.

Beweis:

i) Sei $x \in X$. $b(x,)$ definiert eine Darstellung $\rho : \Gamma \to X \to K^*$.

Wir konstruieren ein ρ-invariantes gebrochenes Ideal $\underline{J} \subset \mathcal{O}_{\hat{C}} \otimes K$, welches lokal prinzipal ist, und dessen Einschränkung auf jede Komponente von $\hat{C}_s$ den Grad Null hat. Die Komponenten C_p von $\hat{C}_s$ werden parametrisiert durch $p \in \tilde{\mathcal{V}}$ = Ecken $(\tilde{\mathfrak{G}})$. Wir wählen $\underline{J}$ so, daß es im Inneren von C_p von einem $g_p \in K^*$ erzeugt wird, und daß für eine Seite $e \in \tilde{\mathcal{E}}$, welche p_1 und p_2 verbindet, $g_{p_2} = f_e^{x_e} g_{p_1}$. Es gibt sicher solche g_p's , und das Ideal $\underline{J}$ ist auch ρ-invariant. Es bleibt zu zeigen, daß man $\underline{J}$ auch in den Doppelpunkten lokal prinzipal wählen kann: Sei $e \in \tilde{\mathcal{E}}$, und betrachte für $n \geq 1$ den Ring $(\mathcal{O}_{\hat{C},e}) \otimes (R/_{\underline{m}}n)$. Er besitzt zwei minimale Primideale $\underline{p}_1$ und $\underline{p}_2$, entsprechend $p_1, p_2 \in \nu$, welche durch e verbunden werden. $\underline{p}_1$ und $\underline{p}_2$ sind Hauptideale, und $\underline{p}_1 \cdot \underline{p}_2 = (f_e)$.

In der Komplettierung wird

$$\mathcal{O}_{\hat{C},e} \cong R[[S,T]]/(ST-f_e) ,$$

und $\hat{\underline{p}}_1$ und $\hat{\underline{p}}_2$ werden durch S bzw. T erzeugt.

Man kann dann $\underline{J}$ so wählen, daß es in $\mathcal{O}_{\hat{C},v} \otimes (R/_{\underline{m}}n)$ isomorph wird zu einer Potenz von $\underline{p}_1$ oder $\underline{p}_2$ (etwa zu $\underline{p}_1^{x_v}$) , und es folgt die Behauptung. Der Grad von $\underline{J}$ auf jedem C_p ist Null, da $\sum_{e \to p} \pm x_e = 0$.

Damit ist Teil i) bewiesen. Die Abbildung $\underline{c}$ ist natürlich eindeutig.

d) Es folgt schon aus den Sätzen 1 und 2, daß man G durch die Mumford-Konstruktion mit Hilfe einer Bilinearform b^* aus $\tilde{G}$ erhält. Wir müssen nun noch nachweisen, daß $\underline{c}$ ein zulässiges System von Isomorphismen definiert, und daß dann $b = b^*$. Man bettet R in einen diskreten Bewertungsring ein, und reduziert sich damit auf den Fall, daß R schon ein solcher ist. Dann ist das Problem für den Fall einer voll degenerierenden Kurve schon in [MD] behandelt worden, und wir folgen den dortigen Ausführungen: Wir können $C, \tilde{C}$, G und $\tilde{G}$ als rigid-analytische Objekte über K auffassen. Es ist dann $C = \tilde{C}/\Gamma$, und $\tilde{G}/i(X)$ ist rigid-analytische abelsche Varietät $(i = b \cdot c : X \to \tilde{G}(K))$. Wir definieren zunächst eine rigid-analytische Abbildung $\phi : C \times C \to \tilde{G}/i(X)$, via $\tilde{\phi} : \tilde{C} \times \tilde{C} \to \tilde{G}$. Dazu müssen wir für jeden endlichen Erweiterungskörper L von K eine Abbildung $\tilde{\phi} : \tilde{C}(L) \times \tilde{C}(L) \to \tilde{G}(L)$ definieren. Da alle unsere Konstruktionen invariant unter Grundkörpererweiterung sein werden, reicht es, $\tilde{\phi}$ auf $\tilde{C}(K) \times \tilde{C}(K)$ zu definieren. Dazu ersetzen wir zunächst C und $\tilde{C}$ durch ihre regulären semistabilen Modelle über R. Dies ändert nichts an allen Definitionen und Behauptungen. Dann ist $\tilde{C}(K) = \tilde{C}(R)$. Seien also $z_1, z_2 \in C(R)$ zwei Punkte. Der Divisor $D = \sum_{\gamma \in \Gamma} \gamma((z_1) - (z_2))$ ist dann Γ-invariant auf $\tilde{C}$, doch hat die Einschränkung von $\mathcal{O}(D)$ auf die irreduziblen Komponenten C_p im allgemeinen nicht den Grad Null. Dies wird nun korrigiert:

e) Wähle ein p_0, und definiere eine Abbildung

$$f : p \longmapsto f(p_0, p)$$
$$V \longrightarrow K^*$$

wie folgt: Orientiere Γ-invariant die Kanten E. Es gilt:

i) $f(p_0, p_0) = 1$

ii) Wenn der kürzeste Weg in $\mathfrak{G}$ von p nach p_0 der Komponente als nächste Ecke p_1 trifft, so sei

$f(p_0, p) = f(p_0, p_1)$ wenn die Orientierung der Kante e zwischen p und p_1 so ist, daß p Anfangs- und p_1 Endpunkt ist.

$f(p_0, p) = f(p_0, p_1) f_e$, bei anderer Orientierung.

Definiere

$$g(p_1, p_2; p) = \prod_{\gamma \in \Gamma} \left(\frac{f(\gamma p_1, p)\ f(\gamma p_2, p_0)}{f(\gamma p_2, p)\ f(\gamma p_1, p_0)} \right) \in K^* \quad .$$

Dabei seien $p_1, p_2 \in \tilde{V}$. Im Produkt ist ein Faktor nur dann verschieden von Eins, wenn die kürzesten Verbindungen (in $\tilde{g}$) von γp_1 und γp_2 mit dem Weg $\overrightarrow{p_0 p}$ verschiedene Fußpunkte auf $\overrightarrow{p_0 p}$ haben. Dies gilt aber nur für endlich viele γ's . (Die Distanz von p_1 nach p_2 muß größer sein als Konstante + Distanz $(p_0, \gamma p_1)$) .

Es ist $g(p_1,p_3;p) = g(p_1,p_2;p)\, g(p_2,p_3;p)$.

Wenn man von p zu einem benachbarten $p' \in \tilde{V}$ übergeht, und e die Kante zwischen p und p' bezeichnet, mit p=Anfangspunkt(e), und p'=Endpunkt(e), so ist

$$\frac{g(p_1,p_2,p')}{g(p_1,p_2,p)} = \prod_{\gamma\in\Gamma} \frac{f(\gamma p_1,p')\, f(\gamma p_2,p)}{f(\gamma p_2,p')\, f(\gamma p_1,p)} .$$

Die Faktoren sind verschieden von Eins nur dann, wenn e auf dem Weg von γp_1 nach γp_2 liegt, und zwar erhält man dann f_e , wenn die Orientierung von e mit der des Weges übereinstimmt, sonst f_e^{-1} .
Es folgt:

i) Sei $\delta \in \Gamma$, entsprechend $x = (x_e) \in X$.

Dann ist

$$\frac{f(p_1,p_2;\delta(p))}{f(p_1,p_2;p)} = \prod_{e\in\overrightarrow{p_1 p_2}} f_e^{\pm x_e} .$$

Der Exponent ist $+1$, wenn die Orientierungen von e und $\overrightarrow{p_1 p_2}$ übereinstimmen, sonst -1 .

ii) Sei $\underline{J}(p_1,p_2) \subseteq \mathcal{O}_{\hat{C}} \otimes K$ das invertierbare Ideal, welches auf C_p von $f(p_1,p_2;p)$ erzeugt wird. (Die Existenz folgt ähnlich wie bei der Konstruktion von $\underline{c}$) .
Dann ist

$$\mathrm{grad}(\underline{J}(p_1,p_2) / C_p) =$$

$$\begin{cases} 0 , & \text{falls } p \notin \Gamma\cdot p_1 \cup \Gamma\cdot p_2 , \text{ oder } p \in \Gamma\cdot p_1 \cap \Gamma\cdot p_2 \\ -1 , & \text{falls } p \in \Gamma\cdot p_1 , p \notin \Gamma p_2 \\ +1 , & \text{falls } p \in \Gamma\cdot p_2 , p \notin \Gamma p_1 \end{cases} .$$

Somit hat für $z_1 \in C_{p_1}$, $z_2 \in C_{p_2}$ die Einschränkung von $\underline{L}(z_1,z_2) = \underline{J}(p_1,p_2) \otimes \mathcal{O}(D)$ Grad 0 auf allen Komponenten. Für $\delta \in \Gamma$, entsprechend $s = (x_e) \in X$, ist

$$\delta^*(\underline{L}(z_1,z_2)) = (\prod_{e \in \overrightarrow{p_1 p_2}} f_e^{\pm x_e}(p_1,p_2))\underline{L}(z_1,z_2) \quad .$$

Wenn man also einen Morphismus

$$\rho : \Gamma \to X \to K^*$$

definiert durch $\rho(\delta) = \prod_{e \in \overrightarrow{p_1 p_2}} f_e^{\pm x_e}$, entsprechend einem $\rho \in T(K)$, so kann man Γ via ρ äquivariant operieren lassen und erhält ein Geradenbündel aus $G(R) = \mathrm{Pic}^0(C)$. Wenn $p_1 = p_2$ ist, so wird dieses Geradenbündel durch den Divisor $z_1 - z_2$ auf C gegeben. Wenn $z_2 = \gamma(z_1)$ mit $\gamma \in \Gamma$, entsprechend $y = (y_e) \in X$, so erhält man das Geradenbündel zu $\underline{c}(y) \in \tilde{G}(R)$. Wenn man das obige Element aus $G(R)$ noch mit $\rho \in T(K)$ multipliziert, ergibt sich schließlich eine Abbildung

$$\tilde{\phi} : \tilde{C}(R) \times \tilde{C}(R) \to T(K)\tilde{G}(R) = \tilde{G}(K) \quad .$$

mit

i) $\tilde{\phi}(z_1,z_2) + \tilde{\phi}(z_2,z_3) = \tilde{\phi}(z_1,z_3)$

ii) $\tilde{\phi}(z_1,\gamma(z_1)) = i(y)$ $\quad (y = \mathrm{Bild}(\gamma) \in X)$

iii) Wenn z_1, z_2 in derselben Komponente liegen, so ist $\tilde{\phi}(z_1,z_2) \in \tilde{G}(R) = \mathrm{Pic}^0(C)$ gegeben durch $\mathcal{O}(z_1 - z_2)$.

f) Außerdem ist $\tilde{\phi}$ verträglich mit Erweiterungen des Grundkörpers. Andererseits hat man über K eine kanonische Abbildung

$$\phi : C_\eta \times C_\eta \to G_\eta = \mathrm{Pic}^0(C_\eta)$$
$$(z_1,z_2) \longmapsto \mathcal{O}(z_1 - z_2) \quad .$$

Es ist G der rigid-analytische Quotient $\tilde{G}/_{i^*(X)}$, mit einer Gruppe von Perioden $i^*(X) \subseteq \tilde{G}(K)$. Es ergibt sich dann eine rigid-analytische Abbildung der universellen Überlagerungen

$$\tilde{C} \times \tilde{C} \to \tilde{G} \quad .$$

Da diese Abbildung mit $\tilde{\phi}$ nahe der Diagonale übereinstimmt, ist sie nach dem Identitätssatz gleich $\tilde{\phi}$.

Wir erhalten also ein kommutatives Diagramm

$$\begin{array}{ccc} \tilde{C} \times \tilde{C} & \xrightarrow{\phi} & \tilde{G} \\ \downarrow{\scriptstyle \Gamma\times\Gamma} & & \downarrow{\scriptstyle i^*(X)} \\ C \times C & \xrightarrow{\phi} & G \end{array} \quad .$$

Da $\tilde{\phi}(z_1, \delta z_1) \in i(X)$, ist $i(X) \subseteq i^*(X)$. Aus einer Betrachtung der Bewertung folgt, daß $i(X)$ endlichen Index in $i^*(X)$ hat. ϕ faktorisiert dann über $\tilde{G}/i(X)$. Dies ist eine endliche Überlagerung von G , damit algebraisch, und wegen der bekannten Eigenschaften der Jacobi'schen ist notwendigerweise $i(X) = i^*(X)$. Wir müssen nun noch zeigen, daß $i = i^*$. Auf jeden Fall stimmen sie schon überein bis auf einen Automorphismus von X . Da i^* ebenso wie i von einer definiten symmetrischen Bilinearform stammt, wird dieser Isomorphismus in einer geeigneten Basis durch eine positiv definite symmetrische Matrix definiert.

g) Alles in allem haben wir den Satz 8 bewiesen bis auf die Tatsache, daß man für ii) und iii) i durch einen Automorphismus von X abändern muß. Wir wollen zeigen, daß dieser Automorphismus die Identität ist.

Es reicht, dies im "universellen" Fall zu tun, das heißt,wenn R die Basis einer versellen Deformation der speziellen Faser C_s ist. Dann ist R regulär, und die f_e für $e \in E$ bilden einen Teil eines regulären Parametersystems. Für jedes e sei $\mathfrak{p}_e \subseteq R$ das zugehörige Primideal der Höhe 1 und G_e die Faser von G über $k(\mathfrak{p}_e)$. Dann ist G_e semiabelsch, mit einem Torusteil der Dimension ≤ 1 . Die Dimension ist genau dann gleich 1 , wenn die Linearform $x = (x_e) \longmapsto x_e$ auf X nicht verschwindet. Falls dies der Fall ist, so besteht der Kern dieser linearen Abbildung aus den x , die auf dem Torusteil von G_e verschwinden. (Der Torusteil von G_e ist in natürlicher Weise ein

Untertorus von T). Aus der Mumford-Konstruktion folgt, daß dies genau dann für x zutrifft, wenn $i^*(x)$ ganz ist in $R_{\mathfrak{p}_e}$.

Wenn $\lambda: X \xrightarrow{\sim} X$ der Automorphismus mit $i^* = i \circ \lambda$ ist, so gilt dann

$$x_e = 0 \Leftrightarrow \lambda(x)_e = 0$$

λ respektiert also alle Hyperebenen $\{x_e = 0\}$, kann trigonalisiert werden und hat somit Eigenwerte ± 1 . Da man λ aber auch durch eine symmetrische positiv definite Matrix darstellen kann, ist $\lambda = \mathrm{id}$. Dies beendet den Beweis von Satz 8.

Wir erhalten auch Informationen über das Verhalten der Abbildung $M_g \to A_g$ am Rande: Bei vorgegebenem g gibt es nur endlich viele Möglichkeiten für den Graphen $\mathbb{G}$ einer stabilen Kurve vom Geschlecht g . Zu jedem solchen $\mathbb{G}$ erhält man symmetrische Bilinearformen $(x,y) \to x_e y_e$ auf $X = H_1(\mathbb{G},Z)$ für $e \in E$. Dann bildet sich die verselle Deformation einer stabilen Kurve mit Graph $\mathbb{G}$ genau dann in die mit der Kegelzerlegung $\{\sigma\}$ definierte Kompaktifizierung $\bar{A}_g$ ab, wenn es ein σ gibt, welches alle Bilinearformen $x_e y_e$ auf X enthält. Wenn dies nicht der Fall ist, so muß man zunächst noch die verselle Deformation durch eine Modifikation ersetzen.

§ 9 DIE KOMPLEXE THEORIE

a) Nach Basiserweiterung zu den komplexen Zahlen erhält man aus $\bar{A}_g$, $\underline{S}$ u.s.w. komplexe Räume. Es ergeben sich die toroidalen Kompaktifizierungen aus [AMRT] oder auch [N] . Da sich auch die Mumford-Konstruktion ins analytische übersetzt, kann man auch die komplex-analytische Version der universellen semiabelschen Varietät beschreiben:

Der Torus S wird über $\mathbb{C}$ gegeben durch

$$S(\mathbb{C}) = \mathbb{C}^{g(g+1)/2} / \mathbb{Z}^{g(g+1)/2} = B(\mathbb{C}^g)/B(\mathbb{Z}^g) \quad .$$

Wenn $\mathbb{H}_g \subset \mathbb{C}^{g(g+1)/2}$ die Siegel'sche Halbebene bezeichnet ($\mathbb{H}_g = \{Z = X+iY / Z = {}^tZ,\ Y > 0\}$) , so erhält $S(\mathbb{C})$ als offene

Teilmenge $\mathbb{H}_g/B(\mathbb{Z}^g)$.

Wenn $\sigma\subset B^+(\mathbb{R}^g)$ ein konvexer rationaler Polyeder ist, so bestimmt σ eine Torus-Einbettung $S\subseteq S_\sigma$. S_σ ist affin algebraisch, und eine Basis des Ringes der algebraischen Funktionen wird gegeben durch

$$\chi_M(Z) = e^{2\pi i \, \mathrm{Spur}(MZ)} ,$$

M eine halbganze symmetrische Matrix, mit ganzen Diagonalelementen, welche im Dual $\sigma^\vee$ liegt $(Y\in\sigma \Rightarrow \mathrm{Sp}(MY)\geq 0)$. Weiter existiert auf S eine "universelle" Bilinearform

$$b : \mathbb{Z}^g \times \mathbb{Z}^g \to \mathcal{O}_S$$
$$b(x,y) = \chi_M ,$$

wobei M die Matrix ist mit Einträgen $m_{jk} = \frac{1}{2}(x_j y_k + x_k y_j)$. b(x,x) setzt sich fort zu einer regulären Funktion auf S_σ .

Sei $\underline{X}\subseteq\mathbb{Z}^g$ die étale Untergruppe, deren Faser über $s\in S_\sigma$ aus den $x\in\mathbb{Z}^e$ besteht mit $b(x,x)(s)\neq 0$. Dann setzt sich b fort zu einer Bilinearform

$$b : \underline{X}\times\mathbb{Z}^g \to \mathcal{O}^*_{S_\sigma} ,$$

und definiert damit $b : \underline{X}\to(\mathbb{C}^*)^g = T$. b ist über $\mathbb{H}_g/B(\mathbb{Z}^g)$ eine Einbettung, und der Quotient $G = T/b(\underline{X})$ ist eine semiabelsche Varietät über einem offenen Stück von S_σ . Dort ist S_σ lokal isomorph zu $\overline{A}_g(\mathbb{C})$, und G liefert die universelle semiabelsche Varietät.

b) Das obige G besitzt ein Geradenbündel $\underline{L}$, welches über $\mathbb{H}_g/B(\mathbb{Z}^g)$ eine Polarisation definiert. Das Pullback von $\underline{L}$ nach T ist kanonisch trivial, und ein globaler Schnitt von $\underline{L}$ wird geliefert durch die θ-Reihe auf T :

$$\theta(e^{2\pi i z_1},\dots,e^{2\pi i z_g}) = \\ = \sum_{\underline{m}=(m_1,\dots,m_g)\in\mathbb{Z}^g} e^{i\pi(\underline{m}Z{}^t\underline{m}+\sum_{j=1}^g m_j Z_{jj})} e^{2\pi i \underline{m}{}^t\underline{z}}$$

b ist die zu den Koeffizienten von θ gehörige Bilinearform (im Sinne von § 2).

c) Das Quadratintegral von g-Formen liefert eine kanonische hermite'-sche Metrik auf $\omega = \Lambda^g \underline{t}^*_G$:

$$||\alpha||^2(s) = \int_{G_s} ||\alpha||^2 .$$

Man rechnet aus, daß für $Z = X+iY \in \mathbb{H}_g$, entsprechend $s \in S$,

$$\left|\left|\frac{dz_1}{z_1} \wedge \ldots \wedge \frac{dz_g}{z_g}\right|\right|^2$$

bis auf einen konstanten Faktor gegeben ist durch $\det(Y)$. Da die Einträge von Y sich aus den Logarithmen der absoluten Beträge der χ_M berechnen, hat die Metrik auf ω_G am Rande nur eine logarithmische Singularität.

ANHANG:

Dieses Manuskript gibt den Kenntnisstand zur Zeit der Arbeitstagung wieder (Juni 1984). Inzwischen (September 1984) gab es die folgenden Entwicklungen:

1.) Die Thesis von C.L. Chai liegt mir vor.

2.) Die minimale Kompaktifizierung läßt sich auch in Charakteristik 2 behandeln:

Betrachte $A_{g,n}$ über $\mathbb{Z}\left[{}^1/n, e^{2\pi i/n}\right]$.

Sei $R = \bigoplus_{m \geq 0} \Gamma(\overline{A}_{g,n}, \omega^{\otimes m})$, $A^*_{g,n} = \mathrm{Proj}(R)$.

Dann gilt:

i) Eine geeignete Potenz von ω wird von globalen Schnitten erzeugt, so daß man eine Abbildung des groben Modulraums (zu $\overline{A}_{g,n}$) $\overline{A}_{g,n}$ nach $A^*_{g,n}$ erhält:

$$\phi : \overline{A}_{g,n} \to A^*_{g,n}$$

ii) $\phi | A_{g,n}$ ist eine offene Einbettung, und $A_{g,n} = \phi^{-1}(\phi(A_{g,n}))$

iii) $A^*_{g,n} - A_{g,n}$ hat Kodimension g in $A^*_{g,n}$.

Der Beweis benutzt θ-Funktionen. Einige Andeutungen: Man betrachtet die Überlagerung $M \to A_g$, welche die symmetrischen Geradenbündel in der Polarisationsklasse liefert (nicht zu verwechseln mit level-2-Strukturen) Für m ungerade (der Einfachheit halber) erhält man dann ein Geradenbündel $\underline{L}_m$ auf $M \times_{A_g} A_{g,m}$, welches für $m \geq 3$ von seinen globalen Schnitten erzeugt wird, und so daß $\underline{L}_m = \frac{m}{2} \cdot \omega$ in $\mathrm{Pic} \otimes \mathbb{Q}$.

Dies liefert globale Erzeugtheit über A_g . Man muß nun noch den Rand betrachten, sowie zeigen, daß die Fasern von $\phi : \overline{A}_g \to A^*_g$ endlich über A_g sind.

Literatur

[A] M. Artin, Algebraization of Formal Moduli I,in Global Analysis, papers in honor of K. Kodaira, 21 - 71, Princeton University Press, Princeton 1969.

[AMRT] A. Ash, D. Mumford, M. Rapoport, Y. Tai, Smooth Compactification of Locally Symmetric Varieties, Math. Sci. Press, Brookline 1975.

[C] C.-L. Chai, Thesis, Harvard 1984.

[DM] P. Deligne, D. Mumford, The irreducibility of the space of curves of a given genus, Publ. Math. IHES 36 (1969), 75 - 110.

[DR] P. Deligne, M. Rapoport, Les schémas de modules de courbes elliptiques, Springer Lecture Notes 349 (1973) 143 - 316.

[F] G. Faltings, Endlichkeitssätze für abelsche Varietäten über Zahlkörpern, Invent. Math. 73 (1983), 349 - 366.

[SGA7] A. Grothendieck, Groupes de monodromie en Géometrie Algebrique (SGA 7 I), Springer Lecture Notes 288 (1972).

[KKMS] G. Kempf, F. Knudsen, D. Mumford, B. Saint-Donat, Toroidal Embeddings I, Springer Lecture Notes 399 (1972).

[L] G. Laumon, Semi-Continuité du Conducteur de Swan (d'après P. Deligne) Séminaire E.N.S. (1978/79), Exposé 9.

[MD] Yu. Manin, V. Drinfeld, Periods of p-adic Schottky groups, Crelles Journal 262 (1973), 239 - 247.

[MB] L. Moret-Bailly, Familles de varietés abéliennes,Thèse, Orsay (1984).

[M1] D. Mumford, Geometric Invariant Theory, Springer-Verlag, Heidelberg 1965.

[M2] D. Mumford, On the Equations Defining Abelian Varieties, Invent. Math. 1 (1966), 287 - 354.

[M3] D. Mumford, An Analytic Construction of Degenerating Curves over Complete Local Rings, Comp. Math. 24 (1972), 129 - 174.

[M4] D. Mumford, An Analytic Construction of Degenerating Abelian Varieties over Complete Rings, Comp. Math. 24 (1972), 239 - 272.

[N] Y. Namikawa, Toroidal compactification of Siegel spaces, Springer Lecture Notes 812 (1980).

THE SCHOTTKY PROBLEM

Gerard van der Geer

By associating to a (smooth irreducible) curve C of genus $g > 0$ its Jacobian $\mathrm{Jac}(C)$ one obtains a morphism $M_g \to A_g$ from the moduli space of curves of genus g to the moduli space of principally polarized Abelian varieties of dimension g. A well-known theorem of Torelli says that this morphism is injective. The image of M_g in A_g is not closed, it is only closed inside A_g^0, the set of points of A_g that correspond to indecomposable principally polarized abelian varieties (i.e. that are not products). For $g=1,2,3$ the closure of the image of M_g equals A_g. Since $\dim A_g = g(g+1)/2$, $\dim M_g = 3g-3$ (for $g > 1$) one sees that for $g > 3$ $\dim A_g > \dim M_g$, and so the question arises how we can characterize the image of M_g in A_g. This question goes back to Riemann, but is usually called Schottky's problem.

In "Curves and their Jacobians" Mumford treats the Schottky problem and the closely related question how to distinguish Jacobians from general principally polarized abelian varieties. In his review of the situation at that moment (1975) he describes four approaches and their merits. He concludes that none of these seems to him a definitive solution. In the meantime the situation has changed a lot. Some of the approaches have been worked out more completely, while new and successfull approaches have appeared. This paper deals with them. I hope to convince the reader that Mumford's statement that problems in this corner of nature are subtle and worthy of his time still very much holds true.

The ingredients.

To begin with, some standard notations.

$\mathbb{H}_g$: the Siegel upper half space of degree g,

$\Gamma_g = Sp(2g, \mathbb{Z})$ the symplectic group acting on $\mathbb{H}_g$,

$\Gamma_g(n,2n) = \{ \begin{pmatrix} A & B \\ C & D \end{pmatrix} \in \Gamma_g : \begin{matrix} A \equiv D \equiv 1 \pmod{n} \\ C \equiv B \equiv 0 \pmod{n} \end{matrix}, \mathrm{diag}\,{}^tAC \equiv \mathrm{diag}\,{}^tBD \equiv 0 \pmod{2n} \}$

$A_g = \Gamma_g \backslash \mathbb{H}_g$, the moduli space of principally polarized abelian varieties of dimension g over $\mathbb{C}$,

$A_g(n,2n) = \Gamma_g(n,2n)\backslash \mathbb{H}_g$, a Galois cover of A_g.

If X is a principally polarized abelian variety over $\mathbb{C}$ we denote by L_X (or simply L) a symmetric invertible ample sheaf of degree 1 defining the polarization and by Θ the divisor of a non-zero section of L_X. We put $X_n = \{x \in X : nx = 0\}$. If $X = \mathbb{C}^g/\mathbb{Z}^g + \tau\mathbb{Z}^g$ ($\tau \in \mathbb{H}_g$) as a complex torus then we write $X = X_\tau$.

The space $\Gamma(X, L_X^{\otimes 2})$ has dimension 2^g. A basis is defined by the functions

$$\Theta_2[\sigma](\tau,z) = \sum_{m \in \mathbb{Z}^g} \exp 2\pi i({}^t(m+\tfrac{\sigma}{2})\tau(m+\tfrac{\sigma}{2}) + 2(m+\tfrac{\sigma}{2})z)$$

$$z \in \mathbb{C}^g,\ \sigma \in (\mathbb{Z}^g/2\mathbb{Z}^g)$$

Here σ is viewed as a vector of length g with zeroes and ones as entries. A different set of generators of $\Gamma(X, L_X^{\otimes 2})$ is given by the squares of

$$\theta[{}^{\varepsilon}_{\varepsilon'}](\tau,z) = \sum_{m \in \mathbb{Z}^g} \exp \pi i({}^t(m+\tfrac{\varepsilon}{2})\tau(m+\tfrac{\varepsilon}{2}) + 2(m+\tfrac{\varepsilon}{2})(z+\tfrac{\varepsilon'}{2}))$$

with $\varepsilon,\varepsilon' \in (\mathbb{Z}/2\mathbb{Z})^g$, ${}^t\varepsilon\varepsilon' \equiv 0 \pmod 2$. These are related by

$$\theta^2[{}^{\varepsilon}_{\varepsilon'}](\tau,z) = \sum_{\sigma} \langle\sigma,\varepsilon'\rangle \Theta_2[\sigma+\varepsilon](\tau,0)\, \Theta_2[\sigma](\tau,z) \qquad (1)$$

$$\langle\sigma,\varepsilon'\rangle = \exp\pi i\, {}^t\sigma\varepsilon'$$

We call a principally polarized abelian variety indecomposable if it is not a product of two principally polarized abelian varieties, i.e. if its theta divisor is irreducible.

The functions $\Theta_2[\sigma](\tau,z)$ define for $X = X_\tau$ a morphism

$$\Phi_X : X \to \mathbb{P}^N \qquad N = 2^g - 1$$
$$z \to (\ldots, \Theta_2[\sigma](\tau, z), \ldots) = \vec{\Theta}_2(\tau, z)$$

which factors through $z \to -z$ and is of degree 2 for indecomposable X. The image is the Kummer variety of X. By taking $z=0$ and varying X we get a morphism

$$\Phi : A_g(2,4) \to \mathbb{P}^N$$
$$\tau \to (\ldots, \Theta_2[\sigma](\tau, 0), \ldots) = \vec{\Theta}_2(\tau, 0)$$

which is generically of degree 1. We also define

$$\Psi : A_g(2,4) \to \mathbb{P}^M \qquad M = 2^{g-1}(2^g+1)-1$$
$$\tau \to (\ldots, \theta^2[{}^{\varepsilon}_{\varepsilon}](\tau, 0), \ldots).$$

Φ and Ψ are connected by the special Veronese V defined by (1) :

$$\begin{array}{ccc} & \overset{\Phi}{\nearrow} & \mathbb{P}^N \\ A_g(2,4) & & \downarrow V \\ & \underset{\Psi}{\searrow} & \mathbb{P}^M \end{array}$$

The morphisms Φ and Ψ can be extended to morphisms of the Satake compactification $\overline{A}_g(2,4)$ of $A_g(2,4)$.

The functions $\Theta_2[\sigma]$ satisfy the differential equations

$$\frac{\partial^2}{\partial z_i \partial z_j} \Theta_2[\sigma] = 4\pi i\,(1+\delta_{ij}) \frac{\partial}{\partial \tau_{ij}} \Theta_2[\sigma], \qquad 1 \le i,j \le g.$$

(δ_{ij}: Kronecker δ)

which are called the Heat Equations.

If M_g is the moduli space of curves of genus g then the map $M_g \to A_g$ defined by $C \to \mathrm{Jac}(C)$ is injective. The closure of the image in A_g (or $\overline{A}_g$) is called the Jacobian locus. Notation : J_g.

APPROACH 1 : ALGEBRAIC EQUATIONS.

This is Schottky's original approach for characterizing the Jacobian locus. It is based on the construction of Prym varieties. For an excellent treatment of Prym varieties, see Mumford [13].

Suppose we start with a curve C of genus g and a non-zero point η of order 2 on $J = \mathrm{Jac}(C)$. This determines an unramified covering

$\pi: \tilde{C} \to C$ of degree 2 and an induced map $Nm : \tilde{J} = Jac(\tilde{C}) \to J$ and gives us a diagram

$$\begin{array}{ccc} \tilde{J} & \overset{\tilde{\lambda}}{\to} & \tilde{J} \\ \phi \uparrow & & \downarrow \hat{\phi} = Nm \\ J & \underset{2\lambda}{\to} & J \end{array} \qquad (2)$$

where $\phi = \pi^*$, $\hat{}$ denotes transpose and $\tilde{\lambda},\lambda$ are the principal polarizations. One defines the Prym variety of $\pi: \tilde{C} \to C$ as the identity component of the kernel of Nm : $P = (\ker Nm)^o$. It is an abelian variety of dimension g. Mumford showed that from a diagram (2) it follows that there exist a symplectic isomorphism $H_1/\{0,\eta\} \to P_2$ with $H_1 = \{\alpha \in J_2 : e_2(\alpha,\eta) = 1\}$ (e_2: Weil-pairing) such that

$$\tilde{J} = J\times P/\{(\alpha,\psi(\alpha)) : \alpha \in H_1\} .$$

Let $\sigma: J\times P \to \tilde{J}$ be the natural isogeny. Then the polarization $\begin{pmatrix} 2\lambda & 0 \\ 0 & \rho \end{pmatrix}$ is the pull back under σ of the polarization $\tilde{\lambda}$ and this implies that ρ is twice a principal polarization. So P carries a principal polarization ! For these facts, see [13],§2.

Now use the elementary

(1.1) <u>Lemma</u>. If D is a divisor of degree $g-1$ on C, then $h^o(\pi^*(D)) \neq 0$ if and only if $h^o(D) \neq 0$ or $h^o(D+\eta) \neq 0$.

One finds (using that for Jacobians the theta divisor in Jac^{g-1} consists of the effective divisor classes of degree $g-1$)

$$\sigma^{-1}(\tilde{\Theta}_o) \cap (Jac(C) \times (0)) = \Theta_o + \Theta_{o,\eta} .$$

Here $\tilde{\Theta}_o$ (resp. $\tilde{\Theta}_o$) denotes the theta divisor on $Jac^{g-1}(C)$ (resp. $Jac^{2g-2}(C)$, i.e, $\Theta_o = \{x \in Jac^{g-1}(C) : h^o(x) > 0\}$. If one now chooses $\alpha \in Jac(C)_4$ such that $2\alpha = \eta$ and a theta characteristic ζ on C, then $\tilde{\zeta} = \pi^{-1}(\zeta + \alpha)$ is a theta characteristic on $\tilde{C}$. If $\tilde{\Theta} = \{x \in \tilde{J} : h^o(x+\tilde{\zeta}) > 0\}$, $\Theta = \{x \in J : h^o(x+\zeta) > 0\}$ are the theta divisors on $\tilde{J}$ and on J then (since $\sigma|J\times(0) = \pi^*$):

$$(\pi^*)^{-1}\tilde{\Theta} = \Theta_\alpha + \Theta_{-\alpha}. \qquad (3)$$

(1.2) The link between the Kummer variety of P and that of J is obtained as follows. There is a morphism

$$\begin{array}{rcl} \delta: P & \to & |2\Theta_J| \\ p & \to & (\pi^*)^{-1}(\tilde{\Theta}_{\tilde{J},-p}). \end{array}$$

Mumford shows in [M1] that δ is the usual Kummer map followed by an inclusion

$$\begin{array}{ccc} & \xrightarrow{\Phi_P} & \mathbb{P}(H^o(L_P^{\otimes 2})^\vee) \\ P & & \downarrow i \\ & \xrightarrow[\delta]{} & |2\Theta_J| \end{array}$$

For any principally polarized abelian variety X the Riemann theta formula

$$\theta(u+v)\theta(u-v) = \sum c_{\alpha\beta}s_\alpha(u)s_\beta(v)$$

with θ a non-zero section of L_X and $\{s_\alpha\}$ a basis of $\Gamma(X,L_X^{\otimes 2})$ gives us a non-degenerate form B on $\Gamma(X,L_X^{\otimes 2})$ via the $(c_{\alpha\beta})$ and gives rise to a diagram

$$\begin{array}{ccc} & \xrightarrow{\Phi_X} & \mathbb{P}(H^o(X,L_X^{\otimes 2})^\vee) \\ X & \sim & \wr\,\, B' \\ & \xrightarrow[\Phi'_X]{} & |2\Theta_X| \end{array}$$

where $\Phi'_X(x) = \Theta_{X,x} + \Theta_{X,-x}$ and B' is induced by B.

Formula (3) thus implies the fundamental relation

$$i(\Phi_P(0)) = B'(\Phi_X(\alpha)). \qquad (4)$$

(1.3) For any indecomposable principally polarized X the theta group $G(L_X^{\otimes 2})$ acts on $\Gamma(X,L_X^{\otimes 2})$ and this defines an action of $G(L_X^{\otimes 2})$ modulo scalars $\cong X_2$ on $\mathbb{P}(\Gamma(X,L_X^{\otimes 2})^\vee)$. If $\alpha \in X_2, \alpha \neq 0$, then α defines a projective involution i_α of $\mathbb{P}^N$ with

$$i_\alpha(\Phi_X(x)) = \Phi_X(x+\alpha). \qquad (5)$$

It is a classical fact that the involution i_α on $\mathbb{P}^N = \mathbb{P}(\Gamma(X,L_X^{\otimes 2})^\vee)$ has as its fixed point set two linear subspaces V_α^+, V_α^-, each of dimension $2^{g-1}-1$ and each intersecting the Kummer variety of X in $2^{2(g-1)}$ points; moreover,

$$(V_\alpha^+ \cup V_\alpha^-) \cap \Phi_X(X) = \Phi_X(\{x \in X_4 : 2x = \alpha\}).$$

The linear spaces $V_\alpha^\pm$ cut out on the modular variety $\Phi(\bar{A}_g(2,4))$ the boundary components. To be precise,

(1.4) <u>Proposition</u>. Let $A_{g-1}(2,4)$ be one of the $2(2^{2g}-1)$ boundary components of $\bar{A}_g(2,4)$ of maximal dimension. The image $\Phi(\bar{A}_{g-1}(2,4))$ in $\mathbb{P}^{2^g-1}$ is the intersection of $\Phi(\bar{A}_g(2,4))$ with one of the linear spaces $V_\alpha^\pm$.

It follows from (4) and proposition (1.4) that for a Jacobian X the intersection $\Phi_X(X) \cap \Phi(A_{g-1}(2,4))$ is not empty (Here we view $A_{g-1}(2,4)$ as a boundary component of $A_g(2,4)$.) : the intersection contains the image of a point of order 4 of X.

(1.5) <u>Definition</u>. The Schottky locus $S_g \subsetneq \bar{A}_g$ is the smallest closed subset of $\bar{A}_g$ containing the points $[X]$ with X indecomposable for which $\Phi_X(X) \cap \Phi(A_{g-1}(2,4)) \neq \emptyset$ for all boundary components $A_{g-1}(2,4)$.

By construction S_g contains J_g, the Jacobian locus (cf. (4)). S_g can be described in terms of theta constants as well. The point is that P can be written as

$$P = \mathbb{C}^{g-1}/\mathbb{Z}^{g-1} + \rho_{g-1}\mathbb{Z}^{g-1} \qquad \text{for some } \rho_{g-1} \in \mathbb{H}_{g-1}$$

and that after suitable normalizations

$$\theta^2[{}^{\varepsilon}_{\varepsilon'}](\rho_{g-1},0) = c\,\theta[{}^{\varepsilon\ 0}_{\varepsilon'\ 0}](\tau_g,0)\,\theta[{}^{\varepsilon\ 0}_{\varepsilon'\ 1}](\tau_g,0) \tag{6}$$

with a constant $c \in \mathbb{C}^*$ independent of $\varepsilon,\varepsilon' \in (\mathbb{Z}/2\mathbb{Z})^{g-1}$. Thus (6) is a translation of (4).

Let

$$T_g \subset \mathbb{C}[\,X_{[{}^{\varepsilon}_{\varepsilon'}]} : \varepsilon,\varepsilon' \in (\mathbb{Z}/2\mathbb{Z})^g,\ {}^t\varepsilon\varepsilon' = 0\,]$$

be the ideal of $\Psi(\bar{A}_g(2,4))$. To an element $f \in T_{g-1}$ we associate

$$\sigma(f) = f(\ldots,\theta[{\varepsilon \atop \varepsilon'}{0 \atop 0}]\theta[{\varepsilon \atop \varepsilon'}{0 \atop 1}](\tau_g,0),\ldots)$$

by substituting $\theta[{\varepsilon \atop \varepsilon'}{0 \atop 0}]\theta[{\varepsilon \atop \varepsilon'}{0 \atop 1}](\tau_g,0)$ for $X_{[{\varepsilon \atop \varepsilon'}]}$. The group $\Gamma_g/\Gamma_g(4,8)$ acts on $\mathbb{C}[\ \theta[{\varepsilon \atop \varepsilon'}](\tau_g,0) : {}^t\varepsilon\varepsilon' = 0]$. Let Σ_g be the smallest $\Gamma_g/\Gamma_g(4,8)$-invariant ideal of this ring containing all $\sigma(f)$ with f in T_{g-1}. Then S_g is the zero-locus of Σ_g in $\overline{A}_g$.

Of course, this description is explicit only if we know T_{g-1} and in general the structure of this ideal is not known.

For $g=4$ one finds that Σ_g is the ideal generated by a Siegel modular form of weight 8 as Schottky showed.

The important question about S_g is whether $S_g = J_g$ and if not, what the components of S_g are. For $g=4$ Igusa proved that S_4 is irreducible. This implies $S_4 = J_4$. Recently van Geemen proved

(1.6) <u>Theorem</u>. (van Geemen [6]) J_g is an irreducible component of S_g.

His proof uses an induction argument and an analysis of the intersection of the Schottky locus with blow-up of a boundary component of $\overline{A}_g(4,8)$.

It is a recurring phenomenon in the history of the Schottky problem that one finds algebraic subsets of A_g that contain J_g as as irreducible component but that may have other components as well. Another example is the Andreotti-Mayer approach. Since it is known that for a Jacobian one has $\dim \operatorname{Sing}\Theta \geq g-4$ one looks at

$$N_g^m = \{[X] \in A_g : \operatorname{Sing}\Theta \neq \emptyset,\ \dim \operatorname{Sing}\Theta \geq m\}$$

Andreotti and Mayer proved that J_g is an irreducible component of N_g^{g-4}, $g \geq 4$. However, N_g^{g-4} contains other components.

APPROACH 2 : TRISECANTS.

One of the remarkable features of Jacobians is that their Kummer varieties possess trisecants :

(2.1) Proposition. Let C be a non-singular curve and let a,b,c,d be points of C. If $r \in X = \mathrm{Jac}(C)$ is such that $2r = a+b-c-d$, then $\Phi_X(r)$, $\Phi_X(r-b+c)$ and $\Phi_X(r-b+d)$ are collinear.

Fay's trisecant identity [3] implies this fact. Gunning [9] has generalized this identity. The idea behind it is essentially the following.

Let N be a line bundle on $X \times X$ such that

$$N_{|X\times t} \cong T_{-t}^{*}(\mathcal{O}(2\Theta)) \qquad (T_t : \text{translation by } t).$$

Fix a point p of C. This defines $\phi: C \to X = \mathrm{Jac}(C)$ by $c \mapsto c-p$. Let Δ be a divisor of degree g on C such that $\phi^*\mathcal{O}(\Theta) \cong \mathcal{O}(\Delta)$. We let M be the vector bundle on X whose fibre at t is $H^0(C,\mathcal{O}(2\Delta+2t))$. Pull back of sections via

$$H^0(X,T_{-t}^{*}(\mathcal{O}(2\Theta)) \xrightarrow{\phi^*} H^0(C,\mathcal{O}(2\Delta+2t)) \tag{7}$$

gives rise to a bundle map

$$\psi: (p_1)_* N \longrightarrow M.$$

(2.2) Lemma. The map ψ is surjective.

Proof. The map $H^0(X,T_{-t}^{*}\mathcal{O}(\Theta)) \to H^0(C,\mathcal{O}(\Delta+t))$ is surjective if the divisor $\Delta+t$ is non-special. Therefore, if $D \in |2\Delta+t|$ can be written as $D = D_1 + D_2$, $D_i \in |\Delta+t_i|$ with $\Delta+t_i$ non-special, then D is the zero divisor of a section in the image of (7). Define a non-empty open set in the symmetric product $C^{(g)}$ by (κ = canonical divisor)

$$U = \{z_1+\ldots+z_g \in C^{(g)} : h^0(D-\textstyle\sum z_i)=1,\ h^0(\kappa-\sum z_i)=0\}.$$

If $D = z_1+\ldots+z_g+z_1'+\ldots+z_g'$ with $\sum z_i \in U$, then $h^0(\sum z_i') = 1$, hence $\sum z_i$ and $\sum z_i'$ are both non-special. This shows that the image of

$\mathbb{P}(H^o(X,T^*_{-t}(\mathcal{O}(2\Theta)) \to \mathbb{P}(H^o(C,\mathcal{O}(2\Delta+t)))$ contains a non-empty open set.

We put as usual

$$W^r_d = \{ x \in \mathrm{Jac}^{(d)}(C) \ : \ h^o(x) \geq r+1 \}.$$

(2.3) <u>Theorem</u>. (Gunning [10]) If $z_1,\ldots,z_n$ are distinct points of C, then

$$W^{n-\mu}_{n-2} - \sum_{i=1}^{n} z_i + 2p = \{2t \in \mathrm{Jac}(C) \ : \ \mathrm{rank}\ \vec{\theta}_2(t+\phi(z_i))^n_{i=1} < \mu \} .$$

<u>Proof</u>. By the lemma, the rank of this $2^g \times n$-matrix is less than $\mu \iff h^o(2\Delta+2t-\sum z_i) > g+1-\mu \iff 2t+2\Delta-\sum z_i \in W^{g+1-\mu}_{2g-n}$. Applying Serre duality $\kappa - W^{g+1-\mu}_{2g-n} = W^{n-\mu}_{n-2}$ and the fact that $\phi(\kappa)=\phi(2\Delta)$ gives the result.

The special case $n=\mu=3$ gives proposition (2.1).

We can generalize this by allowing the points z_i to coincide. If $z_1+\ldots+z_n = m_1x_1+\ldots+m_ex_e$ with $x_i \neq x_j$ if $i \neq j$ then in the rank condition the m_j vectors $\vec{\theta}_2(t+\phi(x_j))$ have to be replaced by

$$\vec{\theta}_2(t+\phi(x_j)) \quad \Delta_1\vec{\theta}_2(t+\phi(x_j)) \quad \ldots \quad \Delta_{m_j-1}\vec{\theta}_2(t+\phi(x_j)),$$

where the Δ_k are differential operators defined as follows. The curve $\phi(C)$ contains at $\phi(x_j)$ an artinian subscheme $\mathrm{Spec}\ \mathbb{C}[\varepsilon]/(\varepsilon^{m_j})$ and this is given by a local homomorphism

$$\mathcal{O}_{X,\phi(x_j)} \to \mathbb{C}[\varepsilon]/(\varepsilon^{m_j})$$

$$f \to f(y_j) + \Delta_1 f(y_j)\varepsilon + \ldots + \Delta_{m_j-1} f(y_j)\varepsilon^{m_j-1}, \qquad y=\phi(x_j)$$

The special case $n=\mu=3$ is important since it gives us back the curve C : Note that $W^o_1 \cong C$ and

$$W^o_1 - \sum_{i=1}^{3} z_i + 2p = \{2t \in \mathrm{Jac}(C) \ : \ \mathrm{rank}\ (\ \vec{\theta}_2(t+\phi(z_i))(t) \leq 2 \} \ !$$

Gunning's idea in [8] was to use this property to characterize Jacobians. Gunning used distinct points z_i but Welters has infinitesimalized Gunning's case to include the case of coinciding points and transformed it into the following beautiful criterion :

(2.4) Theorem. (Gunning-Welters [19]) Let X be an indecomposable principally polarized abelian variety and let $Y \subset X$ be an artinian subscheme of length 3. Assume that

$$V = \{2t \in X : t+Y \subset \Phi_X^{-1}(\ell) \text{ for some line } \ell \subset \mathbb{P}^N\}.$$

has positive dimension at some point. Then V is a smooth irreducible curve and X is its Jacobian.

(2.5) The property of having flexes is closely related to the Kadomčev-Petviashvili equation (K-P-equation), a fourth order partial differential equation satisfied by the theta functions of Jacobians. In [16] Mumford noticed that if the points a,b,c,d in proposition (2.1) coincide, Fay's trisecant identity leads to the K-P-equation.

To get the link, note that an inclusion $\operatorname{Spec} \mathbb{C}[\varepsilon]/(\varepsilon^{N+1}) \to (X,0)$ is given by a local homomorphism

$$\begin{aligned} \mathcal{O}_{X,0} &\longrightarrow \mathbb{C}[\varepsilon]/(\varepsilon^{N+1}) \\ f &\to \sum_{i=1}^{N} \Delta_i(f)\varepsilon^i \end{aligned}$$

where the Δ_i are differential operators satisfying

$$\Delta_o = \text{id}, \quad \Delta_i(gh) = \sum_{k+\ell=i} \Delta_k(g)\Delta_\ell(h)$$

One can show that this is equivalent to the existence of translation invariant vector fields $D_1,\ldots,D_N$ on X such that

$$\Delta_\nu = \sum_{h_1+2h_2+\ldots+\nu h_\nu > 0} (h_1!\ldots h_\nu!)^{-1} D_1^{h_1}\ldots D_\nu^{h_\nu},$$

or formally

$$e^{\sum_{j=1} D_j\varepsilon^j} \equiv \sum_{k=0} \Delta_k\varepsilon^k \pmod{\varepsilon^{N+1}}.$$

We apply this to criterion (2.4). Note that V is defined by the vanishing of the 3×3 minors f_ν, $\nu \in (\mathbb{Z}^g/2\mathbb{Z}^g)^3$, of $(\vec{\theta}_2\ \Delta_1\vec{\theta}_2\ \Delta_2\vec{\theta}_2)$ at some point. If we assume that this point is the origin and that $Y =$

$\operatorname{Spec}\mathbb{C}[\varepsilon]/(\varepsilon^3) \to (X,0)$ is given by D_1, D_2 one finds (using the fact that the rank of $(\vec{\theta}_2\ (\partial_i\partial_j\vec{\theta}_2)_{i,j})$ equals $g(g+1)/2+1$ at $(\tau,0)$; $\partial_i = \partial/\partial z_i$)) that $(V)_2 = Y$. Then, as Welters noticed, one has

$$(V)_3 = \operatorname{Spec}\mathbb{C}[\varepsilon]/(\varepsilon^4) \iff \exists\ D_3 \text{ such that } (\tfrac{1}{3}D_1^3 + D_1D_2 + D_3)f_\nu = 0 \text{ (all } \nu)$$

$$\iff \operatorname{rank}((1\ \ D_1^2\ \ D_1^4 + 3D_2^2 - 3D_1D_3)\vec{\theta}_2)(\tau,0) \le 2.$$

Without changing $\operatorname{Spec}\mathbb{C}[\varepsilon]/(\varepsilon^4) \to (X,0)$ one may effect the change $D_1 \to aD_1$, $D_2 \to a^2D_2 + bD_1$, $D_3 \to a^3D_3 + a^2bD_2 + cD_1$, $a \neq 0, b, c$, hence we can rewrite this as

$$((D_1^4 - D_1D_3 + \tfrac{3}{4}D_2^2 + d)\vec{\theta}_2)(\tau,0) = 0 \qquad (8)$$

This is the K-P equation. By (2.3) the theta functions $\theta_2[\sigma]$ of a Jacobian yield solutions. (Usually, the K-P equation is written $u_{yy} + (u_t + u_{xxx} + uu_x)_x = 0$. It is satisfied on a Jacobian by $u = D_1^2\log\theta(z + xa_1 + ya_2 + ta_3) + c$ for some $a_1, a_2, a_3 \in \mathbb{C}^g$, $c \in \mathbb{C}$, see [15]. Dubrovin formulated the equivalent form (8).) That theta functions yield solutions was noticed by Krichever, who arrived at it in a completely different way. Novikov conjectured then that this should characterize Jacobians :

(2.6) Novikov's Conjecture. An indecomposable principally polarized abelian variety X is a Jacobian if and only if there exist constant vector fields D_1, D_2, D_3 on X and a constant d such that

$$((D_1^4 - D_1D_3 + \tfrac{3}{4}D_2^2 + d)\vec{\theta}_2)(\tau,0) = 0. \qquad (9)$$

Dubrovin proved in [2] that the locus of $[X]$ in A_g for which (9) holds for some D_1, D_2, D_3 and d contains the Jacobian locus as an irreducible component.

(2.7) Soon after a weaker version of (2.4) had appeared Arbarello and De Concini realized that one does not need the positive dimensionality of V, but only the fact that $\mathcal{O}_{V,0}$ contains an artinian subscheme of sufficiently big length, i.e. the condition is that there exist constant vector fields $D_1,\ldots,D_M$ for some big M such that

$$e^{\sum_{j=1}^{M} D_j \varepsilon^j} (\vec{\theta}_2 \wedge \Delta_1\vec{\theta}_2 \wedge \Delta_2\vec{\theta}_2) \equiv 0 \pmod{\varepsilon^M} \quad \text{at} \quad (\tau,0).$$

In this way they were the first to write down equations that characterize the Jacobian locus, see [1]. Using the version of (2.4) given here one can take $M = 6^g g! + 1$.

Recently, Shiota showed that if one makes a minor technical assumption on X then Novikov's conjecture is true, see section 4.

APPROACH 3 : THE GEOMETRY OF THE MODULI SPACE.

The approach here, worked out in joint work with van Geemen [6], is based on the observation that under Φ and Φ_X both the moduli space $A_g(2,4)$ and the Kummer variety of X are mapped to the same projective space, so that we can compare their positions in this space. It was motivated by the special case $g=2$ studied in [7] and a paper of Frobenius dealing with $g=3$, [4].

(3.1) We first look at the tangent space to $\Phi(A_g(2,4))$ at $\Phi([X])$. i.e. we look at the hyperplanes

$$\sum_\sigma \alpha_\sigma \theta_2[\sigma](\tau,0) = 0 \qquad (X=X_\tau) \qquad (10)$$

satisfying

$$\frac{\partial}{\partial\tau_{ij}} \Big(\sum_\sigma \alpha_\sigma \theta_2[\sigma]\Big)(\tau,0) = 0 \qquad \text{for all } i,j.$$

By applying the Heat Equations this is transformed into

$$\Big(\sum_\sigma \alpha_\sigma \frac{\partial^2}{\partial z_i \partial z_j} \theta_2[\sigma]\Big)(\tau,0) = 0. \qquad (11)$$

So let us look at the sections of $\Gamma(X,L_X^{\otimes 2})$ satisfying (10) and (11), i.e. define

$$\Gamma_{00}(X,L_X^{\otimes 2}) = \{ s \in \Gamma(X,L_X^{\otimes 2}) \; : \; m_0(s) \geq 4 \} .$$

with m_0 the multiplicity of a section at zero. Note that for $s \neq 0$

$m_o(s)$ is even). If X is indecomposable, then

$$\mathrm{rk}\ (\ \vec{\Theta}_2\ \ \frac{\partial^2}{\partial z_1 \partial z_1}\vec{\Theta}_2, \dots, \frac{\partial^2}{\partial z_g \partial z_g}\vec{\Theta}_2\ )$$

in $(\tau,0)$ equals $\frac{1}{2}g(g+1)+1$, so the codimension of Γ_{oo} in $\Gamma(X,L_X^{\otimes 2})$ equals $\frac{1}{2}g(g+1)+1$. If $\Phi(A_g(2,4))$ is non-singular at $\Phi([X])$ and if $I_g \subset \mathbb{C}[\dots,X_\sigma,\dots]$ is the ideal of $\Phi(A_g(2,4))$, then $\Gamma_{oo}(X,L_X^{\otimes 2})$ equals

$$\{\ \sum \frac{\partial f}{\partial X_\sigma}(\dots,\Theta_2[\sigma](\tau,0),\dots)\ \Theta_2[\sigma](\tau,z)\ :\ f \in I_g\ \}.$$

(3.2) As an example we take $g=3$. The theory of theta functions gives us a relation

$$\theta[{}^{000}_{000}]\theta[{}^{000}_{100}]\theta[{}^{000}_{010}]\theta[{}^{000}_{110}] - \theta[{}^{001}_{000}]\theta[{}^{001}_{100}]\theta[{}^{001}_{010}]\theta[{}^{001}_{110}] +$$
$$- \theta[{}^{000}_{001}]\theta[{}^{000}_{101}]\theta[{}^{000}_{011}]\theta[{}^{000}_{111}] = 0$$

between the $\theta[{}^{\varepsilon}_{\varepsilon'}](\tau,0)$. We write this as $r_1 - r_2 - r_3 = 0$. This implies the relation

$$r_1^4 + r_2^4 + r_3^4 - 2r_1^2r_2^2 - 2r_1^2r_3^2 - 2r_2^2r_3^2 = 0$$

between the squares of the even thetas. Using (1) this gives an equation

$$F(\dots,\Theta_2[\sigma](\tau,0),\dots) = 0$$

of degree 16 defining a hypersurface in $\mathbb{P}^7$. Hence

$$\phi = \sum \frac{\partial F}{\partial X_\sigma}(\dots\Theta_2[\sigma](\tau,0),\dots)\ \Theta_2[\sigma](\tau,z)$$

belongs to Γ_{oo} and one can check that for indecomposable X_τ it is non-zero. It generates Γ_{oo}. In fact, when expressed in the theta squares this is the function studied by Frobenius in [4].

The first question about Γ_{oo} is its zero locus. Define

$$F_X = \{\ x \in X\ :\ s(x) = 0\ \text{ for all }\ s \in \Gamma_{oo}(X,L_X^{\otimes 2})\}\ .$$

(3.3) <u>Proposition</u>. If $X = \mathrm{Jac}(C)$ then $F_X \supseteq \{(x-y) \in \mathrm{Jac}(C): x,y \in C\}$.

<u>Proof</u>. Use (2.1) and put $a=b$, $c=d$ there. One finds a relation

$$\vec{\theta}_2(a-b) = \lambda\vec{\theta}_2(0) + \sum \mu_{ij}\partial_i\partial_j\vec{\theta}_2(0).$$

(3.4) For a Jacobian one can use the geometry of C to construct elements of Γ_{oo}. Let $|2\Theta|_{oo} = \{ D \in |2\Theta| : m_o(D) \geq 4 \}$. If $x \in \text{Sing}\,\Theta$ then $\Theta_x \cup \Theta_{-x} \in |2\Theta|_{oo}$. Define

$$\Theta_o = \{ \alpha \in \text{Jac}^{g-1}(C) : h^o(\alpha) > 0 \}$$

$$\text{Sing}\,\Theta_o = \{ \alpha \in \Theta_o : h^o(\alpha) > 1 \}$$

and define for $\alpha \in \text{Jac}^{g-1}(C)$:

$$\Theta_\alpha = \{x \in \text{Jac}(C) : \alpha - x \in \Theta_o \}.$$

Then obviously,

$$F_X \subseteq \bigcap_{\alpha \in \text{Sing}\,\Theta} (\Theta_\alpha \cup \Theta_{\kappa-\alpha}). \qquad (\kappa\text{: canonical divisor})$$

If C is hyperelliptic then

$$\text{Sing}\,\Theta_o = g_2^1 + W_{g-3}^o ,$$

hence if $f \in F_X$ one has $\pm f + g_2^1 + W_{g-3}^o \subset W_{g-1}^o$, so $\pm f + g_2^1 \in W_2^o$ and this implies $f = (a-b)$ for some $a,b \in C$. So for hyperelliptic C one finds

$$F_X = \{(x-y) \in \text{Jac}(C) : x,y \in C\}.$$

By semi-continuity it follows that for general X $\dim F_X \leq 2$ and for a general Jacobian $\dim F_X = 2$. We conjectured in [6] that for every C

$$F_{\text{Jac}(C)} = \{(x-y) \in \text{Jac}(C) : x,y \in C\}$$

and provided a lot of evidence for it. Independently, the conjecture was formulated by Mumford [15] (in a dual form) and by Gunning [10]. This conjecture has now been proved by Welters. However, there is one exceptional case, namely $g=4$, where

$$F_{\text{Jac}(C)} = \{(x-y) \in \text{Jac}(C) : x,y \in C\} \cup \{\pm(f-f')\} ,$$

with f, f' the two g_3^1's on C, see [20].

(3.5) <u>Conjecture</u>. Let X be a principally polarized abelian variety of dimension $g \geq 2$. Then X is a Jacobian if and only if $\dim F_X \geq 2$.

An infinitesimal form of this conjecture is related to the Novikov conjecture. Note that

$$x \in F_X \iff \exists\, \lambda, \mu_{ij} \in \mathbb{C} \text{ such that } \vec{\theta}_2(x) = \lambda \vec{\theta}_2(0) + \sum \mu_{ij} \partial_i \partial_j \vec{\theta}_2(0)$$

Now, if $Y = \operatorname{Spec} \mathbb{C}[\varepsilon]/(\varepsilon^{N+1})$ is contained in X at 0 via the local homomorphism $\mathcal{O}_{X,0} \to \operatorname{Spec} \mathbb{C}[\varepsilon]/(\varepsilon^{N+1})$, $f \to \sum_{i=0}^{N} \Delta_i(f)\varepsilon^i$, we have

$$Y \subset F_X \iff \exists\, \lambda, \mu_{ij} \in \mathbb{C}[\varepsilon] \text{ such that}$$

$$\sum_{k=0}^{N} \Delta_k \vec{\theta}_2(\tau, 0)\varepsilon^k = \lambda \vec{\theta}_2(\tau, 0) + \sum \mu_{ij} \partial_i \partial_j \vec{\theta}_2(\tau, 0).$$

Working out the condition for $N=4$ gives

$$\left(\left(\tfrac{1}{24} D_1^4 + \tfrac{1}{2} D_2^2 - D_1 D_3\right) \vec{\theta}_2\right)(\tau, 0) = d\, \vec{\theta}(\tau, 0) + \sum e_{ij} \partial_i \partial_j \vec{\theta}_2(\tau, 0),$$

where e_{ij} is the coefficient of ε^4 in μ_{ij}. If $\sum e_{ij} \partial_i \partial_j \vec{\theta}_2(\tau, 0)$ is a multiple of $D_1^2 \vec{\theta}_2(\tau, 0)$, then we can change coordinates such that this relation becomes the K-P equation.

Gunning has studied the generalizations of Fay's identity. This leads to interesting analogues of (3.3) involving higher derivatives at zero, cf [11].

Instead of intersecting the Kummer variety with the tangent space of the moduli space we can also intersect the Kummer variety with the moduli space itself. As an analogue to (3.3) we find

(3.6) <u>Proposition</u>. If $X = \operatorname{Jac}(C)$ then $\Phi_X(X) \cap \Phi(A_g(2,4))$ contains $\Phi_X(\{\tfrac{1}{4}(x-y) : x, y \in C\})$.

Here $\frac{1}{4}$ means the inverse image under multiplication by 4.

Proof. A divisor class $\underline{a}$ with $2\underline{a} = x+y$ defines a (2:1)-covering $\pi: \tilde{C} \to C$. The Prym variety $P = \ker\{Nm: Jac(\tilde{C}) \to Jac(C)\}$ is a principally polarized abelian variety of dimension g for general $x,y \in C$. There exist theta structures on P and X such that

$$\Phi(\tfrac{1}{4}(x-y)) = \Phi([P]),$$

cf. [13],p.340, where $\frac{1}{4}(x-y) \in Jac(C)$ is such that $2(\frac{1}{4}(x-y)) = \frac{1}{2}(x-\underline{a})$. By symmetry it then follows that all of $\Phi_X(\frac{1}{4}(x-y))$ lies in $\Phi(\overline{A}_g(2,4))$ for general x,y, hence for all x,y.

We made two conjectures in relation to this. First, for a Jacobian we conjectured that

$$\Phi_X(X) \cap \Phi(A_g(2,4)) = \Phi_X(\{\alpha \in X : 4\alpha = x-y,\ x,y \in C\})$$

and we proved this for $g=3$. Secondly, we hope that this characterizes Jacobians :

(3.7) Conjecture. Let X be an indecomposable principally polarized abelian variety of dimension $g \geq 2$. Then X is a Jacobian if and only if $\dim \Phi_X(X) \cap \Phi(\overline{A}_g(2,4)) \geq 2$.

(3.8) The preceding sections suggest to look at the morphism

$$\Xi: \Gamma_g(2,4) \ltimes \mathbb{Z}^{2g} \backslash \mathbb{H}^g \times \mathbb{C}^g = U_g(2,4) \to \mathbb{P}^N$$
$$(\tau,z) \to (\ldots,\vec{\theta}_2[\sigma](\tau,z),\ldots)$$

Is it everywhere of maximal rank ? Since the Kummer variety of an indecomposable X is singular at the images of the points of order 2 of X the rank is certainly not maximal at those (τ,z) for which $2z \in \mathbb{Z}^g + \tau\mathbb{Z}^g$. Using the Heat Equations the question becomes whether the rank of

$$\left(\partial_i \partial_j \vec{\theta}_2 \quad 1 \leq i \leq j \leq g \qquad \partial_k \vec{\theta}_2 \quad 1 \leq k \leq g \qquad \vec{\theta}_2 \right)$$

at (τ,z) is maximal.

Suppose that $2^g \geq \frac{1}{2}g(g+1) + g + 1$, i.e. $g \geq 4$ and that $X = X_\tau$ is indecomposable. If there exist a relation

$$((\sum \alpha_{ij}\partial_i\partial_j + \sum \beta_k\partial_k + \gamma)\vec{\theta}_2)(\tau,z) = 0$$

for z such that $2z \notin \mathbb{Z}^g + \tau\mathbb{Z}^g$ then $\Phi_X(X)$ possesses a flex at $\Phi_X(z)$.

A Jacobian is known to possess a lot of such flexes : if $X = \mathrm{Jac}(C)$, then applying (2.3) with $z_1=z_2=z_3=p$ we find that all points of $\{\frac{1}{2}(x-y) \in X : x,y \in C\}$ are flex points. Hence the rank of Ξ is not maximal at these points.

(3.9) Question. In view of (2.4) we can ask whether for an indecomposable Jacobian $X = X_\tau$ with $g \geq 4$ the only points (τ,z) where Ξ is not of maximal rank are those corresponding to $\{\frac{1}{2}(x-y) : x,y \in C\}$ and whether one could use this to characterize Jacobians.

APPROACH 4 : RINGS OF DIFFERENTIAL OPERATORS

As mentioned above Shiota has settled Novikov's conjecture up to a technical assumption.

(4.1) Theorem. (Shiota [17]) An indecomposable principally polarized abelian variety X of dimension g is the Jacobian of a complete smooth non-singular curve C over $\mathbb{C}$ of genus g if and only if

i) the vector $\vec{\theta}_2(\tau,z)$ satisfies the K-P equation (8) for some D_1, D_2, D_3 and d, and

ii) no translate of the theta divisor of X contains an abelian subvariety of X which is tangent to $D_1(0)$.

Shiota's approach incorporates ideas of Mulase and is based on Krichever's dictionary. Let D be the non-commutative ring which as an additive group equals $\mathbb{C}[[x]][\partial]$ with $\partial = \frac{d}{dx}$ and with multiplication such that

$$\partial \cdot f = f\partial + f' \qquad \text{for} \quad f \in \mathbb{C}[[x]][\partial]. \tag{12}$$

If R is a commutative subring of D containing $\mathbb{C}$ and two elements A,B with $A=\partial^n + \ldots$ ($\ldots$ = lower order terms), $B=\partial^m+\ldots$ with $(n,m)=1$, then any element of R can be written as $C=\alpha\partial^r+\ldots$ with $\alpha\in\mathbb{C}$, $r\in\mathbb{Z}_{\geq 0}$ (Proof: work out the commutator $[A,C]$). If $R_n = \{ C\in R : C=\alpha\partial^r+\ldots \text{ with } r\leq n \}$ one has $\dim R_n/R_{n-1}\leq 1$ and $=1$ for $n >> 0$.

(4.2) Theorem. (Krichever) There is a natural bijection between the following two sets of data:

1) C an irreducible curve, P a smooth point of C, a tangent vector at P and a torsion free rank 1 $\mathcal{O}_C$-module F with $h^o(F)=h^1(F)=1$.

2) $R\subset D$ a commutative subring containing $\mathbb{C}$ and two elements A, B as above.

Let us sketch how to go from 1) to 2). Choose a neighbourhood U of P such that the local coordinate z at P is a unit on $U-P$. Let x be the standard coordinate on $\mathbb{C}$. We now glue $F\otimes\mathcal{O}_{\mathbb{C}}$ on $U\times\mathbb{C}$ and $F\otimes\mathcal{O}_{\mathbb{C}}$ on $(C-P)\times\mathbb{C}$ by multiplication with $e^{x/z}$. This defines a sheaf F^* on $C\times\mathbb{C}$. If V is a suitably chosen neighbourhood of $0\in\mathbb{C}$ then $H^i(C\times V, F^*) = 0$ $i=0,1$. Define now

$$\nabla : F^*(\ell P) \rightarrow F^*((\ell+1)P)$$

by taking $\frac{d}{dx}$ on $C-P$ and $\frac{1}{z}+\frac{d}{dx}$ on U. A non-zero section s_o $H^o(F^*(P))$ generates $H^o(F^*(P))$ as a $H^o(V,\mathcal{O}_{\mathbb{C}})$-module. We normalize s_o such that $s_o = 1+O(z)$ at $p\times V$, i.e. $\frac{d}{dx}s_o = (z^{-1}+O(z))s_o$. Put $s_n=\nabla^n s$. The sections $s_o,\ldots,s_n$ generate $H^o(F^*(n+1)P)$.

If $a\in\Gamma(C-P,\mathcal{O}_C)$ then $as_o\in H^o(F^*(nP))$, hence

$$as_o = \sum_{i=0}^{n-1} a_i(x)\nabla^i s_o.$$

This gives us a map

$$\Gamma(C-P,\mathcal{O}_C) \to D$$
$$a \to \sum_{i=0}^{n-1} a_i(x)\partial^i.$$

The image is a commutative subring R of D.

(4.3) In order to obtain Jacobians one observes that F defines a point of $\mathrm{Jac}(C)$. So let us deform F. Choose variables $t_1,\ldots,t_N$ and consider instead of $F\otimes\mathcal{O}_{\mathbb{C}}$ now $F\otimes\mathcal{O}_{\mathbb{C}^N}$ on $U\times\mathbb{C}^N$ and $(C-P)\times\mathbb{C}^N$ and glue now by $\exp(\sum_{j=1}^{N} t_j z^{-j})$. We introduce formally a variable x by replacing t_1 by t_1+x. This now gives us F^* as above. Define ∇ as above and define

$$\nabla_n : F^* \to F^*(nP)$$

by taking $\frac{\partial}{\partial t_n}$ on $C-P$. We choose a normalized s_o again as above. We now obtain

$$\Gamma(C-P,\mathcal{O}_C) \to D$$

and the image is a commutative subring R_t depending on $t=(t_1,\ldots,t_n)$. The question arises : how does R_t deform with t ? If T denotes the the tangent space we get a map

$$\Phi: T_t\mathbb{C}^N \to D$$
$$\frac{\partial}{\partial t_n} \to B_n(t)$$

where $B_n(t)$ is defined as follows. By the normalization $\frac{\partial}{\partial t_n} s_o = (z^{-n}+O(z))s_o$, so $\frac{\partial}{\partial t_n} s_o = \sum b_i \nabla^i s_o$. We put $B_n(t) = \sum b_i \partial^i$.

We need some notation. Let Ψ be the non-commutative $\mathbb{C}$-algebra whose elements are formal Laurent series in $(\frac{d}{dx})^{-1}$ with coefficients from $\mathbb{C}[[x]]\otimes\mathcal{O}_{\mathbb{C}^N}$. The multiplicative structure is defined by extending the rule (12). Let

$$\Psi^- = \{P\in\Psi : \operatorname{ord} P \le -1\}.$$

So the elements of Ψ^- are expressions $\sum_{j=-\infty}^{-1} a_i(x)\partial^i$, where we suppress the dependence of t in the notation. By extending the map $\Gamma(C-P,\mathcal{O}_C) \to D$ to $Q(\Gamma(C-P,\mathcal{O}_C) \to \Psi$ (Q : quotient field) we see that $1/z$ corresponds to an element of $\frac{d}{dx} + \Psi^-$ which we call L. From the normalization we obtain $B_n = (L^n)_+$, where $(\ )_+$ means taking the differential operator part (non-negative powers of ∂).

The dependence of R_t on t is now expressed by the following deformation equations for $L \in \frac{d}{dx} + \Psi^-$:

$$\left(\frac{\partial}{\partial t_n}\right) L = [(L^n)_+, L] \qquad n=1,\ldots,N$$

Take now infinitely many variables $t_1, t_2, \ldots$, i.e. $t \in \mathbb{C}^\infty = \varinjlim \mathbb{C}^N$ and consider the equations for $L \in \frac{d}{dx} + \Psi^-$:

$$\left(\frac{\partial}{\partial t_n}\right) L = [(L^n)_+, L] \qquad n=1,2,\ldots$$

This set of equations is called the K-P hierarchy. We do not explain here the translation of solutions to this hierarchy of equations into differential equations satisfied by theta functions, but we refer to Shiota's paper and the references there.

If L is a solution to the K-P hierarchy then consider

$$dL : T_t\mathbb{C} \to \Psi^- \qquad \sum c_n \frac{\partial}{\partial t_n} \to [\sum c_n B_n, L] ,$$

the tangent map of the map $t \to L(t)$ at t. We call L a finite dimensional solution if dL is of finite rank. Shiota considers for a finite dimensional L

$$R_L = \Phi(\ker dL) \oplus \mathbb{C}$$

with $\Phi: T_o\mathbb{C} \to D$, $\frac{\partial}{\partial t_n} \to B_n$. He proves

$$R_L = \{P \in D : [P,L] = 0\} .$$

and that R_L is a maximal commutative subring of D if $R_L \neq \mathbb{C}$. Thus a finite dimensional solution to the K-P hierarchy yields a curve by (4.2). Moreover, it turns out that $\ker dL$ can be identified with the tangent space of the Jacobian of this curve at a certain point.

Basically, this is the way Mulase arrived at his theorem which states that the whole K-P-hierarchy characterizes Jacobians. Both Mulase and Shiota then noticed that in fact finitely many equations from this hierarchy suffice,arriving thus at a theorem very similar to the result of Arbarello and De Concini. Shiota then continued by showingthat under condition 2) of (4.1) one can extend a solution to the K-P-equation (the first of the K-P-hierarchy) to a solution of the whole hierarchy, see [17]. Namikawa informed me that Mulase now also obtained such a reduction.

A FINAL REMARK.

Our summaryof recent attacks on the Schottky problem is not intended to be complete. One of the approaches that should be mentioned also is the approach that uses the reducubility of $\Theta \cap \Theta_a$. It is closely related to approach 2 and was suggested by Mumford in [14]. For a Jacobian X with theta divisor Θ one has : if $x \in X$ and $x \neq 0$ then there exist u,v in X with $\{0,x\} \cap \{u,v\} = \emptyset$ such that $\Theta \cap \Theta_x \subset \Theta_u \cup \Theta_v$ if and only if x belongs to $\{(a-b) \in X = \mathrm{Jac}(C) : a,b \in C\}$. (Note that one implication follows from (2.1) by using $X \to \mathbb{P}^N$, $x \mapsto \Theta_x \cup \Theta_{-x} \in |2\Theta|$.) Welters proved the following theorem : Let X be a complex principally polarized abelian variety of dimension g. Assume 1) $\dim \mathrm{Sing}\Theta \leq g-4$, 2) there exist a one-dimensional subset $Y \subset X$ such that for generic $y \in Y$ one has : $\Theta \cap \Theta_y \subset \Theta_u \cap \Theta_v$ for some $u,v \in X$ with $\{0,y\} \cap \{u,v\} = \emptyset$. Then X is the polarized Jacobian of a non-hyper-elliptic curve, see [21]

References.

1 Arbarello, E.,De Concini, C.: On a set of equations characterizing Riemann matrices. Preprint 1983.

2 Dubrovin, B.A.: theta functions and non-linear equations. Uspekhi Mat. Nauk. 36:2 (1981),11-80 = Russian math. surveys 36,(1981),11.

3 Fay, J.D.: Theta functions on Riemann surfaces. Lecture Notes in Math. 352. Springer Verlag,Berlin etc. 1973.

4 Frobenius, F.: Über die Jacobischen Functionen dreier Variabelen. Journal für die reine und angewandte Mathematik 105(1889),35-100.

5 van Geemen, B.: Siegel modular forms vanishing on the moduli space of curves. Preprint Univ. of Utrecht, to appear in Invent. Math.

6 van Geemen, B.,van der Geer, G.: Kummer varieties and the moduli spaces of abelian varieties. Preprint Univ. of Utrecht 1983.

7 van der Geer, G.: On the geometry of a Siegel modular threefold. Math. Annalen 260 (1982),317-350

8 Gunning, R.C.: Some curves in abelian varieties. Invent. Math. 66 (1982),377-389.

9 Gunning, R.C.: On generalized theta functions. Amer. J. Math. 104 (1982),183-208.

10 Gunning, R.C.: Riemann surfaces and their associated Wirtinger varieties. Preprint 1983.

11 Gunning, R.C.: Some identities for abelian integrals. Preprint 1983.

12 Mulase, M.: Cohomological structure of soliton equations, isospectral deformations of ordinary differential operators and a characterization of Jacobian varieties. Preprint MSRI 003-84.

13 Mumford, D.: Prym varieties I. In: Contributions to Analysis, p. 325-350. Academic press, London, New-York 1974.

14 Mumford, D.: Curves and their Jacobians. Univ. of Michigan Press, Ann Arbor 1975.

15 Mumford, D.: Tata lectures on theta II, to appear.

16 Mumford, D.,Fogarty, J.: Geometric invariant theory. Ergebnisse der Mathematik 34, Springer Verlag 1982.

17 Shiota, T.: Characterization of Jacobian varieties in terms of soliton equations. Preprint 1984.

18 Welters,G.: On the flexes of the Kummer variety. Preprint 1983.

19 Welters, G.: A criterion for Jacobi varieties. Preprint 1983.

20 Welters, G.: Preprint 1984.

21 Welters, G.: A characterization of non-hyperelliptic Jacobi varieties. Invent. Math. 71 (1983),437-440.

Gerard van der Geer
Mathematisch Instituut
Universiteit van Amsterdam
Roetersstraat 15
1018 WB Amsterdam.

VOJTA'S CONJECTURE

Serge Lang
Department of Mathematics
Yale University
New Haven, CT 06520
U.S.A.

§ 1. Nevanlinna theory

Let $f : \mathbb{C}^d \longrightarrow X$ be a holomorphic map, where X is a complex non-singular variety of dimension d. Let D be an effective divisor on X, with associated invertible sheaf $\mathcal{L}$. Let s be a meromorphic section of $\mathcal{L}$, with divisor $(s) = D$. We suppose that f is non-degenerate, in the sense that its Jacobian is not zero somewhere. For positive real r we define

$$m(D,r,f) = \int_{\mathrm{Bd}\,B(r)} -\log|f^*s|^2 \sigma .$$

where σ is the natural normalized differential form invariant under rotations giving spheres area 1. When $d = 1$, then $\sigma = d\theta/2\pi$. Actually, $m(D,r,f)$ should be written $m(s,r,f)$, but two sections with the same divisor differ by multiplication with a constant, so $m(s,r,f)$ is determined modulo an additive constant. One can select this constant such that $m(s,r,f) \geq 0$, so by abuse of notation, we shall also write $m(D,r,f) \geq 0$.
We also define

$N(D,r,d)$ = normalized measure of the analytic divisor in the ball of radius r whose image under f is contained in D; (Cf. Griffiths [Gr] for the normalization.)

$$N(D,r,f) = \int_0^r [N(D,r,f)-N(D,0,f)] \frac{dr}{r} + N(D,0,f)\log r .$$

$$T(D,r,f) = m(D,r,f) + N(D,r,f) .$$

Remark. If $d = 1$ then $N(D,r,f) = n(D,r,f)$ is the number of points in the disc of radius r whose image under f lies in D.

One formulation of the FIRST MAIN THEOREM (FMT) of Nevanlinna theory runs as follows. *The function* $T(D,r,f)$ *depends only on the linear equivalence class of* D, *modulo bounded functions* $O(1)$.

The first main theorem is relatively easy to prove. More important is the SECOND MAIN THEOREM (SMT), which we state in the following form:

Let D *be a divisor on* X *with simple normal crossings (SNC, meaning that the irreducible components of* D *are non-singular, and intersect transversally). Let* E *be an ample divisor, and* K *the canonical class. Given* ε, *there exists a set of finite measure* $Z(\varepsilon)$ *such that for* r *not in this set,*

$$m(D,r,f) + T(K,r,f) \leqq \varepsilon T(E,r,f) .$$

This is an improved formulation of the statement as it is given for instance in Griffiths [Gr], p. 68, formula 3.5.

§ 2. Weil functions

Let X be a projective variety defined over $\mathbb{C}$ or $\mathbb{C}_p$ (p-adic complex numbers = completion of the algebraic closure of $\mathbb{Q}_p$). Let $\mathcal{L}$ be an invertible sheaf on X and let ρ be a smooth metric on $\mathcal{L}$. If s is a meromorphic section of $\mathcal{L}$ with divisor D, we define the associated Weil function (also called Green's function)

$$\lambda(P) = -\log|s(P)| \quad \text{for} \quad P \notin \text{supp}(D) .$$

If we change the metric or s with the same divisor, λ changes by a bounded smooth function, so is determined mod $O(1)$. We denote such a function by λ_D. It has the following properties:

The association $D \longmapsto \lambda_D$ is a homomorphism mod $O(1)$.

If $D = (f)$ on an open set U (Zariski) then there exists a smooth function α on U such that

$$\lambda_D(P) = -\log |f(P)| + \alpha .$$

If D is effective, then $\lambda_D \geq -O(1)$ (agreeing that values of λ_D on D are then ∞).

If v denotes the absolute value on $\mathbb{C}_v$ then we write

$$v(a) = -\log|a|_v$$

for any element $a \in \mathbb{C}_v$, so we can write

$$\lambda_D = v \circ f + \alpha .$$

In the sequel, metrics will not be used as such; only the associated Weil functions and the above properties will play a role. Note that these Weil-Green functions need not be harmonic. In some cases, they may be, for instance in the case of divisors of degree 0 on a curve. But if the divisor has non-zero degree, then the Green function is not harmonic.

In the sequel, we shall deal with global objects, and then the Weil functions and others must be indexed by v , such as $\lambda_{D,v}$, α_v, etc.

§ 3. Heights (Cf.[La])

Let K be a number field, and let $\{v\}$ be its set of absolute values extending either the ordinary absolute value on $\mathbb{Q}$, or the p-adic absolute values such that $|p|_v = 1/p$. We let K_v be the completion, and K_v^a its algebraic closure. Then we have the product formula

$$\sum_v d_v v(a) = 0$$

where $d_v = \dfrac{[K_v : \mathbb{Q}_v]}{[K : \mathbb{Q}]}$ and $a \in K$, $a \neq 0$. We let $\|a\|_v = |a|_v^{d_v}$

Let $(x_o,\ldots,x_n) \in \mathbb{P}^n(K)$ be a point in projective space over K . We define its height

$$h(P) = \sum_v \log \max_i \|x_i\|_v$$

If $K = \mathbb{Q}$ and $x_o, \ldots, x_n \in \mathbb{Z}$ are relatively prime, then

$$h(P) = \log \max |x_i|$$

where the absolute value is the ordinary one. From this it is immediate that there is only a finite number of points of bounded height and bounded degree.

Let

$$\varphi : X \longrightarrow \mathbb{P}^n$$

be a morphism of a projective non-singular variety into projective space. We define

$$h_\varphi(P) = h(\varphi(P)) \quad \text{for} \quad P \in V(K^a) .$$

The basic theorem about heights states:

There exists a unique homomorphism $c \longmapsto h_c$

$$\mathrm{Pic}(X) \longrightarrow \text{functions from } X(K^a) \text{ to } \mathbb{R} \text{ modulo bounded functions}$$

such that if D *is very ample, and* $\mathcal{O}(D) = \varphi^* \mathcal{O}_{\mathbb{P}}(1)$, *then*

<u>H 1</u>. $\qquad h_c = h_\varphi + O(1)$.

In the above statement, we denote by h_c any one of the functions in its class mod bounded functions. Similarly, if D lies in c, we also write h_D instead of h_c. This height function also satisfies the following properties:

<u>H 2</u>. If D is effective, then $h_D \geq -O(1)$.

H 3. If E is ample and D any divisor, then

$$h_D = O(h_E) .$$

In particular, if E_1, E_2 are ample, then

$$h_{E_1} \gg \ll h_{E_2} .$$

We are using standard notation concerning orders of magnitude. Since according to our conventions, a given height h_E is defined only mod bounded functions, the notation $h_D = O(h_E)$ or $h_D \ll h_E$ means that there exists a constant C such that for all points P with $h_E(P)$ sufficiently large, we have $|h_D(P)| \leq Ch_E(P)$.

Essential to the existence and uniqueness of such height functions h_c is the property of elementary algebraic geometry that given any divisor D , if E is ample, then $D + mE$ is very ample for all $m \geq m_o$.

A fundamental result also states that one can choose metrics ρ_v "uniformly" such that

$$h_D = \sum_v d_v \lambda_{D,v} + O(1)$$

The right hand side depends on Green-Weil functions $\lambda_{D,v}$, and so is a priori defined only for P outside the support of D . Since h_D depends only on the linear equivalence class of D mod $O(1)$, we can change D by a linear equivalence so as to make the right hand side defined at a given point.

Now let S be a finite set of absolute values on K . We define, relative to a given choice of Weil-Green functions and heights:

$$m(D,S) = \sum_{v \in S} d_v \lambda_{D,v}$$

$$N(D,S) = \sum_{v \notin S} d_v \lambda_{D,v}$$

Then

$$h_D = m(D,S) + N(D,S) ,$$

and one basic property of heights says that h_D depends only on the linear equivalence class of D. This is Vojta's translation of FMT into the number theoretic context, with the height h_D corresponding to the function $T(D)$ of Nevanlinnna theory.

Remark. The properties of heights listed above also hold for T, as well as others listed for instance in [La], e.g. if D is algebraically equivalent to 0, then $T(D) = O(T(E))$ for E ample. As far as I can tell, in the analytic context, there has been no such systematic listing of the properties of T, similar to the listing of the properties of heights as in number theory.

Vojta's translation of SMT yields his conjecture:

Let X *be a projective non-singular variety defined over a number field* K. *Let* S *be a finite set of absolute values on* K. *Let* D *be a divisor on* X *rational over* K *and with simple normal crossings. Let* E *be ample on* X. *Give* ε. *Then there exists a proper Zariski closed subset* $Z(S,D,E,\varepsilon) = Z(\varepsilon)$ *such that*

$$m(D,S,P) + h_K(P) \leq \varepsilon h_E(P) \quad \textit{for} \quad P \in X(K) - Z(\varepsilon) .$$

Or in other words,

$$\sum_{v \in S} d_v \lambda_{D,v} + h_K \leq \varepsilon h_E \quad \text{on} \quad X(K) - Z(\varepsilon)$$

where K is the canonical class.

EXAMPLES

Example 1. Let $X = \mathbb{P}^1$, $K = \mathbb{Q}$, $E = (\infty)$. Let α be algebraic, and let

$$f(t) = \prod_{\sigma} (\sigma\alpha - t)$$

where the product is taken over all conjugates $\sigma\alpha$ of α over $\mathbb{Q}$. Let D be the divisor of zeros of f. The canonical class $\mathbb{K}$ is just $-2(\infty)$. A rational point P corresponds to a rational value $t = p/q$ with $p,q \in \mathbb{Z}$, $q > 0$, and p,q relatively prime. We let S consist of the absolute value at infinity. If $|f(p/q)|$ is small, then p/q is close to some root of f. If p/q is close to α, then it has to be far away from the other conjugates of α. Consequently Vojta's inequality yields from the definitions:

$$-\log|\alpha - p/q| - 2h_{\infty}(p/q) \leqq \varepsilon h_{\infty}(p/q)$$

with a finite number of exceptional fractions. Exponentiating, this reads

$$\left|\alpha - \frac{p}{q}\right| \geqq \frac{1}{q^{2+\varepsilon}},$$

which is Roth's theorem.

Remark. Some time ago, I conjectured that instead of the q^{ε} in Roth's theorem, one could take a power of $\log q$ (even possibly $(\log q)^{1+\varepsilon}$). Similarly, in Vojta's conjecture, the right hand side should be replaced conjecturally by $O(\log h_E)$. If one looks back at the Nevanlinna theory, one then sees that the analogous statement is true, and relies on an extra analytic argument which is called the lemma on logarithmic derivatives. Cf. Griffiths [Gr].

Example 2. Let $X = \mathbb{P}^n$ and let $D = L_o+\ldots+L_n$ be the formal sum of the hyperplane coordinate sections, with L_o at infinity, and $E = L_o$. Let φ_i be a rational function such that

$$(\varphi_i) = L_i - L_o \ .$$

Let S be a finite set of absolute values. Note that in the case of $\mathbb{P}^n$, the canonical class K contains $-(n+1)L_o$. Consequently, Vojta's inequality in this case yields

$$\prod_i \prod_{v\in S} \| \varphi_i(P) \|_v \geq \frac{1}{H(P)^{n+1+\varepsilon}}$$

for all P outside the closed set $Z(\varepsilon)$. This is Schmidt's theorem, except that Schmidt arrives at the conclusion that the exceptional set is a finite union of hyperplanes. In order to make Vojta's conjecture imply Schmidt strictly, one would have to refine it so as to give a bound on the degrees of the components of the exceptional set, which should turn out to be 1 if the original data is linear.

Example 3. Let X be a curve of genus ≥ 2 . Take S empty. The canonical class has degree $2g-2$ where g is the genus, and so is ample. Then Vojta's inequality now reads

$$h_K \leq \varepsilon h_E \quad \text{on} \quad X(K) \ ,$$

except for a finite set of points. Since K is ample, such an inequality holds only if $X(K)$ is finite, which is Falting's theorem.

Example 4. This is a higher dimensional version of the preceding example. Instead of assuming that X is a curve, we let X have any dimension, but assume that the canonical class is ample. The same inequality shows that the set of rational points is not Zariski dense.

This goes toward an old conjecture of mine, that if a variety is hyperbolic, then it has only a finite number of rational points. The effect of hyperbolicity should be to eliminate the exceptional Zariski set in Vojta's conjecture. For progress concerning this conjecture in the function field case, cf. Noguchi [No], under the related assumption that the cotangent bundle is ample, and that the rational points are Zariski dense.

To apply the argument of Vojta's inequality it is not necessary to assume that the canonical invertible sheaf is ample, it suffices to be in a situation when for any ample divisor E, $h_E = O(h_K)$. This is the case for varieties of general type, which means that the rational map of X defined by a sufficiently high multiple of the canonical class gives a rational map of dimension $d = \dim X$. Then we have $h^0(mK) \gg m^d$ for m sufficiently large, and we use the following lemma.

<u>Lemma</u>. *Let* X *be a non-singular variety. Let* E *be very ample on* X, *and let* D *be a divisor on* X *such that* $h^0(mD) \gg m^d$ *for* $m \geq m_0$. *Then there exists* m_1 *such that* $h^0(mD-E) \gg m^d$, *and in particular,* $mD-E$ *is linearly equivalent to an effective divisor, for all* $m \geq m_1$.

<u>Proof</u>. First a remark for any divisor D. Let E' be ample, and such that $D + E'$ is ample. Then we have an inclusion

$$H^0(mD) \subset H^0(mD + mE'),$$

which shows that $h^0(mD) \leqq h^0(mD + mE') = \chi(m(D + E'))$ for m large because the higher cohomology groups vanish for m large, so $h^0(mD) \gg m^d$.

Now for the lemma, without loss of generality we can replace E

by any divisor in its class, and thus without loss of generality we may assume that E is an irreducible non-singular subvariety of X. We have the exact sequence

$$0 \longrightarrow \mathcal{O}(mD-E) \longrightarrow \mathcal{O}(mD) \longrightarrow \mathcal{O}(mD)|E \longrightarrow 0 ,$$

whence the exact cohomology sequence

$$0 \longrightarrow H^0(X,mD-E) \longrightarrow H^0(X,mD) \longrightarrow H^0(E,(\mathcal{O}(D)|E)^{\otimes m})$$

noting that $\mathcal{O}(mD)|E = (\mathcal{O}(D)|E)^{\otimes m}$. Applying the first remark to this invertible sheaf on E we conclude that the dimension of the term on the right is $\ll m^{d-1}$, so $h^0(X,mD-E) \gg m^d$ for m large, and in particular is positive for m large, whence the lemma follows.

For $mD - E$ effective, we get $h_E \leqq h_{mD} + O(1)$ as desired.

Example 5. Let A be an abelian variety, and let D be a very ample divisor with SNC. Let S be a finite set of absolute values of K containing the archimedean ones. Let $\varphi_1,\ldots,\varphi_n$ be a set of generators for the space of sections of $\mathcal{O}(D)$. Let $\mathfrak{o}_S$ be the ring of S-integers in K (elements of K which are integral at all $v \notin S$). A point $P \in A(K)$ is said to be S-integral relative to these generators if $\varphi_i(P) \in \mathfrak{o}_S$ for $i = 1,\ldots,n$. On the set of such S-integral points, we have

$$\sum_{v \in S} d_v \lambda_{D,v} = h_D + O(1)$$

immediately from the definitions. The canonical class is 0. Then again Vojta's inequality shows that the set of S-integral points as above is not Zariski dense.

This is in the direction of my old conjecture that on any affine open subset of an abelian variety, the set of S-integral points is finite. However, in this stronger conjecture, we again see the difference between finiteness and the property of not being Zariski dense.

Example 6. Hall's conjecture Marshall Hall conjectured that if x, y are integers, and $x^3-y^2 \neq 0$ then

$$|x^3-y^2| \geqq \max(|x^3|, |y^2|)^{\frac{1}{6}-\varepsilon}$$

with a finite number of exceptions. Actually, Hall omitted the ε, but Stark and Trotter for probabilistic reasons have pointed out that it is almost certainly needed, so we put it in.

Vojta has shown that his conjecture implies Hall's. We sketch the argument. Let

$$f : \mathbb{P}^2_1 \longrightarrow \mathbb{P}^2_2$$

be the rational map defined on projective coordinates by

$$f(x,y,z) = (x^3, y^2z, z^3) .$$

Then f is a morphism except at $(0,1,0)$. We have indexed projective 2-space by indices 1 and 2 to distinguish the space of departure and the space of arrival. We let $L = L_1$ be the hyperplane at infinity on $\mathbb{P}^2_1$, and L_2 the hyperplane at infinity on $\mathbb{P}^2_2$.

Let C be the curve in $\mathbb{P}^2_1$ defined by $x^3-y^2 = 0$. Let φ be the rational function defined by

$$\varphi(x,y) = x^3-y^2 .$$

Then the divisor of φ is given by

$$(\varphi) = C - 3L .$$

In terms of heights, Hall's conjecture can be formulated in the form

$$\log |\varphi(x,y)| \geq \frac{1}{6} h_{L_2} \circ f(x,y) + \text{error term},$$

or if v denotes the ordinary absolute value on $\mathbb{Q}$,

$$v \circ \varphi(x,y) \leq -\frac{1}{6} h_{L_2} \circ f(x,y) + \text{error term}. \tag{1}$$

Note that $v \circ \varphi = \lambda_{(\varphi)}$ is a Weil function associated with the divisor (φ) . Thus Hall's conjecture amounts to an inequality on Weil functions. By blowing up the point of indeterminacy of f and the singularity of C at $(0,0)$, one obtains a variety X and a corresponding morphism $f_1 : X \longrightarrow \mathbb{P}^2$ making the following diagram commutative:

$$\begin{array}{ccc} X & & \\ \downarrow & \searrow^{f_1} & \\ \mathbb{P}^2_1 & \longrightarrow & \mathbb{P}^2_2 . \end{array}$$

The blow ups are chosen so that the exceptional divisor and C have simple normal crossings. By taking D to be their sum together with the hyperplane at infinity, Vojta shows that his conjecture implies Hall's. By a similar technique, Vojta shows that his conjecture implies several other classical diophantine conjectures. I refer the reader to his forthcoming paper on the subject.

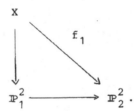

BIBLIOGRAPHY

[Gr] P. Griffiths, *Entire holomorphic mappings in one and several variables*: Hermann Weyl Lectures, Institute for Advanced Study, Institute for Advanced Study, Princeton Univ. Press, Princeton NJ, 1976.

[La] S. Lang, *Fundamentals of Diophantine Geometry*, Springer Verlag, 1984.

[No] J. Noguchi, *A higher dimensional analogue of Mordell's conjecture over function fields*, Math. Ann. 258 (1981) pp. 207-212.

[Vo] P. Vojta, *Integral points on varieties*, Thesis, Harvard, 1983.

A COUNTEREXAMPLE IN 3-SPACE TO A CONJECTURE OF H. HOPF

Henry C. Wente
Department of Mathematics
The University of Toledo
Toledo, Ohio 43606, U. S. A.

In this article we produce a counterexample to the following conjecture of H. Hopf. We shall carefully state the theorems involved in the construction and also provide a geometric description (with suggestive sketches) of the surfaces giving the counterexample. An expanded version complete with proofs is to appear in a paper of the author [8].

Conjecture of Heinz Hopf: If Σ is an immersion of an oriented closed hypersurface in R^n with constant mean curvature $H \neq 0$, then the hypersurface is the standard embedded $(n-1)$-sphere.

If the immersed surface is known to be embedded then a well-known result of A. D. Alexandroff [1] asserts that the conjecture is true. H. Hopf himself [4] showed that if Σ is an immersion of S^2 into R^3 with constant mean curvature then the conjecture is still true. Recently Wu-Yi Hsiang [5] produced an immersion of S^3 into R^4 with constant mean curvature which is not isometric to the standard sphere. However, his construction does not work in the classical dimension (=3) and the conjecture has remained open in this case. We have the following.

Counterexample Theorem: There exist closed immersed surfaces of genus one in R^3 with constant mean curvature. (In fact, we exhibit a countably infinite number of isometrically distinct examples.)

We shall exhibit the surface by producing a conformal mapping of the plane R^2 into R^3 with constant mean curvature which is doubly periodic with respect to a rectangle in the plane. Let $w = (u,v) = u + iv$ represent a typical point in $R^2 = C$ while $\bar{x} = (x,y,z)$ denotes a point in R^3 so that our immersion is given by a function $\bar{x}(u,v)$. We let

$$d\bar{x}\cdot d\bar{x} = ds^2 = E\ (du^2 + dv^2) = e^{2\omega}\ (du^2 + dv^2) \tag{1a}$$

$$-d\bar{x}\cdot d\bar{\xi} = L du^2 + 2M du dv + N dv^2 \tag{1b}$$

be the first and second fundamental forms for the surface. We shall set the mean curvature $H = \frac{1}{2}$. The Gauss and Codazzi-Mainardi equations in this case become (see [4] for details)

$$\Delta\omega + Ke^{2\omega} = 0\ ,\quad K = \text{Gauss curvature} = (LN - M^2)/\,E^2 \tag{2a}$$

$\Phi(w) = (L - N)/2 - iM$ is a complex analytic function. (2b)

Now suppose that $\omega(u,v)$ is a solution to the differential equation

$$\Delta\omega + \sinh\omega\cosh\omega = 0 . \tag{3}$$

If we set $E = e^{2\omega}$, $L = e^{\omega}\sinh\omega$, $M = 0$, and $N = e^{\omega}\cosh\omega$, then it follows that the Gauss and Codazzi-Mainardi equations are satisfied and by a theorem of Bonnet the system can be integrated to yield a surface $\bar{x}(u,v)$, unique up to a Euclidean motion in R^3, having the given fundamental forms. The equations to be integrated are

$$\begin{aligned}
\bar{x}_{uu} &= \omega_u\bar{x}_u - \omega_v\bar{x}_v + L\bar{\xi} \\
\bar{x}_{uv} &= \omega_v\bar{x}_u + \omega_u\bar{x}_v + M\bar{\xi} \\
\bar{x}_{vv} &= -\omega_u\bar{x}_u + \omega_v\bar{x}_v + N\bar{\xi} \\
\bar{\xi}_u &= -k_1\,\bar{x}_u \\
\bar{\xi}_v &= -k_2\,\bar{x}_v
\end{aligned} \tag{4}$$

Here $k_1 = L/E = e^{-\omega}\sinh\omega$, $k_2 = e^{-\omega}\cosh\omega$ so we see that the lines of curvature correspond to lines parallel to the coordinate axes in R^2. Furthermore, the surface is free of umbilic points.

If $\bar{x}(u,v)$ is to be a doubly periodic mapping then so must $\omega(u,v)$. However the converse need not be true. Suppose that $\omega(u,v)$ is a positive solution to the differential equation (3) on a rectangular domain Ω_{AB} lying in the first quadrant with two of its sides on the coordinate axes and the vertex opposite the origin at (A,B). Suppose also that the solution $\omega(u,v)$ vanishes on the boundary of the rectangle. Following the argument used in [3], one can show that $\omega(u,v)$ satisfies the following symmetry properties.

a) $\omega(u,v)$ is symmetric about the lines $u = A/2$ and $v = B/2$. (5)

b) For a fixed v, $0 < v < B$, $\omega(u,v)$ is an increasing function of u, $0 \leqslant u \leqslant A/2$. For a fixed u, $0 < u < A$, $\omega(u,v)$ is an increasing function of v, $0 \leqslant v \leqslant B/2$.

c) $\omega_u(u,0)$ is strictly increasing for $0 \leqslant u \leqslant A/2$.
$\omega_v(0,v)$ is strictly increasing for $0 \leqslant v \leqslant B/2$.

Furthermore, $\omega(u,v)$ can be extended as a solution of the differential equation (3) on all of R^2 by odd reflections across the grid lines $u = mA$, $v = nB$ (m, n integers).

<u>Theorem 2</u>: Suppose $\omega(u,v)$ is a solution to the differential equation (3) on R^2 which is positive on the fundamental rectangle Ω_{AB}, vanishing on the boundary and satisfying the properties (5). The mapping $\bar{x}(u,v)$ obtained by integrating the system (4) is an immersed surface of constant mean curvature $H = \frac{1}{2}$ and satisfying the following symmetry properties.

(6)

a) The curve $\bar{x}((m + \frac{1}{2})A, v)$ lies in a normal plane Π_m with $\bar{x}_u$ as a normal vector to Π_m. If R_m is the reflection map about Π_m in R^3 then $\bar{x}((m + \frac{1}{2})A + u,v) = R_m \circ \bar{x}((m + \frac{1}{2})A - u,v)$.

b) The curve $\bar{x}(u, (n + \frac{1}{2})B)$ lies in a normal plane Ω_n with $\bar{x}_v$ as a normal vector to Ω_n. If R'_n is the reflection map about Ω_n in R^3 then $\bar{x}(u,(n + \frac{1}{2})B + v) = R'_n \circ \bar{x}(u,(n + \frac{1}{2})B - v)$. Each Ω_n is orthogonal to each Π_m.

c) The curve $\bar{x}(u,0)$ is a planar curve lying in a plane Γ_0 which is a tangent plane to the surface at each point. This curve intersects each plane Π_m orthogonally. $\bar{x}_u(u,0)$ is an even function of u. This allows us to conclude that all of the planes Π_m are parallel.

d) The curve $\bar{x}(0,v)$ satisfies the condition $(\bar{x} + \bar{\xi})(0,v) = \bar{c}_o$ a constant vector. Therefore $\bar{x}(0,v)$ lies on a sphere $S(\bar{c}_o,1)$ with center $\bar{c}_o$ and radius one. Similarly $\bar{x}(kA,v)$ lies on a sphere $S(\bar{c}_k,1)$. The points $\bar{c}_k$ lie in every plane Ω_n.

e) $\bar{x}(u + 2A,v) = \bar{x}(u,v) + \bar{b}$ where $\bar{b} = \bar{c}_2 - \bar{c}_o$ is a vector normal to the planes Π_m carrying Π_0 to Π_2.

f) $\bar{x}(u,v + 2B) = \Theta\, \bar{x}(u,v)$ where Θ is a rotation from Ω_o to Ω_2 about their line of intersection, l.

The surface will close up if we can select the rectangle Ω_{AB} so that the translation $\bar{b} = \bar{0}$ (i.e. all the planes Π_m are identical) and so that the rotation angle Θ is a rational multiple of 2π. We use a continuity argument to show that this is possible. The procedure is as follows. Map (via a homothety) all rectangles of similar shape onto a representative rectangle which we select by the standard Schwartz-Christoffel mapping of rectangles onto the unit disk.

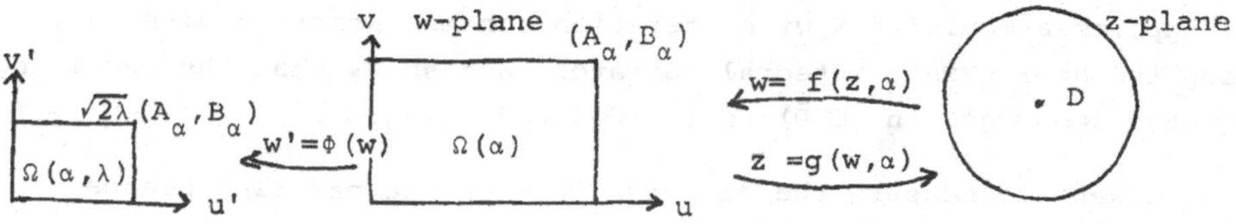

<u>Figure 1</u>: The Fundamental Domain.

We have the following identities satisfied by the various functions defined on the domains pictured in Figure 1.

(7)

a) On $\Omega(\alpha,\lambda)$.

$\Delta\omega + \sinh\omega\cosh\omega = 0$

$\Delta\sigma + \sinh\sigma = 0$ where $\sigma = 2\omega$.

b) On $\Omega(\alpha)$.

$\Delta W + 2\lambda\sinh W\cosh W = 0$ where $W = \omega\circ\Phi$.

$\Delta\Sigma + 2\lambda\sinh\Sigma = 0$ where $\Sigma = 2W$.

c) On the disk D.

$\Delta\Psi + \lambda|f'(z,\alpha)|^2(e^{\Psi} - e^{-\Psi}) = 0$, where $\Psi = \Sigma\circ f$.

$$w = f(z,\alpha) = \int_0^z (t^4 + 2(\cos 2\alpha)t^2 + 1)^{-\frac{1}{2}}\,dt$$

The proof of the existence of positive solutions to the system (7c) on D which vanish on the boundary (and such that small values for λ correspond to large solutions Ψ) is based on a method developed by V.K. Weston [7] and R.L. Moseley [6].

<u>Theorem 3</u>: There exists an open set $O\subset(\alpha,\lambda)$-plane where for each α_1,α_2 with $0 < \alpha_1 < \alpha_2 < \pi/2$ there exists $\tilde{\lambda} = \tilde{\lambda}(\alpha_1,\alpha_2) > 0$ so that $[\alpha_1,\alpha_2]\times(0,\tilde{\lambda}]\subset O$, and a mapping from O to $C(\bar{D})$ denoted by $\Psi(z,\alpha,\lambda)$ such that

a) $\Sigma(w,\alpha,\lambda) = \Psi(g(w,\alpha),\alpha,\lambda)$ is a positive solution to (7b) which vanishes on the boundary.

b) The functions Σ, Σ_u, Σ_v depend continuously on (α,λ) down to $\lambda = 0$ with $\Sigma(w,\alpha,0) = \Sigma_0(w,\alpha) = 4\log(1/|g(w,\alpha)|)$.

c) For $\lambda > 0$ the mapping $(\alpha,\lambda)\longrightarrow\Psi(z,\alpha,\lambda)$ is a continouosly differentiable mapping of O into $C(\bar{D})$.

<u>Remark on the proof</u>: One first constructs a good approximate solution $U_0(z,\lambda)$ with the correct asymtotic limit as λ approaches 0 by using the Liouville form of the exact solution to the differential equation $\Delta V + \lambda e^V = 0$, namely $\lambda e^V = |F'(z)|^2/(1 + |F(z)|^2)^2$ where $F(z)$ is a complex analytic function with at most simple zeros and poles. Then one applies a modified Newton iteration scheme, starting with $U_0(z,\lambda)$ using the appropriate integral operator, and shows that the resulting sequence converges in $C(\bar{D})$ to the desired solution.

We want to measure the distance between the parallel planes Π_0 and Π_1 and wish to show that for certain (α,λ) the distance is zero. It is better to look at the surfaces $\bar{y}(w,\alpha,\lambda) = \bar{x}\circ\Phi(w,\alpha,\lambda)/\sqrt{2\lambda}$ defined relative to the fundamental domain $\Omega(\alpha)$ and to measure the distance between the parallel planes Π'_0 and Π'_1 which correspond to the mapping $\bar{y}$. We do this by looking at the curve $\bar{y}(u,0,\alpha,\lambda)$, a planar curve which cuts through the planes Π'_m orthogonally and has the symmetry indicated in Figure 2.

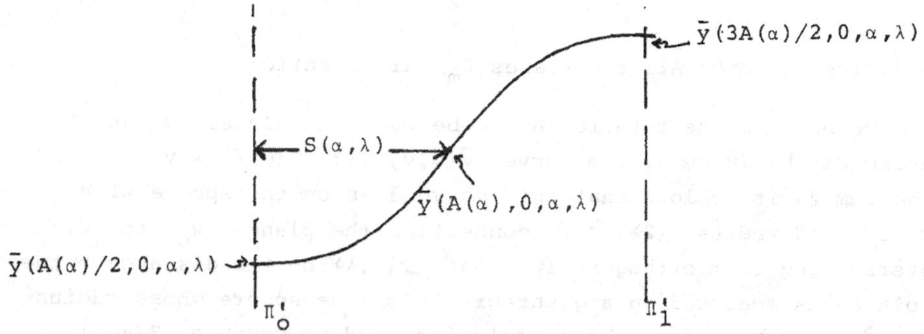

Figure 2: Measuring the Distance between the Parallel Planes Π'_0 and Π'_1

The functions $\bar{y}(u.v,\alpha,\lambda)$ are conformal immersions into R^3 with constant mean curvature $H = \sqrt{2\lambda}$, so that as λ approaches 0 the mean curvature approaches 0 and the mapping tends to a planar map. The functions $\bar{y}$ satisfy a system just like (4) with ω replaced by $W = \Sigma/2$, L is replaced by $\tilde{L} = \sqrt{2\lambda}\ L$ and so on. Since by Theorem 3b the function $W(u,v,\alpha,\lambda)$ approaches $W(u,v,\alpha,0) = 2\log(1/|g(w,\alpha)|)$ as λ approaches 0, the curve $\bar{y}(u,0,\alpha,\lambda)$ approaches a limit curve $\bar{y}(u,0,\alpha,0)$ as λ approaches 0. It follows that the distance function $S = S(\alpha,\lambda)$, as indicated in Figure 2, is continuous down to $\lambda = 0$ and differentiable if λ is positive. Since $W(u,v,\alpha,0)$ is known explicitly one can calculate $S(\alpha,0)$, obtaining

$$S(\alpha,0) = \int_0^{\beta} (\cos 2\theta/(2\cos 2\theta - 2\cos 2\beta)^{\frac{1}{2}})\, d\theta\ ,\ \beta = (\pi/2)-\alpha. \tag{8}$$

We immediately have the following conclusions.

a) $S(\alpha,0)$ is strictly increasing for $0 < \alpha < \pi/2$.

b) $S(\alpha,0)$ approaches $-\infty$ as α approaches 0.

c) $S(\alpha,0)$ is positive for α greater than $\pi/4$.

It follows that there is exactly one value α^*, $0 < \alpha^* < \pi/4$, for which $S(\alpha^*,0) = 0$. We have the following picture (see Figure 3). There is a small rectangle $[\alpha_1,\alpha_2] \times [0,\tilde{\lambda}]$ with $S(\alpha_1,\lambda)$ negative, $S(\alpha_2,\lambda)$ positive, and $S(\alpha^*,0) = 0$. There is a connected set X included in this small rectangle on which S vanishes and which separates the left side of the rectangle from the right side. In particular $(\alpha^*,0)$ is in the set X and every line λ = constant slices into X.

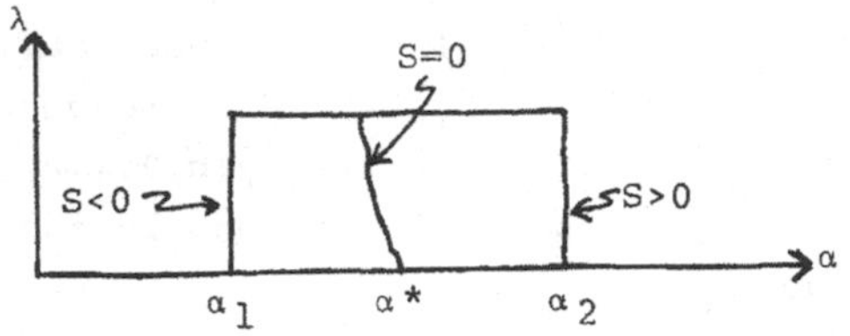

Figure 3: The Set S=0: All the Planes Π_m are Identical.

Now we measure the rotation angle between the planes Ω_o and Ω_1 by looking at the image of the curve $\bar{y}(0,v,\alpha,\lambda)$, $B(\alpha)/2 < v < 3B(\alpha)/2$. From Theorem 2d it follows that this curve lies on the sphere with center $\tilde{c}_o$ and radius $(2\lambda)^{-\frac{1}{2}}$, connecting the planes Ω_o to Ω_1 and intersecting them orthogonally. Let $T(\alpha,\lambda)$ be the distance between these planes as measured on a great circle of the sphere whose radius is $(2\lambda)^{-\frac{1}{2}}$. By repeating the calculation used to compute $S(\alpha,\lambda)$, one finds that for small λ, and α less than $\pi/4$, $T(\alpha,\lambda)$ is positive down to the limit $\lambda = 0$ with the expression for $T(\alpha,0)$ being similar to that for $S(\alpha,0)$. However, for the angle function $\Theta(\alpha,\lambda)$ we have the identity $\Theta(\alpha,\lambda) = (2\lambda)^{\frac{1}{2}} T(\alpha,\lambda)$. This gives us the following:

a) $\Theta(\alpha,\lambda)$ is positive for λ positive.

b) $\Theta(\alpha,\lambda)$ approaches 0 as λ approaches 0.

Since X is a connected set with more than one point (see Figure 3), it follows by continuity that on the set X the function $\Theta(\alpha,\lambda)$ takes on a continuum of values $[0,\varepsilon]$ where ε is positive. Whenever $\Theta(\alpha,\lambda)$ is a rational multiple of 2π the surface will close up. This establishes the existence of a countable number of isometrically distinct immersions of a torus into R^3 with constant mean curvature.

A View of the Immersed Tori.

Let $\Omega = \Omega_{AB}$ be a representative rectangle chosen so that the smallest eigenvalue of the Laplace differential equation

$$\Delta v + \gamma v = 0 \text{ on } \Omega, \quad v = 0 \text{ on boundary } \Omega \tag{9}$$

is $\gamma_1 = 1$. This means that $1 = \gamma_1 = \pi^2((1/A^2)+(1/B^2))$ and in particular A and B are both greater than π. We are to solve the differential equation

$$\Delta W + 2\lambda \sinh W \cosh W = 0 \text{ on } \Omega, \quad W = 0 \text{ on } \delta\Omega. \tag{10}$$

We have the following facts regarding solutions to the differential equation (10).

a) There exists a branch of positive solutions to (10) which bifurcate from the zero solution at $2\lambda = \gamma_1 = 1$ or $\lambda = 1/2$.

b) For any positive solution (W,λ) we must have $0 < \lambda < 1/2$, and for any λ in this interval there exists at least one positive solution.

c) As λ approaches 0 there is a curve of large positive solutions (W,λ) obtained by applying Theorem 3.

It is tempting (but not yet proven) to conjecture that the branch bifurcating from the zero solution at $\lambda = 1/2$ connects up with the branch of large solutions established in Theorem 3. Even more tempting is the following conjecture.

Conjecture: Let (W_1,λ_1) and (W_2,λ_2) be two positive solutions to the system (10). If $0 < \lambda_1 < \lambda_2 < 1/2$ then W_1 is greater than W_2 at every point inside Ω.

For each solution of the system (10) we may apply (7) to get a solution $\omega(u,v)$ to the differential equation (3) and then apply our recipe to construct an immersion $\bar{x}(u,v)$ with constant mean curvature. In the limit case where $W = 0$ the resulting immersion is simply a conformal mapping of the plane onto a circular cylinder whose cross section is a circle of radius one.

In the figures that follow we shall sketch the image $\bar{x}(u,v)$ of a portion of the fundamental rectangle $\sqrt{2\lambda}\ \Omega_{AB}$ as indicated in the first figure and labeled {1,2,3,4,5,6}. A + sign indicates that $\omega(u,v)$ is positive and hence the Gauss curvature of the image surface $K = e^{-2\omega}\sinh\omega\cosh\omega$ is positive, while a - sign indicates that both functions are negative. The rest of the surface is obtained by rotating the surface 180° about the normal line at the image of 2 followed by a series of reflections about the appropriate planes.

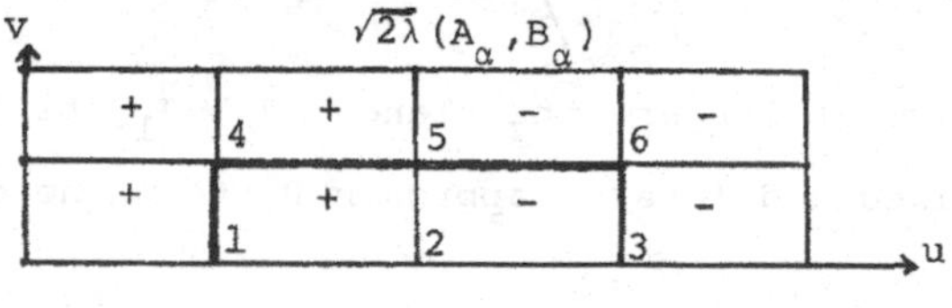

Figure 4: The Fundamental Domain $\sqrt{2\lambda}\ \Omega(\alpha) = \Omega(\alpha,\lambda)$.

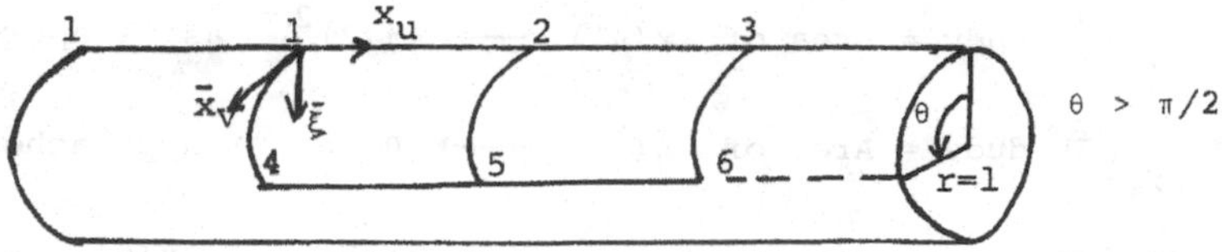

Figure 5: Case 1, W=0, A Pure Cylinder.

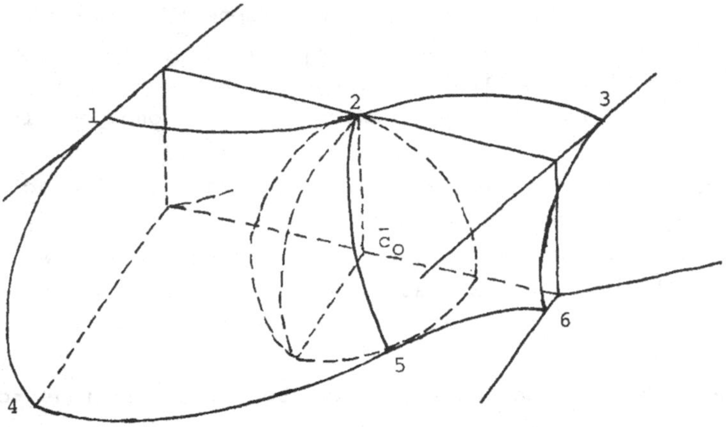

Figure 6: Case 2. W is positive on $\Omega(\alpha,\lambda)$ but not too Large.

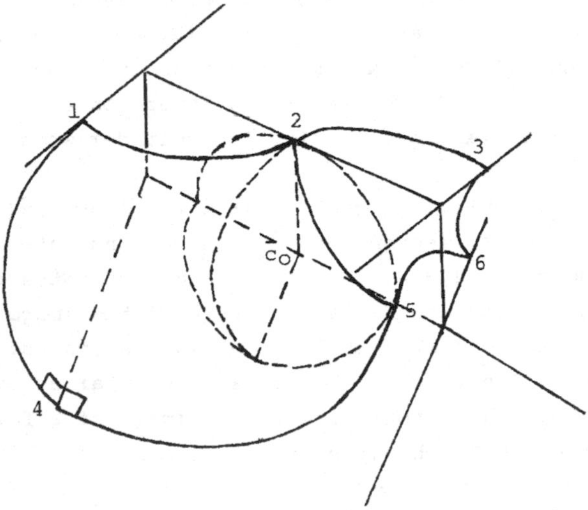

Figure 7: Case 3. W somewhat larger, the Planes Π_o, Π_1 still separated.

If one keeps α fixed and lets λ approach 0, then one can easily show the following.

1) $\int_{\bar{x}(\Omega^+)} K\, dA =$ area of the Gauss map $\longrightarrow 4\pi$ as λ approaches 0.

2) $\int_{\Omega^+} e^{2\omega}\, dudv =$ Area of $\bar{x}(\Omega^+) \longrightarrow 4\pi(2)^2$ as λ approaches 0.

3) $\int_{\Omega^-} e^{2\omega}\, dudv =$ Area of $\bar{x}(\Omega^-) \longrightarrow 0$ as λ approaches 0.

These calculations suggest that as λ approaches 0, $\bar{x}(\Omega^+)$ takes on the shape of a sphere of radius 2.

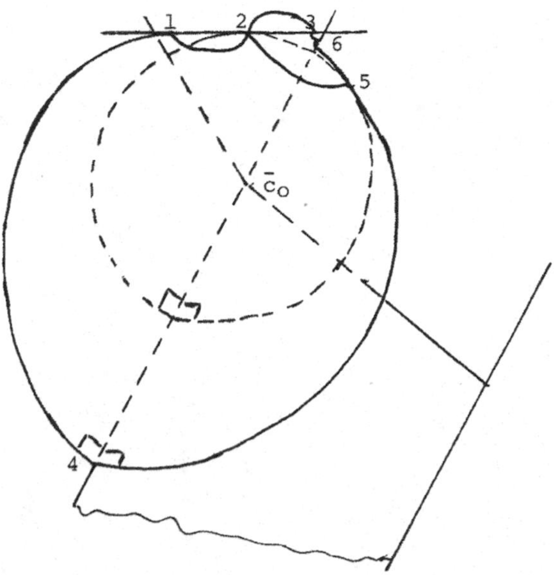

Figure 8: Case 4. The Parallel Planes Π_o, Π_1 are Identical.

If one reflects the sketched Figure 8 about the plane of the paper ($\Pi_o = \Pi_1$) you obtain a surface which resembles a clam shell. Upon rotating this shall 180° about the vertical line c_o-(2) one obtains the other shell. The combined figure is now a clam with the shells opened a bit.

References:

1. A.D. Alexandroff, Uniqueness Theorems for Surfaces in the Large, V. Vestnik, Leningrad Univ. No. 19 (1958) 5-8: Am. Math. Soc. Transl. (Series 2) 21, 412-416.

2. L.P. Eisenhart, A Treatise on the Differential Geometry of Curves and Surfaces, Dover Reprint (1960).

3. B. Gidas, W. Ni, L. Nirenberg, Symmetry and Related Properties via the Maximum Principle, Comm. Math. Physics 68 (1979) No. 3, 209-243.

4. H. Hopf, Differential Geometry in the Large, (Seminar Lectures New York Univ. 1946 and Stanford Univ. 1956) Lecture Notes in Mathematics No. 1000, Springer Verlag, 1983.

5. Wu-Yi Hsiang, Generalized Rotational Hypersurfaces of Constant Mean Curvature in the Euclidean Space I, Jour. Diff. Geometry 17(1982)337-356.

6. J.L. Moseley, On Asymtotic Solutions for a Dirichlet Problem with an exponential Singularity, Rep Amr I, West Virginia University (1981)

7. V.H. Weston, On the Asymtotic Solution of a Partial Differential Equation with an Exponential Nonlinearity, SIAM J. Math Anal 9(1978) 1030-1053.

8. H. C. Wente, Counterexample to a Conjecture of H. Hopf, (to appear) Pac. Jour. of Math.

THE TOPOLOGY AND GEOMETRY OF THE MODULI SPACE OF RIEMANN SURFACES

Scott A. Wolpert*
Department of Mathematics
University of Maryland
College Park, MD 20742

I would like to describe a sampling of recent results concerning the moduli space M_g of Riemann surfaces. My plan is to present several of the ideas underlying the recent work of John Harer on the topology of M_g and of myself on the Hermitian and symplectic geometry. My purpose is not to give a survey; for instance the reader is referred to the papers [7, 9] for the recent progress on the question of whether $\overline{M}_g$, the moduli space of stable curves, is unirational.

The discussion will be divided into two parts: the topology of M_g and $\overline{M}_g$, especially the homology of the mapping class group Γ_g and the geometry of the Weil-Petersson metric. As background I shall start with the basic definitions and notation.

1. Definitions and Notation .

1.1. Let F be a compact topological surface of genus g with r boundary components and s distinguished points in $F - \partial F$; set $S = \partial F \cup \{\text{points}\}$. I shall always assume that $2g - 2 + s + r > 0$ or equivalently that $F - S$ admits a complete hyperbolic metric. Consider $\text{Homeo}^+(F,S)$, the group of orientation preserving homeomorphisms of F restricting to the identity on S and the normal subgroup $I(F,S)$ of homeomorphisms isotopic to the identity fixing S.

<u>Definition</u> 1.1. $\Gamma^s_{g,r} = \text{Homeo}^+(F,S)/I(F,S)$ is the mapping class group for genus g, r boundary components and s punctures.

I shall use the convention that an omitted index is set equal to zero. For genus g and s punctures the mapping class group Γ^s_g acts properly discontinuously on the Teichmüller space T^s_g via biholomorphisms. The quotient M^s_g, the classical moduli space of Riemann surfaces, is a complex V-manifold. To be more specific start by considering triples (R,f,P), where f is a homeomorphism of the topological surface F to a Riemann surface R with $f(S) = P$. An equivalence relation (the marking) is introduced by defining:

---

*Partially supported by the National Science Foundation, Max Planck Institute for Mathematics and Alfred P. Sloan Foundation.

$(R_0,f_0,P_0) \sim (R_1,f_1,P_1)$ provided there is a conformal map k with

$$\begin{array}{ccc} (F,S) & \xrightarrow{f_0} & (R_0,P_0) \\ & \searrow^{f_1} & \downarrow k \\ & & (R_1,P_1) \end{array}$$

commutative modulo a homotopy fixing S and P.

<u>Definition 1.2.</u> T_g^s, the Teichmüller space for genus g and s punctures, is the set of $\sim$ equivalence classes of triples (R,f,P).

Briefly T_g^s is a complex manifold and is homeomorphic to $\mathbb{R}^{6g-6+2s}$. The mapping class group Γ_g^s acts naturally on T_g^s: to the equivalence classes $\{h\} \in \Gamma_g^s$ and $\{(R,f,P)\} \in T_g^s$ assign the class $\{(R,f \circ h,P)\} \in T_g^s$. The action represents Γ_g^s as biholomorphisms of T_g^s.

<u>Definition 1.3.</u> $M_g^s = T_g^s/\Gamma_g^s$ is the moduli space for genus g and s punctures.

As an example the reader will check that for genus 1 and 1 puncture T_1^1 is the upper half plane $H \subset \mathbb{C}$ and Γ_1^1 is the elliptic modular group $SL(2;\mathbb{Z})$ acting on H by linear fractional transformations.

1.2. Now I shall review the definition of the complex structure on T_g^s. For a Riemann surface R with $2g - 2 + s > 0$ consider the hyperbolic metric $\lambda = ds^2$, of constant curvature -1. Associated to R are the L^2 (relative to λ) tensor spaces $H(R)$ of harmonic Beltrami differentials (tensors of type $\frac{\partial}{\partial z} \otimes d\bar{z}$) and $Q(R)$ of holomorphic quadratic differentials (tensors of type $dz \otimes dz$). Of course harmonic is defined in terms of the Laplace Beltrami operator for the hyperbolic metric. A pairing $H(R) \times Q(R) \to \mathbb{C}$ is defined by integration over R: for $\mu \in H(R)$ and $\phi \in Q(R)$ define $(\mu,\phi) = \int_R \mu\phi$. The Ahflors and Bers description of the complex structure of T_g^s is summarized in the diagram

$$\begin{array}{ccc} T^{1,0}T_g^s \times (T^{1,0})^*T_g^s & \longrightarrow & \mathbb{C} \\ & \nearrow_{(\ ,\ )} & \\ H(R) \times Q(R) & & \end{array}$$

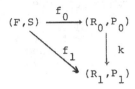

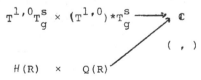

where $T^{1,0}$ is the holomorphic tangent space, $(T^{1,0})^*$ its dual and these spaces are naturally paired [1, 5]. The hyperbolic metric induces a natural inner product on every space of tensors for R. In particular Weil was the first to consider the Hermitian product on $H(R)$ as a metric for T_g^s.

Definition 1.4. Given $\mu,\nu \in H(R)$ the Hermitian product $\langle\mu,\nu\rangle = \int_R \mu\bar{\nu}\lambda^{-1}$ is the Weil-Petersson metric.

The Weil-Petersson metric is Kähler [2,3,10,21] and its Hermitian and symplectic geometry is the subject of Chapter 3.

2. The Homology of the Mapping Class Group.

2.1. In this chapter I shall concentrate on three recent exciting results of John Harer: i) the computation of $H_2(\Gamma_{g,r}^s)$, in brief $H_2(\Gamma_{g,r}^s) \approx \mathbf{Z}^{s+1}$, for $g \geq 5$, ii) the stability theorems, in brief $H_k(\Gamma_{g,r}^s)$ is independent of g and r when $g \geq 3k+1$ and iii) the virtual cohomological dimension of $\Gamma_{g,r}^s$ is $d = 4g-4+2r+s$ for $r+s > 0$, $4g-5$ for $r = s = 0$, in particular $H_k(M_g^s;\mathbf{Q}) = 0$ for $k > d$, [11, 12, 13]. Of course the reader will consult the references for the complete statements especially for the cases of punctures and boundary components.

2.2. A useful technique for computing the homology of a group G is to construct a cell complex C on which G acts cellularly, i.e. cells are mapped to cells. Then the homology of G may be computed from the homology of the quotient C/G and a description of the cell stabilizers. Now $\Gamma_{g,r}^s$ is comprised of isotopy classes of homeomorphisms; $\Gamma_{g,r}^s$ acts on the isotopy invariants of the surface F. An obvious such invariant is the isotopy class of a union of simple loops. Cell complexes with vertices representing unions of simple loops, satisfying appropriate hypotheses, appeared previously in the work of Hatcher-Thurston [15] and Harvey [14]. I shall describe three such complexes (and simple variants) which are at the center of Harer's considerations.

2.3. The cut system complex CS, [11], A cut system $\langle C_1,\ldots,C_g\rangle$ on F is the isotopy class of a collection of disjoint simple closed curves $C_1,\ldots,C_g$ such that $F - (C_1 \cup\ldots\cup C_g)$ is connected. A *simple move* of cut systems is the replacement of $\langle C_i\rangle$ by $\langle C_i'\rangle$ where $C_j = C_j'$ for $j \neq k$ and C_k intersects C_k' once (all intersections are positive). I shall use the convention below that any loop omitted from the notation remains unchanged. Now consider the following sequences of simple moves (see Figure 1).

$$\langle C_i\rangle \;—\; \langle C_i'\rangle \;—\; \langle C_i''\rangle \;—\; \langle C_i\rangle \qquad (R_1)$$

$$\begin{array}{ccc} \langle C_i, C_j\rangle & — & \langle C_i, C_j'\rangle \\ | & & | \\ \langle C_i', C_j\rangle & — & \langle C_i', C_j'\rangle \end{array} \qquad (R_2)$$

$$\langle C_i, C_j\rangle \;—\; \langle C_j, C_j'\rangle \;—\; \langle C_j, C\rangle \;—\; \langle C, C'\rangle \;—\; \langle C_i, C\rangle \;—\; \langle C_i, C_i'\rangle \;—\; \langle C_i, C_j\rangle. \qquad (R_3)$$

Figure 1

The three sequences can be described in terms of the relevant loops (see Figure 2)

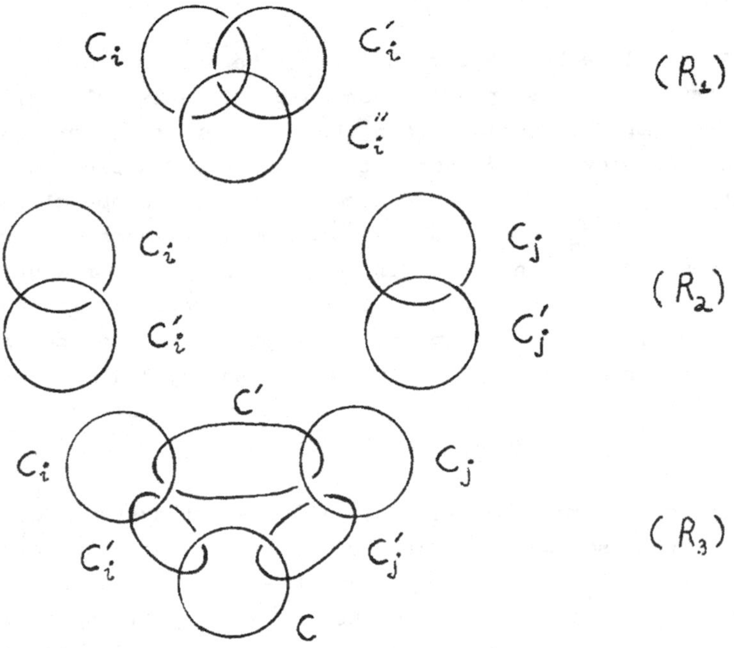

Figure 2

Now define a 0-complex CS_0 with one vertex for each cut system on F; a 1-complex CS_1 by attaching a 1-cell to CS_0 for each simple move; and a 2-complex CS_2 by attaching a 2-cell to CS_1 for each occurence of the cycles R_1, R_2 and R_3. Hatcher and Thurston prove that CS_2 is connected and simply connected. Harer simplifies the description of CS_2 and then attaches 3 cells to obtain a 2-connected 3 complex CS_3, [11]. The stabilizer of a cut system (a vertex of CS_3) is essentially a braid group and Harer analyzes the cell stabilizers of CS_3. Combining this information with an analysis of the homology of the quotient CS_3/Γ_g he obtains the following theorem, [11].

Theorem 2.1. $H_2(\Gamma^s_{g,r};\mathbb{Z}) = \mathbb{Z}^{s+1}$, $g \geq 5$.

The reader will find a slight difference between the above statement and that found in [11]. John Harer has assured me that the above is indeed the correct statement.

The homology group $H_2(\Gamma)$ also admits an interpretation as bordism classes of fibre bundles $F \to W^4 \to T$ with T a closed oriented surface. In particular two bundles are bordant if they cobound a 5-manifold fibering over a 3-manifold with fibre F. The bundle $F \to W^4 \to T$ has s canonical sections $\sigma_1,\ldots,\sigma_s$ defined by the distinguished points of F. In this setting H_2 is spanned by the $s + 1$ natural invariants of W: $\sigma(W)/4$, σ the signature and $\sigma_j \# \sigma_j$, $j = 1,\ldots,s$ the self-intersection numbers of the sections.

2.4. Certainly a basic question is whether or not the homology of the mapping class group falls into any pattern. As an example consider the analogous question for A_n, the coarse moduli space of principally polarized n dimensional abelian varieties. Borel in a fundamental paper computed the rational cohomology of A_n and found that $H^i(A_n;\mathbb{Q})$ is independent of n for n large relative to i, [31]. Recently Charney and Lee have extended these results to $\bar{A}_n$, the Satake compactification of A_n, [8]. Harer establishes the analogous result for $H_k(\Gamma^s_{g,r})$: the answer is independent of g and r provided $g \geq 3k + 1$. In fact the reader will find in Theorem 2.2 that Harer establishes much more but first I would like to mention the work of E. Miller, [18]. Starting with the work of Harer as a basis Miller observes that the boundary connected sum for surfaces $F_{g,1} \# F_{h,1} = F_{g+h,1}$ induces the structure on $A = \varinjlim H_*(\Gamma_{g,1};\mathbb{Q})$ of a polynomial algebra on even generators with an exterior algebra on odd generators. Furthermore by generalizing an example of Atiyah for the generator of $H_2(\Gamma_g)$ Miller is able to find a generator in each even dimension.

To give a precise statement of Harer's stability theorem it is necessary to consider the following three maps of surfaces

$$\Phi: \ F^s_{g,r} \to F^s_{g,r+1} \ , \quad r \geq 1$$

$$\Psi: \ F^s_{g,r} \to F^s_{g+1,r-1}, \quad r \geq 2 \quad \text{and}$$

$$\Xi: \ F^s_{g,r} \to F^s_{g+1,r-2}, \quad r \geq 2$$

where Φ and Ψ are given by sewing on a pair of pants (a copy of $F^0_{0,3}$) (see Figure 3).

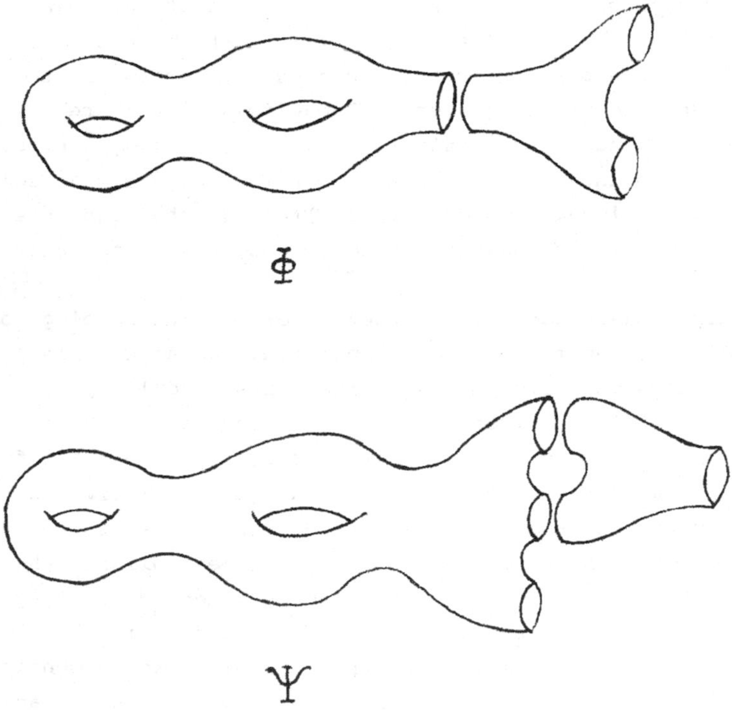

Figure 3

along one boundary for Φ, two for Ψ and Ξ is given by sewing together two boundary components (see Figure 4)

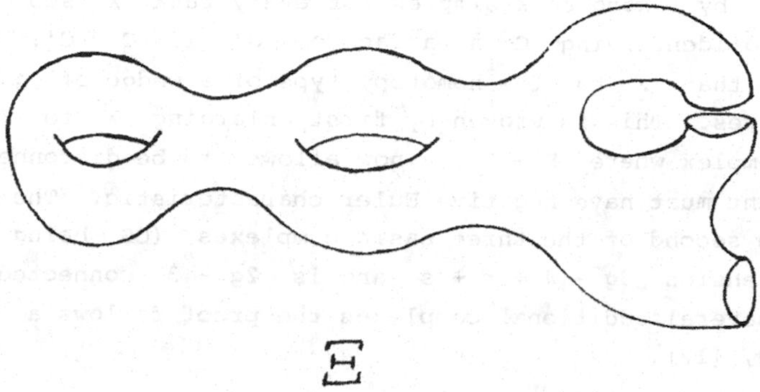

Figure 4

Certainly the maps induce homomorphisms of mapping class groups. The following results for the induced maps on homology with integer coefficients can be found in [12].

Theorem 2.2.

$\Phi_*: H_k(\Gamma^s_{g,r}) \to H_k(\Gamma^s_{g,r+1})$ is an isomorphism

for $k > 1$, $g \geq 3k - 2$, $r \geq 1$

$\Psi_*: H_k(\Gamma^s_{g,r}) \to H_k(\Gamma^s_{g+1,r+1})$ is an isomorphism

for $k > 1$, $g \geq 3k - 1$, $r \geq 2$

$\Xi_*: H_k(\Gamma^s_{g,r}) \to H_k(\Gamma^s_{g+1,r-2})$ is an isomorphism

for $g \geq 3k$, $r \geq 2$.

Corollary 2.3. $H_k(\Gamma^s_{g,r})$ is independent of g and r for $g \geq 3k + 1$.

The proof of a stability theorem for $\Gamma^s_{g,r}$ requires suitable

Γ complexes whose connectivity increases with g. Harer starts with the complex X a variant of the basic complex Z. In order to define X consider the *sub cut system* of rank k i.e. the isotopy class of $k + 1$ disjoint simple loops $C = \{C_0, \ldots, C_k\}$ such that $F - \{C_0, \ldots, C_k\}$ is connected. Define the simplicial complex X of dimension $g - 1$ by taking a k-simplex for every rank k sub cut system of F and identifying C as a face of C' if $C \subset C'$. The first theorem is that X has the homotopy type of a wedge of $g - 1$ dimensional spheres. This is proven by first enlarging X to Z, the analogous complex where $F - C$ is now allowed to be disconnected but each component must have negative Euler characteristic. The complex Z, the second of the three basic complexes (CS being the first), has dimension $3g - 4 + r + s$ and is $2g - 3$ connected. After studying several additional complexes the proof follows a standard outline, [12].

2.5. For the sake of simplicity I shall only discuss one part of Harer's results on the virtual cohomological dimension: that M_g^1 has the homotopy type of a $4g - 3$ dimensional spine. The discussion starts with the description of a Γ_g^1 invariant ideal triangulation of T_g^1. The triangulation arises from the following result of Strebel, [19].

<u>Theorem 2.4</u>. Given a compact Riemann surface R and a point p there exists a unique meromorphic quadratic differential ϕ on R such that

i) ϕ has exactly one pole,

ii) in terms of an appropriate complex coordinate z in a neighborhood of p $\phi = \frac{dz^2}{z^2}$ and

iii) the real trajectories of ϕ are closed.

The differential ϕ may also be described by starting with the differential $\frac{dz^2}{z^2}$ on the disc $D = \{|z| \leq 1\}$ and identifying arcs on ∂D (linearly in radian measure) to obtain the pair (R,ϕ). To see this consider the following simple example (see Figure 5) where one obtains a surface of genus 2; in general any pattern for a genus g surface will occur.

The data for a pair (R,ϕ) is merely the combinatorial pattern for identifying arcs on ∂D, as well as their lengths. In order to record this information for each pair of arcs consider a loop γ

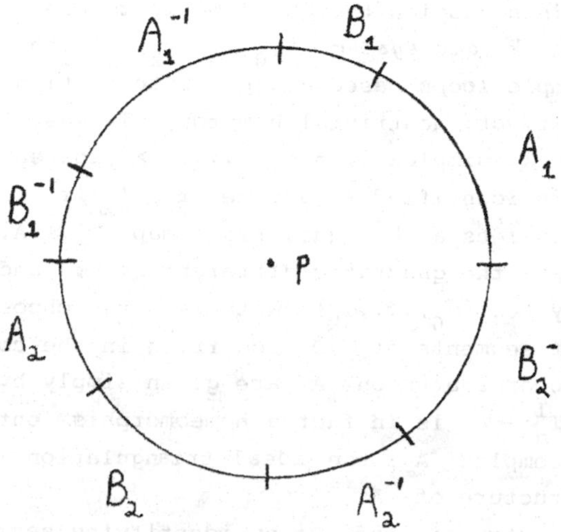

Figure 5

based at p, formed by rays from the origin with endpoints on ∂D identified under the arc pairing and assign to γ a weight w equal to the length of the arcs on ∂D; to (R,ϕ) assign the tuple (γ_j, w_j). Specifically for the loop intersecting A_1 in the above example the picture is (see Figure 6)

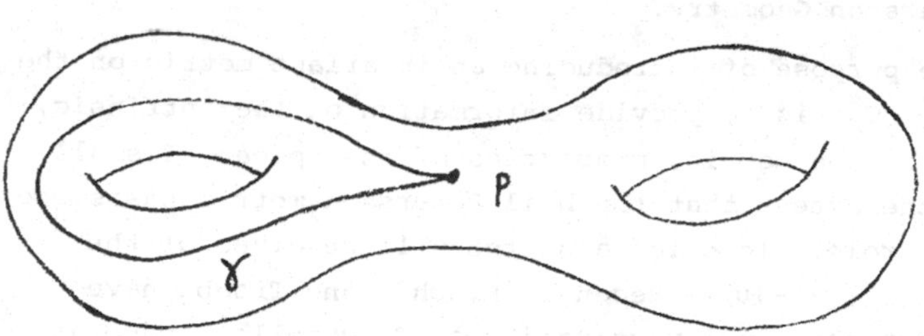

Figure 6

To interpret this situation the third basic complex A is introduced. A *rank* k *arc system* $\{\alpha_0,\dots,\alpha_k\}$ is the isotopy class of k + 1 simple loops based at p, intersecting only at p, and representing distinct, nontrivial homotopy classes. A is the cell complex with a k simplex $\alpha = \langle\alpha_0,\dots,\alpha_k\rangle$ for each rank k arc system and α is identified as a face of α' if $\alpha \subset \alpha'$. Strebel's theorem provides a Γ equivariant map $T_g^1 \to A$. To the pair (R,p) associate the quadratic differential ϕ and to ϕ associate the arc system $\{\alpha_0,\dots,\alpha_k\}$ with weights, choosing one arc for each pair of segments of ∂D occurring in the construction of ϕ. Simplicial coordinates on A are given simply by the weights. The map $T_g^1 \to A$ is in fact a homeomorphism onto the complement of a subcomplex A_∞; the ideal triangulation is the pull-back of the cell structure of A.

Recalling the construction of R by identifying segments on ∂D, observe that every pattern must have at least 4g segments. In particular the rank k arc systems arising from Strebel's theorem have rank $\geq 2g - 1$; the 2g - 2 skeleton of A is contained in A_∞. Now introduce the dual complex Y of $A - A_\infty$. Accordingly Y will not have cells for the 2g - 2 skeleton of A. Thus the dimension of Y is $\dim T_g^1 - (2g-1) = 4g - 3$. Harer shows that T_g^1 may be Γ equivariantly retracted onto Y.

The case of T_1^1 provides a helpful example. A is the standard $SL(2;\mathbb{Z})$ tessellation of H, A_∞ is the rational points of $\mathbb{R} = \partial H$ and Y is Serre's tree for $SL(2;\mathbb{Z})$.

3. The Weil-Petersson Geometry.

3.1. Ideally the purpose of introducing an invariant metric on the Teichmüller space T_g^s is to provide information on the intrinsic, i.e. independent of the metric, properties of the space. I shall try to indicate the extent that the Weil-Petersson metric has successfully filled this role. In brief a sketch will be given of the results in [2, 3, 17, 22-30]. Recently Fischer and Tromba have independently undertaken an investigation of Teichmüller space and the W-P metric, substituting the viewpoint of Riemannian geometry for the classical theory of quasiconformal maps, [20, 21].

As background the reader may check [2, 3, 10, 21] for proofs that the metric is Kähler. Recall that the W-P metric is invariant for the action of Γ_g^s on T_g^s and thus descends to the moduli space of Riemann surfaces M_g^s. On both T_g^s and M_g^s the metric is not complete, [22]. Now this result is best understood in terms of the latter results [17, 26] that the metric has an extension to $\overline{M^s_g}$, the Deligne-Mumford stable curve compactification of M^s_g.

The study of the W-P geometry can be divided into three major topics: curvature considerations, extension of the metric to the compactification $\overline{M^s_g}$, and the symplectic geometry of the Kähler form. As a sample of the results I shall start by sketching three applications. The Kähler form ω_{WP} extends to the compactification $\overline{M^s_g}$; the extension $\overline{\omega}_{WP}$ defines a cohomology class in $H^2(\overline{M^s_g})$, [26, 27, 28]. In [27] it is found that $\frac{1}{2\pi^2}\overline{\omega}_{WP}$ is actually the first Chern class of a known line bundle $\overline{\kappa}_1$ (discussed below) on $\overline{M^s_g}$. In particular $\frac{1}{2\pi^2}\overline{\omega}_{WP}$ is a rational class and the line bundle $\overline{\kappa}_1$ is positive. At this point the Kodaira theorem may be quoted to obtain a purely analytic proof that $\overline{M^s_g}$ is projective algebraic. As a second application consider the Nielsen conjecture: every finite subgroup of Γ^s_g fixes a point of T^s_g. An important ingredient of Kerckhoff's proof of the conjecture is that the *geodesic length functions* ℓ_* are convex along Thurston's earthquake paths, [16]. I have recently found that the length functions ℓ_* are strictly convex along the W-P geodesics [30]. This result provides the basis of an independent but similar proof of the conjecture. The proof starts with an observation of Fricke-Klein: that a suitable sum $S = \sum_j \ell_j$ of length functions will be a proper function on T^s_g. Given $\tilde{\Gamma} \subset \Gamma^s_g$, a finite subgroup, then the sum $S_0 = \sum_{\gamma\in\tilde{\Gamma}} S(\gamma)$ is $\tilde{\Gamma}$ invariant and is also a sum of length functions. Now S_0 is strictly convex along W-P geodesics and thus a critical point is necessarily a relative minimum (S_0 is an index 0 Morse function). Since S_0 is proper it follows that it has a unique minimum and finally the $\tilde{\Gamma}$ invariance of S_0 guarantees that the minimum is fixed by $\tilde{\Gamma}$, the desired conclusion. And finally since the W-P metric is Kähler it follows immediately that the length functions ℓ_* are in fact plurisubharmonic; this observation leads to a new proof that T^s_g is a Stein manifold, [30].

3.2. Ahlfors was the first to consider the curvature of the metric; he obtained singular integral formulas for the Riemann curvature tensor, [3]. As an application he found that the Ricci, holomorphic sectional and scalar curvatures are all negative. Royden later showed that the holomorphic sectional curvature is bounded away from zero and more recently Tromba has found that the general sectional curvature is indeed negative. I shall now present a simple formula for the curvature tensor, [29]. Recall first that the holomorphic tangent space $T^{1,0}\,T_g^s$ at $\hat{R} \in T_g^s$ is identified with the space of harmonic Beltrami differentials $H(R)$ and that dA denotes the hyperbolic area element on R and D the hyperbolic Laplacian. The Riemann curvature is a 4-tensor in particular for $\mu_\alpha, \mu_\beta, \mu_\gamma, \mu_\delta \in H(R)$

$$R_{\alpha\bar{\beta}\gamma\bar{\delta}} = -2\int_R (D-2)^{-1}(\mu_\alpha\bar{\mu}_\beta)(\mu_\gamma\bar{\mu}_\delta)\,dA$$

$$= -2\int_R (D-2)^{-1}(\mu_\alpha\bar{\mu}_\delta)(\mu_\gamma\bar{\mu}_\beta)\,dA$$

where $(D-2)^{-1}$ is the indicated self adjoint operator and observe for $\mu, \nu \in H$ that the product $\mu\bar{\nu}$ is a function. Starting with the above formula it follows that the metric has negative sectional curvature, that the holomorphic sectional and Ricci curvatures are bounded above by $\frac{-1}{2\pi(g-1)}$ and that the scalar curvature is bounded above by $\frac{-3(3g-2)}{4\pi}$. In fact the arguments show that the curvatures are governed by the spectrum of the Laplacian: the negative curvature is a manifestation of the nonpositivity of the Laplacian. These last results have also been obtained by Royden.

As a further application of the techniques I wish to consider the characteristic classes of the Teichmüller curve $\mathcal{T}_g$. $\mathcal{T}_g$ is the natural fibre space over T_g; the fibre above $\hat{R} \in T_g$ is a compact submanifold isomorphic to R. If $\pi : \mathcal{T}_g \to T_g$ is the projection then the kernel of the differential $d\pi : T^{1,0}\mathcal{T}_g \to T^{1,0}T_g$ defines a line bundle (v) on $\mathcal{T}_g$, the vertical bundle of the fibration. The restriction of (v) to a fibre of π is simply the tangent bundle of the fibre; by the uniformization theorem the hyperbolic metric induces a metric on (v). I have computed the curvature 2-form for this metric and found that it is negative, a pointwise version of Arakelov's result that the dual $(v)^*$ is numerically effective, [4]. Once again the curvature is governed by the spectrum of the Laplacian. By integrating powers of the Chern class $c_1(v)$ over the fibers of π one obtains Γ_g invariant characteristic classes $\kappa_n(p) = \int_{\pi^{-1}(p)} c_1(v)^{n+1}$ defined on T_g. Mumford has many intriguing results on the behavior of these classes in the cohomology ring: in particular the κ_n, $n \le 3g-3$ are nontrivial and many geometrically interesting cycles may be written in terms of the κ_n. Mumford guesses that the low dimensional part of the cohomology ring may actually be polynomial in the κ_n. As an example in [29] I find that in fact $\kappa_1 = \frac{1}{2\pi^2}\omega_{WP}$, ω_{WP} the W-P Kähler form.

3.3. Masur was the first to consider the extension of the W-P metric from $\mathcal{M}_g$ to $\overline{\mathcal{M}}_g$, $\overline{\mathcal{M}}_g$ the moduli space of stable curves. In the paper [17] Masur develops the foundations of this topic. Recall now that the compactification locus $\mathcal{D} = \overline{\mathcal{M}}_g - \mathcal{M}_g$ is a divisor with normal crossings. Briefly Masur shows if $\mathcal{D}$ is given locally as $z_1 = 0$ in a coordinate chart $z = (z_1,\ldots,z_n)$ then

$$ds^2_{WP} = \frac{dz_1 d\bar{z}_1}{|z_1|^2(\log 1/|z_1|)^3} + O\left(\frac{1}{|z_1|(\log 1/|z_1|)^3}\right), \qquad [17].$$

In particular the W-P length of a differentiable curve in $\overline{\mathcal{M}}_g$ is finite and the metric extends to a complete metric on $\overline{\mathcal{M}}_g$. Starting from Masur's results it follows that the Kähler form ω_{WP} extends to a closed, positive, (1,1) current $\bar{\omega}_{WP}$ on $\overline{\mathcal{M}}_g$, [26]. In fact the above singularity is sufficiently mild for $\bar{\omega}_{WP}$ to be written as $\partial\bar{\partial}F$, F a *continuous* function (as an example consider $F = \frac{1}{\log 1/|z|}$), [28]. Standard approximation techniques may then be applied to conclude that $\bar{\omega}_{WP}$ is the limit of smooth Kähler forms in its cohomology class. Thus even though the Kähler form is not smooth it is suitable for applications in particular the Kodaira theorem may be quoted to conclude that there is a projective embedding. Now the discussion will continue with the results on describing the cohomology class of $\bar{\omega}_{WP}$.

3.4. The divisor $\mathcal{D} \subset \overline{\mathcal{M}}_g$ is reducible $\mathcal{D} = \mathcal{D}_0 \cup \ldots \cup \mathcal{D}_{[\frac{g}{2}]}$, where the generic surface represented in $\mathcal{D}_k$ has one node separating it into components of genus k and genus $g - k$ (for $k = 0$ the node is nonseparating). Certainly the divisors $\mathcal{D}_k$ define cohomology classes in $H_{6g-8}(\overline{\mathcal{M}}_g)$ and by Poincaré duality (over $\mathbb{Q}$, $\overline{\mathcal{M}}_g$ is a V-manifold) $\bar{\omega}_{WP}$ also defines a class in H_{6g-8}. The first result is contained in the following, [27].

<u>Theorem 3.1</u>. $\{\mathcal{D}_0,\ldots,\mathcal{D}_{[\frac{g}{2}]},(\bar{\omega}_{WP})\}$ is a basis for $H_{6g-8}(\overline{\mathcal{M}}_g;\mathbb{Q})$.

The sketch of the proof is simple enough. By the result of Harer on $H_2(\mathcal{M}_g)$ and an application of Mayer-Vietoris one verifies at the outset that $H_2(\overline{\mathcal{M}}_g;\mathbb{Q})$ has rank $2 + [\frac{g}{2}]$. A candidate basis is then presented for each of H_2 and H_{6g-8} and the intersection pairing is evaluated. The pairing is found to be nonsingular and the proof is complete. The trick for evaluating the integrals of $\overline{\omega}_{WP}$ is to perform a single integration [27] and then deduce the remaining integrals by formal properties.

As an example a 2-cycle E for $\overline{M}_g$ is obtained by considering the family of (stable) curves given as the one point sum of a *fixed* surface S_0 of genus $g - 1$ and an elliptic curve E, which will vary over all (even degenerate) structures represented in its moduli space $\overline{M_1^1}$ (see Figure 7).

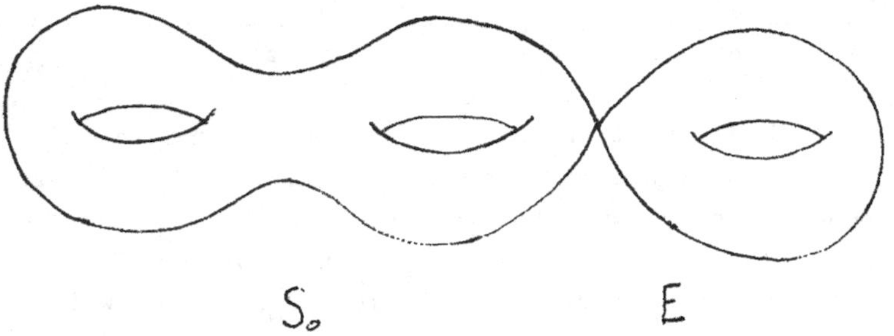

Figure 7

The 2-cycle E is parametrized by $\overline{M_1^1}$. Now to state the desired formal property of ω_{WP} let ω_g^s be the W-P Kähler form for $\overline{M_g^s}$. Then briefly $\omega_g^0|_E = \omega_1^1|_{\overline{M_1^1}}$; the restriction of ω_g^0 to E is naturally identified with the Kähler form for $\overline{M_1^1}$. And so the integral $\int_E \overline{\omega}_{WP}$ is reduced to the $\overline{M_1^1}$ case. This last integral may be evaluated directly; the value is $\frac{\pi^2}{6}$, [25]. The formal properties of the Kähler form will be discussed further in section 3.6 as consequences of the Fenchel-Nielsen coordinates.

After evaluating the intersection pairing one finds that

$$\frac{1}{2\pi^2}\,\overline{\omega}_{WP} = \overline{\kappa}_1$$

where $\overline{\kappa}_1$ is the extension of the class κ_1 discussed in section 3.2. Indeed the above is the generalization to $\overline{M}_g$ of the earlier result $\frac{1}{2\pi^2}\omega_{WP} = \kappa_1$ for M_g. Finally the basic techniques for constructing cycles and computing intersections may be applied to the higher homology groups. For instance in [27] it is shown that $H_{2k}(\overline{M}_g)$, $k < g$ has rank at least $\frac{1}{2}\binom{g-1}{k}$.

3.5. The last two sections will be devoted to the symplectic geometry of the Kähler form ω_{WP}. The symplectic geometry of the triple $(\omega_{WP}, t_*, \ell_*)$, t_* the Fenchel-Nielsen vector fields and ℓ_* the geodesic length functions, is *dual* to the trigonometry, as will be described below, of geodesics in the hyperbolic metric of a surface.

A construction of Fenchel-Nielsen provides for natural flows on Teichmüller space. Fix the free homotopy class of a nontrivial simple loop $\{\alpha\}$ on F and an increment δ of time. If $\hat{R} = \{(R,f)\} \in T_g$ then $\{f(\alpha)\}$ is represented on $\hat{R}$ by a unique geodesic $\alpha_{\hat{R}}$. Cut R open along $\hat{\alpha}_R$, rotate one *side* of the cut relative to the other (by a distance of δ) and then glue the sides in their new position. The hyperbolic structure in the complement of the cut extends naturally to define a hyperbolic structure on the new surface. A geodesic γ intersecting $\alpha_{\hat{R}}$ is deformed to a broken geodesic γ_b with endpoints separated δ units along α. As δ varies a flow, the F-N flow, is defined on the Teichmüller space T_g. The infinitesimal generator of the flow is the F-N vector field t_α. The free homotopy class $\{\alpha\}$ also determines a function ℓ_α, the geodesic length function, on T_g. In brief define $\ell_\alpha(\hat{R})$, $\hat{R} \in T_g$ to be the length of $\alpha_{\hat{R}}$ and the exterior derivative $d\ell_\alpha$ will also be discussed. The basic formula of the symplectic geometry is a duality formula

$$\omega_{WP}(t_\alpha, \ ) = -d\ell_\alpha, \qquad [23, 24, 26].$$

An immediate consequence is that the symplectic form is invariant under the F-N flows on T_g, in particular the flows are W-P volume preserving. There are also formulas for the Lie derivatives $t_\alpha \ell_\beta$ and $t_\alpha t_\beta \ell_\gamma$, [24].

$$\omega(t_\alpha, t_\beta) = t_\alpha \ell_\beta = \sum_{p \in \alpha \# \beta} \cos \theta_p \tag{3.1}$$

$$t_\alpha t_\beta \ell_\gamma = \sum_{(p,q)\in\alpha\#\gamma\times\beta\#\gamma} \frac{e^{\ell_1}+e^{\ell_2}}{2(e^{\ell_\gamma}-1)} \sin\theta_p \sin\theta_q \qquad (3.2)$$

$$-\sum_{(r,s)\in\alpha\#\beta\times\beta\#\gamma} \frac{e^{m_1}+e^{m_2}}{2(e^{\ell_\beta}-1)} \sin\theta_r \sin\theta_s .$$

The right hand side of (3.1) evaluated at $\hat{R} \in T_g$ is the sum of cosines of the angles at the intersections of the geodesics $\alpha_{\hat{R}}$ and $\beta_{\hat{R}}$. Similarly the right hand side of (3.2) is a sum of trigonometric invariants for pairs of intersections; ℓ_1 and ℓ_2 are the lengths of the segments of γ defined by p, q and likewise, for m_1 and m_2 relative to β. Recently W. Goldman has generalized these formulas to the representation space $\mathrm{Hom}(\pi_1(F),G)/G$, G a Lie group with nondegenerate symmetric bilinear form on its Lie algebra, [10]. Consequences of the above formulas are considered in [25, 26]. In particular if $\alpha \# \beta \neq \emptyset$ then $t_\alpha t_\alpha \ell_\beta > 0$ and (3.2) represents a quantitative version of Kerckhoff's observation that the geodesic length functions are convex along earthquake paths, [16]. Finally note that the infinitesimal generators of Thurston's earthquake flows form the completion (in the compact-open topology) of the F-N vector fields.

3.6. Introducing coordinates on Teichmüller space is a question of parametrizing Riemann surfaces. Fenchel-Nielsen suggested a particularly simple solution to this problem. It starts with the observation that the lengths of alternating sides of a right hexagon in the hyperbolic plane may be chosen arbitrarily. Given such a hexagon, form its metric double across the remaining sides to obtain the basic object P, a pair of pants (see Figure 8).

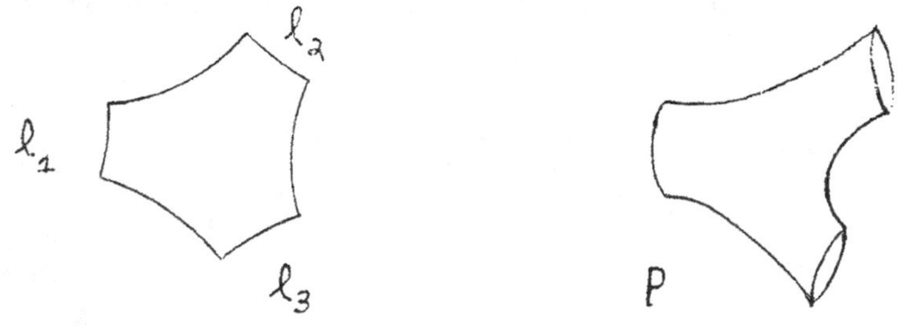

Figure 8

Topologically P is the complement of three disjoint discs in S^2; metrically P is a hyperbolic surface with geodesic boundary. The key observation is that pants P and P' may be metrically summed along their boundaries provided merely that the boundaries are of equal length. Now fixing a combinatorial pattern, then summing the pants $P_1, \ldots, P_{2g-2}$, one obtains the general genus g surface (see Figure 9).

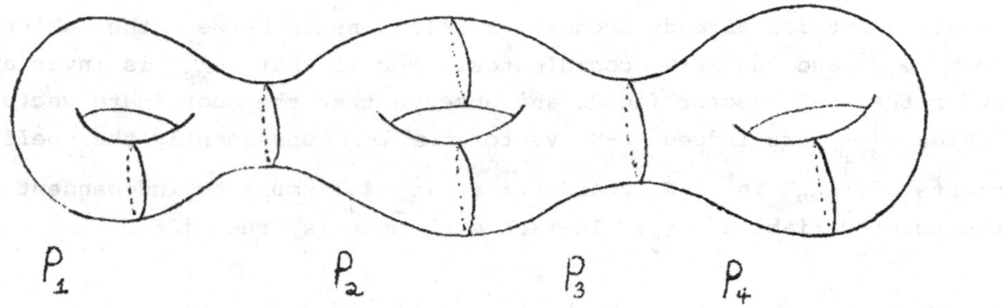

Figure 9

The coordinates for T_g are simply the free parameters for this construction. There are exactly two parameters at each summing locus. Of course the first is simply the length ℓ of the locus, this varies freely in $\mathbb{R}^+$. The second, the twist parameter τ, measures the net displacement between the boundaries. The parameter τ is defined to be the hyperbolic distance between the feet of perpendiculars dropped from appropriate boundaries (see Figure 10).

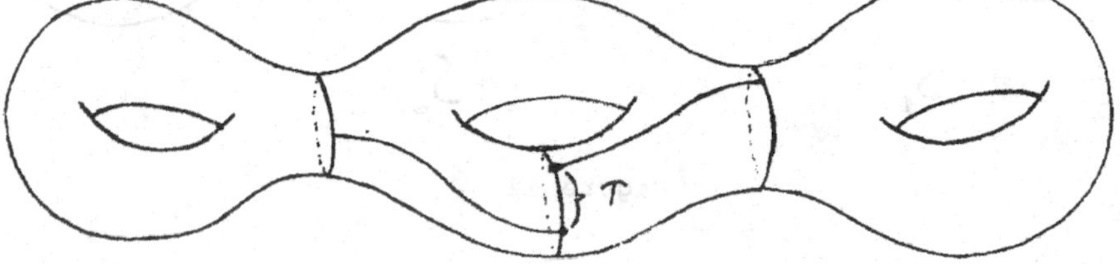

Figure 10

After an initial choice τ is determined by analytic continuation and varies freely in $\mathbb{R}$. In this way Fenchel-Nielsen established the following result, [1, 26].

Theorem 3.2. The map $T_g \to (\mathbb{R}^+ \times \mathbb{R})^{3g-3}$ given by $\hat{R} \to (\ell_j, \tau_j)_{j=1}^{3g-3}$ is a homeomorphism.

In particular T_g is a cell. Furthermore the Deligne-Mumford compactification $\overline{M}_g$ can be constructed from M_g by simply allowing the length parameters ℓ_* to vanish, [1,26]. The discussion of the previous section already suggests a relationship between the Kähler form ω_{WP} and the F-N coordinates. Recall that ω_{WP} is invariant under the F-N vector fields and observe that the coordinate vector fields $\frac{\partial}{\partial \tau_j}$ are indeed F-N vector fields. Consequently the coefficients of ω_{WP} in F-N coordinates (τ_j, ℓ_j) must be independent of the twist variables τ_j. In fact much more is true, [26].

Theorem 3.3. $\omega_{WP} = \sum_j d\ell_j \wedge d\tau_j$.

Briefly (ω, τ_j, ℓ_j) is a completely integrable Hamiltonian system. The Kähler form ω_{WP} is Γ_g invariant in particular the 2-form $\sum_j d\ell_j \wedge d\tau_j$ is independent of the combinatorial pattern for combining pants. Finally the discussion will be concluded with two applications of the above formula. Set a length parameter ℓ_k equal to zero to obtain a degenerate surface S (see Figure 11)

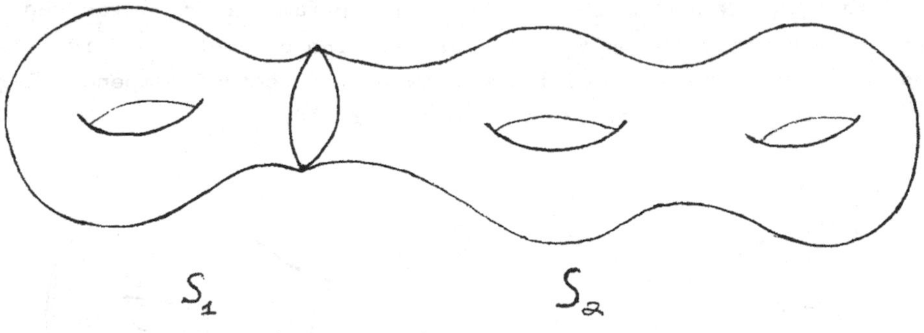

Figure 11

the sum at punctures of surfaces S_1 and S_2. It follows immediately from the above formula that ω_S converges to the sum $\omega_{S_1} + \omega_{S_2}$ as $\ell_k \to 0$. Briefly stated the Kähler form for a sum of surfaces is the sum of the component forms, [27]. To demonstrate the second formal property, behavior for an unramified covering $R \to S$ of surfaces, consider the following example (see Figure 12).

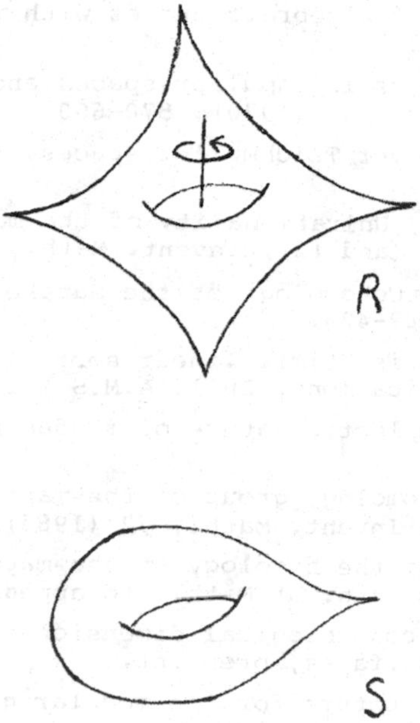

Figure 12

R is a 4 punctured torus, S is a one punctured torus and the covering transformation is the order 4 rotation about the axis of the hole. Introducing F-N coordinates relative to the indicated loops the reader will easily check that $\omega_R = 4\omega_S$; the Kähler form multiplies under unramified coverings, [27].

REFERENCES

1. W. Abikoff, Topics in the Real Analytic Theory of Teichmüller Space, L.N.M. 820 Springer-Verlag, New York, 1980.
2. L. V. Ahlfors, Some remarks on Teichmüller's space of Riemann surfaces, Ann. of Math., 74 (1961), 171-191.
3. L. V. Ahlfors, Curvature properties of Teichmüller space, J. Analyse Math., 9 (1961), 161-176.
4. S. Arakelov, Families of algebraic curves with fixed degeneracies, Izv. Akad. Nauk., 35 (1971).
5. L. Bers, On boundaries of Teichmüller spaces and on Kleinian groups, I, Ann. of Math., 91 (1970), 570-600.
6. L. Bers, Fibre spaces over Teichmüller spaces, Acta. Math., 130 (1973), 89-126.
7. M. C. Chang and Z. Ran, Unirationality of the moduli spaces of curves of genus 11, 13 (and 12), Invent. Math., to appear.
8. R. Charney and R. Lee, Cohomology of the Satake compactification, Topology, 22 (1983), 389-423.
9. D. Eisenbud and J. Harris, Limit linear series, the irrationality of M_g and other applications, Bull. A.M.S., 10 (1984), 277-280.
10. W. M. Goldman, The symplectic nature of fundamental groups of surfaces, preprint.
11. J. Harer, The second homology group of the mapping class group of an orientable surface, Invent. Math., 72 (1983), 221-239.
12. J. Harer, Stability of the homology of the mapping class groups of orientable surfaces, Ann. of Math., to appear.
13. J. Harer, The virtual cohomological dimension of the mapping class groups of orientable surfaces, preprint.
14. W. Harvey, Boundary structure for the modular group, Ann. of Math. Studies, 97 (1978), 245-251.
15. A. Hatcher and W. Thurston, A presentation for the mapping class group of a closed orientable surface, Topology, 19 (1980), 221-237.
16. S. P. Kerckhoff, The Nielsen realization problem, Ann. of Math., 117 (1983), 235-265.
17. H. Masur, The extension of the Weil-Petersson metric to the boundary of Teichmüller space, Duke Math. J., 43 (1976), 623-635.
18. E. Miller, The homology of the moduli space and the mapping class group, preprint.
19. K. Strebel, On quadratic differentials with closed trajectories and second order poles, J. Analyse Math., 19 (1967), 373-382.
20. A. E. Fischer and A. J. Tromba, On a purely "Riemannian" proof of the structure and dimension of the unramified moduli space of a compact Riemann surface, Math. Ann., 267 (1984), 311-345.
21. A. E. Fischer and A. J. Tromba, On the Weil-Petersson metric on Teichmüller space, preprint.
22. S. A. Wolpert, Noncompleteness of the Weil-Petersson metric for Teichmüller space, Pac. J. Math., 61 (1975), 573-577.
23. S. A. Wolpert, The Fenchel-Nielsen deformation, Ann. of Math., 115 (1982), 501-528.

24. S. A. Wolpert, On the symplectic geometry of deformations of a hyperbolic surface, Ann. of Math., 117 (1983), 207-234.

25. S. A. Wolpert, On the Kähler form of the moduli space of once punctured tori, Comment. Math. Helv., 58 (1983), 246-256.

26. S. A. Wolpert, On the Weil-Petersson geometry of the moduli space of curves, Amer. J. Math., to appear.

27. S. A. Wolpert, On the homology of the moduli space of stable curves, Ann. of Math., 118 (1983), 491-523.

28. S. A. Wolpert, On obtaining a positive line bundle from the Weil-Petersson class, Amer. J. Math., to appear.

29. S. A. Wolpert, Chern forms and the Riemann tensor for the moduli space of curves, preprint.

30. S. A. Wolpert, Geodesic length functions and the Nielsen problem, preprint.

31. A. Borel, Stable real cohomology of arithmetic groups, Ann. Sci. Ecole Norm. Sup. 4e(7) (1974), 235-272.

Anhang

Programme

der

25 Arbeitstagungen von 1957 - 1984

1. Mathematische Arbeitstagung 1957

Die Akten enthalten kein offizielles Programm. Aus der Korrespondenz konnten folgende Vortragsthemen festgestellt werden.

1) Vortragsserie A. Grothendieck: Kohärente Garben und verallgemeinerte Riemann-Roch-Hirzebruch-Formel auf algebraischen Mannigfaltigkeiten

2) M.F. Atiyah sprach für einen größeren Kreis über das Thema

"Some examples of complex manifolds"

Dieser Vortrag von Atiyah wurde in den Bonner Mathematischen Schriften Nr. 6 (1958) veröffentlicht.

Das Programm der Tagung sah ungefähr wie folgt aus:

Samstag, den 13.7.:	
17.30 Uhr	A. Grothendieck
Sonntag, 14.7.:	
10.15 Uhr	Ausflug mit Dampferfahrt
Montag, 15.7.:	
9.30 Uhr	A. Grothendieck
15.15 Uhr	M.F. Atiyah
17.15 Uhr	N. Kuiper
Dienstag, 16.7.:	
11.30 Uhr	A. Grothendieck
15.15 Uhr	F. Hirzebruch
Mittwoch, 17.7.:	
11.30 Uhr	A. Grothendieck
15.15 Uhr	M.F. Atiyah
Freitag, 19.7.:	J. Tits
Samstag, 20.7.:	M.F. Atiyah
	H. Grauert

2. Mathematische Arbeitstagung 1958

Die Akten enthalten kein gedrucktes Programm. Die Tagung fand vom 9. bis 16. Juli 1958 statt. Nach privaten Aufzeichnungen haben die folgenden Herren vorgetragen:

S. Abhyankar
R. Bott
H. Grauert
A. Grothendieck
M. Kervaire
J. Milnor
D. Puppe
R. Remmert
J.-P. Serre
K. Stein
R. Thom

3. Mathematische Arbeitstagung 1959

Die Akten enthalten nur folgende Angaben:

Samstag, den 11.7.:

17.30 - 18.30 Uhr A. Borel: Something on transformation groups

Sonntag, den 12.7.:

10.30 - 11.30 Uhr J.-P. Serre: Fundamental groups

12.00 - 13.00 Uhr D. Puppe: Semisimplicial monoid complexes

Montag, den 13.7.:

9.30 - 10.30 Uhr S. Lang: "....

11.00 - 12.00 Uhr J. Tits: "....

Dienstag, den 14.7.:

10.00 - 11.00 Uhr K. Stein: On proper maps

11.30 - 12.30 Uhr J.-P. Serre: Proalgebraic groups

17.00 - 17.30 Uhr Pictures

17.30 - 18.30 Uhr M.F. Atiyah: Fibre homotopy type

Mittwoch, den 15.7.:

10.00 - 11.00 Uhr A. Borel: On torsion in Lie-groups

11.30 - 12.30 Uhr H. Grauert: How to blow down

17.30 - 18.30 Uhr J. Milnor: Scissor and paste arguments

Nach privaten Aufzeichnungen haben außerdem die Herren J.F. Adams, A. Dold und N. Kuiper vorgetragen.

4. Mathematische Arbeitstagung 1960 (11. - 17. Juni)

M.F. Atiyah:	Sheaves and vector bundles

J. Milnor:	Spherical modifications
S. Smale:	Poincaré's conjecture in higher dimensions (I)
M. Kervaire:	A manifold without differentiable structures (I)

J. Stallings:	Combinatorial homotopy-spheres
S. Smale:	Poincaré's conjecture in higher dimensions (II)
M. Kervaire:	A manifold without differentiable structure (II)

H. Grauert:	Some problems concerning non-compact Kähler manifolds
S. Lang:	Cross-sections on algebraic families of curves
	Pictures
A. Dold:	Representable functors and tensor-products

R. Thom:	Singular homology and sheaf theory
R. Remmert:	Rigid complex manifolds

J.F. Adams:	Old-fashioned topology
A. Borel:	Non-singular homology and sheaf theory
J.P. Serre:	L-series

N. Kuiper:	Curvature of index k
R. Bott:	Cauchy formula
M. Hirsch:	Embedding theorems

5. Mathematische Arbeitstagung 1961 (16.- 23. Juni)

A. Borel:	Arithmetic subgroups of Lie groups I

J. Milnor:	The handle body theorem I
M.F. Atiyah:	Finiteness theorem for comapct Lie groups
J. Stallings:	A 5-dimensional example against the Hauptvermutung

---

A. Borel:	Arithmetic subgroups of Lie groups II
J. Milnor:	The handle body theorem II
J. Milnor:	Handle body theorem and results of Wall

A. Andreotti:	Vanishing theorems
M. Kneser:	Approximation in algebraic groups
M. Kneser:	Siegel's theorem and Tamagawa numbers
A. Grothendieck:	Schemes of moduli

B. Eckmann:	Lusternik-Schnirelmann category
J. Tits:	Flags and Bruhat's theorem
R. Palais:	Conjugacy of compact diffeomorphism groups

J. Kohn:	Potential theory on non compact complex manifolds
A. Grothendieck:	Duality theorems in algebraic geometry
R. Remmert:	On homogeneous compact Kähler manifolds

N. Kuiper:	Manifolds admitting functions with few critical points
A. Shapiro:	Graded Clifford modules

F. Peterson:	Squaring operations in a sphere bundle
I. Porteous:	Homomorphisms of vector bundles

6. Mathematische Arbeitstagung 1962 (13. - 20. Juli)

S. Lang:	On the Nash embedding theorem à la Moser

S. Smale:	Stable manifolds of a diffeomorphism
H. Hironaka:	Resolution of singularities
R. Swan:	The Grothendieck ring of a finite group

M.F. Atiyah:	Harmonic spinors
C.T.C. Wall:	Classification problems in differential topology
R. Remmert:	On homogeneous compact complex manifolds

J. Stallings:	Piecewise linear approximation of stable homeomorphisms
M. Kervaire:	2-spheres in 4-manifolds
	Pictures
K. Stein:	Extension of meromorphic mappings

A. Haefliger:	Links
R. Abraham:	Transversality of mappings
H. Grauert:	Rigid singularities

J. Eells - J.H. Sampson:	Harmonic maps

W. Browder:	Homotopy type of differentiable manifolds
M.F. Atiyah:	Explanations of my preceding lecture
N. Kuiper:	Smoothing problems

A. Kosinski:	Piecewise linear functions on combinatorial manifolds
J.C. Moore:	Hopf algebras

7. Mathematische Arbeitstagung 1963 (14. - 21. Juni)

S. Lang:	Transcendental numbers

M.F. Atiyah:	Elliptic operators
J. Eells:	Deformations of maps
M. Kervaire:	Higher dimensional knots

F. Oort:	On automorphisms of varieties
H. Grauert:	On Super-Cocycles
S. Lang:	On Manin's theorem

A. Haefliger:	Combinatorial manifolds
J. Cerf:	$\Gamma_4 = 0$
	Pictures
R. Palais:	Integro-differential operators

F. Hirzebruch:	Elementary proof of Bott's periodicity theorem
S. Abhyankar:	Jungian singularities
A. Dold:	Obstruction theory for cohomology functors

R. Thom:	On generic singularities of envelopes (with slides)

R. Palais:	Integro-differential operators (II)
M.F. Atiyah:	Boundary value problems
N. Kuiper:	Smoothing of combinatorial manifolds

R. Palais:	Morse theory
V. Poenaru:	Thickening and unknotting
Á. Császár:	Allgemeine Approximationssätze

8. Mathematische Arbeitstagung 1964 (15. - 19. Juni)

A. Borel:	Introduction to automorphic forms

N. Kuiper:	The unitary group of Hilbert space is k-connected
R. Wood:	Generalization of Bott's periodicity theorem
S.S. Chern:	Holomorphic vector bundles

W. Haken:	Poincaré's conjecture in dimension three
M. Kneser:	Galois cohomology of p-adic linear groups
W. Browder:	Introduction to cobordism theory

D. Gromoll:	Exotic spheres and metrics of positive curvature
J. Eells:	Variational theory on manifolds
	Pictures
R. Sacksteder:	Foliated manifolds

K. Jänich:	(X,Fredholm) = K(X)

E. Thomas:	Enumeration of vector bundles
J.F. Adams:	Some applications of K-theory to homotopy theory
D. Anderson:	Several aspects of K-theory

R. Wood:	Pre-Palais
R. Palais:	Index theorem for elliptic boundary problems
A. Borel:	Pseudo-concavity for arithmetic groups

A. Kosinski:	Is the Hauptvermutung true for manifolds?
Shih Weishu:	Characteristic classes in K-theory
R. Palais:	The homotopy type of some infinite dimensional manifolds

9. Mathematische Arbeitstagung 1965 (19. - 25. Juni)

M.F. Atiyah:	Elliptic operators on manifolds and generalized Lefschetz fixed point theorem I

A. Douady:	An hour of counterexamples
M.F. Atiyah:	Elliptic operators on manifolds and generalized Lefschetz fixed point theorem II
S. Lang:	Division points on curves

W. Klingenberg:	Closed geodesics
R. Thom:	Topological models for morphogenesis in biology
H. Hironaka:	Projectiveness criterions of Kleiman

A. Grothendieck:	Riemann-Roch I
J.P. Serre:	Formal groups

A. Grothendieck:	Riemann-Roch II
J. Milnor:	Projective class groups in topology
N. Kuiper:	Piecewise linear microbundles are bundles

G. Harder:	Galois cohomology of semi-simple groups
I.G. Macdonald:	Spherical functions on p-adic groups
D. Epstein:	Duality theorems for abelian schemes

A. Van de Ven:	Almost complex manifolds
D. Husemoller:	Cohomology theory
G. Segal:	Equivariant K-theory
R. Palais:	Symplectic manifolds (im Rahmen des Mathematischen Kolloquiums)

10. Mathematische Arbeitstagung 1966 (16. - 22. Juni)

M.F. Atiyah: Global aspects of elliptic operators

H. Grauert: Non-archimedean analysis
J. Eells: Deformations of structures
M. Kervaire: Congruence subgroups after H. Bass

J.P. Serre: l-adic Galois groups
L. Siebenmann: Applications of Wall's invariant
S. Smale: On the structure of diffeomorphisms

M.F. Atiyah: Index theorem (I)
J. Tate: p-adic Galois representations
R. Bott: Vector fields and characteristic numbers

M.F. Atiyah: Index theorem (II)
F. Hirzebruch: Exotic spheres and singularities

S. Smale: Group constructions in the theory of diffeomorphisms
A. Borel: Rigidity of arithmetic groups
J. Milnor: Singularities of hypersurfaces

J.F. Adams: Mahowald's result on the J-homomorphism
A. Van de Ven: Chern numbers of (almost-) complex surfaces
D. Sullivan: Manifolds with singularities

11. Mathematische Arbeitstagung 1967 (16. - 22. Juni)

M.F. Atiyah:	Hyperbolic equations and algebraic geometry

J.P. Serre:	Congruence subgroups and Coxeter hyperbolic groups
I. Šafarevič:	Simple Lie algebras in finite characteristics
D.V. Anosov:	Dynamical systems

N. Kuiper:	Algebraic equations for combinatorial 8-manifolds
J.W.S. Cassels:	Definite functions as sums of squares
M. Postnikov:	K-theory for infinite complexes

I. Šafarevič:	Algebraic analogue of uniformisation
D.V. Anosov:	Asymptotic theory of some partial differential equations
M. Karoubi:	Real and complex K-theory

H. Levine:	Extending immersions of the circle in the plane
B. Venkov:	Cohomology of some groups

J.F. Adams:	Complex cobordism
A. Brumer:	p-adic L-functions
J.I. Manin:	On rational surfaces

R. Abraham:	Generic properties of Hamiltonian systems
T. Matsumoto:	Congruence subgroups and central group extensions
K. Jänich:	Report on part of the Tulane Conference

12. Mathematische Arbeitstagung 1969 (13. - 20. Juni)

N. Kuiper:	Stable homeomorphisms and the annulus conjecture, Kirby's results

---

P.A. Griffiths:	Algebraic cycles
J.F. Adams:	Quillen's work on cobordism and formal groups
R. Bott:	Topological obstructions for foliations

W.-C. Hsiang:	Manifolds with fundamental group $\mathbb{Z}^k$
W. Schmid:	Langlands' conjecture
P. Deligne:	Hodge theory of singular varieties

A. Borel:	Picard-Lefschetz transformations and arithmetic groups
F. Hirzebruch:	S^1-actions on manifolds

L. Siebenmann:	Topological manifolds and related examples
J. Mather:	Stratification of a generic mapping
D. Quillen:	Homotopy theory of schemes

J. Tate:	K_2 of global fields
R. Thom:	Topological linguistics
D. Gromoll:	Periodic geodesics
R. Gardner:	Geometric solution of the Gauchy Problem and a generalization of characteristics

---

C.T.C. Wall:	Integrable almost piecewise linear structures
J. Eells:	Topology of Banach manifolds
G. Harder:	A remark on Tamagawa numbers

A. Bak:	Quadratic modules and unitary K-theory
A. Weinstein:	Symplectic manifolds
W.C. Hsiang:	Falsity of the s-cobordism theorem for lower dimensional manifolds

13. Mathematische Arbeitstagung 1970 (12. - 19. Juni)

C.T.C. Wall:	A survey of free actions on spheres

S. Lang:	Transcendental mappings
R. Kirby:	Triangulation of manifolds
M. Kervaire:	Projective class groups and class field theory

S. Hildebrandt:	Boundary values in capillarities
D. Burghelea:	Homotopy groups of spaces of diffeomorphisms
T. tom Dieck:	Bordism of commuting involutions

S. Smale:	Report on work of Moulton (celestial mechanics)
D.V. Anosov:	New examples of smooth ergodic systems

R. Kiehl:	Satz von Grauert (direkte Bilder von Garben)
T.A. Springer:	Discrete series of finite Chevalley groups
E. Brieskorn:	Singular elements in simple Lie groups

R. Takens:	Partially hyperbolic fixed points
H. Bass:	The Milnor ring of a field
J. Eells:	Fredholm structures and Wiener integrals

J.-L. Verdier:	Serre's duality theorem
H.V. Pittie:	The representation ring of compact Lie groups
F. Waldhausen:	Attempt on higher Withehead groups

R. Gardner:	Rigidity and uniqueness of convex hypersurfaces
U. Koschorke:	Pseudo-kompakte Teilmengen unendlich dimensionaler Mannigfaltigkeiten

14. Mathematische Arbeitstagung 1971 (10. - 14. Juni)

M.F. Atiyah:	The Riemann-Roch theorem for multihomogeneous varieties

---

W. Schmid:	The singularities of Griffith's period mapping
I.G. Macdonald:	Affine root systems and theta-identities
P. Deligne:	The Weil conjecture for surfaces of degree 4 in P_3

S. Lang:	Frobenius automorphisms of modular function fields
G. Lusztig:	The Novikov higher signatures and families of elliptic operators
F. Waldhausen:	Applications of infinite simple homotopy types

---

G. Segal:	Algebraic K-theory
C.T.C. Wall:	Quadratic forms
H. Grauert:	Deformation of singularities

D. Mumford:	Degeneration of curves and non-archimedean uniformisation
R. Bott:	The classifying space of foliations

R. Thom:	The four vertex theorem
W.-Ch. Hsiang:	A reduction theorem of differentiable actions
J. Tits:	Free groups in linear groups

M. Artin:	Lüroth's theorem
E. Bombieri:	On pluricanonical models of algebraic surfaces
A.I. Kostrikin:	Deformation of Lie algebras

R. Langlands:	On arithmetically equivalent representations
J. Simons:	Geometric invariants related to characteristic classes
K. Shiohama:	On the differentiable "pinching problem"

15. *Mathematische Arbeitstagung* 1972 (09. - 15. Juni)

W. Schmid:	Degeneration of algebraic manifolds
M. Karoubi:	Hermitian K-theory
T.A. Springer:	Steinberg functions on a finite Lie algebra
J. Cheeger:	Manifolds with non-negative curvature
T. Petrie:	Real algebraic actions on projective spaces
D.I. Liebermann:	n^{th} order de Rham theory
E. Winkelnkemper:	Open book decomposition of manifolds
A. Bak:	Computation of surgery obstruction groups
A. Michenko:	On infinite dimensional representations of discrete groups
M.F. Atiyah:	Invariants of odd dimensional manifolds
S. Lichtenbaum:	Values of zeta-functions at negative integers
F.J. Almgren:	Geometric measure theory and elliptic variational problems
M. Platonov:	Conjectures of Artin and Kneser-Tits
J.W. Robbin:	Topological classification of linear endomorphisms of R^n
K. Ueno:	Classifications of algebraic varieties
M. Miranda:	Hypersurfaces in R^n of prescribed mean curvature
M. Shub:	Dynamical systems, filtrations and entropy
J.A. Shaneson:	Codimension -2 problems and homology equivalent manifolds

16. Mathematische Arbeitstagung 1974 (12. - 18. Juni)

I.M. Singer:	η-invariant and its relation to real quadratic fields

---

H. Bass:	Russian progress on Serre's problem
N.A'Campo:	Resolution and deformation of plane curve singularities
J.-P. Serre:	Modular forms and Galois representations

J. Eells:	Introduction to stochastic Riemannian geometry

R. Howe:	The Weyl representation over finite fields
G. Lusztig:	Some discrete series representations of finite classical groups
S. Kobayashi - S. Lang:	Hyperbolic geometry and diophantine problems

---

D. Epstein:	Foliations with compact leaves
D. Quillen:	Finite generation of K-groups in the function field case
D. Zagier:	Modular forms in one and two variables

G. Harder:	Betti numbers of modular spaces of vector bundles
W. Schmid:	On the discrete series of semi-simple Lie groups
E. Ruh:	Equivariant pinching problems
K. Ueno:	Canonical bundle formula for certain fibre spaces and algebraic varieties of parabolic type

---

M.F. Atiyah:	Asymptotic properties of eigenvalues in Riemannian geometry
C.T.C. Wall:	Norms of units in group rings
S. Lang:	Fermat curves and units in the modular function field

17. Mathematische Arbeitstagung 1975 (21. - 27. Juni)

M.F. Atiyah:	Algebras of operators in Hilbert space and K-theory

---

J. Moser:	Isospectral deformations
G. Lusztig:	Macdonald's conjecture on discrete series of finite Chevalley groups
B. Kostant:	The η-function formula of Macdonald

R. MacPherson:	Gelfand's formula for the first Pontrjagin class
W. Ziller:	Closed geodesics and homotopy symmetric spaces
J.-P. Serre:	Lower bounds of discrimimants of number fields

T.tom Dieck:	Burnside ring of a compact Lie group

J.C. Jantzen:	Modular representations of semi simple groups
W. Schmid:	Blattner's conjecture on the discrete series of semi simple real Lie groups
A.N. Varchenko:	Newton diagrams of singularities

A.I. Kostrikin:	Tannaka-Artin's conjecture on the multiplicative group of division algebras
B. Mazur:	Rational points on modular curves
W. Casselman:	The $\mathfrak{n}$-cohomology of representations of semi simple Lie groups

---

A.N. Parshin:	Residues and symbols
J.H.M. Steenbrink:	Mixed Hodge structure on vanishing cycles
E. Calabi:	Nearly flat triangulations of Riemannian manifolds

18. Mathematische Arbeitstagung 1977 (21. - 27. Juni)

M.F. Atiyah:	The classical geometry of Yang-Mills Fields (I)

J.-P. Serre:	Function field analogue of $SL_2(\mathbb{Z})$
H.W. Lenstra:	Euclidean number fields
A. Tromba:	Recent progress in Plateau's problem

M.F. Atiyah:	The classical geometry of Yang-Mills fields (II)
M. Gromov:	Hyperbolicity in dynamical systems
P. Griffiths:	Application of residues to geometry

M. Berger:	Wiedersehensmannigfaltigkeiten (Conjecture of Blaschke)
A. Van de Ven:	Inequalities for Chern numbers of surfaces
G. Zuckerman:	Representations of semi simple lie groups

Ch. Thomas:	Space form problems
R. Finn:	Surface tension phenomena and geometry
J.-P. Bourguignon C.L. Terng:	Solution of the Calabi conjecture

---

C. Procesi:	Ideals of determinants and Young diagrams
F. Sakai:	Kondaira dimension of open complex manifolds
A. Andreotti:	Domain of regularity of solutions of partial differential equations

19. Mathematische Arbeitstagung 1978 (16. - 23. Juni)

M.F. Atiyah:	Yang-Mills instantons and algebraic geometry

E. Calabi:	SU(n)- and Sp(n)-manifolds
H. Jacquet:	From GL(2) to GL(3)
J.-P. Bourguignon:	Differential-geometry of the Yang-Mills equation

---

J. Eells:	Holomorphic and harmonic maps of surfaces
J. Steenbrink:	Non-rationality of the quartic threefold
B. Gross:	The Chowla-Selberg formula

B. Mazur:	Rational points on elliptic curves and congruences of L-series
D. Burghelea:	Computation of homotopy groups of diffeomorphism groups of compact manifolds
P. Schweitzer:	Residues of real foliation singularities

T. Banchoff:	The fourth dimension and computer animated geometry
J. Milnor:	Volume of hyperbolic manifolds

K. Ueno:	Birational geometry of fibre spaces
F. Adams:	Finite H-spaces and algebras over the Steenrod Algebra
N. Hitchin:	Twistor spaces

A. Todorov:	Surfaces with $p_g = 1$ and $K^2 = 1$
P. Baum:	K-homology and Riemann-Roch
J. Brüning:	Representations of compact Lie groups and elliptic operators

---

I. Piatetski-Shapiro:	Automorphic forms on the metaplectic group
F. Waldhausen:	Algebraic K-theory of topological spaces

20. Mathematische Arbeitstagung 1979 (06. - 16. Juni)

J. Tits:	On Leech's lattice and sporadic groups

F. Adams:	G. Segal's Burnside ring conjecture
F. Bogomolov:	Converse Galois problems for some Chevalley groups
Wang Yuan:	Goldbach problem

D. Vogan:	Size of representations
L. Berard-Bergery:	A new example of Einstein manifolds

V. Kac:	Infinite dimensional Lie algebras
G. Mostow:	New negatively curved surfaces
G. Lusztig:	Representations of Hecke algebras

B. Gross:	Conjectures of Stark and Tate
Wu-chung Hsiang:	Topological space form problems
E. Looijenga:	Singularities and generalized root systems
M.F. Vigneras:	Isospectral but not isometric Riemannian surfaces

---

A.N. Parshin:	Zeta functions and K-theory
Min-Oo:	Curvature deformations relating to the Yang-Mills fields
G. Harder:	Cohomology and values of L-functions

A. Todorov:	Moduli of Kählerian K3-surfaces
R.P. Langlands:	On orbital integrals for real groups
J.-P. Serre:	The monster game

21. Mathematische Arbeitstagung 1980 (13. - 19. Juni)

M.F. Atiyah: Vector Bundles on Riemann Surfaces

A. Katok: Counting closed geodesics on surfaces

R. Bott: Equivariant Morse theory

K. Ribet: Mazur and Wiles (cyclotomic fields)

A. Borel: L^2 – cohomology of arithmetic groups

I. Bakelman: Topological methods in the theory of Monge-Ampère equations

H. King: Topology of real algebraic varieties

J. Milnor: Groups of polynomial growth (Gromov's work)

Y.-t. Siu: Andreotti-Fraenkel conjecture

M. Artin: Mori's work

B. Gross: L-series of elliptic curves

F. Takens: Turbulence and strange attractors

W. Ziller: Periodic motions in Hamiltonian systems

P. Slodowy: Simple groups over $\mathbb{C}((t))$ and simple-elliptic singularities

---

S. Kudla: Geodesic cycles and the Weil representation

D. Epstein: A theorem of Thurston with applications to group actions and foliations

F. Adams: Recent work on homotopy theory

22. Mathematische Arbeitstagung 1981 (12. - 18. Juni)

M.F: Atiyah:	Convexity and commuting Hamiltonians

B. Mazur:	Abelian extensions of $\mathbb{Q}$
D. De Turck:	"Manifold" of Ricci curvatures
B. Malgrange:	Vanishing cohomology and Bernstein polynomials

D. Mostow:	Complex reflection groups
W. Fulton:	Complex projective geometry (varieties of small codimension)
R. MacPherson:	Intersection homology and nilpotent orbits

J. Tate:	Stark's conjecture about L-series at $s = 0$
W. Meyer:	Gromov's work on Betti numbers
K. Diederich:	Complete Kähler domains

W.D. Neumann:	Thurston's work

A. Derdzinski:	Einstein metrics
S. Zucker:	L^2-cohomology of arithmetic groups
A. Wiles:	Explicit constructions of class fields

J. Duistermaat:	Asymptotics of spherical functions
M-F. Vigneras:	Works of Waldspurger (automorphic forms of halfintegral weight)
R. Schultz:	Topological similarity of representations

23. Mathematische Arbeitstagung 1982 (15. - 21. Juni)

M.F. Atiyah:	The Yang-Mills equations and the structure of 4-manifolds

---

D. Quillen:	Determinants of $\bar{\partial}$-operators
J. Coates:	Heights on elliptic curves
D. Vogan:	Representations with cohomology

R.S. Palais:	Hamilton's work on positively curved 3-manifolds
R. Hartshorne:	Space curves
M. Berger:	Gromov's filling of Riemannian manifolds

J. Bernstein:	Beilinson-Bernstein construction

S.T. Yau:	Manifolds with positive scalar curvature
S. Mori:	Rational curves in 3-folds and applications
G. Harder:	Tate conjecture for Hilbert modular surfaces

L. Siebenmann:	M. Freedman's work on 4-dimensional manifolds
S.J. Patterson:	Limit sets of Kleinian groups
D. Epstein:	On Chapter I of Thurston

S.S. Chern:	Web geometry
B. Mazur:	$\mathbb{Z}_p$-extensions and heights
F. Adams:	Carisson's prove of Segal's Burnside ring conjecture

24. *Mathematische Arbeitstagung 1983 (16. - 22. Juni)*

M.F. Atiyah:	Instantons, monopoles and rational maps

G. Faltings:	The conjectures of Tate and Mordell I
B. Gross:	Heights and L-series I (On the conjecture of Birch and Swinnerton-Dyer)
S.W. Donaldson:	Stable holomorphic bundles and curvature

C. Procesi:	The solution of the Schottky problem (Characterization of Jacobian varieties)
D. Quillen:	Cyclic homology and Hochschild-homology
N. Hingston:	Equivariant Morse theory

D. Zagier:	Heights and L-series II (and applications to the class number problem)
D. McDuff:	The Arnold conjecture on symplectic fixed points (after Conley and Zehnder)
G. Faltings:	The conjectures of Tate and Mordell II (Moduli spaces of abelian varieties)

J. Milnor:	Monotonicity for the entropy of quadratic maps
E. Friedlander:	On the conjectures of Lichtenbaum and Quillen (after Suslin and others)
D. Quillen:	Arithmetic surfaces and analytic torsion
B. Moonen:	Polar multiplicities and curvature integrals

D. Eisenbud:	Special divisors on curves and Kodaira dimension of the moduli space (mostly after Mumford and Harris and Gieseker)
F. Kirwan:	Cohomology of quotients in algebraic and symplectic geometry
W. Tutschke:	Generalizations of the Cauchy-Kowalewski and Holmgren theorems to the case of generalized analytic functions
G. Wüstholz:	Group varieties and transcendence

25. Mathematische Arbeitstagung 1984 (15. - 22. Juni)

J. Tits:	Groups and group functors attached to Kac-Moody data
M. Atiyah:	The eigenvalues of the Dirac operator
A. Connes:	K-theory, cyclic cohomology and operator algebras
G. Segal:	Loop groups
G. Harder:	Special values of Hecke L-functions and abelian integrals
H. Wente:	A counterexample in 3-space to a conjecture of H. Hopf
G. Faltings:	Compactification of $A_g/\mathbb{Z}$
C.T.C. Wall:	Geometric structures and algebraic surfaces
J. Harris:	Recent work on Hodge structures
Y.T. Siu:	Some recent results in complex manifold theory related to vanishing theorems for the semipositive case
W. Schmid:	Recent progress in representation theory
W. Ballmann:	Manifolds of non-positive curvature
B. Mazur - Ch. Soulé:	Conjectures of Beilinson on L-functions and K-theory
H.-O. Peitgen:	Morphology of Julia sets
S.S. Chern:	Some applications of the method of moving frames
S. Lang:	Vojta's conjecture on heights and Green's function
S. Donaldson:	4-manifolds with indefinite intersection form
D. Zagier:	Modular points, modular curves, modular surfaces and modular forms
G. van der Geer:	Schottky's problem
R. Bryant:	G_2 and Spin(7)-holonomy
S. Wolpert:	Homology of Teichmüller spaces
J.-P. Serre:	l-adic representations
M.F. Atiyah:	On Manin's manuscript "New dimensions in geometry"

Teilnehmer der Arbeitstagung 1984

K. Abe (U. of Connecticut)
J. Abresch (MPI/SFB)
J.F. Adams (Cambridge)
S. Akbulut (MPI/SFB)
H. Andersen (Aarhus)
Y. André (Inst. Poincaré, Paris)
G. Andrzejczak (MPI/SFB)
V. Anosov (Steklov Inst., Moskau)
M. Artin (MIT)
M.F. Atiyah (Oxford)
H. Azad (Bochum)
A. Back (SUNY)
A. Bak (Bielefeld)
A. Baker (Cambridge)
W. Ballmann (Bonn)
T. Banchoff (Brown U.)
B. Banieqbal (Manchester)
W. Barth (Erlangen)
G. Barthel (Konstanz)
H.J. Baues (Bonn)
E. Becker (Dortmund)
K. Behnke (MPI/SFB)
R. Berndt (Hamburg)
B. Birch (Oxford)
S. Böcherer (Freiburg)
C.-F. Bödigheimer (Heidelberg)
A. Bojanowska (ETH Zürich)
M. Bökstedt (Bielefeld)
W. Borho (Wuppertal)
R. Bott (Harvard)
J.P. Bourguignon (Ec. Polyt. Palaiseau)
G. Brattström (Harvard)
E. Brieskorn (Bonn)
E. Brown (Brandeis U.)
R. Bruggeman (Utrecht)
J. Brüning (Augsburg)
R.L. Bryant (Rice U.)
D. Burghelea (Ohio State U.)
E. Calabi (U. of Pennsylvania)
S.S. Chern (MSRI Berkeley)
J. Coates (Orsay)
R. Connelly (Cornell U.)
A. Connes (IHES)
M. Crabb (Wuppertal)
C.B. Croke (U. of Pennsylvania)
R. Cushman (Utrecht)
W. Danielewski (Bochum)
W. Decker (Kaiserslautern)
J. Deprez (Leuven)
A. Derdzinski (MPI/SFB)
T. tom Dieck (Göttingen)
K. Diederich (Wuppertal)
A. Dold (Heidelberg)
P. Dombrowski (Köln)
S.K. Donaldson (Oxford)
S. Donkin (London)
A. Douady (ENS Paris)
W. Ebeling (Bonn)
P. Eberlein (MPI/SFB)
J. Eells (Warwick)
H. Eggers (Regensburg)
J. Ehlers (MPI Phys. & Astrophys.)
P.E. Ehrlich (U. of Missouri)
M. Eichler (Basel)
H. Eliasson (MPI/SFB)
O. Endler (Bonn und IMPA)
D. Erle (Dortmund)
H. Esnault (MPI/SFB)
G. Faltings (Wuppertal)
G. Fischer (Düsseldorf)
H. Flenner (Göttingen)
E. Friedlander (North Western U.)
S. Friedlander (U. of Chicago)
S. Gallot (Paris VII)
G. van der Geer (Amsterdam)
E.U. Gekeler (Bonn)
H. Grauert (Göttingen)
R. Greene (UCLA)
G.M. Greuel (Kaiserslautern)
B. Gross (Brown U.)
F. Grunewald (Bonn)
R. Gupta (Brown U.)
F. v. Haeseler (Bremen)
H. Hamm (Münster)
G. Harder (Bonn)
J. Harris (Brown U.)
F. Hegenbarth (Dortmund)
E. Heintze (Münster)
J. Heinze (Springer Verlag)
G. Helminck (Amsterdam)
R. Henderiks (Rotterdam)
M. Herrmann (Köln)
S. Hildebrandt (Bonn)
U. Hirsch (Bielefeld)
F. Hirzebruch (MPI und U. Bonn)
T. Höfer (MPI/SFB)
J. Huebschmann (Heidelberg)
K. Hulek (Erlangen)
J. Hurrelbrink (Louisiana State U.)
D. Husemoller (Haverford College)
H.-C. Im Hof (Basel)
K. Ivinskis (Bonn)
S. Jackowski (ETH Zürich)
J.C. Jantzen (Bonn)
B. Julia (E.N.S. Paris)
H. Jürgens (Bremen)
W. van der Kallen (Utrecht)
W. Kamber (U. of Illinois)
H. Karcher (Bonn)
M. Karoubi (Sceaux)

U. Karras (Dortmund)
J. Kazdan (U. of Pennsylvania)
I. Kersten (Regensburg)
M. Kervaire (Genf)
F. Kirwan (Harvard)
N. Klingen (Köln)
W. Klingenberg (Bonn)
P. Kluitmann (MPI/SFB)
K. Knapp (Wuppertal)
H. Knörrer (Bonn)
F. Koll (Bonn)
J. Konarski (Bochum)
K. Kopfermann (Hannover)
S. Kosarew (Regensburg)
U. Koschorke (Siegen)
M. Kossowski (Rice U.)
B. Kostant (MIT)
H. Kraft (Basel)
G. Kramarz (Bonn)
J. Kramer (Basel)
M. Krämer (Bayreuth)
M. Kreck (Mainz)
F.-V. Kuhlmann (Heidelberg)
K. Kühne-Hausmann (Bonn)
W. Kühnel (TU Berlin)
N. Kuiper (IHES)
R.S. Kulkarni (Indiana U.)
B.A. Kupershmidt (U. of Tennessee)
N. Kuznetsov (Khabarovsk)
C. Lackschewitz (Berlin)
K.Y. Lam (U. of British Columbia)
R. Lamotke (Köln)
S. Lang (Yale U.)
J. Langer (MPI/SFB)
M. Laska (MPI/SFB)
K. Lebelt (Essen)
R. Lee (Yale U.)
L. Lemaire (Brüssel)
M. Levine (U. of Pennsylvania)
I. Lieb (Bonn)
J. Little (Holy Cross)
M. Lorenz (MPI/SFB)
J. Marsden (UC, Berkeley)
S. Maurmann (Bonn)
K.-H. Mayer (Dortmund)
B. Mazur (Harvard)
J. McCarthy (MPI/SFB)
J. Mesirov (AMS)
W. Meyer (MPI/SFB)
W.T. Meyer (Münster)
H. Miller (U. of Washington)
T. Miller (Bonn)
P. Milman (Toronto)
M. Min-Oo (Bonn)
G. Mislin (Zürich)
V. Moncrief (Yale U.)
B. Moonen (Köln)
R. Moore (Canberra)
B.Z. Moroz (MPI/SFB)
D. Morrison (Princeton)
G.D. Mostow (Yale U.)
P. Mrozik (Siegen)
Y. Namikawa (MPI/SFB)
H.J. Nastold (Münster)
W.D. Neumann (U. of Maryland)
W.-M. Ni (U. of Minnisota)
T. Nishimori (Hokkaido U.)
K. Nomizu (Brown U.)
E. Oeljeklaus (Münster)
A. Ogg (Orsay)
R. Olivier (Bonn)
F. Oort (Utrecht)
E. Ossa (Wuppertal)
J.-P. Otal (MPI/SFB)
M. Otte (Bielefeld)
A. Parshin (Steklov Inst., Moskau)
G. Patrizio (MPI/SFB)
H.-O. Peitgen (Bremen)
A. Pereira do Valle (Bonn)
Th. Peternell (Münster)
Ch. Peters (Leiden)
V.Q. Phong (Bremen)
A. Pickl (Regensburg)
B. Pickl (Regensburg)
R. Piene (Oslo)
U. Pinkall (MPI/SFB)
P. Platonov (Minsk)
H. Popp (Heidelberg)
V. Poupko (Wien)
A. Prestel (Konstanz)
M. Prüfer (Bremen)
V. Puppe (Konstanz)
R.M. Range (SUNY, Albany)
T. Ratiu (UC, Berkeley)
B. Reinhart (U. of Maryland)
R. Remmert (Münster)
D. Repovs (Ljubljana)
K. Ribet (UC, Berkeley)
D. Richter (Bremen)
O. Riemenschneider (Hamburg)
J. Rohlfs (Eichstätt)
G. Roland (Bonn)
M. Rost (Regensburg)
E. Ruh (Bonn)
J.M. Sampson (Baltimore)
E. Sato (MPI/SFB)
D. Saupe (Bremen)
N. Schappacher (MPI/SFB)
A. Scharf (Bonn)
R. Scharlau (Bielefeld)
W. Schelter (U. of Texas)
K.D. Schewe (Bonn)
U. Schmickler-Hirzebruch (Vieweg Vlg.)
W. Schmid (Harvard)
C.-G. Schmidt (IHES)
G. Schneider (Kaiserslautern)

P. Schneider (Heidelberg)
R. Schön (Heidelberg)
R. Schoof (U. of Maryland)
F. Schreyer (Kaiserslautern)
M. Schroeder (Göttingen)
G. Schumacher(Münster)
J. Schwermer (Bonn)
R. Sczech (MPI/SFB)
D. Segal (Manchester)
G. Segal (Oxford)
W. Seiler (Mannheim)
J.-P. Serre (Collège de France)
D. Siersma (Utrecht)
Y.-T. Siu (Harvard)
B. Smyth (Notre Dame)
Ch. Soulé (Paris VII)
E. Spanier (MPI/SFB)
B. Speh (Cornell U.)
C. Spönemann (Hannover)
J. Steenbrink (Leiden)
J. Stienstra (Utrecht)
R. Strebel (Heidelberg)
J. Strooker (Utrecht)
U. Stuhler (Wuppertal)
S. Suter (MPI/SFB)
L. Szpiro (ENS Paris)
Ch. Thomas (Cambridge)
K. Timmerscheidt (Essen)
J. Tits (Collège de France)
A. Todorov (Sofia)
P. Tondeur (U. of Illinois)
G. Trautmann (Kaiserslautern)
G. Triantafillou (U. of Minnesota)
A. Tromba (UC, Santa Cruz)
G. Tsagas (Thessaloniki)
A. Tschimmel (Wuppertal)
T. Tsuboi (Tokyo)
A. Van de Ven (Leiden)
P. Verheyen (Leuven)
E. Viehweg (Essen)
M.-F. Vigneras (ENSJF Paris)
W. Vogel (Halle)
T. Vorst (Rotterdam)
F. Waldhausen (Bielefeld)
C.T.C. Wall (Liverpool)
N. Walter (Mannheim)
M. Wang (McMaster U.)
L. Washington (U. of Maryland)
H.C. Wente (U. of Toledo)
J. Werner (MPI/SFB)
A. Wiles (Princeton)
K. Wirthmüller (Regensburg)
J. Wolfart (Frankfurt)
S. Wolpert (U. of Maryland)
P.-M. Wong (Notre Dame)
J. Wood (Chicago)
G. Wüstholz (MPI/SFB)
M. Yoshida (MPI/SFB)
D. Zagier (MPI und U. of Maryland)
G. Zeidler (Leipzig)